NON-SOLAR X- AND GAMMA-RAY ASTRONOMY

NUCLEAR AND CYTOPLASMIC HETEROTOPY

INTERNATIONAL ASTRONOMICAL UNION

UNION ASTRONOMIQUE INTERNATIONALE

SYMPOSIUM No. 37

HELD IN ROME, ITALY, MAY 8–10, 1969

NON-SOLAR X- AND GAMMA-RAY ASTRONOMY

EDITED BY

L. GRATTON

Dudley Observatory, Albany, N.Y., U.S.A.

D. REIDEL PUBLISHING COMPANY

DORDRECHT-HOLLAND

1970

Published on behalf of
the International Astronomical Union
by
D. Reidel Publishing Company, Dordrecht, Holland

Library of Congress Catalog Card Number 71–115885

ISBN-13:978-94-010-3313-8   e-ISBN-13:978-94-010-3311-4
DOI: 10.1007/978-94-010-3311-4

TABLE OF CONTENTS

* An asterisk means that the author(s) has (have) not delivered the text of his (their) paper.

INTRODUCTORY REMARKS

(Invited Discourse)

BRUNO B. ROSSI

Massachusetts Institute of Technology, Cambridge, Mass., U.S.A.

High flying balloons, rockets and artificial satellites have made available to astronomers the whole electromagnetic spectrum reaching the earth from outer space, most of which was previously denied to them by atmospheric absorption. However, even from above the atmosphere, the astronomer interested in objects beyond the solar system has still to reckon with the attenuation of electromagnetic waves by interstellar matter, i.e., by the interstellar gas and the interstellar dust. For practical purposes,

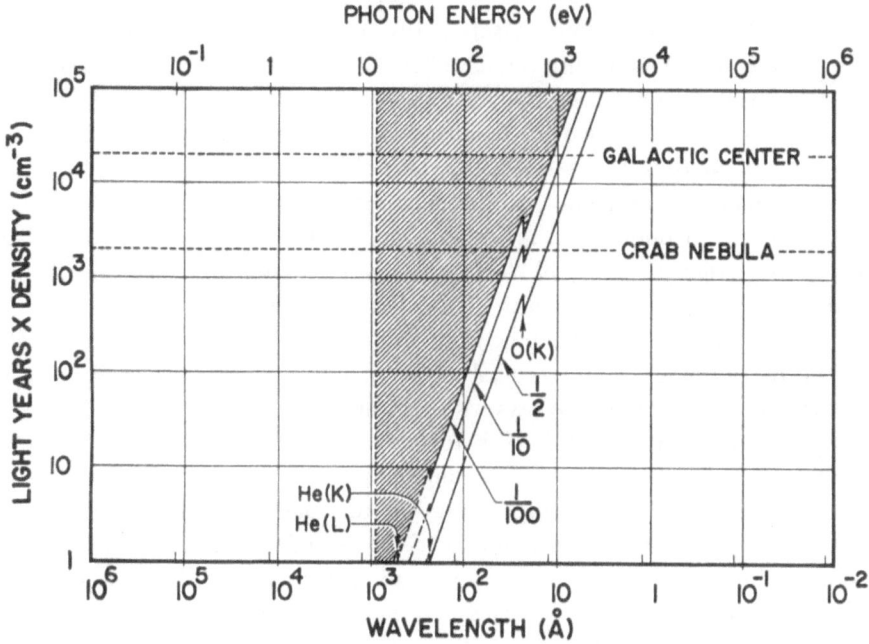

Fig. 1. The absorption of radiations of different wavelengths in interstellar gas. The ordinate is the product of the distance of a given celestial object (in light years) times the average density of interstellar gas (in atoms per cm⁻³). The curves represent the values of this quantity for which radiations of the various wavelengths are reduced in the ratios 2:1, 10:1 and 100:1 in travelling from the object to the earth.

interstellar gas is completely transparent over galactic distances for all wavelengths from radio waves to the far ultraviolet. In this spectral region the visibility is partially impeded only by the unevenly distributed interstellar dust.

But at 912 Å, i.e., at a photon energy equal to the ionization energy of hydrogen

(13.5 eV), the interstellar gas becomes suddenly very opaque. Proceeding toward shorter wavelengths, the absorption decreases gradually (see Figure 1), and interstellar space becomes transparent again at photon energies that vary from about 100 eV for he nearby stars to about 1000 eV for the more distant galactic objects. In all of this spectral region, the absorption of dust is negligible.

Thus interstellar absorption quite naturally divides the electromagnetic spectrum into a *low energy region*, which extends from radio waves to 13.5 eV photons, and a *high-energy region*, which starts somewhere between 100 and 1000 eV, depending on the distance of the object under consideration, and extends to the highest energies at which the photon flux is still above the detection limit of the available instruments.

The subject of our Symposium concerns the high energy region of the electromagnetic spectrum.

From the point of view of the observational requirements, it is convenient to further subdivide this region into the following subregions:

(a) *Soft X-rays* extending from the low-energy cut-off due to interstellar absorption to about 15 keV. Observations of soft X-rays can only be carried out at altitudes above 100 km, and thus require the use of rockets or artificial satellites. Most of the data so far available have been obtained with rockets.

(b) *Hard X-rays* extending from about 15 keV to about 0.5 MeV. Since the spectrum of high-energy photons is very steep, the short observation time provided by rockets is a serious handicap. Satellites, so far, have only been available to a very limited extent for X-ray astronomy. On the other hand, hard X-rays penetrate to a sufficient depth in the atmosphere to be detectable by balloon-borne instruments, which provide observation times of many hours. For these reasons, most of the existing information on hard X-rays has come from balloon observations. These observations are hampered, although not in a crucial way, by a diffuse background of X-rays arising from the interactions of cosmic rays with atmospheric gases.

(c) *γ-rays* extending from 0.5 MeV to several hundreds MeV. The very low intensity of the radiation in this energy range practically rules out the use of rockets. Moreover, since the intensity of high-energy photons of celestial origin decreases with increasing energy more rapidly than that of the secondary photons from cosmic ray interactions in the atmosphere, the background problem that already plagues balloon observations of hard X-rays becomes here even more serious. Thus satellites are essential in γ-ray astronomy.

As you know, high-energy astronomy is a very young science.

A search for extrasolar γ-rays with quantum energies of the order of 100 MeV began in the early 60's, prompted by the prediction that photons in this energy range should originate from the interaction of cosmic rays with interstellar matter. Only two years ago, however, did this search produce unambiguous positive results.

X-ray astronomy had its beginnings in 1962. The early observations revealed the existence of discrete X-ray sources and of a diffuse X-ray background, both many orders of magnitude stronger than anyone had expected to find.

Because of the unexpectedly large intensity of celestial X-rays, X-ray astronomy has evolved with remarkable speed; by now it is well beyond the exploratory stage where γ-ray astronomy still finds itself.

A few words about the program of our Symposium.

The development of new observational techniques of increasing sophistication has played a crucial role in the progress of the new field of astronomy that we are discussing. Thus the organizing committee has considered it appropriate to begin with a brief discussion of these techniques, including not only those that have already been used, but also those that are planned for future experiments. Dr. Gursky will present this discussion immediately after my introductory remarks.

The four subsequent sessions, starting with that of this afternoon, will be devoted to the discrete X-ray sources, which have been the object of most observational programs thus far. One important line of investigation has been a search for new sources and a study of their distribution in the sky. We shall hear a report on this subject from Dr. Friedman, whose group has been responsible for the most extensive surveys, and has been active in other aspects of X-ray astronomy as well. Here I only wish to recall that, while most of the observed sources lie within the Milky Way, a few are located at high galactic latitude. The former sources are clearly galactic objects; the latter are presumably extragalactic objects and one of them, in fact, has been tentatively identified with a radio galaxy.

The general surveys that I have mentioned were performed by means of rockets, which remain above the atmosphere for only a few minutes. Since the flights were designed to cover a large area of the sky, the observation time devoted to each source was very short and the amount of information that could be gathered was correspondingly limited. It was thus necessary to follow the general surveys with a detailed study of individual sources. This has been done by a number of different groups, and we shall hear reports from representatives of these groups, i.e., from Dr. Clark, Dr. Peterson, Dr. Giacconi and Dr. McCracken.

The observations have been directed mainly toward securing the following kinds of information:

(a) Location, for the purpose of attempting an identification with optical or radio objects.

(b) Angular diameter.

(c) Time variations in the X-ray flux.

(d) Gross spectral features.

It is interesting to note that the two sources for which we have most detailed data turned out to be entirely different objects. One of them, of course, is the Crab Nebula; a supernova remnant, whose X-ray emission originates from a region of finite dimensions. These X-rays have a hard spectrum and their flux is constant in time. The other source is Sco X-1. This object appears point-like both to the optical telescopes and to the X-ray detectors with the finest resolution achieved so far (about 20 arc sec). The X-ray spectrum is much softer than that of the Crab and its intensity undergoes large fluctuations.

No reasonable doubt exists today concerning the visible counterpart of Sco X-1 nor concerning the identification of the Crab Nebula as an X-ray source. Other identifications of X-ray sources with visible objects have been suggested, of which some are convincing, others still tentative. These identifications have focussed the interest of optical astronomers upon several faint objects whose peculiar properties had not been noticed previously or had not received much attention. We shall hear about the results of the optical observations of X-ray sources from Dr. Johnson. These observations are of great importance because they place new and very significant limitations on the models of X-ray stars that may be proposed.

A question that still puzzles astronomers is the relation of X-ray stars to other galactic objects. We know that some of the galactic X-ray sources are supernova remnants. But what about the others? Are we to regard them as objects entirely different from all celestial objects previously known? Or are they peculiar members of some known family, such as, for example, old novae? Or do they, perhaps, represent a particular transitory stage in the development of ordinary stars? Some clue to this puzzle may be found in the statistical distribution of the X-ray sources. The observational data concerning this distribution and their possible significance will be discussed by Dr. Gratton at the end of tomorrow's session.

Saturday morning, with the lectures of Dr. Woltjer and Dr. Felten, we shall come to grips with the still unsolved problem of the physical phenomena responsible for the strong X-ray emission by the discrete sources. As far as I know, most theories that have been suggested are based on either one of two assumptions: (1) X-ray emission by optically thin plasmas, locally in nearly thermal equilibrium at temperatures of the order of several tens of millions of degrees; (2) X-ray emission by magnetic bremsstrahlung of very high-energy electrons. Different models have been proposed to account for the large masses of exceedingly hot plasma required by the first assumption, or for the large numbers of exceedingly energetic electrons required by the second assumption. Whether or not any of these models is really convincing is a question about which we may want to reserve judgment until after the discussions that, I am sure, will follow the talks of Dr. Woltjer and Dr. Felten.

The last session, on Saturday afternoon, will be devoted to the background radiation. Dr. Oda will review the observational results concerning the X-ray component of this radiation, Dr. Kraushaar those concerning the γ-ray component. Finally Dr. Setti will discuss the theoretical questions raised by the observations.

I do not wish to anticipate any of the facts or of the conclusions to be presented at this session. I would like, however, to point out that, potentially at least, the diffuse background is of equal, or perhaps even greater cosmological interest than the discrete sources. To support this view I only need to remind you of the suggestion that the X-ray background may be due to inverse compton effect of high-energy electrons colliding, in intergalactic space, with photons of the universal black-body radiation. This radiation is supposed to originate from the big explosion that gave birth to our universe some 15 or 20 billion years ago; and, if the hypothesis is correct, there are reasons to believe that the X-ray photons observed today might have been produced when the Universe was still young.

A SURVEY OF INSTRUMENTS AND EXPERIMENTS
FOR X-RAY ASTRONOMY

(Invited Discourse)

HERBERT GURSKY

American Science and Engineering, Cambridge, Mass., U.S.A.

1. Introduction

A diversity of instruments have been utilized or proposed for use in X-ray astronomy. This in part reflects the fact that the field is very new, that many competing techniques have been devised for any given type of observation and that there has been a progression to increasingly sophisticated instruments. But most of this diversity is real and simply reflects the very broad range of observations that can be performed and the different kinds of observational opportunities that are presented to the several experimental groups. Useful measurements are being performed over three decades of energy extending from 0.25 keV to 500 keV, positional measurements extend from degrees to arc seconds precision, attempts have been made to measure the polarization of the X-ray sources, time variations are being studied and spectra are being measured to resolutions of the order of $\sim 20\%$. The required instruments are being flown on sounding rockets, balloons and satellites. In addition, observation of the diffuse X-ray backgrounds requires even different instruments.

These essential elements of an X-ray astronomy experiment can be reduced to the following items, a collimating device (baffles, wires, focussing devices), a beam conditioner (filters, polarizer, grating) and a detector (proportional counter, photoelectric detector). These units must be mounted rigidly on the carrier (rocket, balloon). Some means must be provided to maneuver or point the carrier in order to acquire or sweep by the X-ray sources or other interesting regions of the sky. Since it is necessary to know where in the sky one is looking, some means of determining celestial attitude must be provided for. Finally, there is the problem of signal conditioning and the return of data to the ground. The essential elements of an X-ray astronomy experiment are shown in Figure 1.

The design of the instrumentation used in X-ray astronomy is very much dependent on some model of the X-ray sky; namely, the distribution and intensity of the sources in the sky, the spectral distribution of the radiation, the effect of the interstellar medium on the radiation and the nature and intensity of the non-X-ray background. Furthermore, the instrumental developments depend a great deal on certain notions we have regarding the nature of the sources; i.e., will the radiation be polarized, will spectral lines be present, are the sources of small or large angular size.

A number of review articles on X-ray astronomy have appeared within the past year; two which deal entirely with instrumentation are by Giacconi *et al.* [1] on

L. Gratton (ed.), Non-Solar X- and Gamma-Ray Astronomy, 5–33. All Rights Reserved
Copyright © 1970 by the IAU

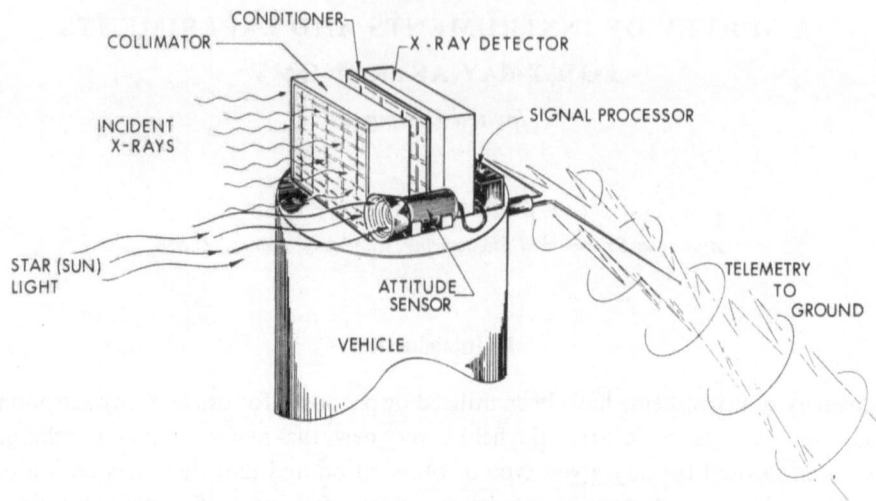

Fig. 1. Elements comprising the X-ray astronomy experiment. In addition to those shown there is normally some means of stabilizing the vehicle.

TABLE I

Anticipated distribution of X-ray sources

	Distance	Number	Intensity (Relative to the Crab Nebula)	Angular Size
Supernova remnants:				
Crab Nebula	1700 pc	1	1	3′
	10 kpc	40	3×10^{-3}	30″
	500 kpc (M31)	100	10^{-5}	0.5″
Sco X-1 like objects:				
Sco X-1	~ 1000 pc	1	10	small
	10 kpc	10^3	10^{-2}	
	500 kpc (M31)	2×10^3	4×10^{-5}	
Stellar corona:				
Sun	5×10^{-6} pc	1	10^7	
	10 pc	10	2×10^{-6}	small
Extragalactic sources:				
M-87	1.1×10^7 pc	1	5×10^{-2}	1′
	1.1×10^8	10^3	5×10^{-4}	6″
	1.1×10^9	10^6	5×10^{-6}	0.6″

conventional techniques and by Giacconi *et al.* [2] dealing exclusively with applications of X-ray telescopes.

2. A Model of the X-Ray Sky

The distribution of X-ray sources on the celestial sphere is shown in Figure 2. The striking feature of this distribution is the concentration of sources along the Milky

Way with the highest concentration being within ~20° longitude of the galactic center. The galactic sources comprise at least two types; in the first place, there are supernova remnants like the Crab Nebula and Cas A which are of large angular size, and secondly there are the star-like objects as exemplified by Sco X-1 and Cyg X-2. The latter are probably highly variable in intensity while the former are probably stable. The range of intensity of the observed galactic sources is between $10\times$ and $0.1\times$ that of the Crab Nebula (20–0.2 photons/cm²-sec in the 2–10 keV energy range). The intrinsic luminosity of these sources is in the range 10^{36}–10^{37} ergs/sec. In addition, we have good evidence for the existence of a class of X-ray galaxies as exemplified by M-87. In Table I, I present what might be expected for a number – space density of these sources. Also, one must consider the sun as an X-ray source, even though the study of solar X-radiation is not normally considered within the province of X-ray astronomy. The reason is simply that at some level of sensitivity, stellar corona in nearby

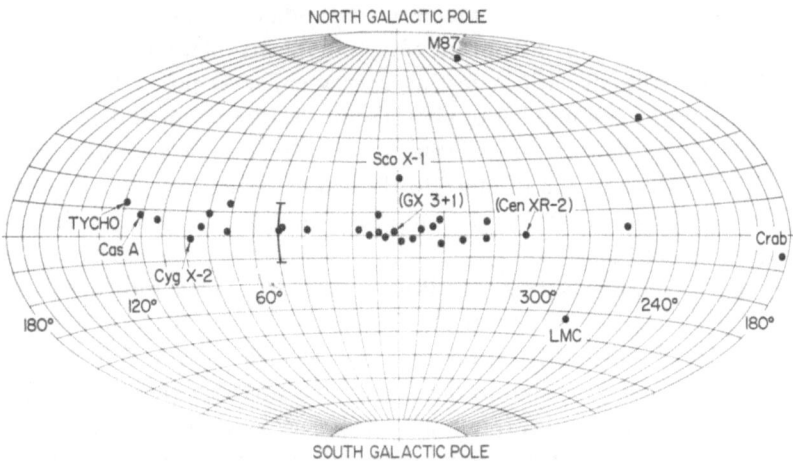

Fig. 2.   Distribution of known X-ray sources in the galactic coordinate system.

TABLE II

Low energy cutoffs due to the interstellar medium

	E-cutoff[a]
Extragalactic objects:	
Pole	0.2 keV
60° latitude	0.3
30° latitude	0.44
Galactic objects:	
in the plane	
10 pc	0.12 keV
1 kpc	0.7
10 kpc (toward galactic center)	3 keV

[a] Energy at which the optical depth is unity.

stars will be observable and that techniques used in solar studies are being adapted for use in X-ray astronomy.

The significance of the data in Table I is that with modest improvements in sensitivity (~factor 10–100) over what is now being achieved one may have a very great increase in the number of observable sources, and when and if sensitivity can be improved 4–5 orders of magnitude we can expect qualitatively new phenomena to reveal themselves, such as the observations of X-ray sources at cosmological distances, of 'common' X-ray sources in nearby galaxies, and of the coronal emission of nearby stars. One point that must be mentioned is that increased sensitivity requires the use of high resolution collimators in order to avoid the problem of source confusion. Even now, mechanical collimators are inadequate to clearly separate the sources at low galactic longitudes. With very high sensitivities it is apparent that instruments with arc second resolution are called for; e.g., the 1000 or so sources in M-31 would be distributed over less than 1 $(\text{deg})^2$.

The spectral composition of the radiation, as modified by the interstellar medium, has an important bearing on instrument design. In the first place, the spectral intensity characteristically falls off with increasing energy either as power laws or as exponentials. Thus, the largest number of photons are always present at the lowest energy. However, the interstellar medium limits how low in energy measurements can be performed, depending on the distance to the sources. Based on calculations of Bell and Kingston [3], the cutoff energies are shown in Table II. These values are uncertain to the extent that we do not yet know the X-ray absorbing characteristics of the interstellar medium. The results of Bowyer et al. [4], which apparently have been confirmed by Kraushaar's group at the University of Wisconsin, indicate that the X-ray absorbing characteristics at 0.28 keV and at high galactic latitudes are lower by about a factor of 3 compared to the calculations in Table II. But the qualitative features of the interstellar absorption are unchanged; namely, that at high galactic latitudes, significant fluxes of radiation can be recorded at 0.3 keV, while in the plane, the cutoffs will lie between 3 or 4 keV and 0.7 keV depending on the location of the source. It is also possible that absorbing material may be present in or near the X-ray sources themselves. Furthermore, if one is content to study the X-ray emission from the nearest 100 or so stellar objects, observations can be extended to at least 0.1 keV, which means that coronal emission can be observed.

3. X-Ray Detectors

A. 1–20 KEV X-RAY

Up to now, this energy range has provided the bulk of the information on the X-ray sources and the background. The principal detector is the gas proportional counter using a window of beryllium or organic films such as mylar. Calculated efficiencies for typical counters are shown in Figure 3. The beryllium window counters have the advantage in that they can be entirely gas tight which gives long shelf-life and the capability of extended life in orbit. The organic films are typically porous to certain gases which limits their life unless special precautions are taken such as continuous renewal of the counter gas.

The state-of-the-art of gas proportional counters has advanced to the point that it is possible to construct counting systems which are virtually free from any counting

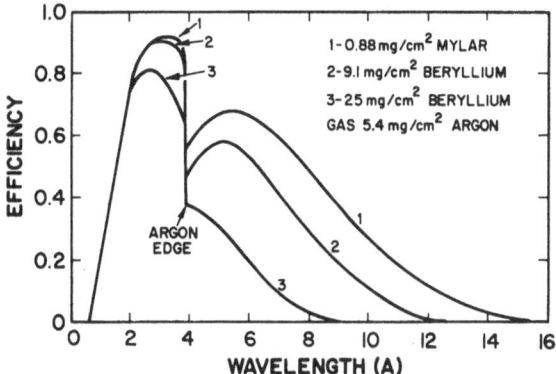

Fig. 3. The efficiency of gas proportional counters using argon as a filling gas. Efficiency can be extended to shorter wavelengths by the use of higher pressure of argon, or the use of heavy gases such as xenon and krypton.

background other than X-rays entering the front window, particularly through the use of pulse shape (or rise time) discrimination, which has allowed a significant improvement in sensitivity for experiments using proportional counters. This technique was first introduced into X-ray astronomy experiments by Dr. Pound's group at Leicester [5], and has been extended by Gorenstein and Mickiewicz [6]. It permits the rejection of counter signals originating from minimum ionizing electrons such as arise from cosmic rays traversing the counter or γ-rays converting in the walls. These events, the latter especially, form the principal non-X-ray background in most experiments.

The basis for pulse shape discrimination is that counter signals arising from X-rays form a narrow spread in pulse rise time; whereas signals coming from γ-rays have a very large spread in rise times, up to 10 times those of X-rays. Thus if one electronically selects events (of a given amplitude) on the basis of their rise time a large fraction of the non-X-ray events can be rejected – up to 99% in certain instances. Figure 4 shows actual performance in a flight experiment of PSD. In this case we estimated that the residual, non-X-ray background was reduced to $\sim 10^{-2}$ cts/cm²-sec for the projected area of the detector. In principle, one should be able to reduce non-X-ray counts even further by the incorporation of several background rejection techniques; e.g., PSD, guard counters and wall-less detectors (the latter are detectors in which the anode is replaced by a grid and the guard counter is incorporated into the primary counter). In practice, however, the limit on the background counting rate is frequently determined by the diffuse X-ray background which of course can not be discriminated against within the detector.

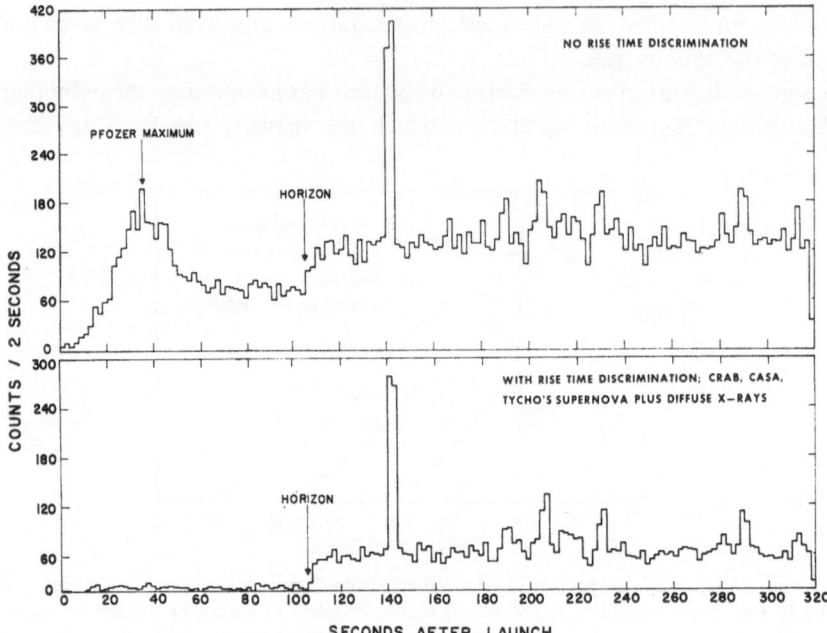

Fig. 4. The effective pulse shape discrimination during an AS & E sounding rocket flight using a system developed by Gorenstein and Mickiewicz. The residual counting rate at times greater than 100 sec is almost entirely the result of the diffuse X-ray background plus individual point sources of X-rays.

B. HARD X-RAY DETECTION

The response of proportional counters can be extended to ~50 keV by using heavy filling gases such as xenon and krypton operating at high pressures. To detect higher energies it is necessary to use scintillation counters with NaI(Tl) being the most common choice of a scintillating crystal. Observations have been made to several hundred keV with NaI scintillation counters on discrete sources and to several MeV on the diffuse background.

The extraneous counting rates represent a very severe problem at these higher photon energies. The cosmic photon fluxes fall off with increasing energies much faster than the extraneous radiation.

The latter is most likely to be photons produced by compton scattering of energetic γ-rays made in cosmic ray interactions. These tend to have a relatively flat energy distribution. Thus intrinsically signal/noise deteriorates with increasing energy. Furthermore, it becomes increasingly difficult to shield against the higher energy photons, since compton scattering, which does not remove the photon but simply degrades it in energy, becomes the dominant form of interaction. The technique of active shielding used by Haymes and by Peterson is effective in reducing background.

C. SOFT X-RAY DETECTION TECHNIQUES

The importance of detecting photons with energies below 1 keV is more than simply the increasing number of photons present at these energies. It may be important to see the low-energy behavior of the X-ray sources. This is best illustrated by the diffuse X-ray background which seems to show an anomalously high intensity at low energy. It is important to establish the number-temperature (power law index) distribution of the sources; i.e., many new sources may be revealed at the very low energies. Finally, the absorbing properties of the interstellar medium and the sources themselves will be revealed at low energies.

In the last few years, a number of experimental groups have concentrated their efforts toward performing measurements below 1 keV using photo-electron detectors and thin-window proportional counters and experiments utilizing the latter devices have achieved some notable results. Physically, the thin-window proportional counters are the same as the beryllium window proportional counter discussed earlier except

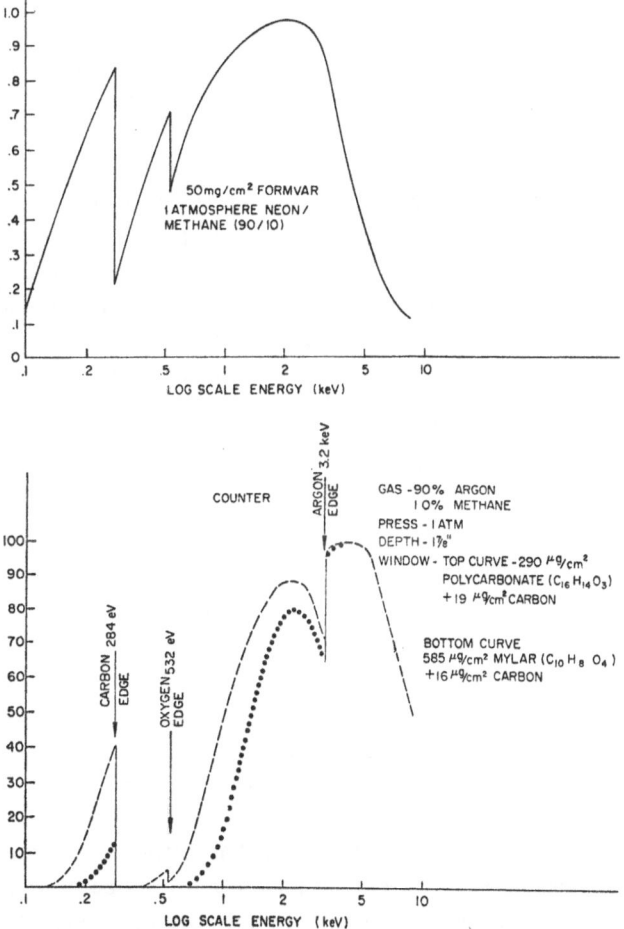

Fig. 5. Calculated efficiency of thin window proportional counters currently being used in sounding rocket experiments.

for the use of the thinner aperture window. Window material that has been used includes 6 and 4μ Mylar, 2μ 'Kimfoil' and Formvar ($0.5-1\mu$). The Lockheed group was the first to attempt flying these detectors. The work of Bowyer et al. [4] on the X-ray background was performed using the thinner Mylar, as was the work of the NRL group [7]. Recent results of the Calgary group [8] have been based on the use of 6μ Mylar. Livermore Radiation Laboratory has successfully flown Formvar window detectors and the University of Wisconsin group has flown 'Kimfoil' and Mylar window detectors. The calculated efficiency of the latter's detectors is shown in Figure 5 (lower half) and the efficiency of the Formvar counters flown by LRL in Figure 5 (upper half).

The great advantage of these detectors over the thicker window devices is the response below the carbon edge at 0.28 keV. However, the experimental problems associated with their use are severe. The windows are porous and a gas-ballast or flow system must be used to maintain pressure during the flight. Since the pulse amplitude is a very sensitive function of the gas pressure, the system must be well regulated and a means must be provided to monitor accurately the gas gain of the counter. Also because of the porosity contaminants, principally oxygen and water vapor, can leak in and the detectors must be carefully flushed before flight or kept evacuated until flight.

Finally, there is apparently more difficulty with spurious effects at these very low energies than in the 1–10 keV energy range. The LRL group reports interference from solar induced X-ray fluorescence in the atmosphere on one occasion and from unexplained sources (probably low energy electrons) on another occasion.

As might not be unexpected the experimental results obtained in this area are not entirely self-consistent. Table III lists the published results of measurements at 0.28 of the diffuse X-ray background. The difference between Bowyer et al. [4] and Henry

TABLE III

Comparisons of results of measurements of
isotropic background at 0.27 KeV

Reported flux (keV/cm²-sec-ster (keV))

	Uncorrected for interstellar attenuation	Corrected
Bowyer et al.	65	73
Henry et al.	168	790
Baxter et al.	583	–

et al. [7] lies mostly in the choice of the interstellar absorption factor. The result of Baxter et al. [8] must be regarded with suspicion, not only because it is larger than the other two, but also because the authors do not observe the expected reduction in intensity at low galactic latitude caused by interstellar absorption. The various measurements on the low energy X-rays from Sco X-1 show even greater inconsistencies.

4. Modulation Collimators

Conventional slat collimators can practically be built to provide ~½° collimator. The modulation or wire-grid collimator on the other hand has been successfully used to much smaller angular widths. The device was invented by Oda [9] and was utilized by the AS&E/MIT collaboration to determine the angular size and position of the Crab Nebula and Sco X-1 with very high precision.

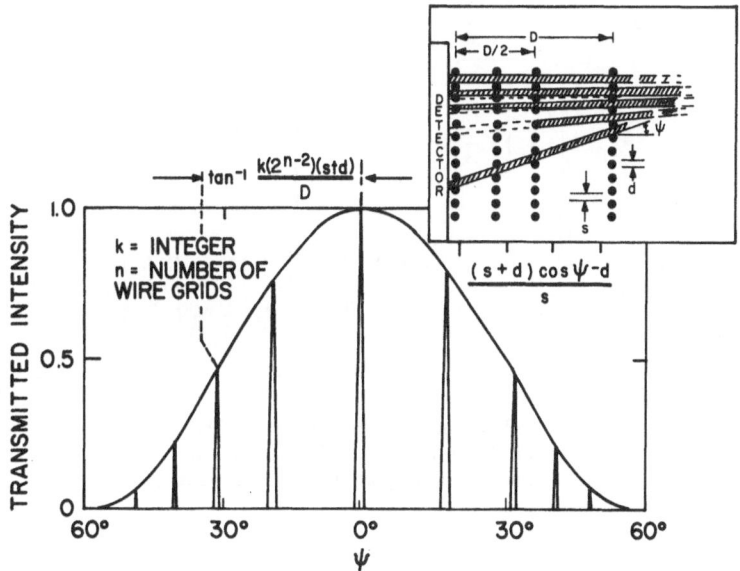

Fig. 6.   Angular response of a four grid modulation collimator.

The functioning of the modulation collimator is shown in Figure 6. The use of wires or electro-formed grids allows obtaining the very fine angular resolution. The planes of wires permit X-rays to arrive at the detector from regularly spaced bands in the sky. The angular width of each band is given by the ratio of the separation of the individual wires and the distances between the outermost wire planes. In the case of the Sco X-1, Crab experiment the wire separation was 0.005″ and the overall length of the collimator was 24″ yielding a basic angular resolution of 2×10^{-4} radians or 40 arc seconds. Data obtained by Gursky *et al.* [10] during the traversal of Sco X-1 is shown in Figure 7.

The modulation collimator can be used in a variety of ways as has been discussed by Bradt *et al.* [11]. A novel application of the device as a 'spatial filter' has been discussed by Schnopper *et al.* [12].

5. Sensitivity for Detection of Faint Sources

The sensitivity of an X-ray astronomy experiment for the detection of faint X-ray

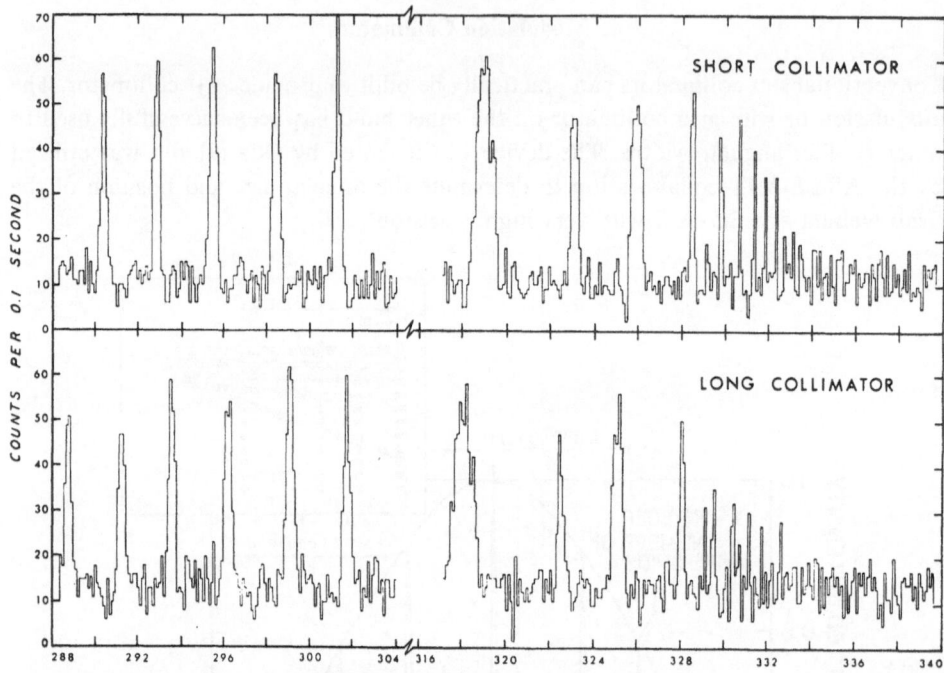

Fig. 7.   Response of a four grid modulation collimator to Sco X-1. Data obtained by Gursky *et al.* [10] during the sounding rocket flight of March 1966. Variations in the width and separation of individual peaks related to the varying drift rate of the rocket.

sources is calculated in this section. The calculation is quite general and can be applied to any scanning type of experiment. The only precaution is that background rates will, of course, be very dependent on the energy range being considered.

The accumulated count N including non-source counts from a source of intensity, I, is given by

$$N = (fI + \Omega B_1 + B_2) \, At \text{ counts} \tag{1}$$

where t, the actual observing time on a source, is given by either

$$t = (\xi/2\pi) \, T \text{ continuously rotating vehicle}. \tag{2}$$

$$t = \xi/\overset{\circ}{\xi} \text{ single slow scan}. \tag{3}$$

T is the total observation time, ξ is the field of view in the direction of the scan, and $\overset{\circ}{\xi}$ is the scan rate.

The other quantities in Equation (1) are as follows: f=vignetting factor (0.75 for mechanical collimators, if ξ is taken to be FWHM); Ω=total field of view (ster); B_1=diffuse X-ray background (counts/cm^2-sec) (Ster)); B_2=non-X-ray background (counts/cm^2-sec); and A=effective detector area.

Both B_1 and I can be obtained from more basic quantities from the following relation

$$I = \int_{E_1}^{E_2} \varepsilon(E)(d\phi/dE)\,dE \text{ counts/cm}^2\text{-sec} \tag{4}$$

where $d\phi/dE$=spectral distribution of the X-radiation (photons/cm^2-sec (keV)), and $\varepsilon(E)$=efficiency of detection as a function of energy E.

The efficiency for proportional counter in turn is calculated from the relation

$$\varepsilon(E) = \exp(-\mu_w x_w)\,[1 - \exp(\mu_g x_g)] \tag{5}$$

where μ=mass absorption coefficient of the detector window (μ_w) or gas (μ_g), and x=thickness (gm/cm^2) of the window (x_w) or gas (x_g).

Equation (4) is only approximate in that it gives the number of photons converting in the detector, but not the number of recorded counts. To obtain the latter one must take into account the details of the response of the detector, principally, the energy resolution and the escape of fluorescent radiation.

The maximum sensitivity is arrived at by the following recipe. In the first place, we assume I is much smaller than $\Omega B_1 + B_2$. Secondly, we state that the weakest observable source is one that gives a number of counts (N_m) equivalent to a 3σ fluctuation of the accumulated count when no source is present.

Thus

$$N_m = f I_m At = 3\sqrt{N}$$
$$= 3\sqrt{(\Omega B_1 + B_2)\,At} \tag{6}$$

or

$$I_m = (3/f)\sqrt{(\Omega B_1 + B_2)/At}.$$

Listed in Table IV are values of At appropriate to particular kinds of experiments and the corresponding minimum observable fluxes. B_1 is taken to be ~ 5 cts/cm^2-sec (ster). The values of B_2 were taken appropriate to experiments performed in the 1–10 keV energy range near the earth's equator, where the effects due to cosmic rays are mini-

TABLE IV
Minimum sensitivity for various experiments

	At(cm^{-2}-sec^{-1})	I_m(counts/cm^2-sec)	In relation to Crab
Recent AS & E Sounding Rocket Experiments (80 sq. deg.)	8000	1.5×10^{-2}	10^{-2}
Ultimate Sounding Rocket ($\Omega B_1 \ll B_2$)	10^5	10^{-3}	5×10^{-4}
X-ray Explorer (1 day, 25 sq. deg.)	4×10^5	10^{-3}	5×10^{-4}
X-ray Explorer (1 week, 25 sq. deg.)	3×10^6	4×10^{-4}	2×10^{-4}
Super Explorer (10 m^2, 1 hr. integration) ($\Omega B_1 \ll B_2$)	2×10^8	3×10^{-5}	1.5×10^{-5}

mized. The condition $\Omega B_1 \ll B_2$ is satisfied for Ω the order of several square degrees. The X-ray Explorer is a payload being prepared for flight in 1970. It will be described in Section 7. The Super Explorer is a payload now in the planning stage at NASA.

The sensitivities presented in Table IV must be examined critically, however, since even the 'ultimate' sensitivity achievable with a sounding rocket has not nearly been achieved. In the case of the sounding rocket, the assumption is that one does a simple on-source, off-source type of scan, which is also the assumption in the case of the Super Explorer example. In the latter case the observing time is taken to be 1 hour. The difficulty with both of these two cases is that they are not suitable for scanning large regions of the sky. They can only be used to sample the sky, or to look at regions of especial interest; otherwise, the total time requirements become prohibitive. This situation occurs as well with major radio and optical observatories; namely, that the ultimate sensitivity of the largest aperture instruments is achieved in only a very limited class of observations. A good example of this is the observation of the radio emission from Sco X-1 by Ables [13], in which at least one order of magnitude greater sensitivity was achieved than is usual for surveys with radio telescopes.

The basic conclusion is that very great improvements in sensitivity can be achieved with conventional instruments, approaching the $10^{-5}-10^{-6} \times$ Crab discussed earlier.

6. Measurements of Source Characteristics

A. LOCATIONS

A single traversal of a source with a mechanical collimator yields a counting rate profile that is typically triangular in shape. The width of the triangle is determined by the collimator and can be as narrow as $\sim \frac{1}{2}°$ for conventional, slat collimators, or as little as a fraction of an arc minute for a wire-grid collimator. The position of the centroid of the counting rate profile determines the X-ray source location to a line on the sky and repeated traversals with varying collimator orientations can be used to obtain intersecting lines of position to further pinpoint the location of source as is shown in Figure 8, which is based on data obtained from a scan of the Cygnus region. Alternately, one can build up count-rate distribution on the sky and find the centroid of that distribution. The precision with which locations can be determined is directly related to the half-width of the collimator and the strength of the source. Also, since it is not possible to reference one X-ray source location to that of other known X-ray sources, it is necessary to determine independently the orientation of the collimators with respect to the star-field.

B. SPECTRAL DATA

The use of proportional counters and scintillation counters allows obtaining spectral data by analyzing the amplitude distribution of the counter signals; essentially, by unfolding the counter response from the amplitude distribution. The technique is described in detail by Gorenstein *et al.* [11a]. The principal complicating factor in these analyses is the fact that not all the energy of an X-rays interacting in the gas of a proportional counter (or in a scintillation crystal) is necessarily deposited in the counter. Secondary fluorescence radiation can escape from the counter and the resulting signals have a lower than expected amplitude.

Spectra obtained in the 1–10 keV can be resolved to between 20 and 30%. At low

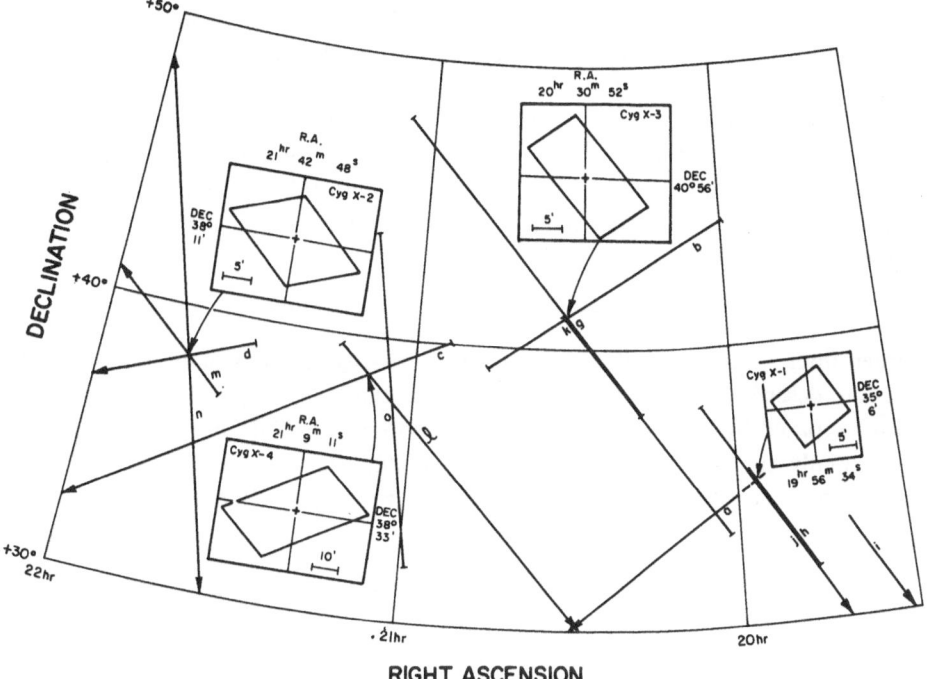

RIGHT ASCENSION

Fig. 8. Location of X-ray sources in Cygnus based on sounding rocket flight of November 1967. Individual lines of positions *a* through *o* are obtained from centroids of peaks in the counting rate curve.

energy, spectral data are best obtained by filter techniques. The presence of absorption edges allows high transmission only below the energy of the edge. The filter can be just the counter window. The very soft X-ray detectors all make use of the carbon edge to obtain spectral data at 0.28 keV. Other counters have been constructed that make use of the aluminum edge at 1.5 keV and the fluorine edge (teflon) at 0.7 keV.

Although spectral data have been used almost exclusively to determine the shape of the energy distribution, it should be possible to detect at least one spectral line, if it is present, namely, the iron line at 7.1 keV. Tucker [14] has calculated that the emission in this line should constitute about 1% of the total X-ray emission from a hot plasma of 5×10^7 K temperature with the normal abundance of iron. This feature has been searched for unsuccessfully in Sco X-1 by Fritz *et al.* [15] during an observation with an exposure (At) of 4000 cm²-sec. The authors estimated that an exposure of at least 16000 cm²-sec was required to observe the iron line strength predicted by Tucker. Such exposures are well within the capabilities of sounding rocket experiments.

18   HERBERT GURSKY

C. POLARIZATION

Novick and his collaborators have searched for polarization of the X-rays from Sco X-1 and from the Crab Nebula using a Thompson scattering polarimeter [16]. The technique involves allowing the X-rays to be incident on blocks of lithium and detecting the X-rays scattered at right angles. Because of the competition between photo-absorption and scattering, the sensitivity of such a polarimeter peaks at about 7 keV. The result on Sco X-1 was that the source was unpolarized to a 1σ limit of 7%, for an exposure of about 10^5 cm²-sec. The actual area was 900 cm² and represents the geometric area of the entire instrument. In order to obtain a significant result on the Crab an exposure of about 10^6–10^7 cm²-sec with this instrument is required. It may be possible to construct a more efficient polarimeter using hydrogen as the scatterer.

7. Description of the X-Ray Explorer

Rather than describe additional elements of X-ray instrumentation individually, I want to describe an entire experiment; namely, the X-ray Explorer Payload which is being prepared by AS&E under NASA sponsorship for flight in 1970. The payload is essentially similar to sounding rocket payloads that we and others have flown. This experiment is the first of NASA's small astronomy satellites (SAS). The second

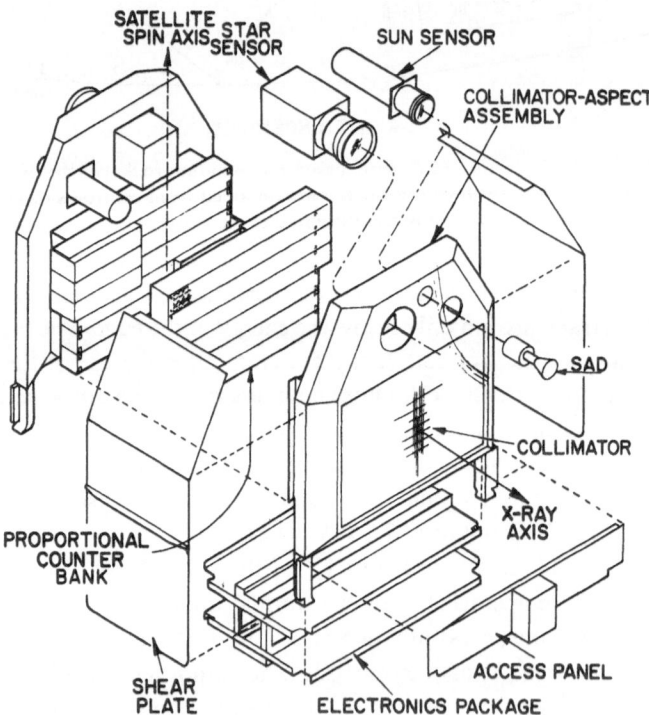

Fig. 9.   Exploded view of the experiment portion of the X-ray Explorer satellite.

SAS payload is to be devoted to γ-ray astronomy using a spark chamber. Later SAS payloads will be pointed and are planned for both X-ray astronomy and optical astronomy.

The principal scientific objective of the Explorer mission is an all-sky survey to a limit of sensitivity of about $10^{-3} \times$ Crab in the energy range 1–10 keV. Spectral data are obtained by the use of proportional counters. Positional accuracy of source locations will be $\sim 1'$ for the stronger sources. Time variations can be studied on a time scale of minutes, days and months and the useful life of the satellite should be at one year.

The payload is shown schematically in Figure 9. The unit is organized into two, almost identical halves. The collimators are built up from extruded aluminum tubing. The field of view on one side $\frac{1}{2} \times 5°$ (FWHM) and the other side is $5° \times 5°$. The detectors on both sides are beryllium window proportional counters. The window thickness is 2 mils (50μ). The effective area on each side is about 800 cm^2. Attitude is determined by a star sensor – sun sensor combination.

The spacecraft portion is being built by the Applied Physics Laboratory of Johns Hopkins University. Its basic characteristics are as follows. The spacecraft is spin-stabilized by a high-speed rotor, the spacecraft itself will spin with about a 10 min period, which will be controllable. The spin axis can be pointed to any position in

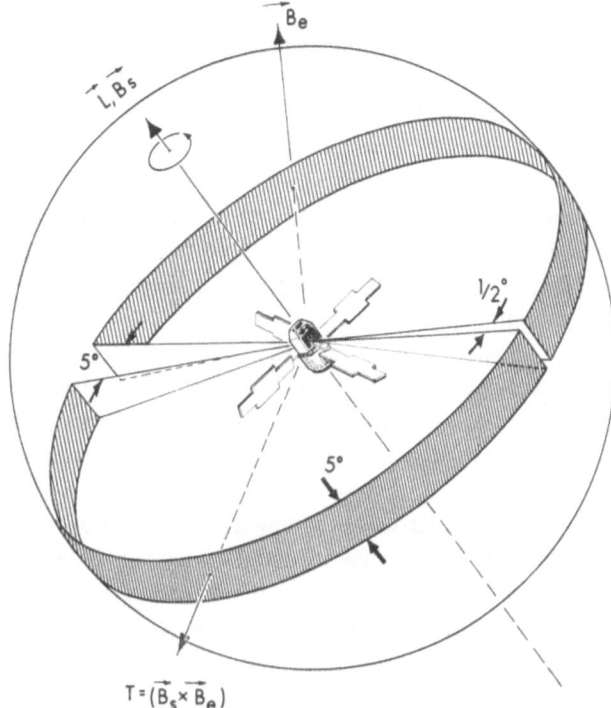

Fig. 10. Scan and maneuver geometry for the X-ray Explorer on the celestial sphere. Maneuvring is performed by internally generating the magnetic field B_s along the spacecraft spin axis which interacts with the earth's magnetic field B_e to produce the Torque T as shown.

the sky using a magnetic torqueing control. As shown in Figure 10 the X-ray detectors will sweep out a 5° band in the sky once per rotation. We plan to spend between $\frac{1}{2}$ and one day per band and then reposition the spin axis to scan out another band. At this rate about a month will be required to scan the entire celestial sphere.

As shown in Table IV, the ultimate limit of sensitivity of the Explorer payload could be about $10^{-4} \times$ Crab. Based then on the extrapolation of source distributions shown in Table II, as many as $\sim 10^3$ sources could be observed compared to the 40 or so now reported. Of course many of these, particularly in the Sgr-Sco region will not be resolved.

In addition to an all-sky survey, the maneuvring capability and the extended lifetime of the Explorer spacecraft allows operating the experiment in a manner analogous to that of a conventional astronomical observatory. Observations can be planned according to a pre-arranged schedule to allow simultaneous acquisition by other groups of optical or radio data. Individual experiments can be planned that call for extended periods of time observing a single source or particular region of the sky and certain transient phenomena (such as novae outbursts) can be investigated for accompanying X-ray emission.

8. Grazing Incidence Optics

I have discussed above certain limitations of 'conventional' X-ray instrumentation; notably, the difficulty of working at long wavelengths and the restriction on angular resolution. Instruments that employ grazing incidence optics do not suffer from these limitations and, additionally, offer the possibility of very high sensitivity for the detection of faint sources, high spectral resolution, and efficient polarimetry. Above all, telescopes allow obtaining pictures in X-rays of cosmic features such as the Crab Nebula, Cas A and M-87 and more ambitiously of M-31. Such pictures have been obtained of the sun by the Solar Physics Groups at American Science and Engineering and the Goddard Space Flight Center.

The fact that X-rays reflect at grazing incidence allows the construction of focussing optics. This was first discussed by Wolter [17] in connection with X-ray microscopy and astronomical applications were proposed by Giacconi and Rossi [18]. Functioning telescopes were first developed at AS & E and their initial application was solar photography.

X-ray optics can be used in a variety of configurations, as shown in Figure 11. The only configuration which I will discuss in detail will be the two-surface focussing arrangement of paraboloid-hyperboloid. This is not to say that the other configurations may not find important applications in certain circumstances, it is rather that this one has the widest range of applications and imposes a minimum of operational restraints.

The most important single parameter characterizing the device is the grazing angle (α) for paraxial radiation on the first surface. The wavelength response, the collecting area and the focal length are determined by this quantity. The calculated reflectivity

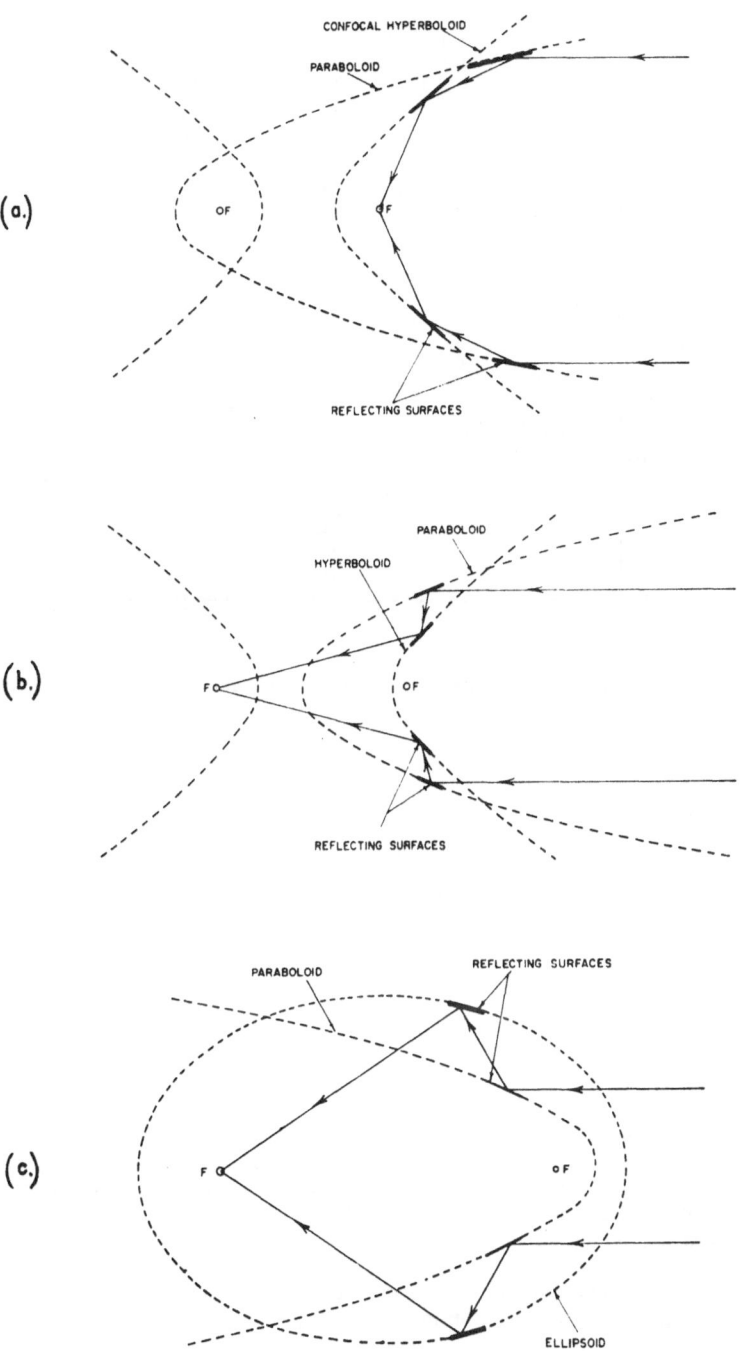

Fig. 11.   Three possible configurations of reflecting services that can be used for producing
X-ray energies.

HERBERT GURSKY

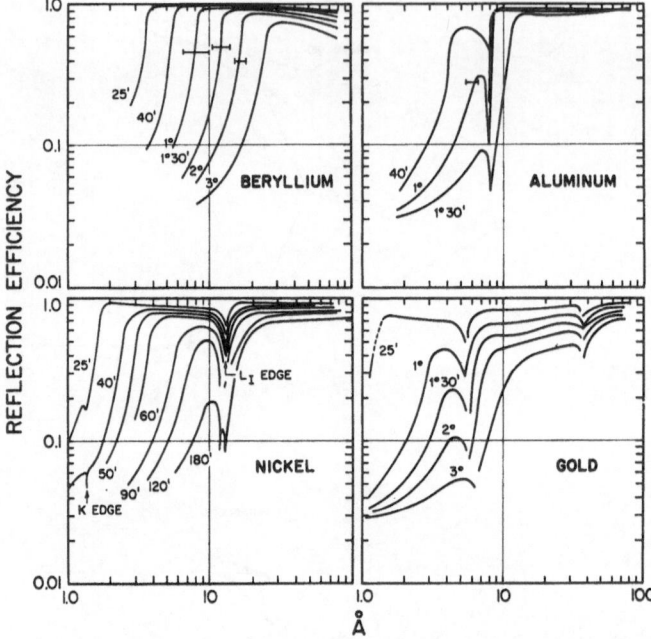

Fig. 12.   Calculated reflectivity for several materials at grazing angles of incidence.

TABLE V

Telescope parameters

	Telescope A[a]	Telescope B[a]
Diameter (cm)	30.5	87
Focal length (meters)	2.1	6.1
Collecting area (cm²)	44	1000
Field of view, diameter (degrees)	1	1
Point-source image distribution	5% of incident flux imaged in 5 arc sec diameter circle	20% of incident flux imaged in 2 arc sec diameter circle
Telescope resolution element (maximum size)	5 arc sec	2 arc sec
Net effective area in tel. resolution element (cm²)[b]	0.22	20
Noise/telescope resolution element (counts/sec)	10^{-3}	10^{-3}
Number of telescope resolution elements in field	5×10^5	3×10^6

[a] The mirrors of Telescope A are identical to those which have been fabricated for the S-054 solar X-ray telescope experiment to be flown on the NASA Apollo Telescope Mount. Telescope B is a system we are designing for a stellar X-ray astronomy experimental program.
[b] Calculated on the basis of a detector efficiency of 10%.

for several materials is shown in Figure 12. One sees the characteristic cut off at short wavelength and the discontinuities at the absorption edges.

Other properties of a telescope are derived (approximately) from the following relations:

Collecting area $A = \pi R^2 l\alpha$

Focal length $f = R/8\alpha$

where R=radius of aperture and l=length of mirror section. The field of view of the telescope is approximately α.

The specifications of mirrors that are being planned for flight use are listed in Table V.

In principle very high angular resolution can be attained with X-ray telescopes. Even for the smallest aperture devices yet built the diffraction limit is less than 0.01″. More practical limits are imposed by mechanical tolerances and surface finish. The results of a ray tracing study are shown in Figure 13, which indicates that a fraction of an arc second resolution can be achieved within the central arc minute of a 'real' telescope; by real I mean one that is figured to state of the arc tolerances. Resolution

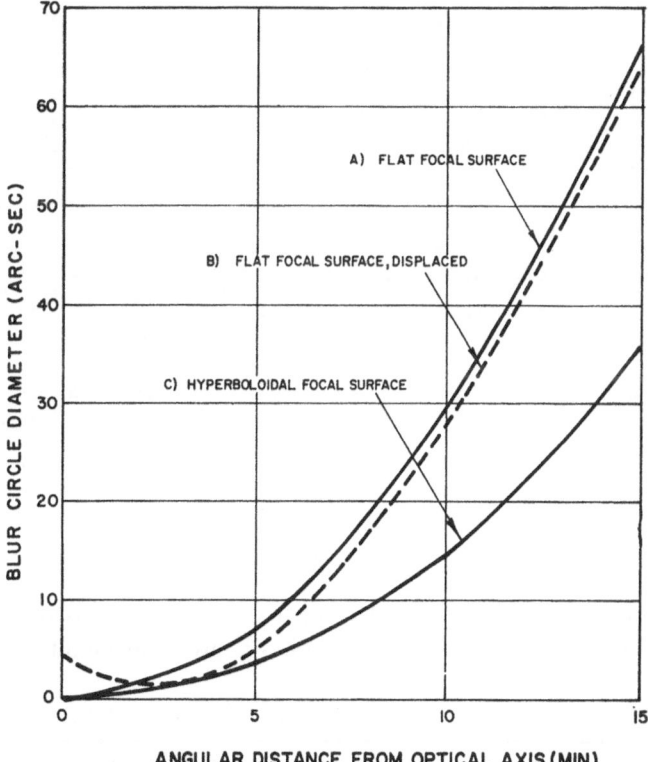

Fig. 13.   Blur circle diameter vs. angular distance for the optical axis of a paraboloid-hyperboloid telescope as obtained by the ray drawing analysis.

of several arc seconds has been achieved in AS & E sounding rocket flights [19], as is shown in Figure 14. Examples of existing telescopes are shown in Figure 15.

In comparison to the angular resolution, only poor efficiency (reflectivity of the surfaces) has been obtained with present telescopes. The telescope used to obtain the picture in Figure 14 had a net reflectivity of only several percent or about a factor of 10 below the calculated value. This is apparently the result of diffuse scattering from surface irregularities. The present AS & E mirrors have been fabricated by Diffraction Limited from Kanigen (an amorphous nickel alloy) which is coated onto beryllium. The surface finish is good only to between 50–100 Å. It is anticipated that more satisfactory performance can be obtained with materials such as cervite or fused silica

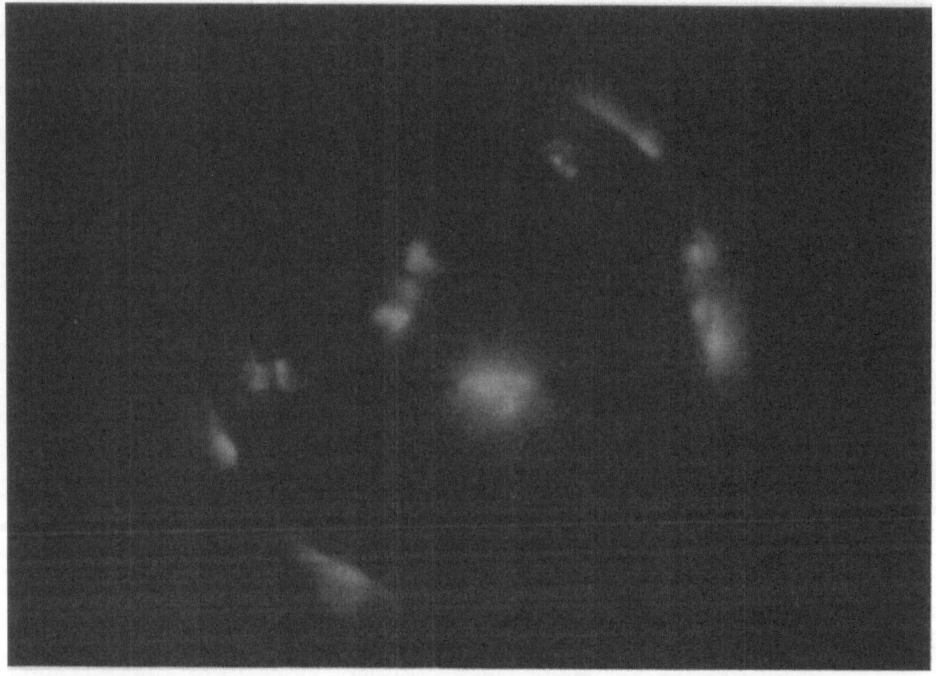

Fig. 14.   X-ray photograph of the sun obtained on June 8, 1968, by the AS & E Solar Physics group using a 9″ diameter telescope. The rocket was flown during a solar flare which shows up as the very bright region near the center of the picture.

which is known to take a superior finish compared to Kanigen. Fused silica telescopes have been successfully flown by the Solar Physics group at Goddard Space Flight Center [20].

Fig. 15.   Existing X-ray telescopes. The top figure shows the sounding rocket telescope used to obtain the photograph in Figure 14; the bottom photograph illustrates a pair of confocal, axial telescopes that will comprise the mirrow system for AS & E solar ATM experiment.   →

A. IMAGING WITH X-RAY TELESCOPES

The classical use of a telescope is, of course, to produce an image of an object. Equally important is the fact that telescopes concentrate; i.e., produce a much higher flux density in the focal plane than is present at the aperture. In view of the high background of X-radiation that is present, this concentration means that telescopes can be used to achieve very high sensitivity. Assuming that one has a 'noiseless' detector, the faintest observable source is one yielding a flux, Φ min, given by,

$$N_n = 10 = \varepsilon A T \Phi_n \text{ counts}$$

or

$$\Phi_n = 10/\varepsilon A T \text{ photons/cm}^2\text{-sec},$$

where 10 counts is taken to be the minimum signal (against a background of zero) indicating the presence of a source. The quantities A and T are as defined previously and ε is the net throughput, including the mirror reflectivity and detector efficiency.

In a 'typical' sounding rocket experiment $A = 40 \text{ cm}^2$, $T = 250$ sec and assuming $\varepsilon = 0.10$ which should be achievable, Φ min $\simeq 10^{-2}$ photons/cm^2-sec which is adequate to image any of the known sources. A more ambitious, but not totally unrealistic case is $A = 10^3 \text{ cm}^2$, $T = 3600$ sec, in which case Φ min $\simeq 3 \times 10^{-5}$ photons/cm^2-sec, which is at least as good as the 'super-explorer'. Furthermore, unlike conventional detection systems, the telescope should not suffer from source confusion at even this sensitivity.

Achieving even modest sensitivity with telescopes requires the use of X-ray image converters that combine good spatial resolution and high quantum efficiency. We do not consider film to be useable in these experiments; not only because of its low efficiency, but also because of the operational constraints that are imposed; namely, the long integrating time while accurately pointed and the awkwardness of data retrieval.

The X-ray imaging device that AS & E has been developing is based on the use of channel-plate multipliers as have been developed by Bendix, Rauland (Div. of Zenith) and Mullard. This device (shown schematically in Figure 16) consists of a bundle of many thousands of individual glass tubes, each comprising a channel electron-multiplier. Photons, converting at the front end, continuously accelerate and multiply down the tube. A net gain of $\sim 10^5$ electrons/photoelectron is achieved. In our devices, the output electrons are additionally accelerated onto a phosphor screen. The net gain is such that each X-ray converted at the front end will yield a visible (or photographable) spot of light on the phosphor screen. The sensitivity to X-rays is achieved by coating a bit of the inner surface of the front end of the tubes with CsI or similar material which is known to have a high efficiency for conversion of X-rays to photoelectrons. The noise in devices which AS & E has been using has been $\sim 10^{-3}$ counts/res. elem-sec and it is not known whether this figure represents any kind of limit. The number is the basis for our statement about noiseless detectors; namely, for an image that occupies a single resolution element, the 'noise' would amount to only a single 'count' in 1000 sec.

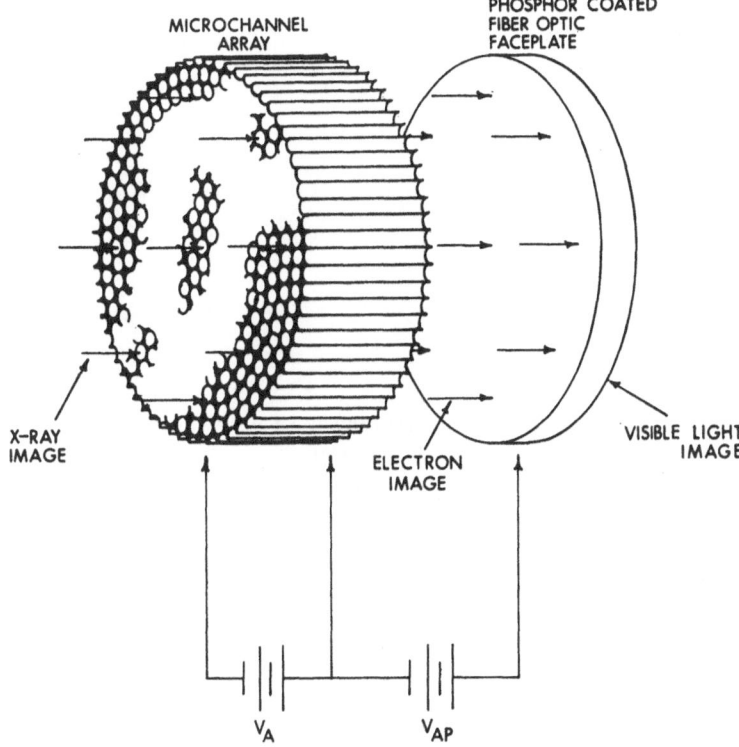

Fig. 16.   Schematic outline for an X-ray image intensifier that makes use of a micro-channel electron multiplier array. Each channel defines a single resolution element.

B. X-RAY SPECTROSCOPY

Two kinds of dispersing spectrometers have been proposed for use with focussing telescopes; one, which has been studied by Schnopper and Kalata [21] makes use of a curved Bragg Crystal in a Johann Mount, as shown in Figure 17. The X-ray image of the telescope is placed on the Rowland circle and forms a line-image on the Rowland circle for X-rays satisfying the Bragg condition. The choice of crystal is determined by the wavelength to be studied. Mica with a 2d spacing of 13 Å and KAP with a 2d spacing of 26.4 Å are typical choices. With these spectrometers, resolution of $(\lambda/\Delta\lambda)$ of $\sim 10^3$ is attainable.

The Bragg spectrometer can be used without a telescope as is commonly done in studies of the sun, in which case a flat crystal is used. There are two principle difficulties with this geometries; namely, since the crystal face constitutes the aperture, very large apertures may be difficult to attain, and also the detector must be as large as the crystal face and one has a very adverse signal/noise situation since only a very small fraction of the incident beam is transmitted to the detector.

A totally novel dispersing spectrometer has been developed for use in X-ray astronomy, which is the slitless spectrometer as proposed by Gursky and Zehnpfennig

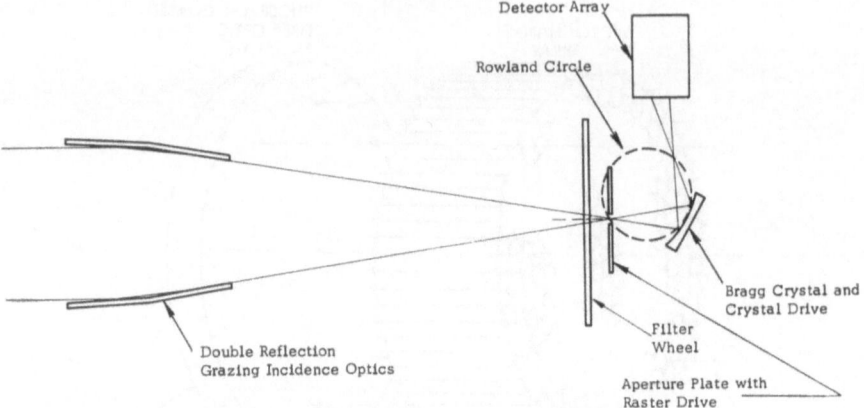

Fig. 17. A focussing high resolution spectroscopy using an X-ray telescope and a spherically bent crystal.

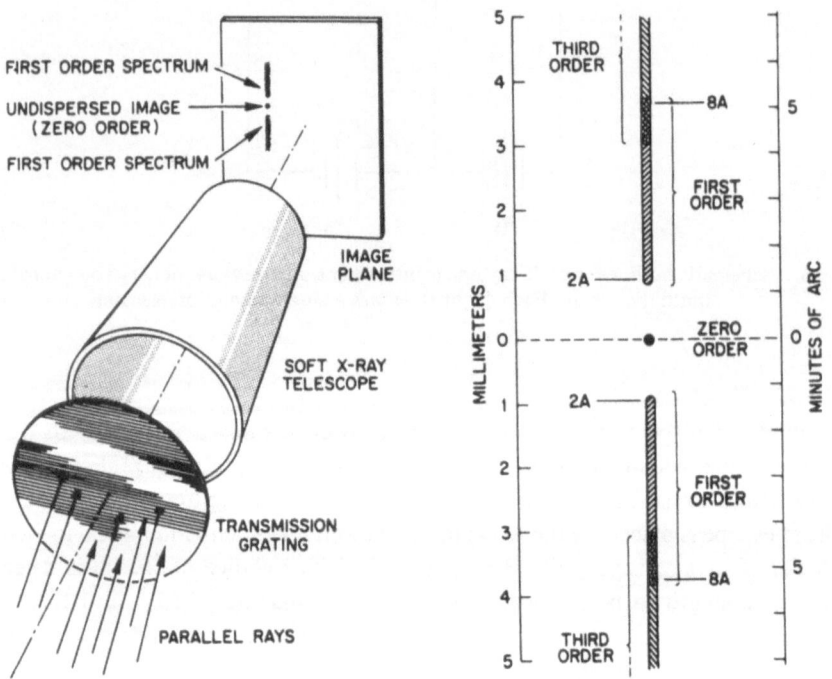

Fig. 18. Schematic representation of the X-ray slitless spectrometer using a transmission grating.

[22]. As shown in Figure 18, a transmission diffraction grating placed behind the telescope mirror, produces a diffracted image in the focal plane. The incident beam is diffracted by the amount

$$\theta = n\lambda/d \quad n = 0, 1, 2, \ldots,$$

where d is the grating spacing. In principle the dispersing power of the spectrometer is determined by the quality of the grating as H is for any grating spectrometer, in practice, the actual angular dispersions are so small ($\sim$ arc minutes) that the angular resolution ($\Delta\theta$) of the mirror determines the dispersing power; i.e.,

$$\lambda/\Delta\lambda \approx \theta/\Delta\theta = n\lambda/d\,(\Delta\theta).$$

A spectrum obtained with this device is shown in Figure 19. $\lambda/\Delta\lambda$ of 50–100 seems to be attainable.

The Bragg spectrometer and grating spectrometer are totally different instruments

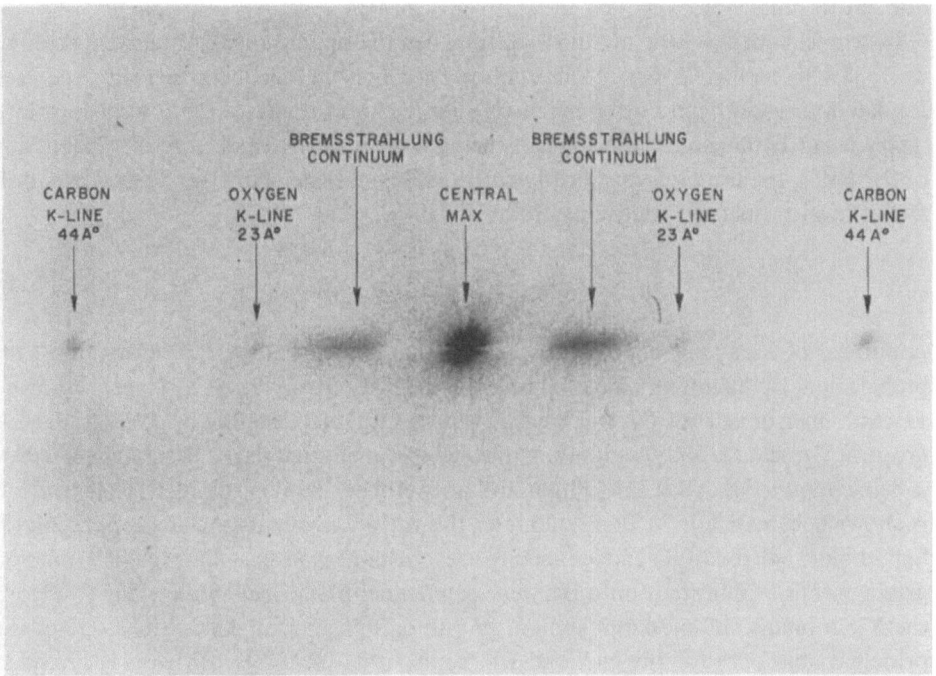

Fig. 19. A spectrum obtained from a thick target bremsstrahlung X-ray source using the slitless spectrometer obtained by Zehnpfennig.

and will be utilized for correspondingly different observations. The Bragg spectrometer is very slow since it is sensitive to only a single wavelength per crystal setting. On the other hand its great resolution allows the possibility of studying line profiles and of observing wavelength shifts as might be caused by doppler or gravitational effects. By comparison the grating spectrometer is a crude device, but does have very high sensitivity since a large fraction of the X-rays ($\sim 40\%$) is dispersed and the entire spectrum can be studied simultaneously. This device will be most useful for determining the general features of X-ray spectra (line emission, absorption features) for even very faint sources.

C. OTHER APPLICATIONS OF GRAZING-INCIDENCE OPTICS

The ability of X-ray optics to concentrate flux allows development of instruments that otherwise do not necessarily make use of the imaging properties of the optics. The concentration factor is just the ratio of the effective telescope collecting area to the arc of the detector in the focal plane and can be as large as several thousand. This means that the signal/noise in the detector can be improved by the same amount. Fisher of Lockheed has used an arrangement of focussing slats in this capacity. A more ambitious instrument developed by the Columbia group used microscope slides arranged in a segmented approximation to a two-surface telescope. These arrangements should be particularly useful at low photon energies where large detectors are difficult to build.

Instruments of this kind are most useful when the signal/noise is otherwise prohibitively low to perform a useful observation, or when the particular detector required for the observation can not be made very large. An example of the former is polarimetry using Thompson scattering as is being performed by the Columbia group, and of the latter, the use of cryogenically cooled solid-state detectors for spectral analysis as has been proposed recently by Boldt of GSFC.

10. Description of Telescope Systems

A number of X-ray astronomy experiments that incorporate X-ray optics are in the preparation or planning stage. These include the University College, London-Leicester instrument for OAO-C and the proposed Dutch satellite by Prof. de Jager's group at Utrecht. A NASA sounding rocket experiment that uses a focussing telescope is being prepared at AS & E for flight and an Aerobee 150 this summer. That payload is shown schematically in Figure 20, and the various elements include a paraboloid-hyperboloid mirror of 34 cm^2 collecting area (Kanigen on beryllium), a transmission grating (gold evaporated) onto parylene, a channel-plate image intensifier to record the X-ray image and a 16 mm camera to photograph the star-field. Since one of the principal objectives for this payload will be accurate locations, a means is provided to cross-reference the X-ray image intensifier and the star-field camera. This is accomplished by having an image of the image intensifier appear on the star-field. The

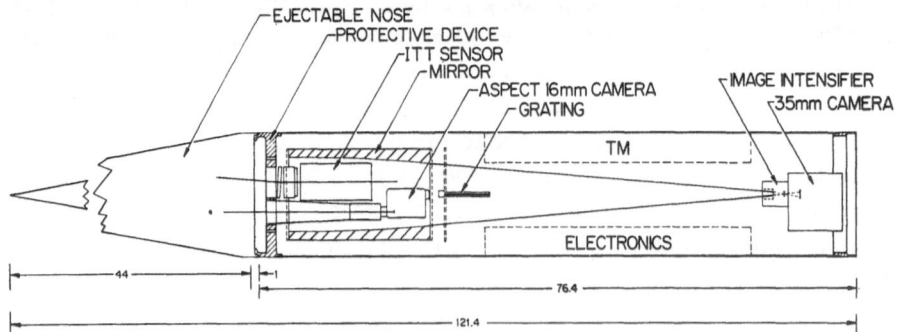

Fig. 20. Layout of a sounding rocket experiment making use of an X-ray telescope for high resolution studies of X-ray sources. This payload is being prepared for flight on an Aerobee 150.

face of the image intensifier will be illuminated and will be viewed through the X-ray telescope by the star camera through a cross-cube prism. This allows direct reference of the X-ray image to the star image.

The sounding rocket group at GSFC is responsible for the remainder of the instrumentation. The guidance system is the stellar-strap, which consists of a gyro-controlled, gas jet course attitude control, and an ITT star-tracker for fine attitude signals. The star-tracker will lock onto a pre-selected bright star ($\leqslant +2.5$ m); since, the X-ray object in general is not coincident with a visually bright object there must be an angular offset between the star-tracker axis and the X-ray optic axis. This system will hold to a target to within $\sim 5''$.

The capability of this payload will be as a slitless spectrometer to obtain a spectrum of Sco X-1, or to image any of the known sources with the possible exception of M-87. As an example of the latter, the recently prepared pulsar in the Crab should show up clearly as a point image within the diffuse nebula, even during the short time duration of a rocket flight. A far more ambitious NASA program calls for the use of a single or a cluster of large area telescopes as a facility with a number of instruments available in the focal plan for observations. One possibility provides for two telescopes, one with high-resolution ($\sim 1''$) and ~ 1000 cm^2 collecting area, to be used for imaging and high resolution spectroscopy and the other of moderate resolution ($\sim 5''$) and 5000 cm^2 collecting area to be used for polarimetry, low resolution spectroscopy and for low noise-high sensitivity applications. A schematic of a possible payload is shown in Figure 21. The experiment could be launched from a Titan-3c and it might be possible to station-keep the experiment from a nearby manned orbiting workshop. Station-keeping allows the use of film and cryogenics, and the possibility of replacing the focal plane instruments.

11. Summary

Based on present knowledge, the instruments available for observations in X-ray astronomy are capable of several orders of magnitude improvement compared to what is actually being attained at the present time. These improvements include detection o f very faint sources, high resolution imaging, spectroscopy and polarimetry. The only real limitation to the application of these instruments is the availability of space-borne platforms. It is also noteworthy that the principal requirements on these space platforms is weight-carrying capability. The stability, maneuvring, and data transmission requirements are typically less stringent than what now seems to be routinely accomplished in many satellite programs. Furthermore, one does not require exotic orbits; the most desirable place for an X-ray astronomy experiment is near the equator and at low altitude where we obtain the maximum shielding from cosmic rays by the terrestrial magnetic field and is still out of the regions of trapped radiation.

Acknowledgements

I wish to acknowledge the assistance that I have had with the astronomy group at

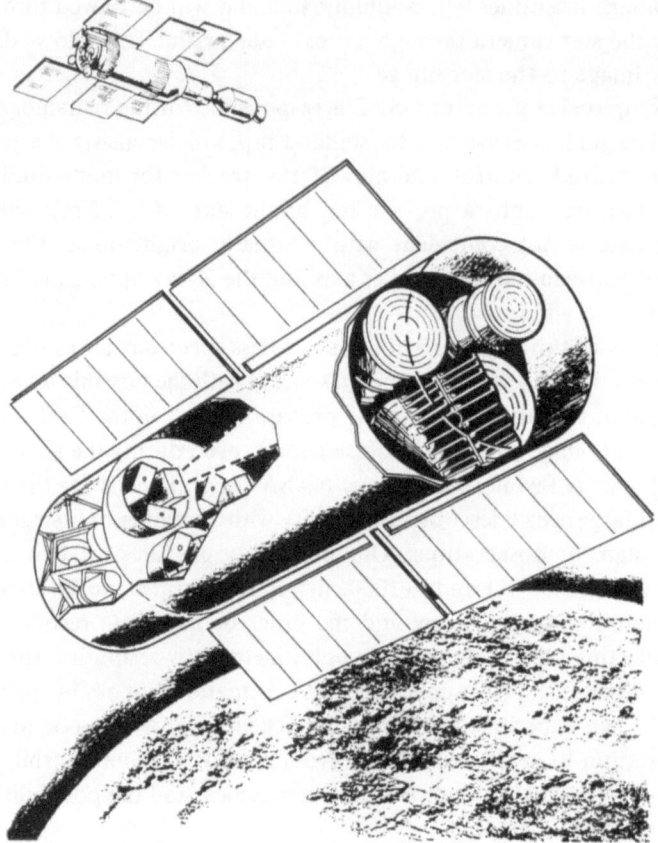

Fig. 21.   Conceptual drawing of a large X-ray telescope facility where telescope is shared
by several users.

AS&E with preparation of the report and in particular with Drs. Gorenstein and
Kellogg for their contributions in the area of conventional X-ray astronomy experiments, and Drs. Vaiana, Van Speybroeck and Zehnpfennig for their contributions in
the area of telescope techniques. I have also benefited from discussions with Dr.
Giacconi and Professor Rossi of MIT regarding the contents of the paper.

Also, I wish to thank Professor Peterson, University of California, San Diego,
Dr. Seward, Lawrence Radiation Laboratory, and Professor Bunner, University of
Wisconsin, for providing material that I have used in this paper.

References

[1] Giacconi, R., Gursky, H., and Van Speybroeck, L.: 1968, *Ann. Rev. Astron. Astrophys.* **6**, 373.
[2] Giacconi, R., Reidy, W., Vaiana, G., Van Speybroeck, L., and Zehnpfennig, T.: 1969, *Space Sci. Rev.* **9**, 3.
[3] Bell, K. L. and Kingston, A. E.: 1967, *Monthly Notices Roy. Astron. Soc.* **136**, 241.
[4] Bowyer, C. S., Field, G. B., and Mack, J. E.: 1968, *Nature* **217**, 32.

[5] Mathieson, E. and Sanford, P. W.: 1964, in *Proceedings of the International Symposium on Nuclear Electronics, Paris, 1963*, ENEA, p. 65.

[6] Gorenstein, P. and Mickiewicz, S.: 1968, *Rev. Sci. Inst.* **39**, 816.

[7] Henry, R. C., Fritz, G., Meekins, J. F., Friedman, H., and Byram, E. T.: 1968, *Astrophys. J.* **153**, L11.

[8] Baxter, A. J., Wilson, B. G., and Green, D. W.: 1969, *Astrophys. J.* **155**, L145.

[9] Oda, M.: 1965, *Appl. Opt.* **4**, 143.

[10] Gursky, H., Giacconi, R., Gorenstein, P., Waters, J. R., Oda, M., Bradt, H., Garmire, G., and Sreekantan, B. V.: 1966, *Astrophys. J.* **146**, 311.

[11] Bradt, H., Garmire, G., Oda, M., Spada, G., Sreekantan, B. V., Gorenstein, P., and Gursky, H.: 1968, *Space Sci. Rev.* **8**, 471.

[11a] Gorenstein, P., Gursky, H., and Garmire, G.: 1968, *Astrophys. J.* **153**, 885.

[12] Schnopper, H. W., Thompson, R. I., and Watt, S.: 1968, *Space Sci. Rev.* **8**, 534.

[13] Ables, J. G.: 1969, *Astrophys. J.* **155**, L27.

[14] Tucker, W. H.: 1967, *Astrophys. J.* **148**, 745.

[15] Fritz, G., Meekins, J. F., Henry, R. C., Bryam, E. T., and Friedman, H.: 1968, *Astrophys. J.* **153**, L199.

[16] Angel, J. R. R., Novick, R., Vanden Bout, P., and Wolf, R.: 1969, *Phys. Rev. Letters* **22**, 861.

[17] Wolter, H.: 1952, *Ann. Phys.* **10**, 94.

[18] Giacconi, R. and Rossi, B.: 1960, *J. Geophys. Res.* **65**, 773.

[19] Vaiana, G., Rossi, B., Zehnpfennig, T., Van Speybroeck, L., and Giacconi, R.: 1968, *Science* **161**, 564.

[20] Underwood, J. H. and Muney, W. S.: 1967, *Solar Phys.* **1**, 129.

[21] Schnopper, H. and Kalata, K.: 1969, *Appl. Phys. Letters* **15**, 134.

[22] Gursky, H. and Zehnpfennig, T.: 1966, *Appl. Opt.* **5**, 875.

GENERAL SURVEY OF X-RAY SOURCES

(Invited Discourse)

H. FRIEDMAN

Naval Research Laboratory, Washington, D.C., U.S.A.

Prof. Friedman did not communicate any text or Summary of his invited discourse.

TECHNIQUES FOR IMPROVING THE SENSITIVITY OF
PROPORTIONAL COUNTERS USED IN X-RAY ASTRONOMY

P. W. SANFORD, A. M. CRUISE, and J. L. CULHANE

Mullard Space Science Laboratory, University College of London, London, England

The discovery and measurement of cosmic X-ray sources has almost invariably been performed with proportional counters which have large window areas. In the energy range from 1 to 50 keV, proportional counters have advantages over other types of detectors; they provide energy resolution and they can be made relatively easily with very large window areas.

The ability to detect cosmic X-ray sources is determined by the collecting area of the detector, the available observing time and the background counting rate of the detector. The observing time and the collecting area are necessarily limited by the performance and the size of the rocket or satellite. However, the background counting rate is within the control of the experimenter and it is this factor which we have studied and been able to improve.

There are two main contributors to the background counting rate; the isotropic cosmic X-ray flux and the response due to higher energy charged particles and γ-rays, which deposit only a small part of their energy in the proportional counter. The contribution from the isotropic flux can be reduced below that from the higher energy radiation by suitable collimation. In this paper we are concerned with reducing the background contribution from the higher energy radiation.

In 1963 Mathieson and Sanford [1] discovered that it was possible to discriminate between X-ray pulses and background pulses from a proportional counter. Although background and X-ray counts may have equal amplitudes, their pulse shapes differed. They were able to show that the response to γ-rays from Co^{60} could be reduced by one order of magnitude, with the aid of pulse shape discrimination, whilst leaving the response to low energy X-rays unchanged.

It is of course possible to reduce the background from high energy charged particles by using anticoincidence shields. However, charged particles only contribute between 30 to 50% of the high energy radiation in the laboratory and about 30% at rocket altitudes. The remaining contribution from γ-rays cannot be effectively eliminated with anti-coincidence counters. Pulse shape discrimination does, however, offer the possibility of significantly reducing the background from γ-rays and charged particles.

The pulse shape discrimination system which we have developed for rocket payloads is shown in Figure 1. Pulses from the preamplifier are differentiated twice and then passed to the Schmitt trigger. The output of the Schmitt trigger is integrated and its amplitude is then a function of the rise time of the pulse from the proportional counter. Pulses due to cosmic rays or γ-rays have a slower rise time than pulses from X-rays. The comparator generates a blocking pulse for the slower rise time pulses.

The proportional counter which has been developed for rocket payloads is shown

L. Gratton (ed.), Non-Solar X- and Gamma-Ray Astronomy, 35–40. All Rights Reserved
Copyright © 1970 by the IAU

in Figure 2. The window has an area of 800 cm^2 of 6 μ mylar and it is supported by
the egg crate collimator. Two anodes are used and these are mounted parallel along
the counter. The performance of the pulse shape discrimination system is shown in
Figure 3. We have found that it is essential to check the pulse shape discrimination

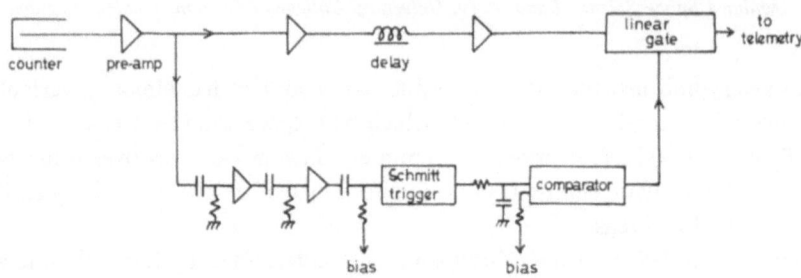

Fig. 1.   Pulse shape discrimination system.

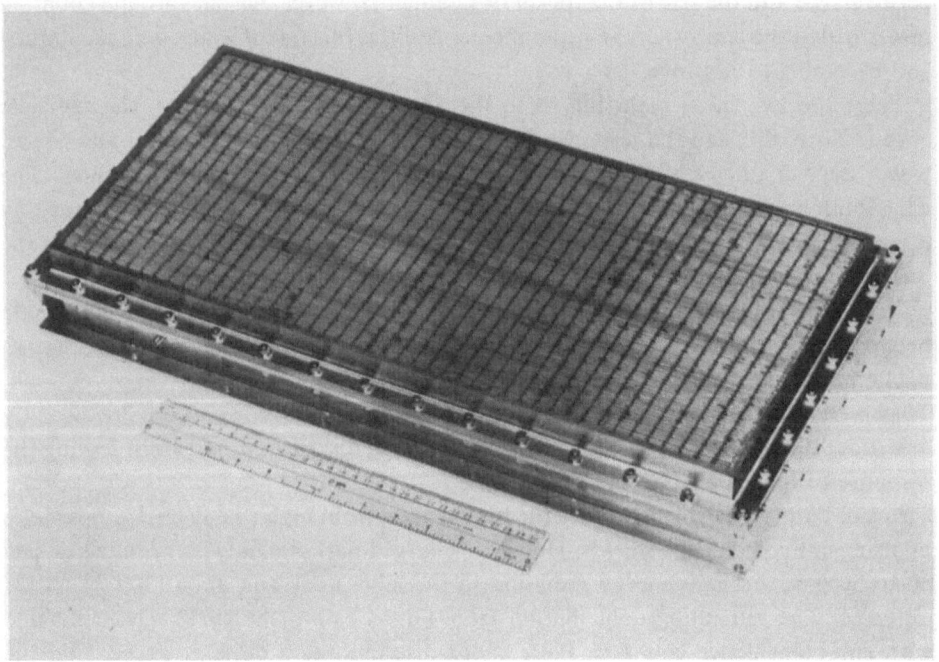

Fig. 2.

system over a wide energy range as it is no trivial problem to ensure that the system is
not energy dependent. The top curves are pulse amplitude spectra with and without
pulse shape discrimination for sources of Fe^{55} and Pu^{238} which give peaks corresponding to 3, 6, 13.5, 16.5 and 20 keV. The lower distributions, from Cs^{137}, demonstrate the efficacy of the system. At the most probable pulse amplitude the Cs^{137}
response is reduced by a factor of 50 while the X-ray response is reduced by only 10%.

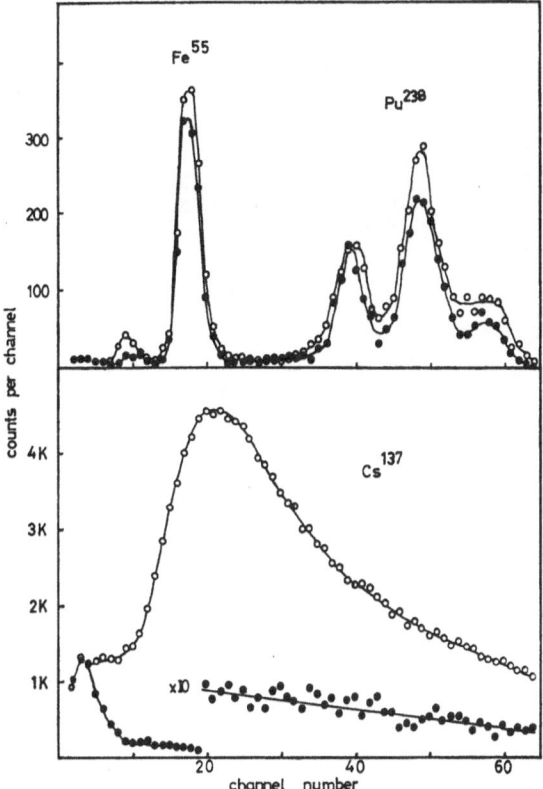

Fig. 3.   Pulse shape discrimination with 800 cm² counter.

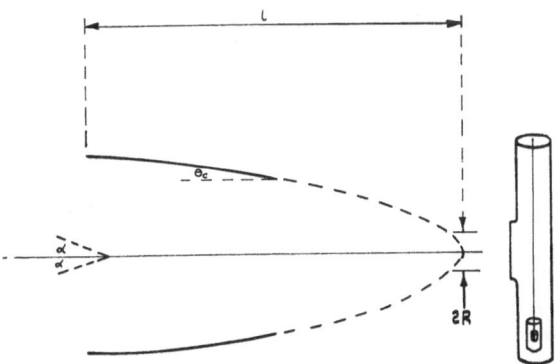

Fig. 4.   A grazing incidence X-ray telescope.

Moving to higher energies, the reduction factor is 45 at 10 keV and 40 at 17 keV. The reduction factor gets less below 6 keV until a 2 keV there is little improvement. Similar reductions are achieved in the background pulse amplitude distributions.

At energies below about 3 keV it becomes possible to reduce the background

Fig. 5.

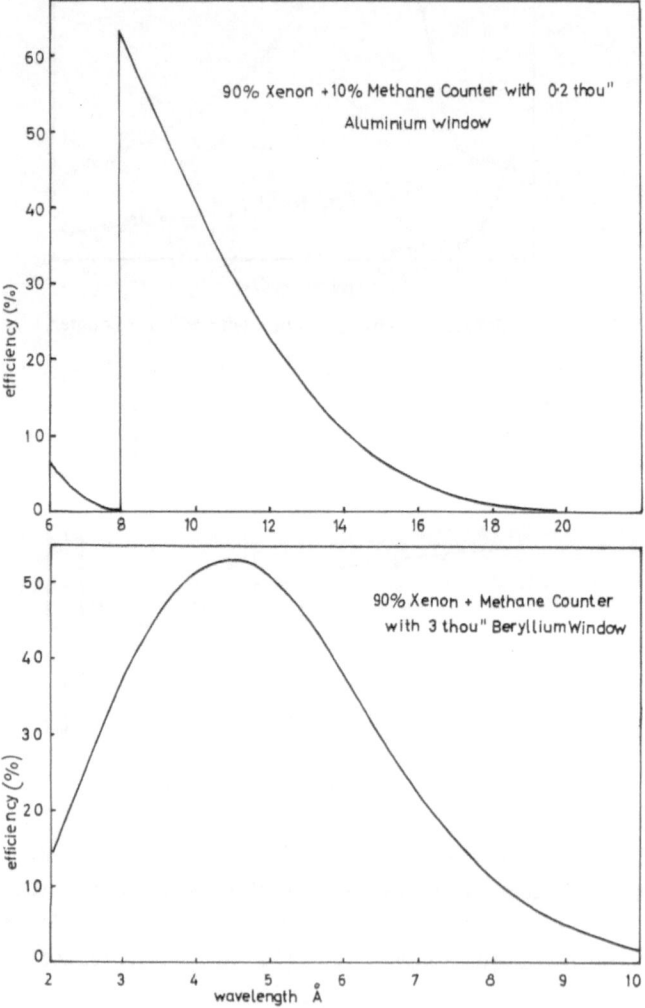

Fig. 6.   Detection efficiencies of small volume counters.

counting rate by the use of grazing incidence collecting mirrors. Figure 4 shows a paraboloid reflector which brings the X-radiation to a focus and is then detected by a proportional counter shown on the right. The improvement in background counting rate is achieved by making the proportional counter as small as possible. A small volume double proportional counter has been developed by 20th Century Electronics for our OAO experiment with grazing incidence mirrors and this is shown in Figure 5. The bottom counter is for calibration purposes and it continuously counts Fe^{55} radiation. The active length of the counter is 1 cm and the cathode internal diameter 0.5 cm.

The photon detection efficiencies of these small counters are shown in Figure 6.

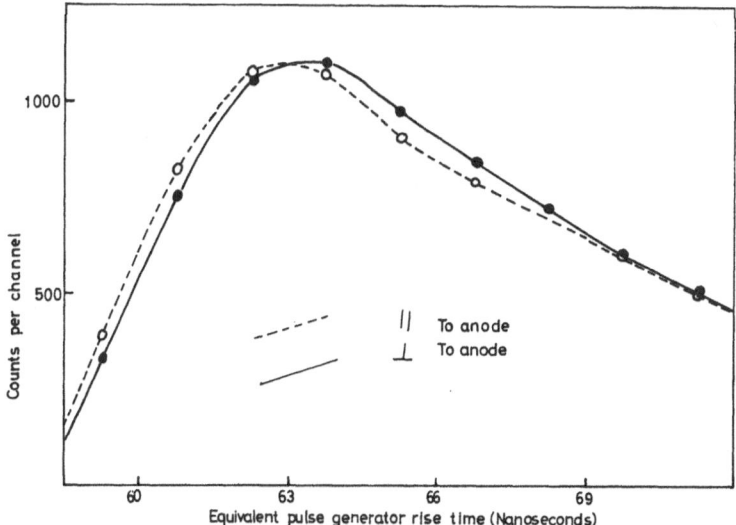

Fig. 7.   Pulse rise time distributions for polarized 16.5 keV X-rays.

Two window materials are used; Aluminium 5 μ thick for the wavelength range 8 to 18 Å, and Beryllium 75 μ thick for the wavelength range 3 to 9 Å. Xenon is used for the main filling gas since it enables the highest possible X-ray absorption. The measured background counting rate for these counters in the laboratory is one count in 100 sec. With mirrors effective collecting area of only 10 cm² the background performance yields an improvement of a factor of 50 over an equivalent 10 cm² window area counter.

We have demonstrated that it is possible to reduce the background counting rate in proportional counters by two methods. These are to some extent complementary since collecting mirrors are useful for X-ray energies less than about 3 keV whilst the Pulse Shape Discrimination system can be employed with large area counters for energies greater than about 2 keV. Pulse Shape Discrimination is however limited to energies where the ejected photoelectron produces a track of ionisation which is small compared with the tracks of background radiation. We suggested (1967) that the directions of the photoelectrons, produced by higher energy X-rays in a gas, could be used to measure the polarisation of an X-ray source, the direction of the photoelectron being determined by measuring the rise time of the pulse from a proportional counter. Figure 7 shows pulse rise time distributions for polarised 16.5 keV X-rays. It can be seen that it is possible to detect polarisation by this technique and with a suitably designed detector it may be possible to achieve a very sensitive polarimeter.

References

[1] Mathieson and Sanford: 1963, Proc. Inst. Symp. Nuclear Electronics, Paris p. 65
[2] University College London Proposal for the ESRO TD1 Satellite.

UPPER-AIR FLUORESCENCE AS A TOOL IN X-RAY ASTRONOMY AND SEARCHES FOR X-RAYS FROM NP 0532 AND OTHER PULSARS

W. N. CHARMAN, R. W. P. DREVER, J. H. FRUIN and J. V. JELLEY

Nuclear Physics Division, Atomic Energy Research Establishment, Harwell, Berkshire, England

and

J. L. ELLIOT, G. G. FAZIO, D. R. HEARN, H. F. HELMKEN,
G. H. RIEKE, and T. C. WEEKES

Smithsonian Astrophysical Observatory, and Harvard College Observatory, Cambridge, Mass., U.S.A.

1. Historical Introduction

The object of this paper, which describes work which has been carried out in close collaboration between members of our two groups, is to present the background of the upper-air fluorescence technique as a viable ground-based tool for studies of cosmic X-rays. After a discussion of the characteristics, merits and limitations of the technique, we describe some preliminary experiments which have been carried out, both separately and also jointly, with our respective installations. These are sited at Mount Hopkins (the SAO field station at 2300 m altitude near Amado, Arizona) and at Sparsholt Firs (approx. 300 m altitude, on the Berkshire Downs, near Harwell).

It has long been known that the upper atmosphere fluoresces in the UV, visible and infrared spectral regions under excitation by X-rays. This phenomenon has already been applied effectively to the detection of atomic bombs in space [1, 2], through their copious yield of X-rays at the instant of detonation.

The technique is being extensively developed by Greisen and his colleagues at Cornell [3, 4, 5] as the basis of a method of detecting very large cosmic-ray air showers. In this application, however, the fluorescence is observed at much lower altitudes, altitudes to which primary X-rays could not reach, and at which the fluorescent conversion efficiency is very low, $\sim 10^{-5}$. It is this very extensive work by Greisen which has stimulated interest in this general field, for other applications of the technique to astrophysics.

It was suggested by Colgate [6] that the shock ejection of the outer layers of a supernova should be accompanied by the emission of X-rays and γ-rays, to a total yield of $\sim 5.10^{47}$ ergs at the source, with an energy spectrum extending to ~ 2 GeV. On the basis of this prediction and estimates of the rate of formation of supernovae in galaxies, Fichtel and Ögelman [7] showed that with relatively simple equipment, it should be possible to detect fluorescent flashes in the upper atmosphere at a sensitivity level yielding a rate perhaps as high as one flash per 8-hour night.

It was with this background of information, and the very thorough laboratory studies of air fluorescence by Hartman [8] that two of the authors of this paper have

L. Gratton (ed.), Non-Solar X- and Gamma-Ray Astronomy, 41–49. All Rights Reserved

suggested [9] that the technique might well be applied to the search for X-rays from pulsars. Delegates of the Meeting should realise that this work started about Sept. (1968) prior to the discovery by Friedman [10] of the pulsed component of the X-rays from NP 0532, and its subsequent confirmation by many other groups, as reported at this Meeting. While our interest has until now been concentrated predominantly on the pulsars, we are now becoming increasingly enthused by the possibilities of the supernova experiment.

To conclude this section, may we mention that Sir Bernard Lovell has in private correspondence shown considerable interest in the possibilities of detecting X-rays from Flare Stars, already under extensive study in the radio and optical bands of the spectrum.

At this meeting the emphasis is on the search for X-rays from pulsars.

2. The Atmospheric Fluorescence and its Basic Sensitivity

The basic idea of the technique is to detect fluorescence in the upper atmosphere generated by cosmic X-rays, the observations to be made with simple optical equipment sited on the ground.

This fluorescence, which occurs predominantly in three wavelength bands, is associated with molecular nitrogen. These bands are the following (i) the 1st P-system of N_2, in the infrared, (ii) the 2nd P-system of N_2, in the UV, and (iii) the 1st N-system, of N_2^+, in the visible. Of these, for a variety of reasons, we have only considered (iii).

The bulk of the emission in this, the optical band, occurs within $\Delta\lambda \sim 20$ Å, at $\bar{\lambda} = 3914$ Å, with a conversion efficiency $\eta = 3.4 \times 10^{-3}$ [8] at an altitude $h \sim 65$ km. The yield of the isotropically emitted light is approximately linear with the X-ray energy deposited. It has been shown [9] that if the collecting area of the light receiver is A and the solid angle of the field of view is Ω, the signal-to-noise of the system varies as $(A\Omega)^{1/2}$, and that with a phototube of given A, the optimum performance is attained if this bare phototube is allowed to view the entire upper hemisphere of sky. For a variety of reasons however, certain advantages are in practice gained by the use of an optical system, consisting for example of a large searchlight mirror with a numerical aperture of typically $f/0.5$.

The mean lifetime of the electrons, between their photoelectric production by the primary X-rays at say 80 km, to their ionization of the N_2 molecules is ~ 1 μsec, and the lifetime of the excited N_2^+ ion for emission of the light in the 3914 Å transition is ~ 100 n.sec. A further time-dispersion, $\sim$ tens of μsec, may occur for an oblique X-ray front to traverse the atmospheric layers encompassed by the field of view of the instrument.

The input energy bandwidth of this type of detector is about three decades, from ~ 100 eV to ~ 100 keV. The lower limit is set by the steep reduction in the excitation cross-section below 100 eV, and the upper limit by collisional de-excitation competing with radiation at the higher pressures, for $E_x > 100$ keV. The overall bandwidth may be further reduced by interstellar and intergalactic absorption, as discussed elsewhere at this Meeting.

The overall sensitivity of an installation must depend on the various contributions to the total background light from the night-sky, and on the electronic bandwidth; the light pulses have to be observed against the fluctuation noise, and the bandwidth required clearly depends on the duration of the X-ray flashes, and hence on the particular application.

Under an ideal sky, for which the night-sky background light flux is ϕ_b, and for which the atmospheric transmission is T, the shot-noise from the photocathode, of quantum efficiency ε_k may be calculated directly from the mean cathode current i_k expressed in photoelectron emission rate (dN_e/dt), so that

$$i_k = (dN_e/dt) = A\Omega \int \phi_b(\lambda)\, T(\lambda)\, \varepsilon_k(\lambda)\, d\lambda \qquad (1)$$

and the mean square value of the shot noise is:

$$< \Delta i_k^2 > = 2ei_k \cdot \delta f \qquad (2)$$

in a bandwidth δf.

To the shot noise must be added other contributions to fluctuations in the light of the night-sky, many of which have already been discussed by us elsewhere [9].

Since most of the fluorescence occurs in a narrow band, the sensitivity can be raised by the use of narrow band filters, though in general interference filters are excluded, by the large values of Ω. Gelatine or glass filters will nevertheless enhance the sensitivity. For example, the combination of an S-11 cathode and a Corning 'Amethyst' filter, type 7-62, provides an admirable combination.

3. Merits and Limitations

It is essential to realise that the technique is limited to X-ray sources for which there are variable components of the flux which occur on time-scales short compared with those associated with fluctuations in the sky background. The applications are therefore restricted to supernovae, pulsars, variable X-ray sources and flare stars. The time-scales involved are approximately as follows:

Supernovae, $\sim$ tens of μsecs [6]
Pulsars, frequency components $\sim$ 10 Hz to 5 kHz
Variable X-ray sources (e.g. Sco X-1) $\sim$ mins
Flare stars, $\sim$ minutes to hours.

In general the lower the frequencies the greater the problems.

The *chief merits* of the fluorescence technique are:

(1) Large input energy bandwidth, $\sim$ 1 keV–100 keV, with an additional decade, 100 eV–1 keV, limited by interstellar and intergalactic absorption [11].

(2) Long integrating times, $\sim$ 6 hr per night, for unlimited periods.

(3) Ground-based, simple, low cost, and

(4) Since the fluorescent light is emitted isotropically, the detectors individually

are non-directional, so that all sources present above the horizon may be observed simultaneously. In the case of the supernova application, widely spaced stations can however be used to obtain directional information to locate galaxies to a precision $\sim 1°$, a resolution small compared with the dimensions ($\sim 10° \times 10°$) of the larger clusters of galaxies [12].

The *disadvantages* of the technique are:

(1) A single detector is non-directional.

(2) The performance is sensitive to site, climate and local weather.

(3) The system is difficult to calibrate.

(4) There is no spectral information.

With these considerations in mind, each application has its own particular problems which are summarised briefly as follows:

Supernovae. Technically straightforward, but the expected relatively low event-rate demands the organisation of a network of widely spaced stations, to ensure avoidance of local optical interference (lightning, meteors and man-made flashes) and to obtain an adequate coincidence probability of clear skies at two or more sites.

Pulsars. By far the greatest problem is that the first harmonic of the line-frequencies, 100 or 120 Hz, lies in precisely the electronic bandwidth required, so that it is essential to site the installations far from sources of A.C. lighting.

Variable X-ray sources and flare stars. Since the time-scales for these phenomena are very long, it is anticipated that the sensitivity will be limited by the constancy of the mean level of the sky-background over the periods required.

4. Preliminary Experiments at Harwell and Mt. Hopkins

We will now describe briefly some preliminary experiments which have been carried out at Harwell, starting in September of 1968, and on Mt. Hopkins, around November, 1968.

A. THE HARWELL EXPERIMENT

A 150-cm dia. $f/0.5$ parabolic searchlight reflector was used with a 30-cm dia. EMI phototube (type 9545B) having an S-11 cathode of sensitivity ~ 50 μA/lumen; no filter was used. The analogue signals from the tube were amplified, filtered ($\Delta f \sim 2k$ Hz), digitised and recorded on magnetic tape, along with a clock reference signal derived from an ovened 5 MHz crystal, stable to 1 in 10^8 and calibrated against the 200 kHz carrier of the B.B.C. Droitwich transmitter. The tapes were read out into a 512-channel Laben analyser cycled from the clock reference via a vernier scaler system [13]. The tapes were read out several times, for periods corresponding to those for the various pulsars, these periods being corrected for orbital and axial rotation Doppler shifts. The installation was sited at Sparsholt Firs, a dark site several kilometers from nearby villages and towns. In spite of this, residual 100 Hz 'hum' was considerable, though this was eventually quenched by hum-bucking, as shown in Figure 1, which illustrates the essential features of the recording system.

Runs were carried out on a number of occasions, under a variety of conditions and selection of the various parameters, and the tapes were analysed for the periods of the pulsars NP 0532, CP 1133, and CP 0950.

The best overall conditions occurred on Jan. 18–19 (1969) when a run was obtained

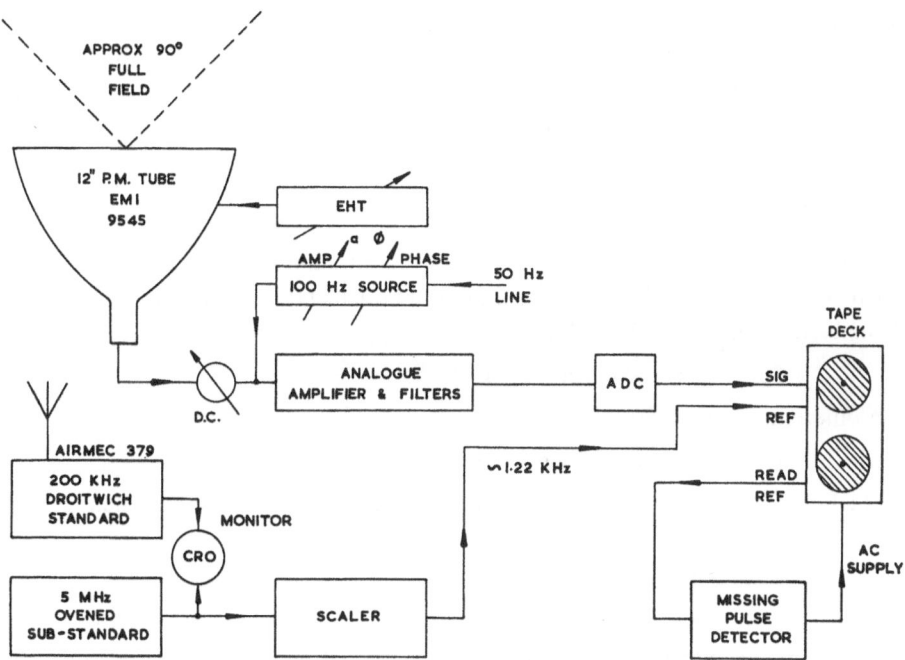

Fig. 1.   The basic features of the recording system used at AERE for a search for X-ray fluorescent pulses from the three pulsars. As used during the runs of Jan. 18–19th, a 120-cm diam. searchlight mirror was used, and a ÷ 10³ scaler was interposed between the ADC and the recorder.

for approximately three hours, broken into three periods for which the phase was preserved during each of these periods.

So far, none of these observations have as yet on analysis yielded any evidence for pulses at a significant level, and it is hoped the observations will be continued next winter, particularly as there is now firm evidence for X-ray pulses from NP 0532 at a variety of energies.

During these observations low frequency components were frequently seen. These, discussed by the SAO group, may have contributed additional noise to the basic shot-noise, thus lowering the sensitivity.

Theoretical sensitivity for NP 0532

For the above installation, $A\Omega = 2200$ cm^2 sr. For the S-11 cathode $\bar{\lambda} \sim 4000$ Å, $\delta\lambda \sim 1500$ Å (3 db points), and $\bar{\varepsilon}_k \sim 0.1$. In this bandwidth, $\int \phi \, d\lambda = 6.4 \times 10^7$ photons cm^{-2} sec^{-1} sr^{-1} [14]. With $T_{4000} = 0.63$ [15], and the above figures, we find from (1) that

$$(dN_e/dt) = 8.8 \times 10^9 \text{ photoelectrons sec}^{-1}.$$

Let us consider NP 0532, for which the main pulse has width $\Delta t \sim 2$ msec, and $P \sim 33$ msec. We therefore require the analyser to have say 20 channels. Consider now an integration period of $t = 1$ hr. In this time we accumulate 3.2×10^{13} counts or 1.6×10^{12} counts per bin, with $\sigma = \pm 1.3 \times 10^6$.

For a 1% confidence level over 20 channels, we require that one of these channels has a fluctuation $\geqslant 3.5\sigma$ or $\geqslant 4.6 \times 10^6$ photoelectrons or $\geqslant 4.6 \times 10^7$ photons in an hour's run. Since $1\ hv.$ at $3914\ \text{Å} \equiv 3.2$ eV, this represents a time-averaged light flux in the pulses, of

$$\phi_L \geqslant 18.6 \text{ eV cm}^{-2} \text{ sec}^{-1} \text{ sr}^{-1}.$$

With a conversion efficiency $\eta = 3.4 \times 10^{-3}$ and assuming the light may be emitted isotropically over 4π, this represents a minimum detectable X-ray flux, over any part of the band approx. 1 keV–100 keV, of

$$\Phi_x = 6.9 \times 10^4 \text{ eV cm}^{-2} \text{ sec}^{-1} \tag{3}$$

or a differential sensitivity of

$$\phi_x \geqslant 4.6 \times 10^{-27} \text{ erg cm}^{-2} \text{ sec}^{-1} \text{ Hz}^{-1}. \tag{4}$$

The light yield from a complete spectrum of form $\phi_x = kE^{-\gamma} \cdot dE$ is then

$$\Phi_L = k \int_{E_1}^{E_2} \cdot \eta(E) \cdot E^{-\gamma} \cdot dE. \tag{5}$$

In Figure 2 are plotted the spectral points obtained by rocket and balloon flights which have been presented at this Meeting, along with others already published. A reasonably good fit is obtained in the energy-range of interest, by taking a power-law with $\gamma = 1.0$ and between $E = 1$ keV and $E = 100$ keV. If we now take our limiting sensitivity (3) and a spectrum of the same *slope* as that of the observed spectrum, we obtain the line bb' in Figure 2, showing that the overall sensitivity lies somewhat outside the errors of existing measurements on NP 0532, for an hour's integration. The addition of filters and an increased integration time should bring the source within detectable range, expecially as some relaxation on the statistics is permitted, when allowance is made for the known double-pulse structure of the expected signal, with the characteristic 14 msec spacing.

That no positive effects have been seen on the experimental runs so far, is attributed to a number of limitations, among which were (i) loss of performance due to residual hum, (ii) degradation of statistics in pre-scaling between the ADC and the recorder, and (iii) non-Gaussian noise at the lower frequencies, from the night-sky, as mentioned below.

B. THE SMITHSONIAN ASTROPHYSICAL OBSERVATORY (SAO) EXPERIMENT

Four 130 mm diameter phototubes, two of type RCA 4522 and two of type Amperex

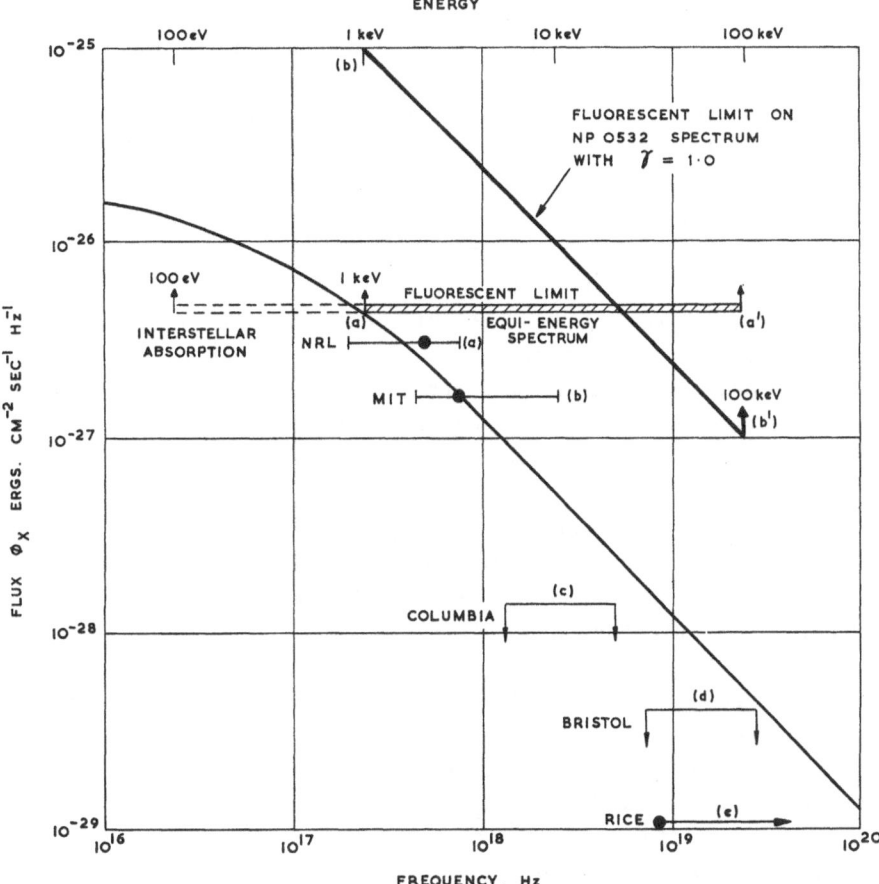

Fig. 2.   The differential spectrum of X-rays from the pulsar NP 0532. (a) The spectral point obtained by N.R.L. and reported by H. Friedman at this Meeting (this volume, p. 34, also ref. [16]). (b) The results of the MIT experiment, as reported by B. Rossi at this Meeting (this volume, p. 1, also ref. [17]). (c) The Columbia results, from the work reported by R. Novick at this Meeting (this volume, p. 145). This upper limit flux, of $\leqslant 5\%$ of the general Crab background, refers however only to the component of the pulsar radiation at the fundamental frequency of 30 Hz. (d) The upper limit flux reported by the Bristol group, ref. [18]. (e) This result, from the group at Rice, was reported at this Symposium by R. C. Haymes (this volume, p. 185, also ref. [19]). aa' represents the theoretical limiting sensitivity calculated by the authors of this paper, for a 1 hour's integration of the fluorescence light of the upper atmosphere, in a 1 kHz bandwidth, with a 150 cm diam. $f/0.5$ searchlight and a 30-cm diam. EMI phototube of quantum efficiency $\sim 10\%$. The curve bb' represents the sensitivity of the same instrument when exposed to a spectrum having the same slope ($\gamma \sim 1.0$) as that observed in the rocket and satellite experiments reported at this Meeting. The curves aa' and bb' are derived assuming a perfect dark sky for which (i) the power spectrum is flat throughout the frequency band considered i.e. ~ 10 Hz–3 kHz, and (ii) that the sky is totally free from line-frequency hum. The optical point on the spectrum, off the graph itself, is derived from the work of Lynds et al. [20] with a time-averaged flux of $\bar{m}_V = 16.5$, $B - V = 0.6$ and $m_B = 17.1$, using the absolute flux at $m_B = 0$ of $\phi_{\text{opt}} = 6.5 \times 10^{-9}$ erg cm^{-2} sec^{-1} Å^{-1} at $\lambda = 4401$ Å obtained by Willstrop (Monthly Notices Roy. Astron. Soc. (1960) **121**, 17). The spectral point, after correction of 1.1 mag kpc^{-1} for absorption, is at a flux level of $\phi_{\text{opt}} = 3.7 \times 10^{-26}$ erg cm^{-2} sec^{-1} Hz^{-1} at $\nu_{\text{opt}} = 6.85 \times 10^{14}$ Hz.

58 AVP were mounted in a light screen to view the zenith-sky, so that the total sensitive area A was 402 cm^2 and the effective solid angle Ω was 2.36 sr. The analogue signals from the tubes were linearly added, amplified and filtered ($\Delta f \sim 1$ kHz), then linearly converted to frequency by a voltage controlled oscillator. The output of the oscillator was recorded on one channel of magnetic tape, while 1 kHz clock pulses were recorded on the second channel. The clock signals were obtained from the SAO satellite tracking station at Mt. Hopkins and are stable to one part in 10^{10}. The magnetic tapes were analyzed on a 100-channel pulse height analyzer operating in the multiscalar mode. The analyzer was recycled every two pulsar periods from the clock pulses via a vernier scaler system [13]. The pulsar period was corrected for earth orbital and axial rotation.

The experiment was located at the 2300 m elevation of Mt. Hopkins, Arizona. The data were recorded under very dark site conditions and during periods of excellent sky transparency. A 120 Hz signal recorded due to distant city lights was present, but at a low level.

Data were taken over four nights from November 18 through November 21, 1968, and approximately 33 hours of data were recorded, with each tape reel lasting about 1.5 hours. Absolute time marks were also recorded on each tape but no effort was made to add successive tapes coherently.

The data were analyzed only for the pulsar NP 0532 (Crab Nebula). Thus far these tapes have yielded no evidence of X-ray pulsations, with an upper limit to the X-ray flux of 2.5×10^4 eV cm^{-2} sec^{-1}.

A most interesting aspect of the data taken at Mt. Hopkins is the power-spectrum (square of the amplitude of the Fourier components in a given frequency interval) of the night airglow over the frequency range from 0 to 600 Hz. The amplitude of the power spectrum is not constant at all frequencies, as would be expected, but is higher in the 0 to 100 Hz region, at a level about 3 to 10 db higher than the rather flat spectrum above 200 Hz. Thus, in the band where most of the pulsar frequency components are expected, the night sky light fluctuations appear to be higher than would be expected from purely statistical effects. The sensitivity is not as great, therefore, as theoretically calculated in the previous section. The origin of the excess components below 100 Hz is completely unknown to us, and must be investigated further if the present upper limits are to be lowered significantly.

5. Conclusions

It is concluded that with care it should be possible to detect the X-rays from NP 0532 by ground-based observations of air fluorescence, with integration times $\sim$ hours.

Of the other suggested applications, the search for supernovae by this technique is particularly attractive. The estimates of expected counting rates [7], $\sim$ a few per week, make it a viable experiment, and the simultaneous operation of a number of stations spaced $\sim$ 1000 km apart would yield a directional accuracy $\sim 1°$, with recording of real time to a precision of $\sim$ 100 μsec. This resolution is adequate for astronomical search purposes and is considerably less than the angular sizes of the larger clusters of galaxies, $\sim 10°$. The recording of real time to ~ 100 μsec is straightforward, using VLF transmissions from stations such as WWV, MSF or HBG.

Since the recordings would be made in a frequency band $\delta f \sim 100$ kHz–1 MHz, the technique should not be plagued by the background of line-frequency components of the light of the night-sky, or by other low-frequency fluctuations.

The application to the detection of supernovae has obvious astrophysical interest, as it presents a direct test of Colgate's model of the shock-wave phenomenon, and, if successful, could be developed to alert astronomers for optical observations of the early phases of the subsequent light-curve.

References

[1] Westervelt, D. R. and Hoerlin, H.: 1965, *Proc. IEEE* **53**, 2067.
[2] Donahue, T. M.: 1965, *Proc. IEEE* **53**, 2072.
[3] Greisen, K.: 1965, in *Proc. 9th Internat. Conf. on Cosmic Rays, London*, p. 609.
[4] Bunner, A. N.: 1966, Cornell-Sydney University Astronomy Center Report No. 62, also Cornell thesis.
[5] Jenkins, E. B.: 1966, Center for Radiophysics and Space Research, Cornell, N.Y., Report No. CRSR 250, also Cornell thesis.
[6] Colgate, S. A.: 1968, *Canad. J. Phys.* **46**, S476.
[7] Fichtel, C. E. and Ögelman, H. B.: 1968, NASA Technical Note No. 4732.
[8] Hartman, P. L.: 1968, Los Alamos Scientific Lab. Report LA-3793.
[9] Fazio, G. G. and Jelley, J. V.: 1969, AERE Rep. No. 6095.
[10] Friedman, H.: IAU Telegramme No. 2141, also this volume, p. 34.
[11] Felten, J. E. and Gould, R. J.: 1966, *Phys. Rev. Lett.* **17** (7), 401.
[12] Abell, G. O.: 1965, *Ann. Rev. Astron. Astrophys.* **3**, 1.
[13] Papaliolios, C., Carleton, N. P., Horowitz, P., and Liller, W.: 1968, *Science* **160**, 1104.
[14] Jelley, J. V.: 1967, *Prog. Elem. Particle Cosmic Ray Phys.* **IX**, 53.
[15] Allen, C. W.: 1955, *Astrophysical Quantities, Univ. of London, Athlone Press*, 1st ed., p. 116.
[16] Fritz, G., Henry, R. C., Meekins, J. F., Chubb, T. A., and Friedman, H.: 1969, *Science* **164**, 709.
[17] Bradt, H., Rappaport, S., Mayer, W., Nather, R. E., Warner, B., McFarlane, M., and Kristian, J.: 1969, *Nature* **222** (5195), 728.
[18] Hillier, R. R., Jackson, W. R., Murray, A., Redfern, R. M., and Standing, K. G.: 1969, *Nature* **222** (5189), 149.
[19] Fishman, G. J., Harnden, F. R., and Haymes, R. C.: 1969, *Astrophys. J.*, in press.
[20] Lynds, R., Maran, S. P., and Trumbo, D. E.: 1969, *Astrophys. J.* **155**, L. 121.

A PROPOSAL FOR AN X-RAY EXPERIMENT FOR COS-B

A. P. WILLMORE

Mullard Space Science Laboratory, Physics Dept., University College London, London, England

1. Introduction

It was proposed that the COS payload should include an experiment to study cosmic γ-rays and one for cosmic X-rays. From a scientific point of view the association of these experiments is clearly valuable, the more so in view of Ogelmann's suggestion that the γ-rays observed by Clark, Garmire and Kranshaar and by Hutchinson represent the high energy extension of the spectra of galactic X-ray sources. It may, however, be argued that the X-ray sources can independently be studied and that, in any case, the most important sources near the galactic plane will be well enough known by the launch date of the COS satellite. This argument may not be valid in view of the observed variability of X-ray sources, but in any case a strong case can be made for the proposed kind of experiment.

The suggested objectives of the experiment, more or less in descending order of weight as far as this design is concerned, would be (a) to measure with very high precision the location of discrete sources, to assist in optical identifications; (b) to improve knowledge of their spectral characteristics; (c) to study short period variations, and (d) to discover new sources.

Only a few of the known sources can be identified with any confidence with optically known objects – in this respect X-ray astronomy presents a close parallel with the early days of radio astronomy. The reason is simply that existing techniques do not in general permit the location of the source to better than about 1°, leaving an area of uncertainty containing many thousands of stars. No experiment is at present planned anywhere on the time scale we are considering which will enable sources to be located with the precision necessary to give only 1–10 known stars (i.e. down to say 20^m) in the field of view. The precision required for this is about 10 arc sec in the worst case. The planned U.S. experiments will probably not achieve better than an arc minute, at which level the ambiguity will still include up to a hundred stars.

An experiment with a rather large field of view (perhaps 45° included angle) and capable of discriminating all the sources in this field, is also valuable for spectral studies and for studies of variability. The planned experiments on SAS-A and SAS-C (with one exception) are not of this character, except for strong sources, because of their limited throughput. Thus, the inclusion of a suitable wide-field experiment on COS-B would give considerable new information on the spectral characteristics of fainter sources, including probably extra-galactic sources, only one of which (M-87) is known at present. Finally, by covering the same region of the sky for long periods – 10–20 days – the short period variability of many sources will be observed, including both flaring of the type exhibited by SCO-X1, if this occurs in other sources, and

L. Gratton (ed.), Non-Solar X- and Gamma-Ray Astronomy, 50–53. All Rights Reserved
Copyright © 1970 by the IAU

possibly also emission from flare stars which has not so far been observed at all in X-rays. The proposed experiment has an extremely high sensitivity, the collecting efficiency being 10 and 100 times greater than the two low angular resolution experiments on SAS-A and when combined with the long pointing time this should enable new sources, many of which are likely to be extra-galactic, to be discovered.

2. Experimental Method

The proposal for COS-A, made by Dr. Pounds, included a novel device in the form of a moving collimator. The effective application of this system required a spacecraft with 3-axis stabilisation and low rates of angular drift. In the case of COS-B which is to be spin-stabilised this system can no longer be used. However, Pound's device is an example of a general class of such collimators which the writer has recently studied theoretically. An alternative form, which is not in any respect inferior to the moving collimator and which has certain advantages, is a rotation collimator. In this case the two grids are maintained at a fixed spacing, and rotated about their common axis. The X-rays from a source off this axis are modulated at a frequency depending on their position. This system is compatible with the proposed COS-B, the spacecraft providing the rotation so that the collimator remains fixed relative to the spacecraft. For both the versions of the experiment, the rate of drift of the pointing direction is the main limiting factor in the determination of source position. The rotating satellite has the advantages that the gyroscopic rigidity of the satellite automatically minimises the steady drift, and that the nutation can readily be damped by conventional means. On the other hand, the spin rate will then be higher than the period associated with the moving collimator, so that the modulating frequencies will be higher in this case and cannot directly be telemetered. Since the modulating waveform has the fundamental period of the spin frequency, the counts can be accumulated in a buffer store, the addresses in the store being determined by a sector generator synchronised to the sun, if sufficient accuracy can be obtained, or to a star. The weight and power figures given assume that the core store and sector generator are included in the experiment although they are also required for the γ-ray experiment and it would obviously be more efficient to combine them. In addition, because of the high stability required, it is assumed that an independent sub-structure is provided for the X-ray experiment and quasi-kinematically attached to the spacecraft.

A detailed description of the experiment is not given, because it is so similar to that proposed by Dr. Pounds. In particular, the use of a photomultiplier to determine the orientation of the collimator is included. However, some comments and explanations may be useful. The orbit selected for COS-B is advantageous for the X-ray experiment also, because the observing time is influenced by similar factors to the γ-ray experiment. As for that experiment, advantage has been taken of this to reduce the size of the detectors without reducing the sensitivity. This factor, together with the elimination of the high precision drive for the collimators reduces the weight requirement. It has been assumed that the experiment would be located in the annular space around the central tube, though a structural re-arrangement might possibly be more sensible. The limiting factors for the angular accuracy of the experiment are the drift rate of the spin axis and the telemetry rate (assuming the requirements for nutation are satisfied). The companion visible star system enables the position of the axis to be determined in about 1 minute of time. A certain quantity of data is thus collected. If sufficient telemetry is available for this to be transmitted every minute, then the effective image degradation is the spin axis movement in 1 min. At a lower telemetry rate, the position is determined less often (the quantity of data to be transmitted increased only quite slowly with the period) and the image degradation is worse. The design presented assumes the figures of the COS-B feasibility study.

A total length available of 70 cm has been assumed, though an extension to the full length of the spacecraft would be desirable and would result in a useful improvement in angular accuracy.

3. Experiment parameters

Collimator and detector package

Size: Approximately a cylinder 35 cm diameter, 70 cm long, constrained as necessary to fit in one of the large equipment compartments.
Detector diameter (active area) 30 cm.
Weight 18.5 kg.
Electronics package (not including cabling) 6 kg.
Total weight 24.5 kg.

Mounting requirements: To be aligned with an accuracy of $0°5$ to the spacecraft spin axis.

Stabilisation: The spacecraft should be spin-stabilised, with a drift rate of less than $0°5$/orbit and an amplitude of nutation not exceeding $5''$ arc. (This may necessitate a fluid rather than a ball-in-tube nutation damper). A spin rate of 10 r.p.m. is suitable.

Spin timing pulse: A timing pulse is required, once each rotation relative to fixed stars. Over a five minute period, no individual pulse should depart from the position corresponding to an exact rotation reference by more than 15 arc sec. On the other hand, neither the actual position of this pulse in rotation nor its long-term stability are of any importance.

Telemetry: The most efficient method of interfacing the experiment with the telemetry via a core store requires a detailed study from a system point of view, and this has not been possible in the time available for producing this proposal. A method has been worked out which may not be an optimum but which would work.

Each spin cycle would be divided by a sector generator synchronised to the spin timing pulse into 8K segments. The counter output would be divided by pulse height analysis into two energy ranges, the counts from each in each sector being stored in a 4 bit count. In addition an 8-bit count would be required for the star tracker. Thus, the total memory storage required is 16K of 8-bit words, half word access being required to half the store. This store would be read out via the telemetry on a continuous cycle, the read-out period being 1280 sec at a bit rate of 100 b/s. Together with 100 b/s this gives the 200 b/s capacity envisaged in the design study, but a small increase would of course be necessary to satisfy the requirements of the spacecraft housekeeping.

Power: A reasonable power allocation for the experiment would be 4 watts.

Angular accuracy of position measurement: The effective image size of the collimator depends on the characteristics of the collimator, together with blurring resulting from the instability of the spin direction:

Image size due to collimator	90 arc sec.
drift in core recycling period	20
nutation (total amplitude)	10
	120 arc sec.

For faint sources which are only just detected this would be the error in position measurement. For bright sources, of which there would be many because of the high sensitivity of the experiment, the position error would be about $\frac{1}{10}$ of the image size, or 12 arc sec, together with an error of about 9 arc sec contributed by the star tracker. Combining these errors quadratically since they are independent we obtain an overall error of 15 arc sec.

Ambiguity in position: The rotation collimator determines two position coordinates for the sources in the view field (in this respect it differs from the original proposal of Dr. Pounds; the difference arises essentially because both frequency and phase information in the modulation are available) but because of its two-fold symmetry about the rotation axis there is a corresponding ambiguity in the azimuth angle of the sources To resolve this ambiguity, two sets of measurements are required. The simplest scheme appears to be to displace either the spacecraft spin axis, or the axis of the experiment, periodically through a small angle. An angle larger than the image size of the experiment, but smaller than is detectable by the γ-ray experiment seems a good choice; we propose 5 arc min. Thus, one orbit might be spent on one position, followed by a displacement for the succeeding orbit. If, however, a careful study shows that the spin axis drift is sure under all circumstances to lie between 0.1 and 0.5 degrees per orbit, this provision would be unnecessary.

STUDIES OF DISCRETE COSMIC X-RAY SOURCES AT M.I.T.

(Invited Discourse)

GEORGE W. CLARK

*Center for Space Research and Department of Physics, Massachusetts Institute of Technology,
Cambridge, Mass., U.S.A.*

The following is a brief summary of some of the recent results obtained at M.I.T. which bear on the properties of various discrete X-ray sources, both galactic and extragalactic. Most of the work referred to has been described in publications which contain more complete descriptions.

1. Galactic Sources

In a July 1968 rocket observation with narrow slat collimators, Bradt *et al.* scanned the region of the Milky Way in the range of galactic longitudes from about $l^{II} = -20°$ to $+17°$ and in the energy range from 2 to 8 keV. They obtained a variety of results on the positions, angular sizes, and spectra of various sources. Their data confirmed the positions for the sources GX3+1, GX5−1, GX9+1, GX13+1, and GX17+2 [1] determined in their July 1967 flight [2]. The agreement between the previously and recently measured intensities of the sources north of $l^{II} = 0°$ indicated that they have not changed by more than 30% in over 2 years. Upper limits of 100 arc sec were placed on the angular sizes of four sources [3].

In a rocket flight on April 26, 1969 Bradt *et al.* observed a pulsating component of X-rays from the Crab Nebula [4]. The period and phase of the X-ray pulsations were the same as the optical pulsations of the pulsar NP 5032 which was observed nearly simultaneously at the MacDonald and Mt. Palomar Observatories. Details of this observation are presented elsewhere in this Symposium.

Rappaport *et al.* have developed a sensitive and reliable method for measuring the effects of interstellar absorption on X-ray spectra near 10 Å [5]. They employ proportional counters with thin ($7.7\,\mu$) aluminum windows that define two distinct spectral bands of sensitivity above and below the Al K-edge near 8 Å. Using data on the ratio of counts in these two bands obtained in their July 1968 flight that scanned the Scorpio-Sagittarius region, and their April 26, 1969 flight that measured the Crab Nebula pulsar, they measured or placed limits on the column density of atoms between the earth and several sources as summarized in Table I. Significant absorption was also observed in the spectra of several other of the Sagittarius sources. This recent work confirms and extends their earlier measurements of low energy spectra based on a July 1967 flight [6].

Bradt *et al.* have examined their rocket data on Sco X-1 for evidence of iron line emission. An upper limit of 1×10^{-8} ergs cm^{-2} sec^{-1} was set on the intensity of the iron K-line from Sco X-1 [6].

L. Gratton (ed.), Non-Solar X- and Gamma-Ray Astronomy, 54–58. All Rights Reserved
Copyright © 1970 by the IAU

TABLE I

Summary of results on
interstellar X-ray absorption

Source	Column density from earth to source (atoms/cm^2)
Sco X-1	$< 10^{21}$
Crab Nebula	$< 3 \times 10^{21}$
GX349 + 2	$\geqslant 1 \times 10^{22}$
GX5 − 1	$\geqslant 1 \times 10^{22}$

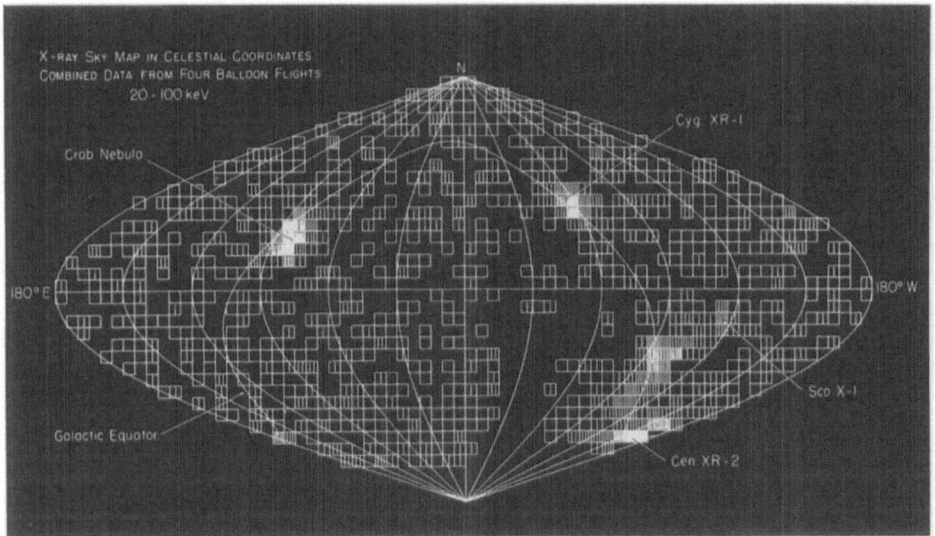

Fig. 1.   Composite of four balloon surveys of cosmic X-rays above 20 keV. The number of bars in any given box is the number of statistical standard deviations by which the observed counting rate exceeds the background rate.

In the higher energy range accessible to balloon observations, Lewin *et al.* completed a general sky survey of X-rays above 20 keV with an angular resolution of approximately 10° FWHM. Figure 1 is a summary of their results from four flights in the form of a sky map which shows that all the prominent high energy sources are concentrated near the galactic equator. Only the Crab Nebula and Cyg X-1 stand out in the northern hemisphere. Detailed examination of the data obtained in flights in the southern hemisphere has led to the following results:

(1) An X-ray flare was observed from Sco X-1 on October 15, 1967 during a flight in Australia when the observed intensity in the energy range from 20 to 30 keV increased by a factor of 4 in less than 10 minutes and then decreased during the next 20 minutes [7]. The time structure of the X-ray flare resembled that of the optical flares previously observed, but its amplitude, measured both in terms of absolute

energy flux and fractional increase, far exceeded the amplitudes of any of the optical flares observed so far.

(2) Cen XR-2 was observed [8] during the October 15 balloon flight in Australia as an intense source above 20 keV shortly after it had gone undetected in a rocket experiment sensitive in the range 2–8 keV. A position determination was made with an error radius of 1.°5. Within this circle Eggen et al. [9] reported the variable blue star WX Centauri which may be the optical counterpart.

(3) A possible decrease in the intensity of Cen XR-2 was noted between the October 15 flight and a subsequent flight 9 days later [10].

(4) Recent results from a balloon observation by Lewin et al. [24] on March 20, 1969 show that the intensity of Cen XR-2 had decreased by at least a factor of 7 since October 15, 1967.

(5) The data on a complex of high energy sources observed between $l^{II} = -40°$ and $+5°$ can be accounted for in terms of four discrete sources [11].

Overbeck and Tanenbaum [12] have conducted a series of balloon flights using the same detector to make precision measurements of X-ray fluxes in several energy bands above 15 keV. In the cases of both Sco X-1 and Cyg XR-2 they reported the occurrence of time variations by factors from 2 to 3 between measurements made approximately one month apart.

In a study of the angular sizes of high energy X-ray sources Floyd [13] has obtained an upper limit of 1.4 arc min for Cyg X-1 at energies above 20 keV in a balloon experiment with a gyrostabilized modulation collimator. Telemetry difficulties prevented full use of the capability of the instrument which has a potential angular resolution of 0.1 arc min. The Crab Nebula has been observed in a more recent flight with the same instrument and the data are now being analyzed.

2. Extragalactic Sources

With the most sensitive balloon and rocket instruments available to date, the minimum detectable X-ray energy flux from a discrete source lies in the range from 10^{-10} to 10^{-9} ergs/cm^2 sec. A flux in this range exceeds the radio energy fluxes of the brightest extragalactic (and galactic) radio sources by factors of 10 or more. Moreover, it exceeds the optical energy flux of a tenth magnitude object which would be among the brighter galaxies of the Virgo Cluster. Thus a valid observation of X-ray emission from even a relatively near exterior galaxy implies an astonishing non-thermal luminosity in the form of X-ray photons. Moreover, the detection of X-rays from any of the quasars (which are comparatively faint objects in the optical and radio regions, presumably because of their great distances) would imply that the emission is completely dominated by X-rays. There is reason, therefore to look very skeptically at any report of an extragalactic source which is based on an excess number of counts above background which is no more than three standard deviations of a Poisson distribution. Below this level of statistical significance the chances are dangerously large that a false signal will be found at one or another of the many potentially interesting positions scanned in a broad sky survey. When there are, in addition, significant systematic errors in the background rate evaluation, even greater caution is obviously necessary.

At the present time the X-ray source in Virgo is the only one which has been identified as an extragalactic object with reasonable certainty on the basis of rocket surveys [14, 15, 16] in the 1–10 keV region. These surveys showed signals with 3σ significance or better from a source located within a small region containing M-87 but otherwise devoid of conspicuous objects. Haymes *et al.* [17] recently reported a balloon observation of M-87 at energies above 30 keV which they interpreted as evidence for a continuation to 100 keV of the power law spectrum that fits the data for the emission of the jet from the visible to the soft X-ray region. In contradiction, McClintock *et al.* at M.I.T., using an oriented detector with 2000 cm² of sensitive area and an 8° FWHM field of view, found no significant evidence of a source of X-rays above 20 keV at the M-87 position [18]. If the results of Haymes *et al.* were correct, the M.I.T. experiment should have reported a 5σ peak. McClintock *et al.* have emphasized the need for great caution and precision in evaluating the background rate in a balloon experiment in which, as in the case of M-87, the signal is at best only a small percentage of the background rate. They did their evaluation by comparing the rates at nearby azimuths on either side of M-87. In contrast Haymes *et al.* evaluated the background only at an azimuth 180° from that of M-87.

There have been several reports of the detection of X-rays from other extragalactic sources, namely the radio galaxy Cyg A [18], the quasar 3C273 [15], and the Large Magellanic Cloud [19]. A subsequent survey of the Cygnus region by Giacconi *et al.* [20] placed an upper limit on Cyg A which contradicts the earlier claim.

The experiment of Bradt *et al.* [16], which placed M-87 in the 6 deg² uncertainty area around the measured position of the X-ray source, also scanned the position of 3C273 and found a 3σ upper limit consistent with the intensity claimed for it by Friedman *et al.* [15]. However, an optimal analysis of the raw data subsequently published by Friedman *et al.* [21], reveals that the counting rate peak attributed to 3C273 contained a number of counts above background which was less than one standard deviation of the Poisson distribution. Other similar criticisms of the 3C273 report have been made [22]. Therefore, at the present time, there is no significant evidence that X-rays have been detected from 3C273.

Finally, with regard to the Large Cloud of Magellan (LCM), consideration of the graphical display of the rocket data of Mark *et al.* [19] together with the information supplied with regard to the rate of scan shows that the excess of counts above background in this experiment was not more than two standard Poisson deviations and that the uncertainty in the location of the alleged source was relatively large. Therefore, their conclusion that the LCM has an X-ray luminosity in the 1–10 keV region equivalent to 100 Crab Nebulae does not appear to be well substantiated. It should be noted also that a 3σ *upper limit* of 120 Crab Nebulae was previously placed on the luminosity in the range 20–60 keV by Lewin *et al.* [23] on the basis of a balloon survey.

In the light of the weakness of the experimental evidence for the detection of extragalactic sources other than M-87 and considering the absolute X-ray luminosities that such sources must have to be detectable in current rocket and balloon experiments, it would appear that further substantial progress in the study of extragalactic X-ray sources may well be made only by the satellite experiments due for launching in 1970 and 1971.

References

[1] Mayer, W., Bradt, H. V., and Rappaport, S.: 1970, *Astrophys. J. (Letters)* **159**, L115.
[2] Bradt, H. V., Naranan, S., Rappaport, S., and Spada, G.: 1968, *Astrophys. J.* **152**, 1005.
[3] Polucci, G., Bradt, H. V., Mayer, W., and Rappaport, S.: 1970, *Astrophys. J. (Letters)* **159**, L109.
[4] Bradt, H. V., Rappaport, S., Mayer, W., Nather, R., Warner, B., MacFarlane, M., and Kristian, J.: 1969, *Nature* **222**, 728.
[5] Rappaport, S., Bradt, H. V., and Mayer, W.: 1969, *Astrophys. J. (Letters)* **157**, L21.
[6] Rappaport, S., Bradt, H. V., Naranan, S., and Spada, G.: 1969, *Nature* **221**, 428.
[7] Lewin, W. H. G., Clark, G. W., and Smith, W. B.: 1968, *Astrophys. J. (Letters)* **152**, L55.
[8] Lewin, W. H. G., Clark, G. W., and Smith, W. B.: 1968, *Astrophys. J. (Letters)* **152**, L49.
[9] Eggen, O. J., Freeman, K. C., and Sandage, A.: 1968, *Astrophys. J. (Letters)* **154**, L27.
[10] Lewin, W. H. G., Clark, G. W., and Smith, W. B.: 1968, *Nature* **220**, 249.
[11] Lewin, W. H. G., Clark, G., Gerassimenko, M., and Smith, W. B.: 1969, *Nature* **223**, 1142.
[12] Overbeck, J. W. and Tanenbaum, H. D.: 1968, *Astrophys. J.* **153**, 899.
[13] Floyd, F.: 1969, *Nature* **222**, 967.
[14] Byram, E. T., Chubb, T. A., and Friedman, H.: 1966, *Science* **152**, 66.
[15] Friedman, H. and Byram, E. T.: 1967, *Science* **158**, 257.
[16] Bradt, H. V., Mayer, W., Naranan, S., Rappaport, S., and Spada, G.: 1967, *Astrophys. J. (Letters)* **150**, L199.
[17] Haymes, R. C., Ellis, D. V., Fishman, G. J., Glenn, S. W., and Kurfess, J. D.: 1968, *Astrophys. J. (Letters)* **151**, L131.
[18] McClintock, J. E., Lewin, W. H. G., Sullivan, R. J., and Clark, G. W.: 1969, *Nature* **223**, 162.
[19] Mark, H., Price, R., Rodrigues, R., Seward, F. D., and Swift, C. D.: 1969, *Astrophys. J. (Letters)* **155**, L143.
[20] Giacconi, R., Gorenstein, P., Gursky, H., and Waters, J. R.: 1967, *Astrophys. J. (Letters)* **149**, L85.
[21] Friedman, H., Byram, E. T., and Chubb, T. A.: 1968, *Science* **159**, 747.
[22] Argyle, E.: 1968, *Science* **159**, 747.
[23] Lewin, W. H. G., Clark, G. W., and Smith, W. B.: 1968, *Nature* **220**, 249.
[24] Lewin, W. H. G., McClintock, J. E., and Smith, W. B.: 1970, *Astrophys. J. (Letters)*, March issue.

PROPERTIES OF INDIVIDUAL X-RAY SOURCES

(Invited Discourse)

LAURENCE E. PETERSON*

Dept. of Physics, Space Physics Group, University of California, San Diego, La Jolla, Calif., U.S.A.

Abstract. Observations to determine the spectra and time variations of hard X-rays from cosmic sources have been made from balloons and from the OSO-III satellite. These data have been obtained using actively collimated scintillation counters with apertures between 6 and 24° FWHM, areas between 10 and 50 cm² and which operate over the 10–300 keV range. The Crab Nebula has been observed on three occasions over a 22-month period between September 1965 and July 1967. The power law spectrum has a number index of 2.0 ± 0.1. No long-term changes were observed over the 30–100 keV range with a limit at 3%/yr. A balloon search with a 10 cm² Ge(Li) detector for X-ray lines at 62.5 keV, 110 keV and 180 keV due to heavy element radioactive decays which would be produced in the initial Crab explosion based on the Cf^{254} hypothesis has resulted in upper limits at about 10^{-3} γ-rays cm²-sec. This is about a factor of 20 above the predicted levels. Simultaneous X-ray and optical observations of SCO XR-1 from OSO-III confirm that X-ray and optical flaring are indeed coincident phenomena, and that although the X-ray intensity increases about a factor of two during the flare, the equivalent temperature of the excess radiation is nearly the same as that of the quiescent object. Upper limits, 95% confidence, on the flux of M-87 at 40 keV have been obtained. These are inconsistent with the flux of 1.2×10^{-4} photons/cm²-sec-keV reported in the literature. CYG X-1 has been observed to have a power law of number index 2.0 ± 0.2. The OSO-III has observed a number of sources in the southern skies including NOR XR-2 and the variable source Centaurus XR-2.

1. Introduction

In this paper, significant previous results are reviewed and some new data presented which have been obtained relative to point X-ray sources by the UCSD Gamma-Ray Astronomy Group. Most of the observations obtained since 1965 from balloons have been concerned with the spectra and time variations of these sources in the northern skies. These observations have been primarily made with scintillation counters and cover 15–300 keV energy range. Although most of the earlier data were obtained with instrumentation flown from high altitude balloons, recent observations have been obtained with the OSO-III satellite, which was launched March 8, 1967, and from which preliminary results are now available.

2. Instrumentation

Detector systems which have been developed at UCSD for these observations over the past five years are based upon the principle of the actively shielded collimated scintillation counter. Use of these detectors was started in a collaboration with Ken Frost (Frost *et al.*, 1966) of the Goddard Space Flight Center in 1962, and many preliminary data necessary to determine background properties of the detector in its radiation environment were obtained in a series of experiments on the OSO-I satellite

---

* Other members of the UCSD Gamma-Ray Astronomy Group who have contributed to this work include H. S. Hudson, A. S. Jacobson, J. L. Matteson, D. McKenzie, R. M. Pelling, and D. A. Schwartz.

L. Gratton (ed.), Non-Solar X- and Gamma-Ray Astronomy, 59–80. All Rights Reserved

(Peterson, 1966) and at balloon altitudes. The detector flown on the OSO-III satellite, which resulted from these early studies, has been described in the literature (Hicks *et al.*, 1965). This 10 cm² detector with a 23° full width half maximum response (FWHM) was identical to the detector which was used during many of the early balloon observations (Peterson and Jacobson, 1966).

The design criterion for these detectors is to construct a collimating shield around the photon sensing element so that the total radiation through the front aperture, which consists of atmospheric X-rays secondary to cosmic rays and cosmic X-rays due to the diffuse sky component, is approximately equal to the remaining, or 'true detector background' consisting of effects due to shield leakage, Compton scattering in the shield, neutron interactions in the central detector, and possibly other interactions which are not presently understood. The most successful way to build a detector with such a property at small solid angles Ω is to design the shield of scintillation material, such as $CsI(Na)$. This is observed by a photomultiplier, which is connected in electrical anticoincidence with a central scintillation counter, usually of $NaI(Tl)$. The shield thickness is optimized according to the relation

$$\frac{\Omega}{4\pi} \cong e^{-\bar{t}/\tau},$$

where $\bar{t}$ is the average thickness of the shield and τ is the photon mean free path in the shield material at the highest energy where operation is desired. Although more background rejection may be obtained by constructing a shield which is thicker, for a given

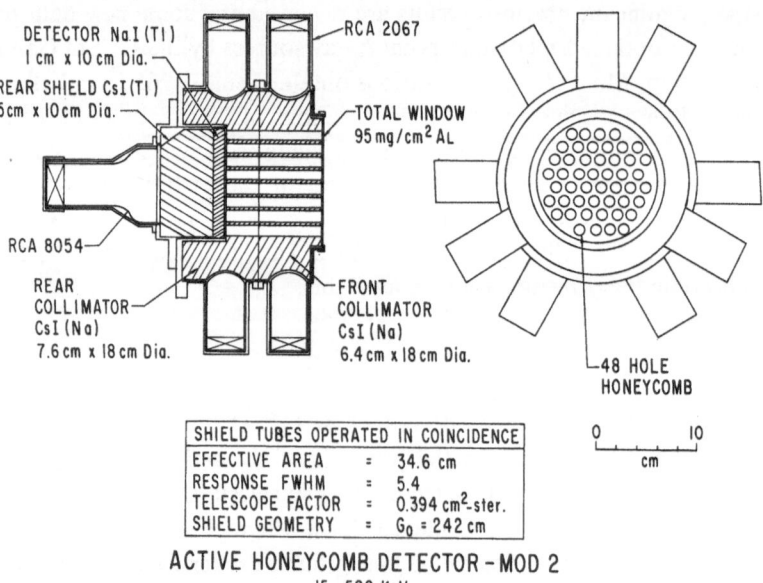

SHIELD TUBES OPERATED IN COINCIDENCE		
EFFECTIVE AREA	=	34.6 cm
RESPONSE FWHM	=	5.4
TELESCOPE FACTOR	=	0.394 cm²-ster.
SHIELD GEOMETRY	=	G_0 = 242 cm

ACTIVE HONEYCOMB DETECTOR – MOD 2
15 – 500 KeV

Fig. 1.   An actively collimated hard X-ray detector developed recently at UCSD. The aperture is formed by a honeycomb hole pattern. Cosmic-ray produced background and higher energy γ-ray events are rejected by anticoincidence circuitry.

aperture the signal-to-noise will be increased by only a factor of $\sqrt{2}$ at a considerable expense in weight, deadtime and cost. Detectors vary from the simple cylindrical geometry designed for the OSO-III to a model which uses a CsI collimator consisting of holes drilled in a honey-comb pattern and a separate 'well' to provide side and rear shielding. This 48 cm^2 detector with a 8.4° FWHM operative detector and its properties have also been described in the literature (Peterson *et al.*, 1968).

The latest version developed at UCSD by J. L. Matteson and R. M. Pelling and shown in Figure 1 has 36 cm^2 area and an aperture of 5.4° FWHM. This detector, which has no inert mass inside the shield, uses a NaI(Tl)/CsI(Tl) 'phoswich' to provide attenuation from the rear. The honeycomb collimator of CsI(Na) is observed with a large number of phototubes in coincidence in order to detect the lowest energy photons scattered into the shield and therefore to improve the background rejection.

Detectors such as these, before being placed in operation, are test flown on a balloon at the altitude where the observations will be eventually made in order to determine the background properties in the correct radiation environment. The background at 3 gm/cm^2 over Palestine, Texas, is also very similar to that encountered in an OSO satellite at 550 km, 33° inclination. The test gondola contains an additional scintillation counter which may be moved in front of the aperture and connected in anticoincidence. Spectra of the central detector are obtained with the aperture

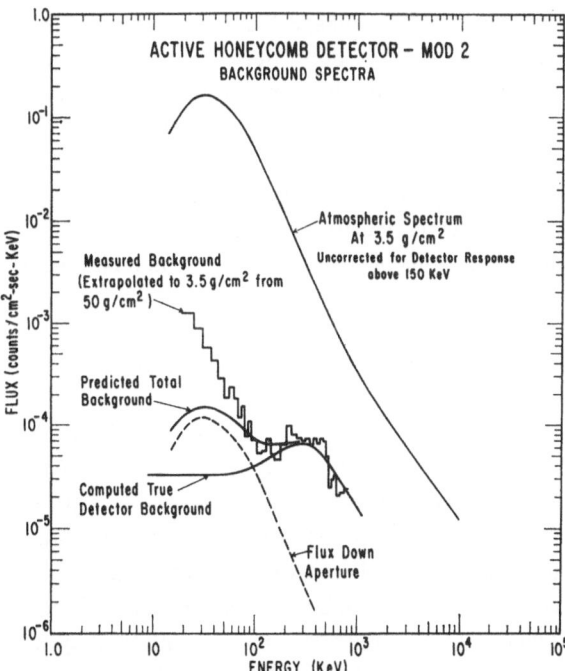

Fig. 2. The aperture and background properties of the detector shown in Figure 1 at balloon altitudes. The isotropic background spectrum is attenuated by the aperture solid angle factor of 1000 and is added onto various leakage and scattering contributions in the detector due to its total radiation environment.

blocked and unblocked, with the anticoincidence turned on and off and with the shield operating in different modes in order to determine various contributions to the background. Figure 2 shows typical test data obtained on the detector shown in Figure 1.

The upper curve indicates the total atmospheric spectrum measured at 3 gm/cm^2 (Peterson *et al.*, 1967). At this altitude the total atmospheric plus diffuse cosmic spectrum forms nearly an isotropic distribution. If the detector shield were perfect, the measured flux through the aperture would therefore be the atmospheric background reduced by the ratio of $\Omega/4\pi$, as indicated by the dotted line. Background not due to external fluxes entering the aperture, which is determined with the collimator blocked, is due mostly to Compton scattering of higher energy γ-rays which leak through the finite shield thickness and scatter in the central detector. A Monte-Carlo computer program for determining this effect has been developed by J. L. Matteson. Using the known atmospheric spectrum, this model technique predicts a true detector background which is indicated by the solid line. The sum of this background plus the flux through the aperture should equal the total background spectrum measured with the aperture opened, assuming no cosmic point sources are within the aperture.

The measured spectrum is shown in the histogram and agrees very well with the predictions at energies above about 80 keV. At the lower energies, however, an effect which is not directly predicted from the interactions of radiation with matter produces an additional exponential background spectrum extending to the lowest energies. This effect is apparently caused by large noise pulses produced by phosphorescent effects in the phototube, photocathode, or the scintillation material (Jerde *et al.*, 1967), and may be reduced by eliminating fast decay events in the central detector. We feel that the background properties of these detectors are sufficiently well understood so that similar systems of much larger area and smaller aperture could be designed and that these techniques could be extended to higher energies.

In order to make observations of cosmic sources from a balloon, a system is required which provides pointing and data transmission (Peterson *et al.*, 1967) facilities. One of the more recent balloon gondolas flown at UCSD is shown in Figure 3. This consists of a bottom structure or base which is suspended through a rigging mechanism and the parachute to the balloon. A large yoke containing an equatorial axis is free to move in azimuth about a vertical axis. This axis is driven in azimuth by a servo which obtains error signals either from a magnetometer system or from a star tracker and aligns the equatorial axis in the N–S meridian plane. This axis is also driven in elevation by a servo which also obtains error signals from the star tracker, or may be operated from ground command, based on the known latitude of the balloon. The detector may be controlled to a number of preselected declination and hour angle positions with reference to the equatorial axis. A command system permits tracking mode selection, real-time star search and acquition, and dec/h.a. selection.

In practice the device is usually operated in a drift scan mode, in which a detector is fixed at the appropriate declination corresponding to the source and at some angle about the polar axis. As the earth rotates, the source then drifts through the aperture

Fig. 3. The equatorial axis gondola used to orient detectors similar to that shown in Figure 1. The bottom structure is suspended from the balloon and the yoke servoed in the North/South meridian plane by error signals from either a magnetometer or star tracker. The detector is commanded to fixed positions in declination and hour angle.

and background is obtained before and after the observation. Then the detector is advanced in hour angle in such a way that a second drift scan may be obtained. This process is continued until as many observations as required have been obtained or until the source is too low in elevation to obtain successful results. Between observations the detector is calibrated by commanding it to a position so that it observes a radioactive source. In order to verify pointing position the sky is photographed with a camera attached to the polar axis. Although the design goal is for a pointing accuracy on the order of 0.1°, about 0.3° rms has been obtained. It is felt that continued refinements of the servo mechanisms and mechanical integrity of the structure will permit attainment of the design goal with this rather simple configuration.

Data are telemetered on an FM/FM/PCM link. Events from the pulse height analyzer are coded on a PCM word which is transmitted, along with data relative to total counting rates, background effects, deadtime, detector mode, etc., on the PCM format. Information required to operate the pointing controls and various other housekeeping functions is transmitted on additional subcarriers. Data are decoded in a standard PCM ground station and spectra are displayed by reading events into the memory core of a laboratory pulse height analyzer. Data are also recorded, along with timing information, on magnetic tapes which are processed through a computerized system after the flight has been performed.

Although the detector and observation system described here represents the present state-of-the-art development at UCSD, most of the observations which are discussed here were performed with relatively simpler systems developed during the evolution of detector, pointing, and communication techniques.

3. The Crab Nebula

The Crab Nebula was the first X-ray source to be positively identified with a known optical or radio object (Bowyer *et al.*, 1964). It was also the first source to be detected via a balloon observation by Clark in 1964 (Clark, 1965). Since then it has been studied extensively from balloons at UCSD. The spectrum obtained from our work (Peterson *et al.*, 1968) together with certain other results at higher (Haymes *et al.*, 1968a) and lower energies (Chodil *et al.*, 1967) are shown in Figure 4. There are also many additional observations, all of which are in general agreement with the spectrum in Figure 4. The data now available on the Crab Nebula form a continuous spectrum defined between 1 and 500 keV, which has a power law shape. Under present thinking, this is indicative of a synchrotron process.

If indeed X-rays are produced by synchrotron radiation in the 10^{-4} G field of the Nebula, the predicted lifetime of the electrons which produces 50 or 100 keV X-rays is only on the order of a few hours. Unless the process which accelerates the electrons to produce these X-rays is rather continuous, one should expect to see time variations in the X-ray spectrum. Therefore, it is germane to observe the Crab Nebula as sensitively and as often as possible in order to search for these variations. Observations at UCSD using essentially the same detector, which avoids systematic effects, have

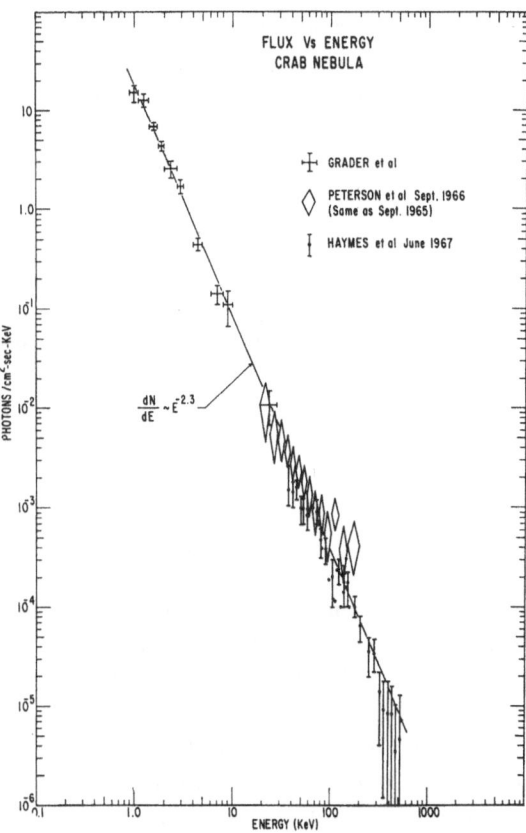

Fig. 4. The spectra of the Crab Nebula determined by a series of rocket and balloon observations over the 1–500 keV range. The data are all consistent with a power law spectrum, suggesting a synchrotron process.

been obtained over the 22-month period from September 1965 through July 1967. On three flights, 22 September 1965, 21 September 1966, and 7 July 1967 (Jacobson, 1968), no significant changes were observed over the 30–100 keV range. At a 95% confidence level, hard X-rays from Crab Nebula must be constant within 3% per year. This small variation presents remarkable constraints upon the accelerator if indeed the synchrotron process is responsible for these X-rays.

In addition to studying the continuous emission from the Crab Nebula, effort has been devoted to search for γ-ray line emission. Such lines have been predicted based upon the production of r-process elements in the supernova explosion (Clayton and Craddock, 1965). The model for these predictions is the production of fissionable Cf^{254}, which was advanced to explain the 55-day decay associated with Type 1 supernovae. If indeed Cf^{254} is produced in the initial explosion, other heavy elements should also have been formed and some of the longer half-life products presently undergoing radioactive decay in the remnant. The activities predicted have been calculated initially by Clayton and Craddock and include lines due to Am^{241}, Cf^{251}, and Cf^{249}.

Jacobson (1968) of UCSD has made an extensive search for these lines using a germanium-lithium drifted cooled detector in an active anticoincidence shield similar to that flown on the OSO-III. These observations were made with a 10 cm^2 area detector, 5 mm thick, which was flown in a gondola which had a simple alt-azimuth pointing control. The upper limits obtained by Jacobson from his observations in July 1967 are shown in Table I. This table also shows the initial predictions of Clayton

TABLE I

Upper limits on predicted γ-ray lines from the Crab Nebula

r-Process element	Energy (keV)	Predicted flux [Clayton and Craddock (1965)] (cm^2-sec)$^{-1}$	Predicted flux (This work) (cm^2-sec)$^{-1}$	Previous upper limits [Haymes et al. (1968)]	Upper limits (This work) (cm^2-sec)$^{-1}$
Am 241	59.6	5.7×10^{-5}	5.1×10^{-5}	3.9×10^{-3}	1.4×10^{-3}
X-rays	104	—	4.0×10^{-5}	–	1.0×10^{-3}
Cf 251	180	1.9×10^{-5}	7.6×10^{-5}	1.5×10^{-3}	1.1×10^{-3}

and Craddock as well as corrections to the predicted fluxes based upon improved branching ratios. An additional component, that of KX-rays of the heavy elements, has also been added by Jacobson. The predictions are about a factor of 20 below the present upper limits.

These upper limits, however, are based upon the assumption that all of the fission energy of Cf^{254} is emitted as light and that this is the principle energy source. Any inefficiency will result in a higher γ-ray intensity. Furthermore, these intensities were predicted under the assumption that the distance to the Crab was approximately 1.3 Kpcs. Recent measurements by Trimble (1968) indicate a new distance of 2.02 Kpcs. At this distance the optical extinction is larger, which results in an increase in the expected γ-ray line intensity. At the new distance the upper limits are only a factor of four above the predicted line intensities for an efficiency of unity (Jacobson, 1969). Thus, the upper limits and the predicted fluxes are bracketed within a rather narrow range, and it seems that the efficiency for conversion of fission energy through ionization into light energy must have been at least 25%, and that this efficiency must have maintained during the large expansion of the nebula that occurred immediately following the explosion.

Therefore, based on these results it seems that the Cf^{254} fission is becoming rather untenable as an explanation for the initial energy source of the Crab Nebula. There are other considerations which involve the dynamics of the expansion, heating, etc. (Sartori and Morrison, 1967), which have also cast doubt upon this idea in recent years. It would seem that only a small increase in sensitivity could absolutely disprove this hypothesis.

The recent discovery of Pulsar NP0532 has, of course, cast an entire new dimension to the study of the Crab Nebula. Since this pulsar has now been observed in the optical range (Cocke et al., 1969) and recently in the rocket X-ray range (Fritz et al., 1969), the search for pulses in the 20–100 keV spectral range becomes a most important effort. Jacobson (private communication) has reanalyzed the data from his July 1967 flight in order to search for such pulses, and has found an upper limit for pulsed X-ray emission over this range of equal to or less than 15% of the continuous emission. This is not in disagreement with the result reported by Haymes at this meeting (Fishman *et al.*, 1970) of $7 \pm 2\%$ for the pulsed emission over the 40–100 keV range.

4. Scorpius XR-1

The original determination of the spectrum of Scorpius over the 20–50 keV range made by UCSD in June 1965 (Peterson and Jacobson) indicated that the major emission was of an exponential character, suggesting thermal emission from a thin gas at 50×10^6 K. Since then many workers have confirmed this spectrum and have extended it to higher energies. SCO XR-1, the strongest X-ray source in the sky at 3 keV, has now been identified with a 13th magnitude blue optical object which undergoes a number of different time variations (Mook, 1967). These include fast flickering, occasional flare-like events where the optical flux approximately doubles, and slow variations in which the optical flux varies over a period of hours or days by about an optical magnitude. If indeed the X-ray emission from Scorpius is due to a hot gas, as the spectral measurements and other evidence indicate, it would seem that the X-ray emission and the optical emission must be coupled, and the variations should be correlated. Attempts to perform simultaneous observations have been made by the Lawrence Radiation Laboratory from short duration rocket flights, in conjunction with photometric observations by Hiltner (Chodil *et al.*, 1968b). UCSD has also attempted simultaneous observations from balloons in order to observe the higher energy component over an interval of several hours.

These observations were performed May 21/22, 1968, in cooperation with Dr. Hugh Johnson, using the KPNO 84-inch telescope. The optical observations were made with approximately a 10-sec integrating time constant and using three different wavelength filters. One was approximately 20 Å wide centered around Hβ, one a wide band filter of 109 Å, also centered at Hβ and therefore included mostly the continuum near Hβ; and the third filter was 84 Å wide and included the features associated with He II at 4686 Å. Observations are also made on a comparison star. The filters and reduction procedures have been described previously (Johnson and Golson, 1968).

The 20–40 keV X-ray flux measured during the balloon flight, shown in Figure 5, was observed with the original 48 cm^2, 8.4° FWHM honeycomb collimated detector. The large increase during the ascent is due to the cosmic-ray transition maximum. Once the balloon reached floating altitude, a rather steady background of both cosmic and atmospheric X-rays was obtained, after which three drift scan transits of the object through the detector aperture were made. To obtain the flux due to Scorpius, after subtracting the background, it is necessary to correct for the position of the source within the aperture using the orientation information and the detailed aperture

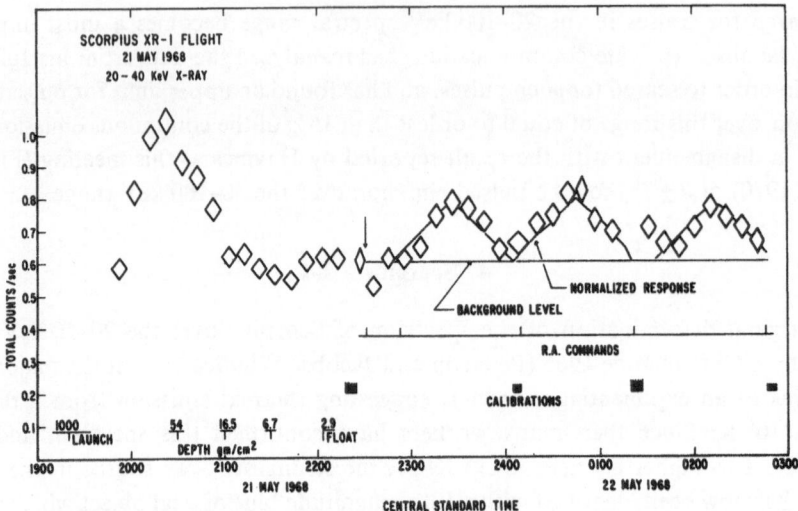

Fig. 5.   Counting rates are a function of time during the SCO XR-1 observations of May 21/22, 1968.
Background was obtained before and after the three transits of the source.

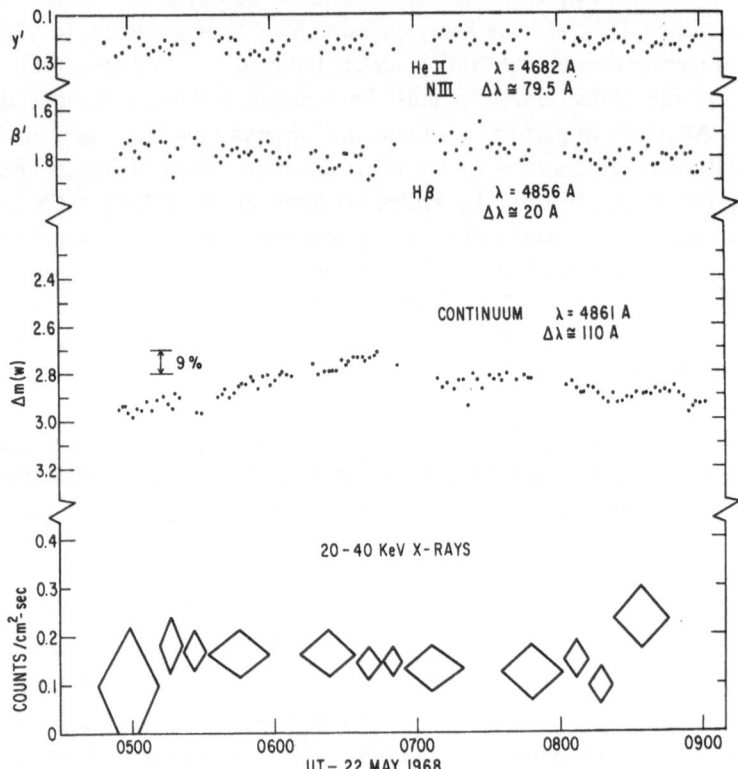

Fig. 6.   Corrected counting rates due to X-rays from Scorpius as a function of time compared with
simultaneous optical observations. Although the optical intensity varied approximately 25%, statistically significant variations in the total X-ray flux were apparently not observed.

response. The results of carrying out these corrections are indicated in Figure 6 along with the results obtained by Johnson over the various wavelength ranges. Clearly, no variations were obtained from the comparison source (y') or the narrow band centered upon Hβ(β'). The wide band, however, showed a slow increase followed by a decrease during the three-hour observing period. Statistically significant variations of the total X-ray flux over the 20–40 keV energy range were not observed.

Significant changes were, however, observed in the spectrum during each of the three transits. In order to relate these changes to a physically meaningful variable, a thermal emitting source was assumed. The expected measured spectrum was then computed, taking account of atmospheric absorption, detector angular response and efficiency, and resolution spreading. These spectra were then fitted to the measured data using a least square method with temperature as a parameter.

The results of this process are shown in Figure 7, where the best fit incident spectrum,

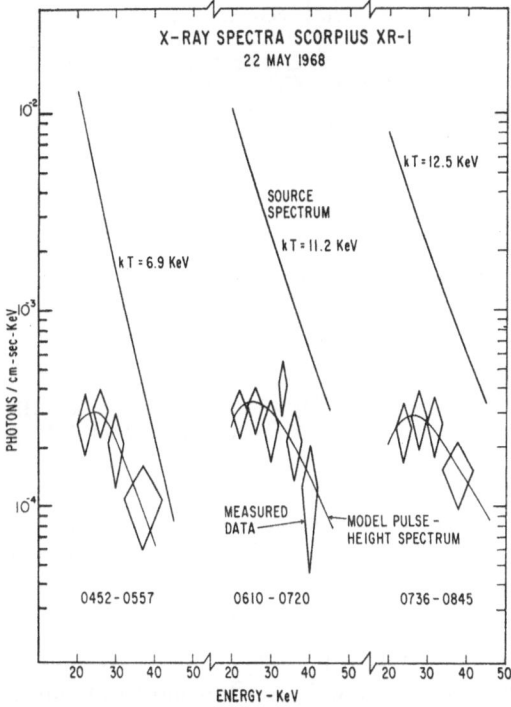

Fig. 7. Spectra of Scorpius determined during each of the three transits. Using a thermal brems-strahlung spectrum as an input, the expected spectrum at the detector has been computed and is least-square fitted to the measured data using temperature as a parameter.

the model pulse height spectrum at the detector, and the data are all shown for each of the three intervals. The calculated standard deviation of temperature as a parameter is ± 3 keV; a chi-squared test, however, permits the temperature to have remained constant. The higher best-fit temperatures are associated with increased optical emission as observed by LRL (Chodil *et al.*, 1968b). The variation of total flux at a given energy is a complex function of the change of temperature and emission measure. It is felt, however, that the lack of observed flux change within 20% is not inconsistent with the LRL interpretation based on a thermal model sufficiently compact to be self-absorbing in the optical range.

Observations of time variations of Scorpius have also been obtained from the OSO-III satellite, whose data reduction and analysis have been accomplished by H. S. Hudson and D. A. Schwartz. The 10 cm² OSO-III detector, which is shown in Figure 8,

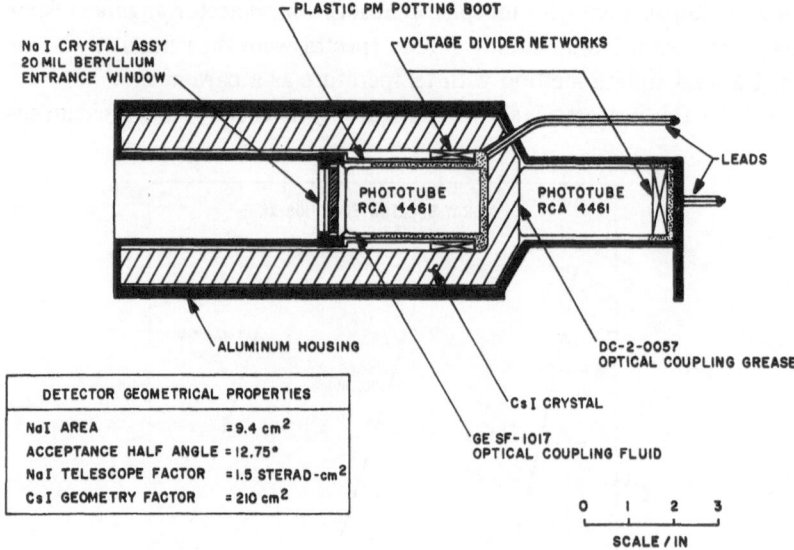

Fig. 8.   This detector, located on the rotating wheel section of the OSO-III satellite measures solar, cosmic and background X-rays over the 7.7–200 keV range.

is the earliest version of the anticoincidence shielded detectors developed at UCSD. This detector is mounted looking radially outward in the rotating wheel section of the OSO satellite, and therefore scans across the sky, the sun, and usually the earth with every two-second rotation period. Because the OSO wheel plane is constrained to lie in the earth-sun line, the scan plane precesses at about a degree per day and, as shown in Figure 9, slowly moves across the sky, drifting mostly in hour angle. A point source near the equatorial plane, therefore, slowly drifts across the aperture over a period of about a month.

The X-ray spectrum between 7.7 and 200 keV is obtained in six logarithmically spaced channels with approximately 1.7 energy ratio. Data are obtained in a number of different satellite-programmed modes. One of these accumulates X-rays from the solar direction and reads out every 15 sec; another reads a direction signature for each non-solar event and sequences through the six channels every 12 min. The third, or background mode, which is relevant to the results described here, consists of simply

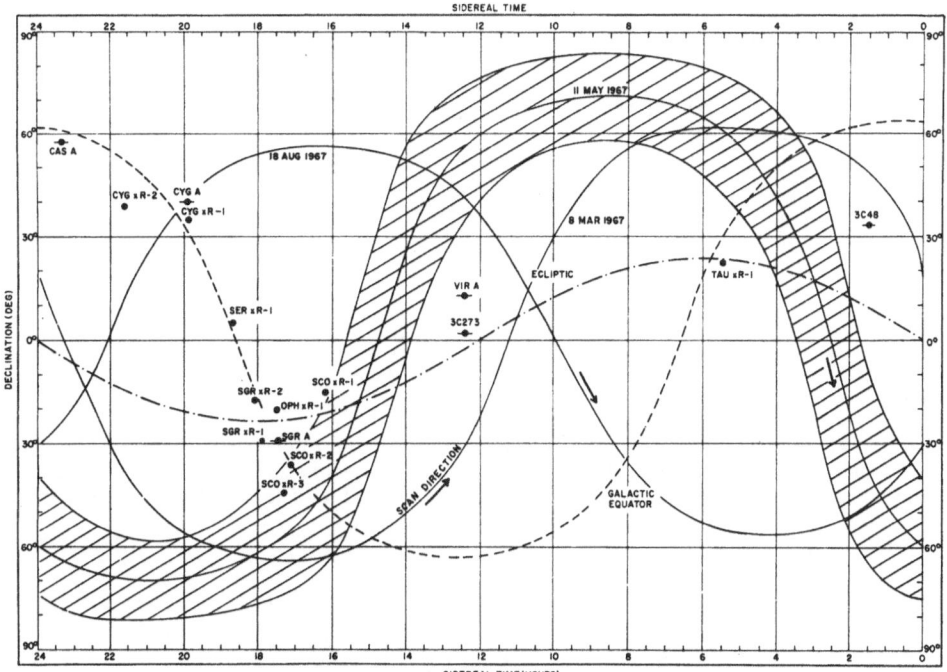

Fig. 9.   The area on the celestial sphere scanned by the OSO-III detector during each two-second wheel rotation. Since the scanned plane precesses approximately one degree per day in r.a., a source such as Scorpius slowly drifts through the aperture.

reading out the total counts accumulated during 1.96 sec in each channel every 15 sec during nighttime satellite operation. These events include a mixture of background, diffuse and point cosmic sources, and earth albedo. Scorpius XR-1 dominates the counting rate in this mode even though the source itself is within the aperture of the detector only 7% of each 2-sec scan. This mode was intended to obtain background information necessary to correct data from the solar and the sky modes; the effect of a strong point source such as Scorpius was not anticipated.

Figure 10 shows the counting rates obtained in the lowest channel during May and June of 1967 when the source made its first drift across the scan plane of the satellite. Data are selected to eliminate trapped radiation effects and averaged so that each point has a 10% or smaller statistical error. Background obtained before and after the scan plane intercepts Scorpius is at a level of 1.3 counts/sec.

Also shown is the predicted response based on the assumption that the background is 1.3 counts/sec and that the flux of Scorpius is that indicated by our earlier balloon observation (Peterson and Jacobson, 1966). Calculations include the model spectrum of Scorpius

$$F(E)\,dE = 110\,e^{-(E/4.3)}\,\text{keV/cm}^2\text{-sec-keV},$$

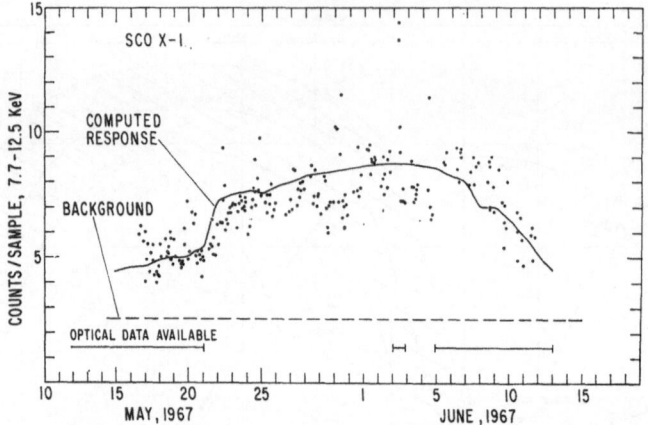

Fig. 10.   The total counting rate observed on the OSO-III in the lowest channel during May–June 1967, as Scorpius was drifting through the detector aperture. This strong source so dominates the total counting rate that it is significantly above the background level, as determined before and after the source entered the aperture.

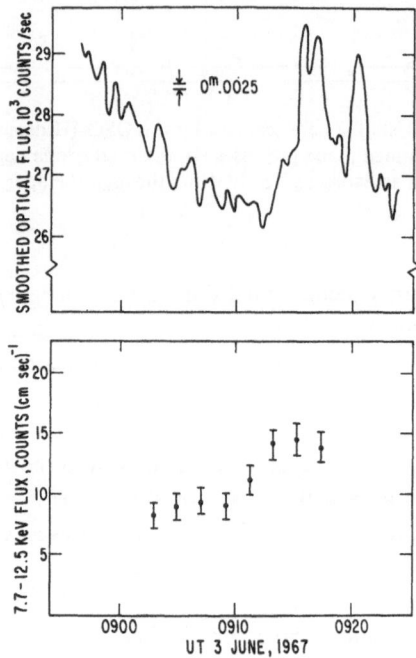

Fig. 11.   Simultaneous X-ray and optical observations of the Scorpius flare 0910 on 3 June 1967. The optical observations were fortuitously obtained at the Mt. Wilson telescope.

the instrument response and efficiency and the detailed aspect solution for the satellite. The observed data follows closely the predicted rates, which indicates that we are observing Scorpius, and that the flux is, on the average, close to that measured previously. The fact that some of the data depart from the predicted value more than the statistics would allow suggests that there are X-ray variations in Scorpius.

Fortuitously, optical observations were also being made during this period, as indicated in Figure 10. These have been obtained from two primary sources: B magnitude photometry obtained on the 36″ Cerro Tololo telescope and some high time resolution broad band photometric observations on the Mt. Wilson/Palomar telescopes. Correlations of the optical and X-ray observations for slow variations have not yet been completed, since these data are in a most preliminary form.

One striking result, however, is the confirmation of the observation by MIT (Lewin *et al.*, 1968a) of X-ray flares on Scorpius and their correlation with the related optical phenomena. The optical observations were made by Westphal *et al.* (1968) on the 100-inch telescope on the night of 3 June 1967. Figure 11 shows the Scorpius flare of 0910 on the 3 June observation by the satellite OSO-III and also the correlated optical data. Over a 5-min period the X-ray flux nearly doubled, while the optical flux increased about 25%. Although the accuracy of the time comparison is only about two minutes, these results indicate that X-ray flares and optical flares on Scorpius are

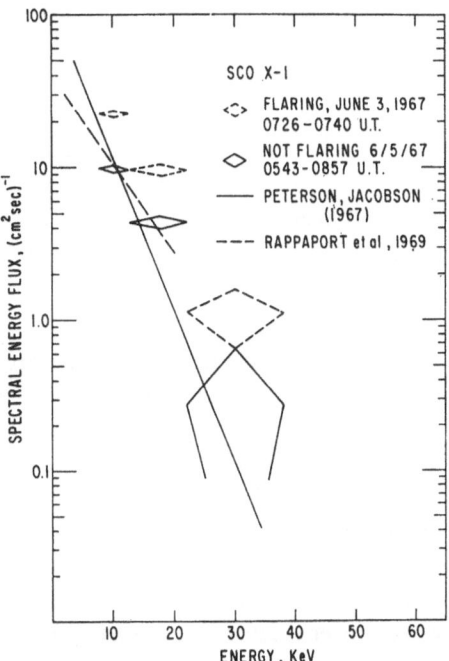

Fig. 12. The Scorpius flare spectrum as determined during another flare on 3 June. According to these data, the spectral shape of the flare is very similar to that of quiescent Scorpius.

indeed related phenomena. Another flare was observed on the same night at 0725. Figure 12 shows the X-ray spectrum from this flare. The flare spectrum as deduced from this observation indicates a uniform increase in all channels over the pre-flare flux. This is in disagreement with the observation of Clark in which he inferred that the flare spectrum was softer than that of the quiescent object. Although additional observations will be required to settle this and other questions, this work has indicated for the first time that X-ray and optical flares do indeed occur simultaneously on Scorpius XR-1.

The combination of optical, radio, rocket and balloon X-ray observations have now indicated a model for Scorpius. Figure 13 shows the fluxes which have been observed in all spectral ranges. These observations and their time variations imply a model for

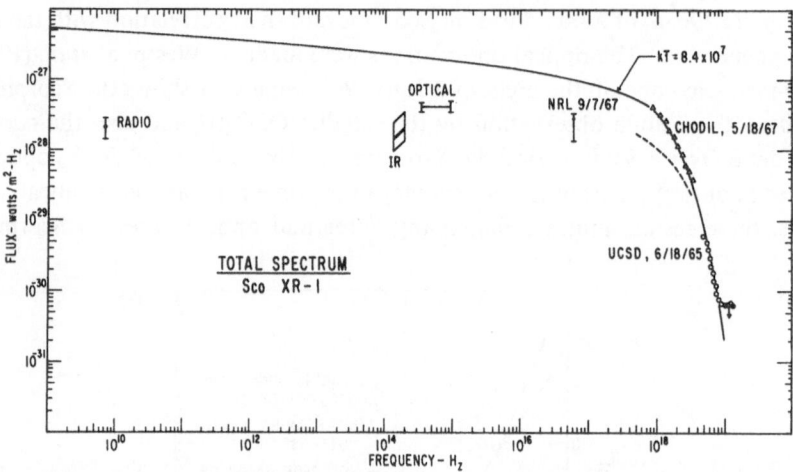

Fig. 13. A summary of radio, optical and X-ray observations of SCO XR-1 compared with the thermal bremsstrahlung expected from an optically thin object uniformly at 84×10^6 K. The data demand a more complex model.

Scorpius in which the object consists of a hot gas, on the order of 50×10^6 degrees, which is rather compact and becomes optically thick in the infrared range, and therefore follows the Rayleigh-Jeans law in emission (Neugebauer *et al.*, 1968). The source must have a diameter of approximately 10^8 cm and a density of $10^{16}/\mathrm{cm}^3$ based upon a distance of 0.17 kpc. (Gatewood and Sofia, 1968). The 'high energy tail' of hard X-rays may be due to an additional component of electrons, and the object must be associated with a cooler region to account for the atomic emission lines. The radio observations are not consistent with a compact thermal source. The radio data and the higher energy X-rays have been related in a model by Riegler and Ramaty (1969), in which an additional region of 10^{12} cm radius and rather lower density which contains a magnetic field is postulated. The electrons which produce the radio emission by synchrotron radiation are responsible for the higher energy X-rays by bremsstrahlung.

5. Other Sources

In addition to the Crab Nebula and Scorpius, which have been studied extensively, a number of other sources have been observed from the OSO-III satellite and from balloons. Because of the design of the detectors used at UCSD, broad sky surveys from balloons are not attempted and observations are generally made on X-ray sources identified from rocket observations. Sky surveys at high sensitivity are best accomplished from satellites.

The spectrum of the X-ray source which has been associated with the object M-87 (Bradt *et al.*, 1967) is of considerable interest since it is the only known extra-galactic X-ray source. This very weak X-ray source, $\frac{1}{30}$ of the Crab Nebula at 3 keV, may be observed at balloon altitude with sufficiently sensitive technique if the spectrum is not too soft. A series of upper limits on this object obtained at UCSD are shown in Figure 14. M-87 is below the level of detectability of the OSO-III satellite; therefore,

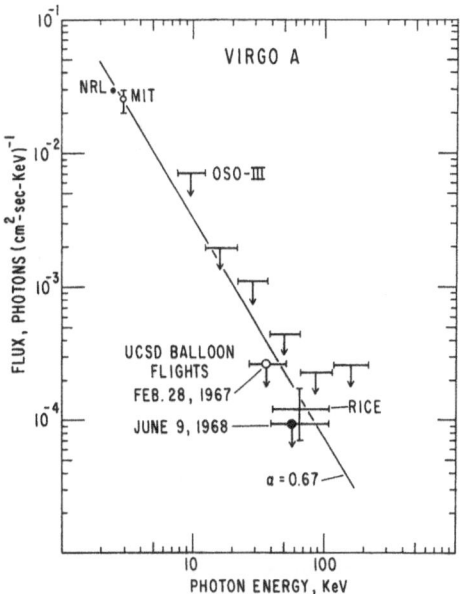

Fig. 14.   A summary of the upper limits on M-87 obtained from the OSO-III satellite and from the series of balloon flights at UCSD. Also shown are the fluxes measured at rocket energies and the reported flux at 40 keV by Rice University.

only upper limits are available. Two attempts have been made to detect this object from balloons, one during February 1967, and again during June of 1968. Although neither of these flights was sufficiently successful so that the ultimate sensitivity of the observation could be realized, significant upper limits have nevertheless been obtained, as shown in Figure 14. These upper limits, which are at a 3σ confidence level, seem to be significantly below the reported flux at 40 keV due to Haymes at Rice University (Haymes *et al.*, 1968c). We therefore question the validity of the Rice observation.

A series of observations have also been obtained of the Cygnus region. Cygnus X-1, which at the higher energies seems to have a power law spectrum of similar intensity and slope to the Crab Nebula, was originally observed by UCSD in September 1966

(Peterson *et al.*, 1968) and has been reobserved by Jacobson and Laros in 1968 (private communication). A summary of selected results is shown in Figure 15, compared with the UCSD observations. Time variations have been reported for this source and the large scatter of the rocket data indicates that this is indeed the case. The UCSD observations do not presently extend to high enough energy to verify the

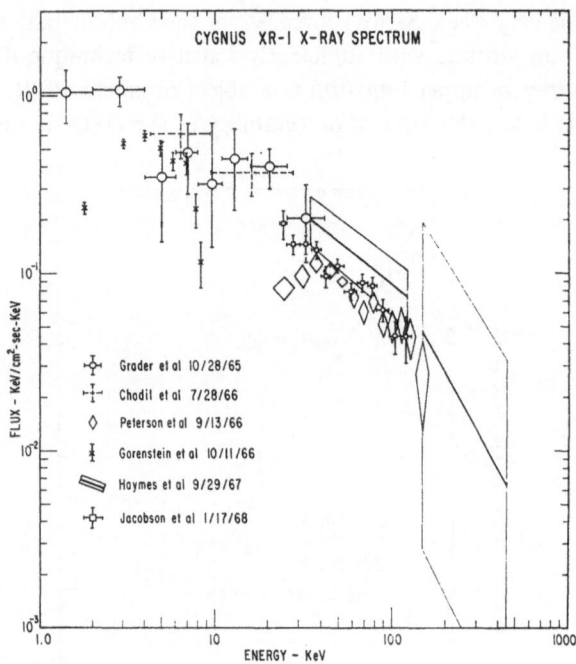

Fig. 15.   The spectrum of Cygnus XR-1 determined from a series of balloon and rocket observations. Although there seems clear evidence of time variations at the lower energies, the two UCSD observations are consistent with a simple power law spectra over the 30–100 keV range.

break in the spectrum at about 150 keV reported by Haymes (Haymes *et al.*, 1968b).

The spectrum of the weak source Cygnus XR-3 has also been determined on a balloon flight on 22 May, 1968. All of the data on this source are shown in Figure 16. Although this object appears to have a thermal spectrum at lower energies (Giacconi *et al.*, 1967), an additional hard component is indicated above about 20 keV. It may be that all sources which appear to be thermal-like over the rocket range have additional hard components associated with them, such as has already been indicated for Scorpius.

The OSO-III has made observations of certain sources in the southern sky, some of which will be presented here in a preliminary form. Figure 17 shows the source Norma XR-2 at 15 hr, 32 min r.a., −57° dec. This source was first identified in a rocket survey (Friedman *et al.*, 1967), observed from a balloon by Lewin and Clark (Lewin *et al.*, 1968b) and was previously identified erroneously with Lupus XR-1 in a UCSD preprint (Schwartz *et al.*, 1968). No significant change was observed in this

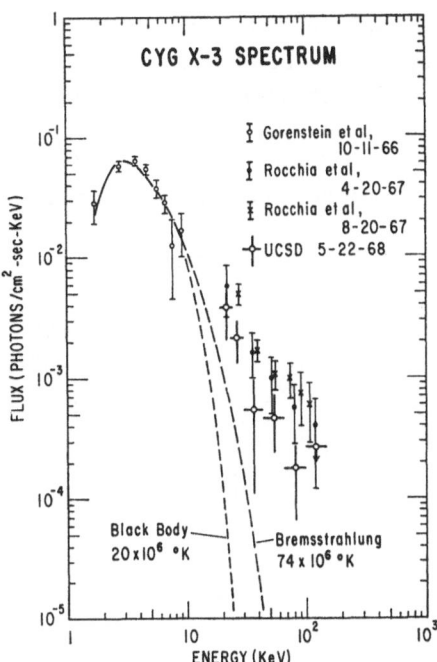

Fig. 16. The Cygnus X-3 spectrum as determined from several observations. Although this source is consistent with thermal bremsstrahlung at lower energies, there is apparently, like Scorpius, an additional component at the higher energies.

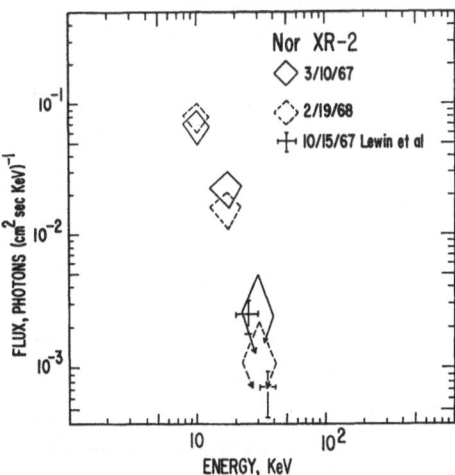

Fig. 17. Spectrum of Norma XR-2 determined on two different occasions by the OSO-III satellite, compared with the balloon observation of Lewin and Clark. Over nearly a year there is no evidence of time variation.

source between 10 March 1967 and 19 February 1968. The spectrum agrees with the observation on 15 October 1967 by Lewin and Clark. Norma XR-2 apparently is not a variable source within about 20% over a one-year period.

The variable X-ray source Centaurus XR-2 has also been observed by the OSO-III. Originally discovered by rockets during May and June 1967, it suddenly appeared and subsequently decayed with a 30-day time constant (Chodil *et al.*, 1968a). It is located

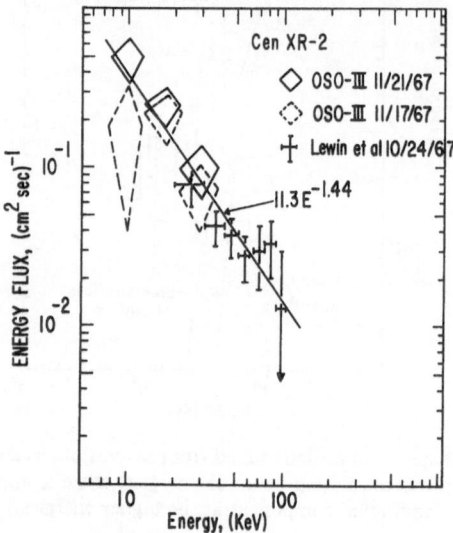

Fig. 18.   The variable source Centaurus XR-2 measured during October and November 1967 by the OSO-III compared with the balloon measurement of Lewin and Clark.

at 196.5° r.a., −64° dec and has been tentatively identified with the Nova WX Centaurus (Eggen *et al.*, 1968). Lewin and Clark observed this source 24 October 1967 to have a hard component, although rocket observations indicated that the source was no longer observable (Chodil *et al.*, 1968a). The OSO-III observations were during November 1967, and by January and February of 1968 the source had decreased below the threshold of detectability over the 10–100 keV range. The spectral data obtained in October and November of 1967 are summarized in Figure 18.

Conclusion

In this brief presentation we have tried to review some of the significant results which have been obtained by the UCSD X-ray and Gamma-ray Astronomy Group on individual point sources during the past several years. Balloon observations of the sources in the Northern Hemisphere are continuing, and the reduction of the OSO-III satellite will be completed very shortly. From these data it is hoped to obtain information on the spectra and time variations of 10 or 15 of the strongest sources in the sky. These results will be published in the literature. In addition this experiment will provide data on the isotropy and spectrum of the cosmic background component (Hudson *et al.*, 1969). This component has also been studied extensively at UCSD using the ERS-18 satellite where the spectrum has been extended to approximately 6 MeV (Vette *et al.*, 1970).

Acknowledgements

This research was supported under NASA Grant NSG-318 and Contract NAS5-3177 with the Goddard Space Flight Center. In addition to those directly mentioned, the author acknowledges the work of many other students, staff and colleagues who have contributed to these experiments.

References

Bowyer, S., Byram, E. T., Chubb, T. A., and Friedman, H.: 1964, 'Lunar Occultation of X-Ray Emission from the Crab Nebula', *Science* 146, 912.

Bradt, H., Mayer, W., Naranan, S., Rappaport, S., and Spada, G.: 1967, 'Evidence for X-Radiation from the Radio Galaxy M-87', *Astrophys. J. Letters* 150, L199.

Chodil, G., Mark, Hans, Rodrigues, R., Seward, F. D., and Swift, C. D.: 1967, 'X-Ray Intensities and Spectra from Several Cosmic Sources', *Astrophys. J.* 150, 57.

Chodil, G., Mark, Hans, Rodrigues, R., and Swift, C. D.: 1968a, 'Nova-Like Behavior of the X-Ray Source Centaurus XR-2', *Astrophys. J. Letters* 152, L45.

Chodil, G., Mark, Hans, Rodrigues, R., Seward, F. D., Swift, C. D., Turiel, Isaac, Hiltner, W. A., Wallerstein, George, and Mannery, E. J.: 1968b, 'Simultaneous Observations of the Optical and X-Ray Spectra of SCO XR-1', *Astrophys. J.* 154, 645.

Clark, George W.: 1965, 'Balloon Observation of the X-Ray Spectrum of the Crab Nebula above 15 keV', *Phys. Rev. Letters* 14, 91.

Clayton, D. D. and Craddock, W. L.: 1965, 'Radioactivity in Supernova Remnants', *Astrophys. J.* 142, 189.

Cocke, W. J., Disney, M. J., and Taylor, D. J.: 1969, *Nature* 221, 525.

Eggen, O. J., Freeman, Kenneth C., and Sandage, Allan: 1968, 'On the Optical Identification of the X-Ray Source CEN XR-2 as WX CEN', *Astrophys. J. Letters* 154, L27.

Fishman, G. J., Harnden, F. R., and Haymes, R. C.: 1970, this volume p. 116.

Friedman, H., Byram E. T., and Chubb, T. A.: 1967, 'Distribution and Variability of Cosmic X-Ray Sources', *Science* 156, 3773, 374.

Fritz, G., Henry, R. C., Meekins, J. F., Chubb, T. A., and Friedman, H.: 1969, 'X-Ray Pulsar in the Crab Nebula', *Science* 164, 3880, 709.

Frost, K. J., Rothe, E. D., and Peterson, Laurence E.: 1966, 'A Search for the Quiet-Time Solar Gamma-Rays from Balloon Altitudes', *J. Geophys. Res.* 71, 17, 4079.

Gatewood, George and Sofia, Sabatino: 1968, 'Physical Characteristics of SCO X-1', *Astrophys. J. Letters* 154, L69.

Giacconi, R., Gorenstein, P., Gursky, H., and Waters, J. R.: 1967, 'An X-Ray Survey of the Cygnus Region', *Astrophys. J. Letters* 148, L119.

Haymes, R. C., Ellis, D. V., Fishman, G. J., Kurfess, J. D., and Tucker, W. H.: 1968a, 'Observation of Gamma Radiation from the Crab Nebula', *Astrophys. J. Letters* 151, L9.

Haymes, R. C., Ellis, D. V., Fishman, G. J., Glenn, S. W., and Kurfess, J. D.: 1968b, 'Detection of Gamma Radiation from the Cygnus Region', *Astrophys. J. Letters* 151, L125.

Haymes, R. C., Ellis, D. V., Fishman, G. J., Glenn, S. W., and Kurfess, J. D.: 1968c, 'Detection of Hard X-Radiation from Virgo', *Astrophys. J. Letters* 151, L131.

Hicks, D. B., Reid, L. Jr., and Peterson, L. E.: 1965, 'X-Ray Telescope for an Orbiting Solar Observatory', *IEEE Trans. Nucl. Sci.* NS-12, 54.

Hudson, H. S., Peterson, L. E., and Schwartz, D. A.: 1969, 'Solar and Cosmic X-Rays above 7.7 keV', *Solar Phys.* 6, 205.

Jacobson, A. S.: 1968, A Search for Gamma-Ray Line Emissions from the Crab Nebula, Thesis.

Jacobson, A. S.: 1969, The Gamma-Ray Line Spectrum of the Crab Nebula (UCSD Preprint).

Jerde, R. L., Peterson, L. E., and Stein, W.: 1967, 'Effects of High Energy Radiations on Noise Pulses from Photomultiplier Tubes', *Rev. Sci. Instr.* **38**, 1387.

Johnson, H. M. and Golson, J. C.: 1968, 'Narrow-Band and UBV Photometry of GX3-1 and Two Wolf-Rayet Stars', *Astrophys. J. Letters* **154**, L77.

Lewin, W. H. G., Clark, G. W., and Smith, W. B.: 1968a, 'Observation of an X-Ray Flare from SCO X-1', *Astrophys. J.* **152**, L55.

Lewin, W. H. G., Clark, G. W., and Smith, W. B.: 1968b, 'Observation of CEN XR-2 and Other High-Energy X-Ray Sources in the Southern Sky', *Astrophys. J. Letters* **152**, L49.

Mook, Delo E.: 1967, 'UBV Photometry of SCO XR-1', *Astrophys. J.* **150**, L25.

Neugebauer, G., Oke, J. B., Becklin, E., and Garmire, G.: 1969, 'A Study of Visual and Infrared Observations of SCO XR-1', *Astrophys. J.* **155**, 1.

Overbeck, James W. and Tananbaum, Harvey D.: 1968, 'Time Variations in Scorpius X-1 and Cygnus XR-1', *Astrophys. J.* **153**, 899.

Peterson, L. E.: 1966, 'Upper Limits of the Cosmic Gamma-Ray Flux from OSO-I', *Space Research*, Vol. VI, p. 53.

Peterson, L. E. and Jacobson, A. S.: 1966, 'The Spectrum of SCO XR-1 to 50 keV', *Astrophys. J.* **145**, 962.

Peterson, L. E., Jerde, R. L., and Jacobson, A. S.: 1967, 'Balloon X-Ray Astronomy', *AIAA J.* **5**, 1921.

Peterson, L. E., Jacobson, A. S., Pelling, R. M., and Schwartz, D. A.: 1968, 'Observations of Cosmic X-Ray Sources in the 10–250 keV Range (Presented at 10th International Cosmic Ray Conference, Calgary, Canada, June 1967) *Can. J. Phys.* **46**, S437.

Riegler, G. and Ramaty, R.: 1969, 'Physical Properties of the Radio-Emitting Region of SCO X-1 (GSFC X-611-69-123)'; Preprint March.

Sartori, L. and Morrison, P.: 1967, 'Thermal X-Rays from Non-Thermal Radio Sources', *Astrophys. J.* **150**, 385.

Schwartz, D. A., Hudson, H. S., and Peterson, L. E.: 1968, 'Satellite Observation of the X-Ray Source in Lupus', Preprint January 1969.

Trimble, Virginia: 1968, 'Motions and Structure of the Filamentary Envelope of the Crab Nebula', *Astron. J.* **73**, 535.

Vette, J. I., Matteson, J. L., Gruber, D., and Peterson, L. E.: 1970, this volume p. 355.

Westphal, J. A., Sandage, Allan, and Kristian, Jerome: 1968, 'Rapid Changes in the Optical Intensity and Radial Velocities of the X-Ray Source SCO X-1', *Astrophys. J.* **154**, 139.

PROPERTIES OF INDIVIDUAL X-RAY SOURCES

(Invited Discourse)

K. G. McCRACKEN

University of Adelaide, Adelaide, South Australia

Prof. McCracken did not communicate any text or Summary of his invited discourse.

ROCKET OBSERVATIONS OF VIRGO XR-1

D. J. ADAMS, B. A. COOKE, K. EVANS and K. A. POUNDS

X-Ray Astronomy Group, Dept. of Physics, University of Leicester, England

Skylark 723, launched at 10.30 GMT on 12th June 1968 was an unstabilised rocket carrying two proportional counter detectors, each of 1385 cm^2 effective area. Detector C1 covered the energy range 1.4 to 2.5 keV and detector C2 the range 2.0 to 18 keV. The detector outputs were analysed into 4 and 9 energy channels respectively. The field of view of each detector was 28° and 4° (FWHM) with the greater collimator extension mounted parallel to the longitudinal rocket axis for C2 and $\frac{6}{10}$ of the C1 detector, the remaining $\frac{4}{10}$ of the C1 collimator being canted at 40° to the major axis. During the flight, the rocket spun at a constant rate of 75° per sec whilst the spin axis slowly precessed about a flat cone, thereby surveying some 80% of the visible sky. On

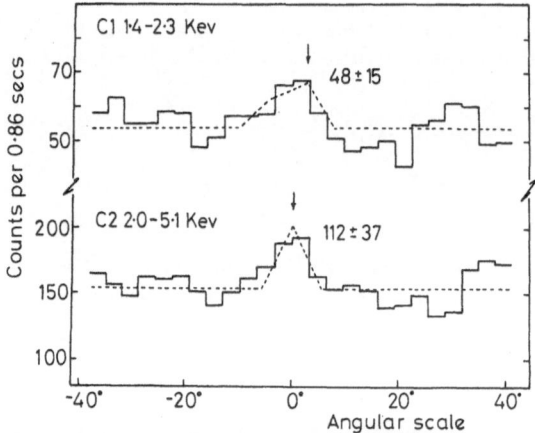

Fig. 1. The count rate histograms of each detector from the flight of 12th June. Data from 20 scans passing through the Virgo region have been added for each trace and the position of the Virgo XR-1 source is arrowed. The apparent 2° difference between the C1 and C2 peaks is in agreement with pre- and post-flight measurements of the relative detector alignment. Each count plotted is a running mean over two 3.2° intervals and corresponds to an accumulation time of 0.86 sec.

25 consecutive scans the Vir XR-1 source, which by chance lay near the precession axis, crossed the field of view of both detectors. The count rate profiles for twenty scans in which the source passed within 21° of the centre of the field of view have been added together for each detector, resulting in the totals shown in Figure 1. Significant peaks are seen in the position of Vir XR-1 for both detectors and these have been fitted with the measured transmission profiles of the respective collimators assuming

L. Gratton (ed.), Non-Solar X- and Gamma-Ray Astronomy, 82–86. All Rights Reserved
Copyright © 1970 by the IAU

parallel incident radiation. The peaks are consistent with the observation of X-rays from a single point source.*

Skylark 403, launched at 0605 GMT on 8th July, was equipped with a sun-pointing attitude control system and roll control was obtained by reference to magnetic field sensors. Two small X-ray proportional counters, each of 25 cm² effective area, were mounted, viewing at 76° to the sun-pointed rocket axis. The detectors were collimated to 20 arc min and 3.5° (FWHM) respectively (in a direction parallel to the longitudinal rocket axis), both having a 10° field of view in the rocket spin plane. During the 250 sec of stabilised flight, the two detectors were scanned four times, back and forth, across the region of sky containing the source Vir XR-1. The correct pointing was verified later by an on-board star camera. A systematic increase in count rate was observed in the direction of the source by the more coarsely collimated detector and the spectral analysis of these source counts has been used in the present paper. The finely collimated detector recorded no significant source counts, a result consistent with the flux measured by the other detector, and therefore no additional information can be given about the precise position of Vir XR-1.**

The spectral results of the two rocket experiments are summarised in Figure 2. The source counts from the June flight have been divided into seven energy channels and those from the July flight into two channels. The error bars on each individual point are quite large, owing to the limited counting statistics in a single channel, but the overall evidence for a continuously rising spectrum at least to 1.5 keV is good. This result is not in agreement with that of Bradt et al. [1] which indicated a flat photon spectrum near 2 keV and suggested that there might be significant absorption at this energy. It is noteworthy that the present data also indicate a source strength about three times as high as that found by Bradt et al., and in better agreement with the measurements of the NRL group [2, 3]. The greater intensity recorded by the four NRL and Leicester experiments is in part consistent with the absence of a low energy cut-off in the Vir XR-1 spectrum since all four experiments had thin plastic window detectors, giving significantly greater sensitivity below 2.5 keV than the MIT groups beryllium window detectors. The intensity of the Vir XR-1 source from the present data is 1.5×10^{-9} ergs cm² sec in the energy range 1.5 to 5 keV.

Finally, it is interesting to speculate briefly on the proposed identification of Vir XR-1 with the radio galaxy M87 and, in particular, on the possibility that the X-radiation arises from synchrotron emission of the relativistic electrons in the optical jet. It is widely believed that both the 'core' radio emission and the polarised, blue continuum of M87 arise from synchrotron emission in the jet and Shklovskii [4] has suggested that the X-ray observation of Vir XR-1 represents an extension of this spectrum into the kilovolt range. The X-ray spectral data presented in Figure 2 are

* It may be noted that the method of addition will tend to smooth out count rate peaks for any sources away from the Virgo region, but separate examination of the data has shown only one other discrete X-ray source in the high latitude sweep covered by the data of Figure 1. This is a weak source in Crater which gives rise to some excess counts to the left of the Virgo peak in Figure 1, particularly on the C1 trace. The Crater observation will be reported in a separate paper.
** This experiment was originally intended to observe the (briefly) much stronger source Cen XR-2.

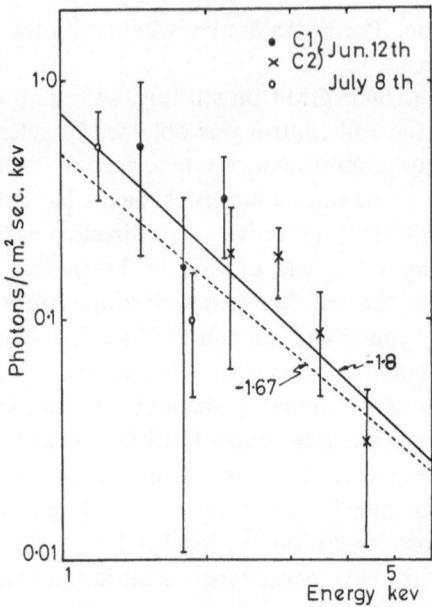

Fig. 2. The differential photon spectrum of Virgo XR-1 using the results of both rocket flights. Solid circles represent results from the C1 detector and crosses from the C2 detector of the June 12th flight. The full lines represent the best fit to all the data points and the dashed line the extrapolation of the core-radio spectrum.

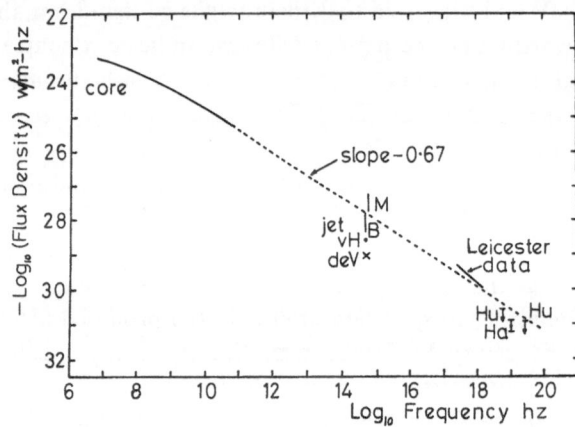

Fig. 3. The X-ray spectrum of Figure 2 is plotted together with the M87 core-radio spectrum extrapolated to high frequencies (after reference [6]), optical results from the M87 jet and some published high-energy X-ray data (see refs. [6–8]).

consistent with a power-law spectrum (though admittedly the energy band is insufficient to *prove* such a fit) as would be likely for a synchrotron source. More convincing, perhaps, is the evidence of Figure 3 in which the X-ray data have been plotted together with the M87 core-radio spectrum and various optical measurements of the jet. The overall fit to the extrapolated radio spectrum, with a power-law index of 0.67, is certainly remarkable over 9 decades of frequency. A well-known difficulty for the synchrotron mechanism, even at optical frequencies, is the short-lifetime of the radiating electrons. Shklovskii [5] and others have discussed a model in which a proton flux in the jet interacts with the ambient gas, producing a continuous supply of energetic electrons by ϱ-π-μ-e decays. As Felten [6] has pointed out, the occurrence of X-ray synchrotron emission would conflict with this model owing to the non-observation of 10^{13} eV γ-rays. Independent evidence against the secondary production model is provided by the absence of absorption in the X-ray spectrum, if this emission does indeed arise in the jet. Examination of Figure 2 indicates an optical depth at 1.5 keV of less than 0.5, corresponding to a neutral gas column of less than 5×10^{21} atoms (of cosmic abundance) per cm^2. The maximum gas density in a column 36 pc deep (0.5 arc sec [7, 8] is thus $100/cm^3$, compared with $700/cm^3$ necessary to prevent the jet being disrupted by its internal cosmic ray pressure [7].

In summary, it appears that the M87 jet is a source of intense optical and radio synchrotron emission, possibly extending up to X-ray frequencies. The latest optical studies [7, 8] show the optical emission arising mainly from three intense 'point sources', probably less than 1 arc sec in diameter and with some indication of fine structure. Moreover, there appears to be no evidence for the large gas densities required to confine the jet, either gravitationally or as the seat of ordered magnetic fields. We therefore propose a model analogous to, though on a much larger scale than that outlined recently by Burbidge and Hoyle for the Crab Nebula supernova remnant [9]. Thus in the case of M87, the visible radiation in the jet is envisaged to arise in the magnetospheres of massive, coherent fragments presumed to be thrown out from the galactic nucleus [10] perhaps a million years before. Continuous ejection from these fragments maintains the supply of energetic electrons and the required magnetic fields are anchored in the condensed objects. With this model, the absence of both optical emission lines and of X-ray absorption effects may be readily explained, whilst the electron lifetime and jet containment problems do not arise. The X-ray emission region might be very localised in each source and it is an intriguing possibility that the differences in measured flux referred to earlier could represent real variability.

Acknowledgements

This work was supported by a grant from the Science Research Council. The assistance of R. Batstone and J. Green in the data analysis and of the skilled support of teams at the R.A.E., Farnborough; Elliott-Automation Ltd.; and W.R.E., Salisbury, South Australia, in preparing the rocket payloads are gratefully acknowledged.

References

[1] Bradt, H., Mayer, W., Naranan, S., Rappaport, S., and Spada, G.: 1967, *Astrophys. J.* **150**, L199.
[2] Byram, E. T., Chubb, T. A., and Friedman, H.: 1966, *Science* **152**, 66.
[3] Friedman, H. and Byram, E. T.: 1967, *Science* **158**, 257.

[4] Shklovskii, I. S.: 1967, *Sov. Astron.* **11**, 45.
[5] Shklovskii, I. S.: 1962, *Sov. Phys.-Usp.* **5**, 365.
[6] Felten, J. E.: 1968, *Astrophys. J.* **151**, 861.
[7] Felten, J. E., Arp, H. C., and Lynds, C. R.: 1969, Contribution from the Kitt Peak National Observatory, No. 390.
[8] de Vaucouleurs, G., Angione, R., and Fraser, C. W.: 1968, *Astrophys. L.* **2**, 141.
[9] Burbidge, G. R. and Hoyle, F.: 1969, *Nature* **221**, 847.
[10] Burbidge, G. R.: 1967, *Nature* **216**, 1287.

INTENSITIES AND SPECTRA
OF SEVERAL GALACTIC X-RAY SOURCES

R. BATSTONE, B. COOKE, R. GOTT, and E. STEWARDSON

University of Leicester, Leicester, England

No text nor Summary was communicated by the Authors.

X-RAY FLUX FROM DISCRETE SOURCES

U. R. RAO, E. V. CHITNIS, A. S. PRAKASARAO, and U. B. JAYANTHI

Physical Research Laboratory, Ahmedabad, India

Abstract. Preliminary results of two rocket flights carrying X-ray payloads conducted from Thumba Equatorial Rocket Launching Station (TERLS), Trivandrum, India, on November 3, 1968, and November 7, 1968, respectively, are presented. The results indicate the first evidence for the existence of low energy X-ray flux in the energy range 2–20 keV from Cen-X2 source since the reported extinction in May, 1967. The energy spectrum and the absolute flux of X-rays from Cen-X2, Sco-X1 and Tau-X1 are presented and compared with other observations.

Two identical payloads for the detection of X-rays were launched from Thumba Equatorial Rocket Launching Station (TERLS), Trivandrum, India, one at 0319 UT on November 3, 1968, and another at 0305 UT on November 7, 1968, respectively. Both the spin stabilized rockets were launched towards the zenith (85° elevation) when Sco-X1, Tau-X1 and Cen-X2 were in the field of view of the detector, mounted perpendicular to the spin axis of the rocket. The detector system consisted of a Xenon-Methane filled, 2 mil. Berylium window (60.8 cm²) proportional counter collimated to 8.7° half transmission in azimuth and 17.2° half transmission in elevation by a slat type mechanical collimator. Inflight calibration of the counters was done using the 6 keV line from Fe^{55} source (15% FWHM), which was explosively ejected at 70 km altitude. Suitably oriented magnetic sensors and Sco-X1 sightings have been utilised to define the rocket attitude to within a $\pm 1°$ accuracy. The preliminary results are presented here and the detailed analysis and results will be published elsewhere (Rao *et al.*, 1969a).

Even though the Centaure rocket launched on November 3, 1968, remained spin

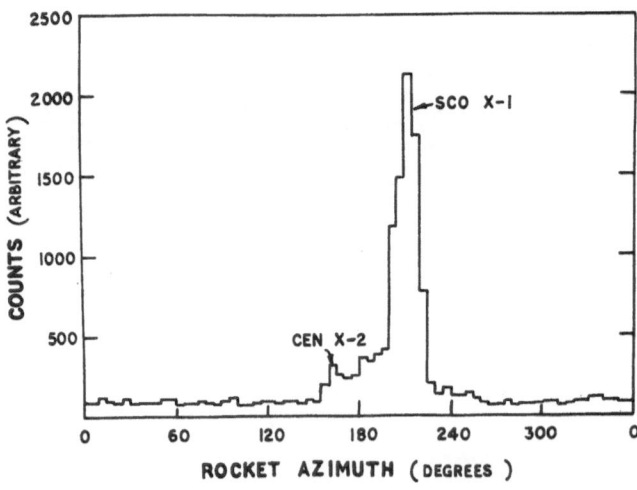

Fig. 1. X-ray count rate from various discrete sources observed on November 3, 1968.

L. Gratton (ed.), Non-Solar X- and Gamma-Ray Astronomy, 88–93. All Rights Reserved
Copyright © 1970 by the IAU

stabilized at 8 RPS, with the rocket axis pointing at $10^h18'$ R.A. and $15.0°$N declination on the celestial sphere, the Nike-Apache rocket launched on November 7, 1968, got into a precession after the split nosecone ejection. The precession axis of the latter rocket was at $10^h8'$ R.A. and $36°$N declination with the half cone precession angle of $54°$. Figures 1 and 2 show the observed flux of X-rays in the energy range

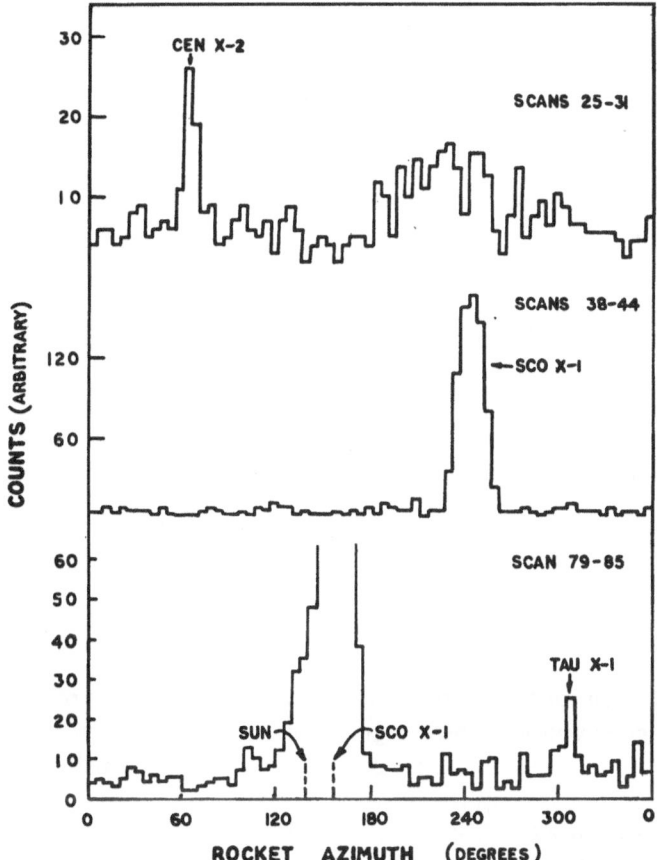

Fig. 2.   X-ray count rate from various discrete sources observed on November 7, 1968.

2–6 keV from different sources for the flights of November 3, 1968, and November 7, 1968, respectively. The absolute flux for each source was determined by fitting a triangular response to the observed data after summing up appropriately.

X-ray Flux from Tau-X1:

The results of both the flights for the absolute flux of Tau-X1 are consistent with a power law spectrum of the type

$$f(E) = 8.0 \, E^{-0.9 \pm 0.2} \, dE.$$

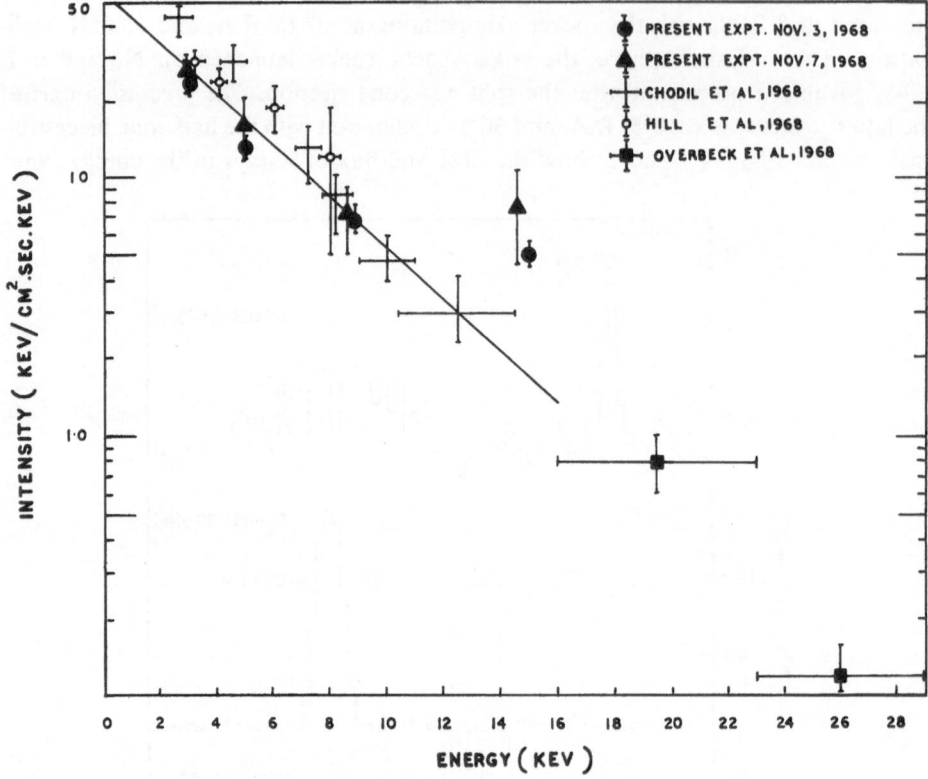

Fig. 3.   Energy spectrum of Sco-X1.

The observed energy of X-rays from Tau-X1 in the range 2–5 keV of $(1.6 \pm 0.3) \times 10^{-8}$ ergs/cm²sec agrees well with the energy estimates observed by others, which adds to the credibility of the absolute values of flux that we have derived.

X-ray Flux from Sco-X1:

Figure 3 shows the energy spectrum of Sco-X1 in the energy range 2–20 keV, the data being obtained for different energy windows of nominal value 2–4 keV; 4–6 keV; 6–12 keV and 12–18 keV. The results are consistent with an exponential spectrum with $E_0 = 4.4 \pm 0.2$ keV corresponding to a temperature of $(5.1 \pm 0.2) \times 10^7$ K.

Since the discovery of fluctuations in the optical intensity from Sco-X1 by Hiltner and Mook (1967), the investigation of time variation of X-ray flux from Sco-X1 has assumed a great importance. An examination of all the available results on the absolute flux of Sco-X1, since 1965, as seen from Figure 4 shows that the X-ray luminosity of Sco-X1 has decreased steadily over the period 1965–68. Sporadic short time variations are superimposed upon this general decrease which is consistent with an exponential decay with a time constant of about 4.1 years. If this is true, then the flux of Sco-X1 would decrease by two orders of magnitude in about 20 years (Rao *et al.*,

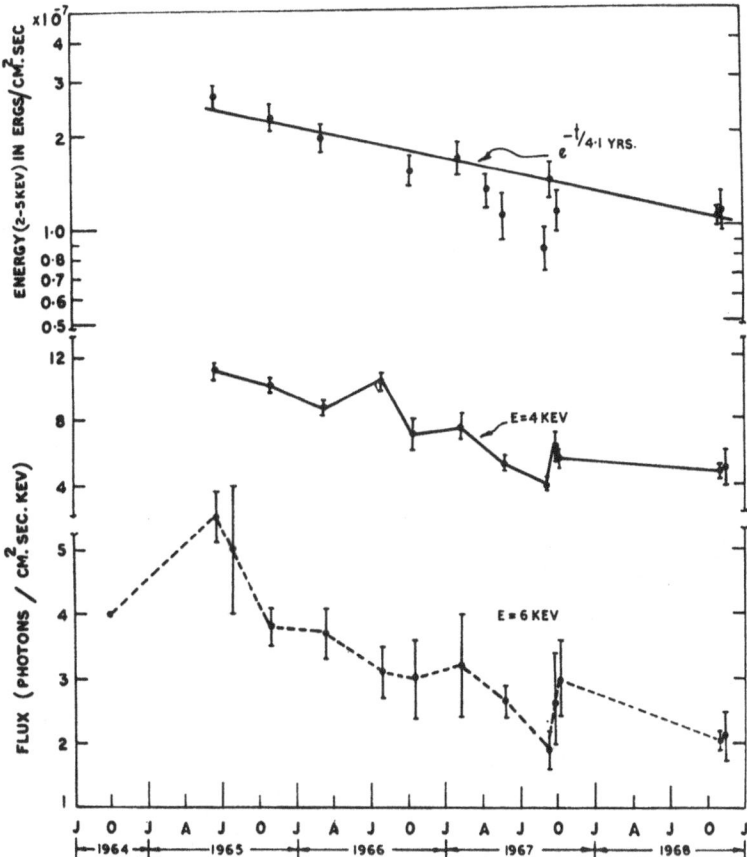

Fig. 4. Time variation of the absolute flux and energy from Sco-X1.

1969b; the references therein), an estimate which is not unreasonable from the theoretical point of view.

X-ray flux from Cen-X2:

The most important result of these two flights is the rediscovery of the low energy X-ray flux from Cen-X2 source. The observed spectrum during both the flights, plotted in Figure 5, is consistent with a power law having an exponent of $-(1.2\pm0.2)$. In the same figure are plotted the data for Cen-X2 in 20–100 keV range observed on October 15, 1967, by Lewin *et al.* (1968). It is evident that a single power law spectrum can satisfactorily explain both the low energy flux observed by us and the high energy flux observed by Lewin *et al.* almost a year earlier.

Cen-X2 was not detectable in October, 1965 (Grader *et al.*, 1966), was observed as a time varying object during April–May, 1967 (Francey *et al.*, 1967; Cooke *et al.*, 1967) and again apparently ceased to exist in September, 1967 (Chodil *et al.*, 1968). The decrease in flux during April–May, 1967 can be represented by an exponential

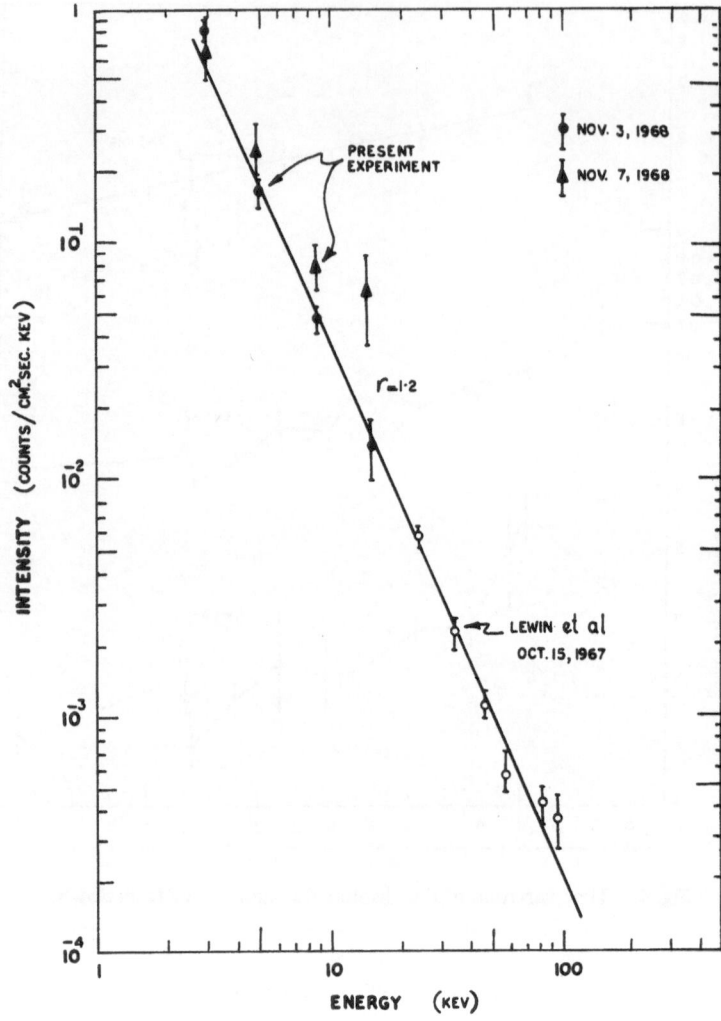

Fig. 5.   Absolute flux of Cen-X2 as a function of energy.

decay with a time constant of 23.4 days. The extrapolated X-ray flux in the energy 2–5 keV using Lewin *et al.*'s spectrum observed on October 15, 1967, is also shown in the same figure. However, no low energy flux was observed on June 12, 1968, by Pounds (1970). They have provided an upper limit of 0.15 photons/cm²sec for the flux in the energy range 2–5 keV. The first evidence for the presence of low energy flux, since May 1967, is reported here from our flights on November 3, 1968 and November 7, 1968.

Acknowledgements

The Nike-Apache rockets for the experiments were made available under the NASA-INCOSPAR agreement. Thanks are due to Mr. H. G. S. Murthy and his colleagues for their able assistance in launching the rockets. The authors are thankful to Professors V. A. Sarabhai K. R. Ramanathan and M. Oda for many helpful discussions.

References

Chodil, G., Mark, H., Rodrigues, R., and Swift, C. D.: 1968a, *Astrophys. J. (Letters)* **152**, L45.

Chodil, G., Mark, H., Rodrigues, R., Seward, F. D., Swift, C. D., Turiel, I., Hiltner, W. A., Wallerstein, G., and Mannery, E. J.: 1968b, *Astrophys. J.* **154**, 645.

Cooke, B. A., Pounds, K. A., Stewardson, E. A., and Adams, D. J.: 1967, *Astrophys. J.* **150**, 189.

Francey, R. J., Fenton, A. G., Harries, J. R., and McCracken, K. G.: 1967, *Nature* **216**, 773.

Grader, R. J., Hill, R. W., Seward, F. D., and Toor, A.: 1966, *Science* **152**, 1499.

Hill, R. W., Grader, R. J., and Seward, F. D.: 1968, *Astrophys. J.* **154**, 655.

Hiltner, W. A. and Mook, D. E.: 1967, *Astrophys. J.* **150**, 851.

Lewin, W. H. G., Clark, G. W., and Smith, W. B.: 1968, *Astrophys. J. (Letters)* **152**, L49.

Overbeck, J. W. and Tananbaum, H. D.: 1968, *Astrophys. J.* **153**, 899.

Pounds, K. A.: 1970, in this volume, p. 82.

Rao, U. R., Chitnis, E. V., Prakasarao, A. S., and Jayanthi, U. B.: 1969a, Communicated to *Proc. Ind. Acad. Sci.*

Rao, U. R., Chitnis, E. V., Prakasarao, A. S., and Jayanthi, U. B.: 1969b, *Astrophys. J. (Letters)*.

Rao, U. R., Jayanthi, U. B., and Prakasarao, A. S.: 1969c, *Astrophys. J. (Letters)*.

ENERGY SPECTRA OF SEVERAL DISCRETE X-RAY SOURCES IN THE 20–120 keV RANGE

P. C. AGRAWAL, S. BISWAS, G. S. GOKHALE,
V. S. IYENGAR, P. K. KUNTE, R. K. MANCHANDA and B. V. SREEKANTAN

Tata Institute of Fundamental Research, Bombay 5, India

1. Introduction

In this paper we report on our observations of hard X-rays from several X-ray sources in the energy range 20–120 keV. The results were obtained from the data collected during two balloon flights made from Hyderabad, India (latitude 17.6 °N, longitude 78.5 °E). The first flight was made on April 28, 1968, and the balloon reached a ceiling of about 5.3 g cm^{-2} residual atmosphere and floated from 0230 to 0800 hrs. IST (Indian Standard Time). The second balloon was launched on December 22, 1968 and floated at about 7.5 g cm^{-2} of residual air from 1000 to 1130 hrs. IST.

The X-ray telescope consisted of a NaI (Tl) crystal of area of 97.3 cm^2 and thickness 4 mm, coupled to a 5″ photomultiplier. The crystal was surrounded by both active and passive collimators. The passive collimator was a cylinder of graded shielding of lead, tin and copper and the active collimator, a cylindrical plastic scintillator. The field of view of the telescope was 18.6° at FWHM. The geometrical factor of the telescope for isotropic radiation was 13.2 cm^2 sr. The pulses from the NaI crystal were sorted out into 10 continuous channels extending from 17 to 124 keV. An Am^{241} source came in the field of view of the telescope periodically and provided in flight calibration of the detector. All the information was recorded on a continuously moving photographic film.

The X-ray telescope was mounted on an oriented platform which was programmed to look at specified directions in the sky. The axis of the telescope was kept inclined at a fixed angle to zenith, 25° in the first flight and 32° in the second. In the first flight it was planned that the azimuth of the telescope would be aligned to the north and the south directions alternately for about 10 min each. Although the telescope looked at the directions close to north and south for a considerable period of time, it also scanned some other directions of the sky for significant period due to oscillations and 'hunting' of orienting system. Fortituously this enabled us to make interesting observations on the X-ray intensities from these directions. A continuous record of the aspect of the telescope was made from the output of a pair of flux gate magnetometers.

In the second flight the oriented platform was programmed to look at four specified directions successively spending about 4 min in each direction during a cycle of about 16 min. The four specified directions were, N ($\phi = 0°$), S ($\phi = 180°$), NE ($\phi = 310°$, with the convention $\phi = 270°$ being due east) and SW ($\phi = 110°$). The performance of the orientor was very good in this flight and the telescope looked at the preselected directions of the sky successively for five cycles, from 0950 to 1115 hrs. IST.

L. Gratton (ed.), Non-Solar X- and Gamma-Ray Astronomy, 94–103. All Rights Reserved
Copyright © 1970 by the IAU

2. Sco X-1

On April 28, 1968 the balloon reached ceiling altitude of 5.3 g cm^{-2} at 0230 hrs. IST. During the period 0230–0330 hrs. IST Sco X-1 was in the field of view of the telescope intermittently for a total time of 1455 sec. The angle between the telescope axis and Sco X-1 was computed every 15 sec by noting the corresponding aspect and taking into account the drift of the balloon in longitude. The effective exposure time corresponding to 100% efficiency was deduced from these data to be 255 sec. The background counting rate was determined from the counting rates during the period when no known sources were in the field of the telescope. The relevant data are given in

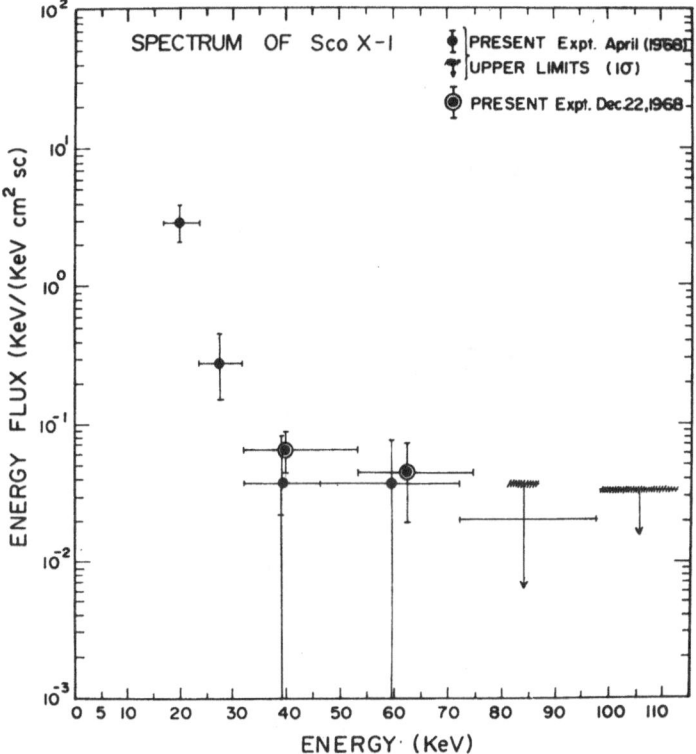

Fig. 1.   X-ray intensities of Sco X-1 measured on April 28, 1968 and December 22, 1968.

Table I. It is seen that finite fluxes from Sco X-1 are available in four energy channels I to IV (16–72 keV) and upper limits can be put in two higher energy channels V and VI (72–125 keV). The fluxes shown in Table I are not corrected for the escape probability of K X-rays and also for the resolution and the efficiency of the detector. This correction is small and amounts to about 10%. This will be taken into account in the final analysis of the data of the X-ray sources.

During the flight on December 22, 1968, Sco X-1 was in the field of view of the

TABLE I

Energy spectrum and intensity of Sco X-1

Date	Energy band (keV)	Mean energy (keV)	Excess counts	Exposure time (sec)	Equivalent exposure at 100% efficiency (sec)	Flux	
						Photons/keV cm² sec	keV/keV cm² sec
28.4.68	16.6–23.4	20.1	268 ± 80	1455	255	$(1.43 \pm 0.43) \times 10^{-1}$	2.86 ± 0.85
	23.4–31.5	27.5	72 ± 40	600	114	$(1.00 \pm 0.55) \times 10^{-2}$	$(2.76 \pm 1.6) \times 10^{-1}$
	31.5–47.7	39.6	89 ± 97	1455	255	$(9.7 \pm 10.6) \times 10^{-4}$	$(3.85 \pm 4.2) \times 10^{-2}$
	47.7–72.0	59.8	116 ± 135	1455	255	$(6.0 \pm 7.0) \times 10^{-4}$	$(3.61 \pm 4.2) \times 10^{-2}$
	72.0–97.4	84.6	$2\sigma = 205$	1455	255	$< 8.9 \times 10^{-4}$	$< 7.5 \times 10^{-2}$
	97.4–123.7	110.5	$2\sigma = 175$	1455	255	$< 6.0 \times 10^{-4}$	$< 6.6 \times 10^{-2}$
22.12.68	29.9–52.3	41.1	178 ± 60	566	450	$(1.54 \pm 0.52) \times 10^{-3}$	$(6.3 \pm 2.1) \times 10^{-2}$
	52.3–74.7	63.5	137 ± 79	566	450	$(7.0 \pm 4.0) \times 10^{-4}$	$(4.5 \pm 2.5) \times 10^{-2}$

telescope periodically (for about 4 min at a time) over a time period of about 90 min, from about 1000 hrs. to 1130 hrs. IST. The detector looked at Sco X-1 for a total time of 566 sec. Since the orientor performed very satisfactory the mean efficiency was as high as ∼80% and the effective exposure time with 100% efficiency was 450 sec. This long exposure enabled us to obtain finite and statistically significant flux in the two energy intervals of 30–52 keV and 52–75 keV (Table I), in spite of the fact that the atmospheric thickness along the line of sight to Sco X-1 was as high as 9.2 g cm^{-2} of air. In estimating the excess counts the background rate was obtained from the north region of sky which included no discrete X-ray sources.

The measured intensities of Sco X-1 during the two flights are plotted in Figure 1. In the December 22 flight rapid variations in the intensity of Sco X-1 within the period of observation of 90 min have been noticed and these results have been reported earlier [1]. However for a period of about 1 hour, the flux of Sco X-1 in the energy range 30–52 keV was fairly constant and near to the values reported by others, as well as to our flux measurements of April 28. In the figure we have plotted the flux values corresponding to this period of observation on December 22 for the two energy channels 30–52 keV and 52–75 keV.

It is seen from the Figure 1, that in the range 15 to 45 keV, the spectrum of Sco X-1

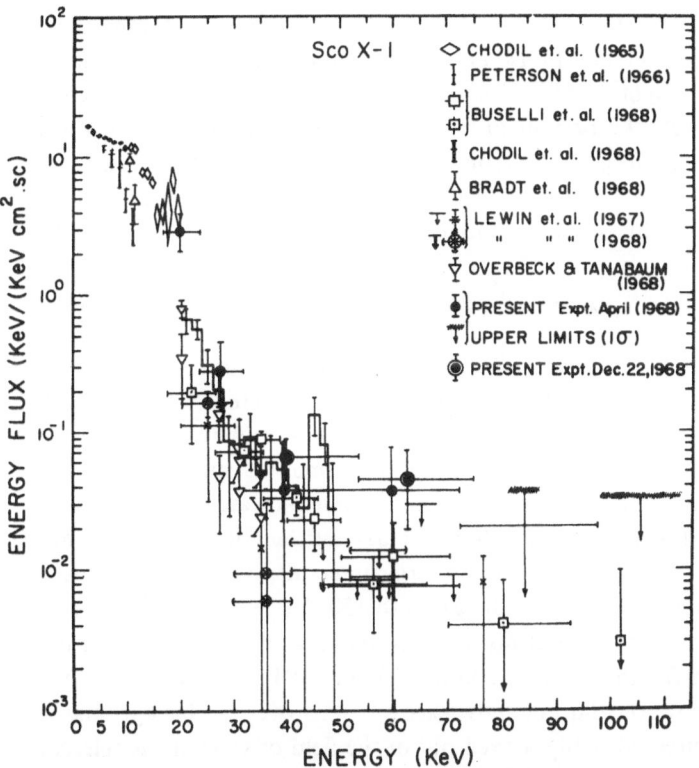

Fig. 2.   X-ray intensities of Sco X-1 measured in the present experiments on 28.4.1968 and on 22.12.1968 together with available spectral information on Sco X-1.

can be represented by an exponential of the form $N(E) = (K/E) \exp - E/E_0$, where $E_0 \sim 5$ keV the corresponding temperature being $T = 5.8 \times 10^7$ K. In the energy interval 45–75 keV, our data indicate a rather flat spectrum $E_0 \sim 20$ keV or more, suggesting that the same spectral exponent is not valid for energies below and above 45 keV.

A comparison of the present data with the available spectral data [2–8] on Sco X-1 is shown in Figure 2. It is clear from the figure that at any given energy the agreement in the absolute flux values is not better than by a factor of 2–3, at energies below 40 keV. At higher energies the spectral information on Sco X-1 is still rather poor. Our flux values at 60 and 63 keV obtained in two different flights agree with others. But the values are considerably higher than those of Buselli *et al.* [6], who also find a flattening of the spectrum at high energies ($KT \sim 15$ keV).

3. Cyg X-1

On December 22, the X-ray telescope looked at Cyg X-1 in the eastern sky during the 3rd, 4th and 5th cycles. The total time of exposure at 100% efficiency was 202 sec. Excess flux was detectable in three energy channels. The background was taken corresponding to the north direction as was done in the case of Sco X-1. The flux values of Cyg X-1 are shown in Table II and plotted in Figure 3. For comparison available spectral data [8–13] are also shown in the same figure. The flux value measured in this flight at 41 keV is within the range of previously measured values. At 64 and 97 keV the present intensities appear to be higher than those measured by some investigators [11, 12, 14] but nearer to those of Overbeck and Tananbaum [8]. The spread in the flux values may be due to time variations in the intensity of Cyg X-1 also as pointed out by Overbeck and Tananbaum [8].

4. A New Source at High Galactic Latitude (TWX)

In April 28 flight, we had indication of some excess flux from the direction of azimuth 90°–120° and zenith angle 25° during the time 0210 to 0300 hrs. IST. This was clearly noticeable in the energy channel 17 to 48 keV and weakly in the energy channel 48–97.5 keV. Although the total exposure time was small (60 sec) and consequently statistical errors large, it was significant that the calculated minimum intensities of this source (assuming to be along the axis of the telescope) was almost the same as that of Sco X-1 for equivalent energy intervals. The observations in the azimuth 90–120°, and of Sco X-1 in the azimuth 150–210° were made intermittently over the same period of time and identical background rates were applied for both. An analysis of the data for the period 0300 to 0330 hrs. IST for the same azimuth and zenith direction, however did not show excess counts over the background level suggesting that the source probably moved out of the field of view of the telescope.

We scanned this region of the sky in the direction (SE($\phi = 110°$)), on December 22, 1968 and detected considerable excess counts in the energy channels 30–52 keV and

TABLE II
Spectrum of Cyg X-1

Date	Energy band (keV)	Mean energy (keV)	Excess counts	Exposure time (sec)	Equivalent exposure at 100% efficiency (sec)	Flux photons/keV cm^2 sec
December 22, 1968	29.9–52.3	41.1	149 ± 80	688	202	$(4.7 \pm 2.6) \times 10^{-3}$
	52.3–74.7	63.5	195 ± 88	688	202	$(3.4 \pm 1.6) \times 10^{-3}$
	74.7–118.7	95.7	403 ± 130	688	202	$(2.6 \pm 0.8) \times 10^{-3}$

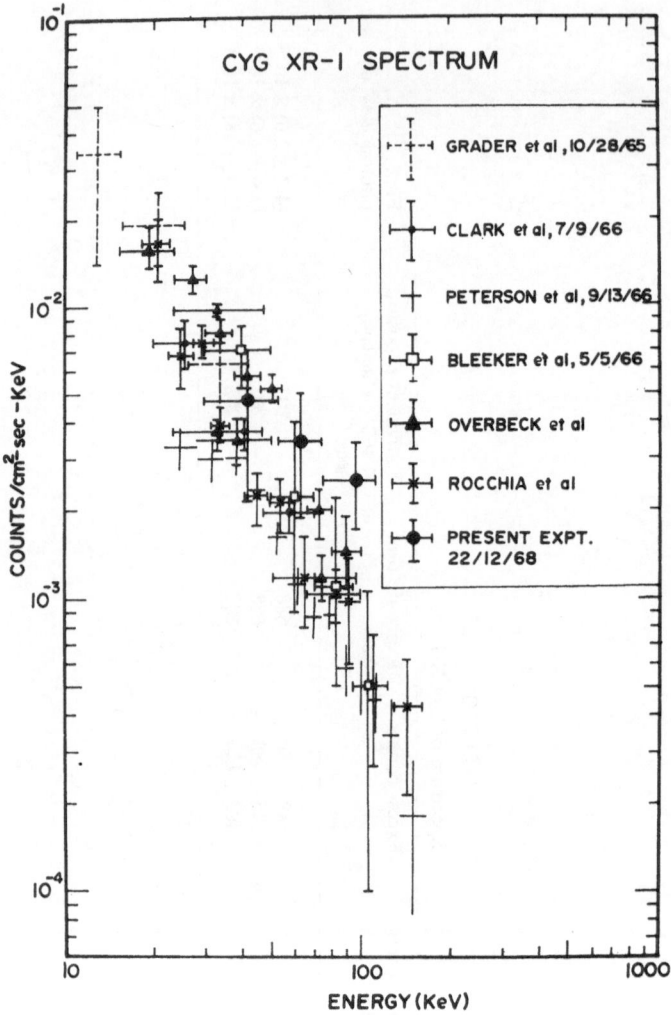

Fig. 3.   Flux values of Cyg X-1 measured on 22.12.1968 together with available flux data of Cyg X-1.

52–86 keV. The exposure time on this source region was 755 sec. During the same rotational cycles of the telescope Sco X-1 was detected in the direction $S(\phi = 180°)$ and Cyg X-1 in the direction $NW(\phi = 310°)$. Figure 4 shows the counting rates for the four directions scanned. Using the same value of the background counting rate as used for Sco X-1 and Cyg X-1, the flux values of the new source were calculated assuming it to be along the axis of the telescope. The flux values are given in Table III. The intensities measured on April 28 and December 22, are plotted in Figure 5 together with measurements on Sco X-1 and Cyg X-1 during these flights. The energy spectrum of the source seems to be similar to that of Sco X-1.

From the observations on April 28, the source location was deduced to be in the celestial region, R.A. = 14.0 hrs. to 15.6 hrs., and $\delta = -5.5°$ to $+24.9°$. In the De-

TABLE III

Intensity and energy spectrum of the new source

(The source is assumed to be along the axis of the telescope)

Date	Energy band (keV)	Mean energy (keV)	Source counting rate-c/sec	Background counting rate-c/sec	Excess counts due to source	Total exposure Time (sec)	Flux photons/keV cm² sec
April 28, 1968	16.6–47.7	32.2	8.08	7.39	40 ± 23	60	$(2.6 \pm 1.5) \times 10^{-3}$
	47.7–97.4	72.5	13.30	12.61	38 ± 28	60	$(5.3 \pm 4.0) \times 10^{-4}$
December 22, 1968	29.9–52.3	41.1	5.29	4.98	231 ± 94	755	$(1.1 \pm 0.5) \times 10^{-3}$
	52.3–74.7	63.5	6.03	5.65	290 ± 94	755	$(8.9 \pm 2.9) \times 10^{-4}$
	74.7–118.7	96.7	11.87	11.60	203 ± 135	755	$(2.4 \pm 1.6) \times 10^{-4}$

P. C. AGRAWAL ET AL.

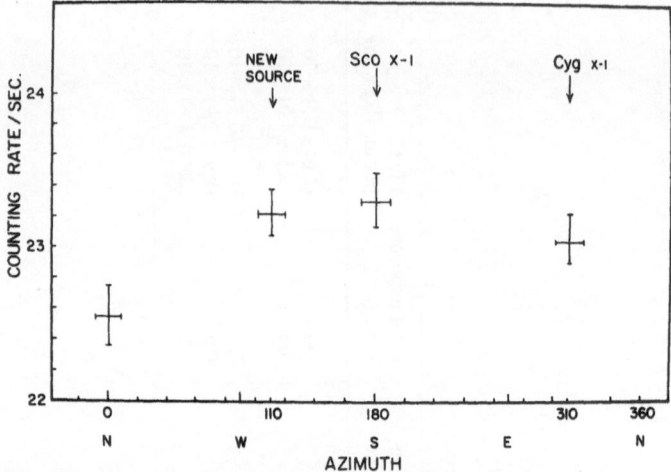

Fig. 4. Counting rates of the X-ray telescope in the energy interval 30–119 keV on 22.12.1968 for the 4 different directions of the sky. This shows the counting rate due to the new source as compared to that of the background ($\phi = 0°$), Sco X-1 ($\phi = 180°$) and Cyg X-1 ($\phi = 310°$).

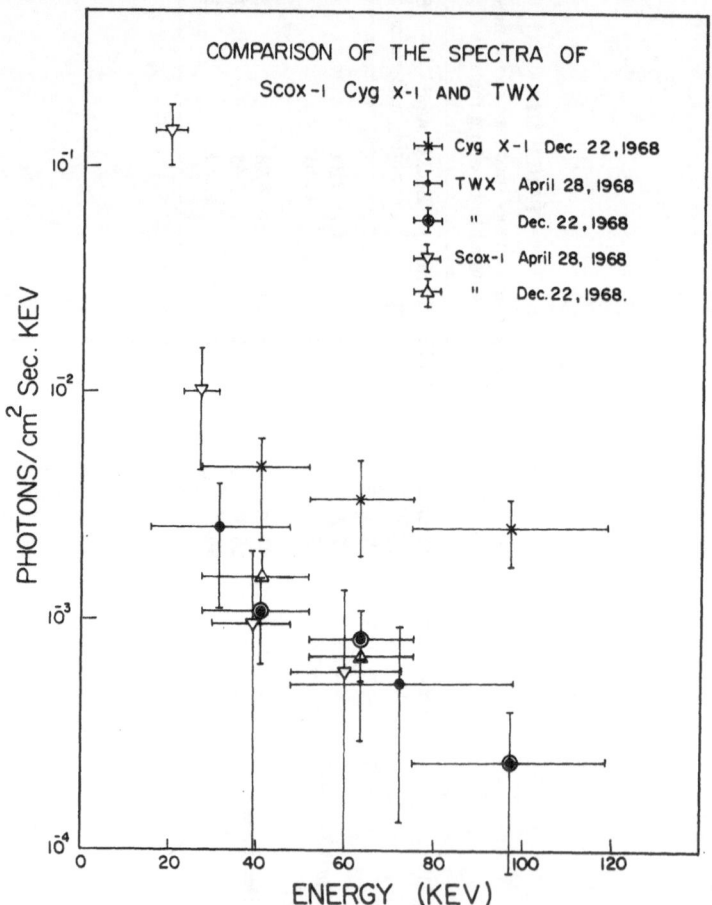

Fig. 5. Comparison of the spectra of the new source (TWX) and Sco X-1 measured on 28.4.1968 and 22.12.1968 and of Cyg X-1 on 22.12.1968.

cember 22 flight, the location of the source was in region, R.A. = 13.5–15.2 hrs. and $\delta = -9.5°$ to $+23.8°$. Fortuitously since the measured intensities in the two flights were nearly the same, it is estimated that the source is located in the overlapping region of the sky at R.A. = 14.0–15.2 hrs. and $\delta = -5.5$ to $+23.8°$, which lie in the constellations of Virgo-Bootes-Serpens. In galactic coordinates the source lies at high galactic latitude, at $l^{II} \approx 359°$ and $b^{II} \approx +58°$.

References

[1] Agrawal, P. C., Biswas, S., Gokhale, G. S., Iyengar, V. S., Kunte, P. K., Manchanda, R. K., and Sreekantan, B. V.: 1969, *Nature* **224**, 51.
[2] Chodil, G., Jopson, R. C., Mark, H., Seward, F. D., and Swift, C. D.: 1965, *Phys. Rev. Letters* **15**, 605.
[3] Peterson, L. E. and Jacobson, A. S.: 1966, *Astrophys. J.* **145**, 62.
[4] Lewin, W. H. G., Clark, G. W., and Smith, W. B.: 1967, *Astrophys. J. (Letters)* **150**, L153.
[5] Chodil, G., Mark, H., Rodrigues, R., Seward, F. D., Swift, C. D., Turiel, L., Hiltner, W. A., Wallerstein, G., and Mannery, E. J.: 1968, *Astrophys. J.* **154**, 645.
[6] Buselli, G., Clancy, M. C., Davison, P. J. N., Edwards, P. J., McCracken, K. G., and Thomas, R. M.: 1968, *Nature* **219**, 1124.
[7] Bradt, H., Naranan, S., Rappaport, S., and Spada, G.: 1968, *Astrophys. J.* **152**, 1005.
[8] Overbeck, J. W. and Tananbaum, H. D.: 1968, *Astrophys. J.* **153**, 899.
[9] Grader, R. J., Hill, R. W., Seward, F. D., and Toor, A.: 1966, *Science* **152**, 1499.
[10] Clark, G. W., Lewin, W. H. G., and Smith, W. B.: 1968, *Astrophys. J.* **151**, 21.
[11] Peterson, L. E., Jacobson, A. S., Pelling, R. M., and Schwartz, D. A.: 1968, *Can. J. Phys.* **46**, S437.
[12] Bleeker, J. A. M., Burger, J. J., Deerenberg, A. J. M., Scheepmaker, A., Swanenburg, B. N., and Tanaka, Y.: 1967, *Astrophys. J.* **147**, 391.
[13] Rocchia, R., Rothenflug, R., Boclet, D., and Durochoux, Ph.: 1969, *Astron. Astrophys.* **1**, 48.

SUDDEN CHANGES IN THE INTENSITY OF HIGH
ENERGY X-RAYS FROM SCO X-1

P. C. AGRAWAL, S. BISWAS, G. S. GOKHALE, V. S. IYENGAR, P. K. KUNTE,
R. K. MANCHANDA, and B. V. SREEKANTAN

Tata Institute of Fundamental Research, Bombay 5, India

In this note we wish to report briefly the observation of sudden changes in the intensity of Sco X-1 by a factor of about 3 recorded in the energy interval 29.9–52.3 keV on December 22, 1968 between 04 h 27 m and 05 h 53 m UT. The observation was made with an X-ray telescope flown in a balloon from Hyderabad, India. The balloon was launched at 0200 hr UT and reached the ceiling of 7.5 g/cm^2 of residual atmosphere at 0435 hr UT. The X-ray telescope consisted of a NaI(T1) crystal with an area of 97.3 cm^2 and thickness 4 mm, surrounded by both active and passive collimators. The telescope was mounted on an oriented platform which was programmed to look in four specified directions successively, of azimuths, $\phi=0°$, 110°, 180° and 310° ($\phi=0°$ being North and $\phi=90°$, West), spending about 4 min in each direction during a cycle of period of about 16 min. The axis of the telescope was inclined at an angle of 32° with respect to the zenith. A pair of crossed flux gate magnetometers provided information every 8.2 sec on the azimuth of the telescope. The pulse heights from the X-ray detector were sorted into several channels extending from 10 to 120 keV. An Am^{241} source came into the field of view of the telescope once in 15 min for about 30 sec to provide in-flight calibration of the detector. The meridian transit of Sco X-1 was at 0454 hr UT. Just before the balloon reached the ceiling Sco X-1 was in the field of view of the telescope for 3 min and 41 sec. After the balloon reached ceiling, Sco X-1 was in the field of view of the telescope on five occasions between 0443 and 0553 hr UT. During the last observation, however, the balloon had lost altitude by about 1 g/cm^2. The excess counts due to Sco X-1 were obtained by subtracting the counting rates corresponding to the North direction which did not include any known X-ray sources. The observation on Sco X-1 in the 1st cycle was made while the balloon was still ascending and consequently the interposed grammage was changing from 10.5 to 9.7 g/cm^2. However, for the energy range under consideration, the change in the background counting rate was not significant and there cannot be any doubt regarding the genuineness of the excess counts recorded.

In Figure 1, we have plotted the observed intensity of Sco X-1 corrected for atmospheric attenuation in the energy interval 29.9–52.3 keV as a function of time. The dashed line (1) represents the weighted mean intensity of all the first five observations and the dashed line (2) represents the weighted mean intensity of the 2nd, 3rd and 4th observations.

It is clear from the figure that the observed intensities deviate considerably from dashed line (1), suggesting that the mean is not a good fit for the data. (The probability level is 0.05.) On the other hand dashed line (2) is a good fit for the 2nd, 3rd and 4th

L. Gratton (ed.), Non-Solar X- and Gamma-Ray Astronomy, 104–106. All Rights Reserved
Copyright © 1970 by the IAU

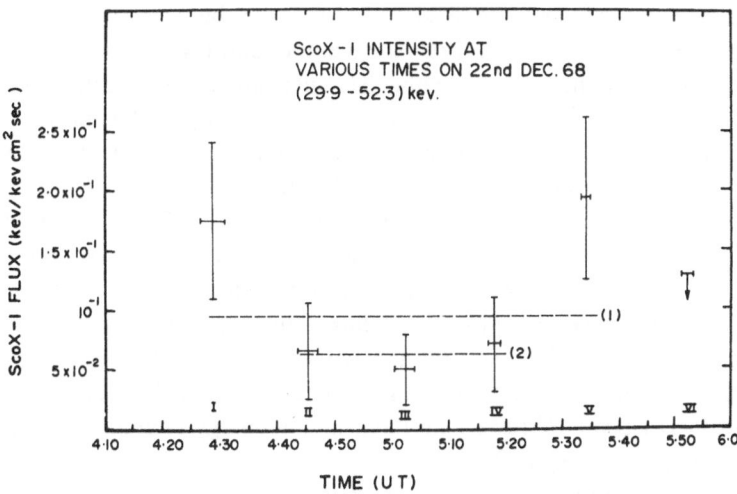

Fig. 1.   Intensity variations of Sco X-1 at various times on December 27, 1968, in the energy interval of 29.9–52.3 keV.

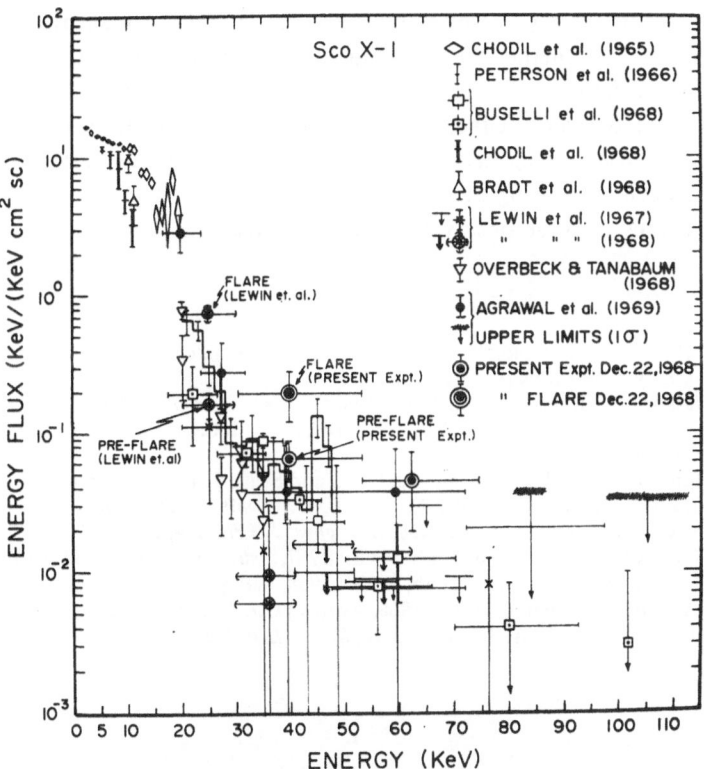

Fig. 2.   The pre-flare and flare-time intensities of Sco X-1 on December 22, 1968, plotted with available spectral data of Sco X-1.

observations. These two features make us believe that the intensity of Sco X-1 in the energy range 29.9–52.3 keV has not remained at the same level during the entire period of our observation extending over 90 min. The intensity has first decreased by a factor of about 3, and remained constant for some time, and again shot up by a factor of 3 and decreased again – the rise and fall occurring in times less than about 15 min. For this reason the 5th observation may be called a flare.

The absolute flux values corresponding to the mean of 2nd, 3rd and 4th observations for two energy channels 29.9–52.3 and 52.3–74.7 keV, have been plotted in Figure 2, where we have summarised the existing spectral data (1–9) on Sco X-1. In the same figure we have also plotted the flux value recorded in the energy channel 29.9–52.3 keV during the 5th cycle (designated – flare present). The flare observed in this experiment is in some respects similar to one reported by Lewin *et al.* (1968). A fourfold increase was observed by them in the energy channel 20–30 keV at an atmospheric depth of 3.5 g/cm^2.

It is clear from Figure 2 that there is considerable spread in the flux values of Sco X-1 recorded by different observers at different times. A few of the observations at least have been attributed to long term (several months) variations in the intensity of Sco X-1 by Overbeck and Tananbaum (1968). While our pre-flare value corresponding to the energy channel 29.9–52.3 keV falls within the range of previously reported values, the flux corresponding to the flare point is higher by almost a factor of 5. We cannot say whether the higher intensity recorded during the 1st observation is again due to a flare or whether the intensity had remained high for a considerable period of time. However, the fact that the intensity dropped within 15 min by a factor of 3 to a value close to the generally reported values, favours the possibility of another flare having occurred during the 1st observation also.

Further details of these observations are reported in *Nature* **224**, 51, 1969.

References

Agrawal, P. C., Biswas, S., Gokhale, G. S., Iyengar, V. S., Kunte, P. K., Manchanda, R. K., and Sreekantan, B. V.: 1969 (this proceedings).
Bradt, H., Naranan, S., Rappaport, S., and Spada, G.: 1968, *Astrophys. J. Lett.* **152**, 1005.
Buselli, G., Clancy, M. C., Davison, P. J. N., Edwards, P. J., McCracken, K. G., and Thomas, R. M.: 1968, *Nature* **219**, 1124.
Chodil, G., Jopson, R. C., Mark, H., Seward, F. D., and Swift, C. D.: 1965, *Phys. Rev. Lett.* **15**, 605.
Chodil, G., Mark, H., Rodrigues, R., Seward, F. D., Swift, C. D., Turiel, I., Hiltner, W. A., Wallerstein, G., and Mannery, E. J.: 1968, *Astrophys. J. Lett.* **154**, 645.
Lewin, W. H. G., Clark, G. W., and Smith, W. B.: 1967, *Astrophys. J. Lett.* **150**, L153.
Lewin, W. H. G., Clark, G. W., and Smith, W. B.: 1968, *Astrophys. J. Lett.* **152**, L55.
Overbeck, J. W. and Tananbaum, H. D.: 1968, *Astrophys. J. Lett.* **153**, 899.
Peterson, L. E. and Jacobson, A. S.: 1966, *Astrophys. J. Lett.* **145**, 62.

PROPERTIES OF INDIVIDUAL X-RAY SOURCES

(Invited Discourse)

RICCARDO GIACCONI

In the past 2 years the group at American Science & Engineering, Inc., including Gursky, Gorenstein, Kellogg and me, has been engaged in a program of rocket launches aimed at the solution of several outstanding questions in X-ray astronomy. Of four rocket launches, one was almost totally unsuccessful due to a mechanical failure; one was only partially successful due to a failure of the ACS (Attitude Control System); and two performed as planned. Notwithstanding the delays induced by this run of bad luck, a considerable amount of data has been gathered and analyzed. The major results obtained in the flights on October 11, 1966, and on February 2, 1968, have been published. It is my intent today to report on preliminary results of a recent flight (December 5, 1968) and briefly summarize conclusions from previous flights, and comment on a few selected topics to which this experimental program may contribute.

(1) The first objective of the program was to accurately locate X-ray sources and was prompted by the state of confusion still existing as late as 1966 on the location, intensity and variability of galactic X-ray sources. The successful identification of Sco X-1 had earlier convinced us of the need for accurate location. We set ourselves to the task of examining with as high a sensitivity and as high an angular resolution as possible several regions where numerous sources had been detected and to extend the survey to regions of the northern hemisphere not yet scanned at high sensitivity. We hoped that improved locations and decreased sources confusion would lead to new optical identifications and consequently to a better understanding of the nature and distance of individual sources.

(2) The second objective was to obtain more accurate spectra of the X-ray sources. We recognized that by use of proportional counters sensitive in the region from 1 to 10 keV we would not in many cases have a wide enough energy range to select the specific spectral shape for each source. We hoped, however, to be able to obtain relative spectral measurements which would allow us to distinguish the essential features of the different sources. Also, by observing the presence of low energy cut-offs, we might be able to ascribe distances to the sources based on the degree of interstellar absorption present.

(3) The third objective was to examine the distribution of the galactic X-ray sources in an effort to decide whether they would be assigned a type I or II distribution; whether in other terms, we were dealing with relatively young or relatively old objects.

(4) The fourth objective was to determine whether several objects which were considered likely candidates for X-ray emission, such as novae, supernovae, and radio sources, would, in fact, prove to be candidate objects when the precision of location of sources was increased by approximately one or two orders of magnitude.

L. Gratton (ed.), Non-Solar X- and Gamma-Ray Astronomy, 107–115. All Rights Reserved
Copyright © 1970 by the IAU

Fig. 1.   X-ray astronomy payload for sounding rocket containing collimated proportional counters and aspect camera.

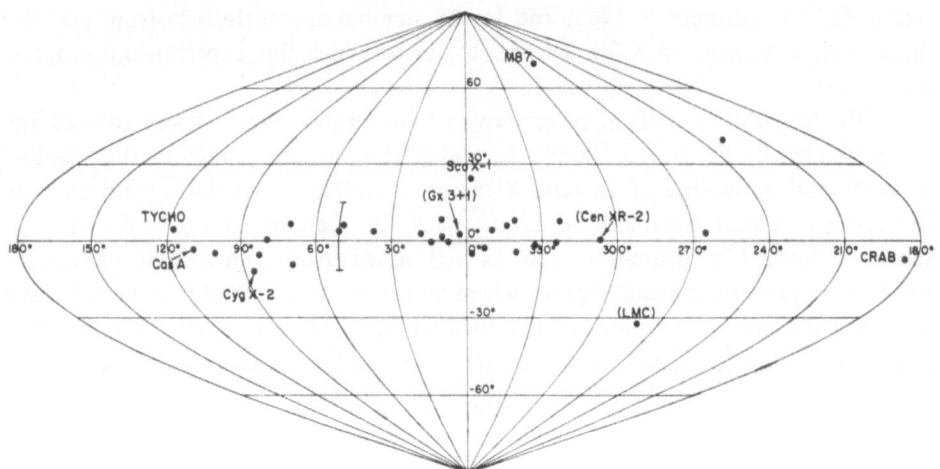

Fig. 2.   The distribution of X-ray sources on a galactic coordinate system.

To accomplish these objectives we designed a rocket payload shown in Figure 1. This identical payload was flown, recovered, and reflown four times with only slight modifications. Details of the instrumentation have been described previously (Gursky *et al.*, 1968; Gorenstein *et al.*, 1969a, b). It consists of two large area banks of proportional counters with slit type collimators and an aspect camera to obtain aspect by photography of the star field in visible light. An attitude control system was used to permit us to scan a selected portion of the sky at a selected rate in the manner first used by Fisher. From the early flight of October 1966 to the last flight of December 1968, the only substantial modifications to the payload were the addition of PSD (pulse shape discrimination), which reduced the residual non X-ray background after anti-coincidence by a factor of 10, the use of a thinner window (1 mil Be, rather than 3) to extend the response to longer wavelength, and the use of two slits rather than one to reduce the maneuvers necessary to obtain positions for each source. Apart from the obvious convenience and economy of re-using the same payload, there is a very real advantage in determining the variability of sources by having a consistent set of data. In Figure 2, a plot is given in galactic coordinates of the distribution of X-ray sources in the sky. I would like to use this plot to indicate the portions of the sky scanned in each flight. In the flight of October 1966 we examined the galactic equator region from $l^{II} = -15$ to $l^{II} = +90$ (the region which includes Scorpio and Cygnus). The flight of February 1968, although intended to scan the Cepheus-Lacerta regions, resulted in a fast scan including $l^{II} = 150$ to $l^{II} = 190$ and $l^{II} = 250$ to $l^{II} = 270$ over the regions of the Crab and Vela. A recent flight in December 1968 included the scan of the region of Cepheus, Lacerta (containing Cas A and Tyco supernovae) from $l^{II} = 100$ to $l^{II} = 140$ and extended to $l^{II} = 190$. In this manner we have examined most of the galactic equator regions visible from White Sands, New Mexico, from $l^{II} = -15$ to about $l^{II} = +270°$.

The results of the flight prior to December 1968 have been reported in detail in a series of articles which have appeared in the literature: Giacconi *et al.* (1967a, b),

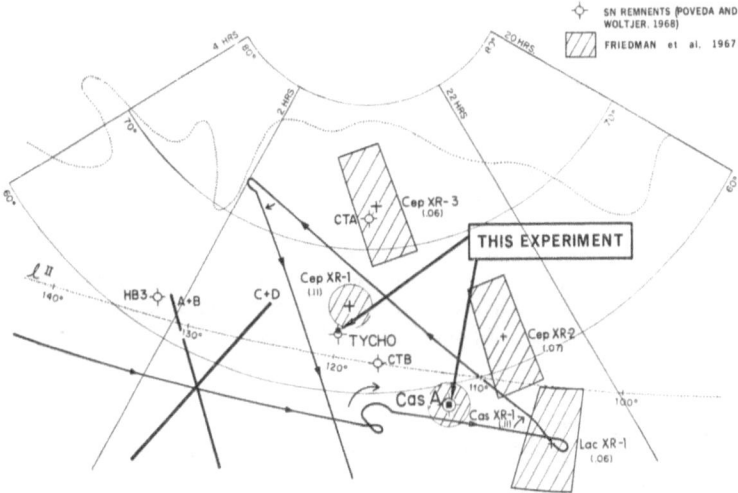

Fig. 3. Motion of the center of the field of view (fov) on the celestial sphere for the Cassiopeia portion of the flight. At earlier times the center moves approximately along the galactic equator starting at $l^{II} \approx 200°$. The line segments labelled $(A + B)$ and $(C + D)$ show the orientation of the fov and are drawn smaller than the actual fov which is $2° \times 45°$. The numbers along the scan line correspond in time to those in Figure 2. The locations of the X-ray sources observed in this experiment as well as sources reported from a previous survey (Friedman *et al.*, 1967) with their strength relative to the Crab (plus known supernova remnants) are shown.

Gorenstein *et al.* (1967), and Gursky *et al.* (1967, 1968). I would like to restrict my remarks in this paper to aspects of those earlier publications which, together with the results of our most recent flight, have a bearing on (1) the X-ray emission from supernova remnants, and (2) the distribution of X-ray sources in the galaxy.

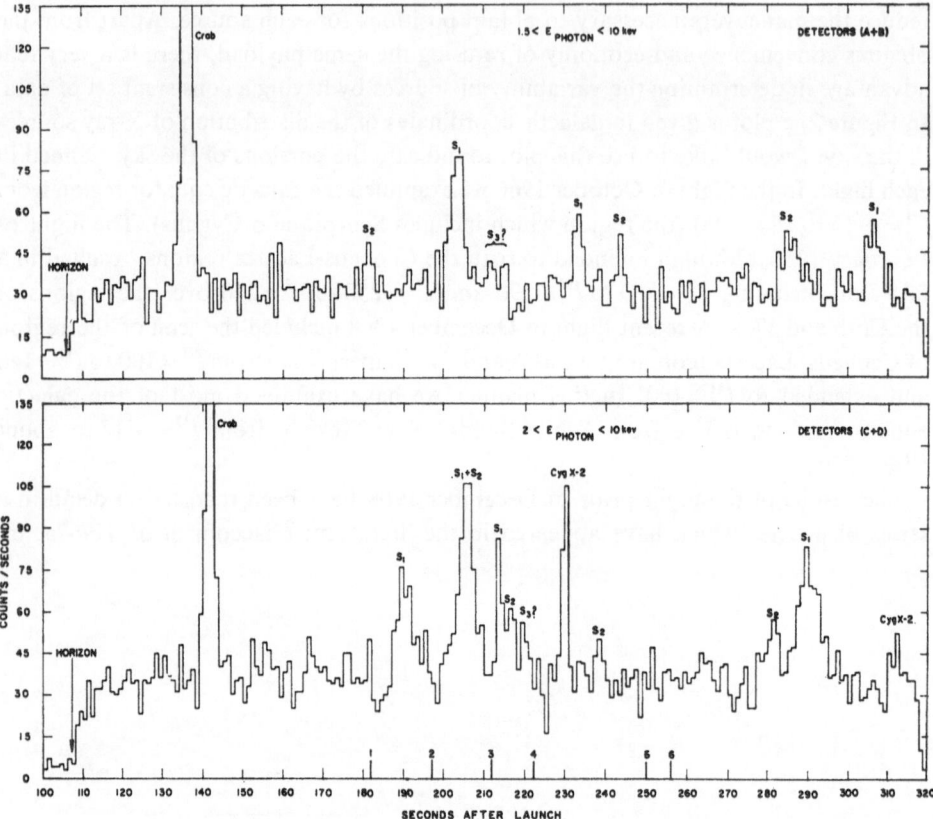

Fig. 4. Actual counts per one second interval vs. time. All the peaks labelled 'S_1' ('S_2') are due to the same source. The numbers along the time scale correspond in celestial position to those of Figure 1. The width of a peak is determined by the instantaneous rate of scan which is either
3°/sec, 1°/sec, or 0.5°/sec.

(1) *Supernovae X-Ray Emission*

In the recent flight of December 1968 we scanned the region of the sky from Crab to Cepheus Lacertae along the galactic equator. The results of this flight, giving greater detail, will shortly by submitted for publication by our group. Briefly the scan mode is shown in Figure 3. The scan was carried out essentially along the galactic equator from $l^{II} \sim 200\text{–}120°$. Complex maneuvers were then carried out to sweep the region containing five previously reported sources: Cep XR-1, XR-2, and XR-3; Cas XR-1 and Lac XR-1 (Friedman *et al.*, 1967).

In Figure 4 the counting rate is shown vs. time. Several obvious peaks in the distribution are observed; in addition, several significant peaks not yet analyzed may also be present. Upon analysis of the aspect photography, all of the obvious peaks except Crab were found to define the position of only two sources corresponding in position to the location of the two supernova remnants Tyco and Cas A. In Figures 5 and 6 the X-ray positions and associated region of uncertainty are shown super-

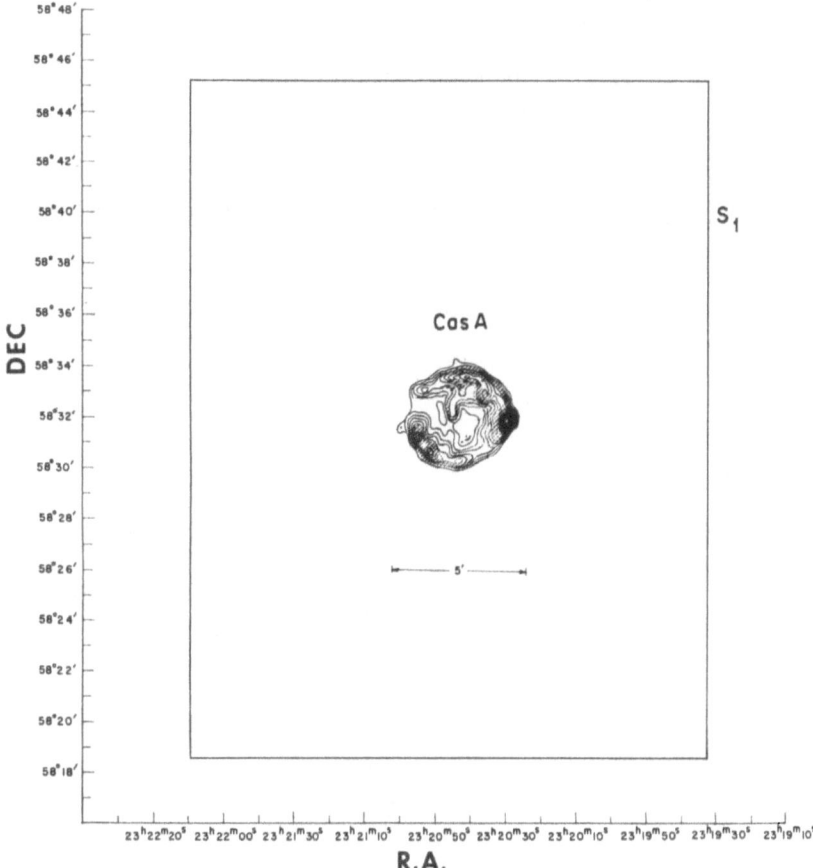

Fig. 5. Position of source S_1 as determined from 'best' intersection of 8 great circle segments. The box shows the region corresponding to a one standard deviation uncertainty in the source position. A radio contour map of Cas A (Ryle *et al.*, 1965) is superimposed upon the position of S_1.

imposed on a radio contour map of the supernova remnants. Although Cep XR-2 and Cep XR-3 were reported to have a strength as large as $\frac{1}{2}$ of that of Cep XR-1 and Cas XR-1, we see at first glance no evidence of their existence. Further detailed analysis may reveal additional sources, but none can be much brighter than $\frac{1}{3}$ of Cas A and Tyco. In particular, we find no evidence for the source Cep XR-3 whose closeness to the location of the supernova remnant CTA-1 had suggested the possibility of an identification with that supernova. Its distance was estimated by Poveda and Woltjer to be about 2000 pc, spectral index $\alpha = 0.2$, a diameter of 120'. On the other hand, we believe that our data confirm very convincingly the emission of X-ray radiation from Cas A and firmly establish the existence of X-ray emission from Tyco SN.

Crab, Tyco and Cas A have thus been established to be X-ray emitters. Kepler SN, CTA-1, and Vela X have been shown not to be X-ray emitters at the present level of sensitivity. While the greater distance from us of Kepler SN (1000 pc) leads us to expect a factor 10 smaller emission from Kepler than from Tyco (3500 pc), CTA-1 and Vela X are estimated to be closer to us than either Tyco or Cas A.

We note that only supernova remnants with an angular diameter less than several minutes of arc have been definitely observed to be X-ray sources. If the diameter is taken to be an indication of the age of the remnant, one would be tempted to conclude that only young supernova remnants are known to be X-ray emitters. On the other hand, the angular diameter may be determined by physical conditions in the interstellar medium surrounding the source or by specific characteristics of the supernova event itself, and not by age.

In any case, it is interesting to note that while other parameters, such as flux density

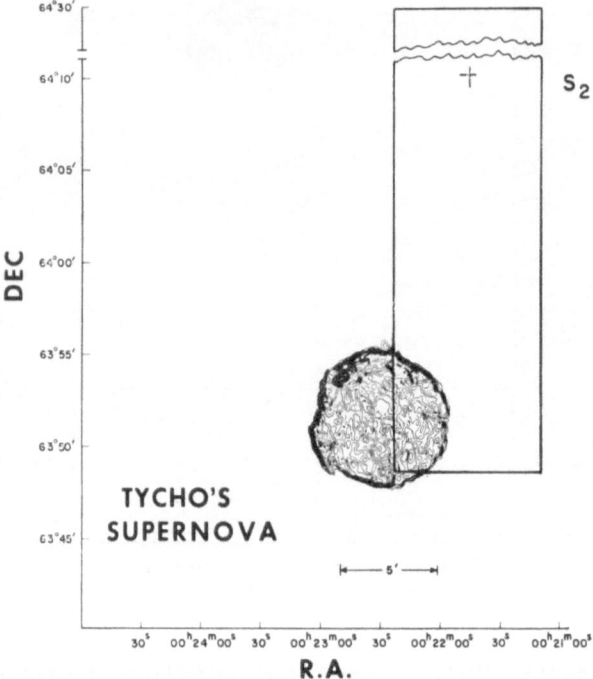

Fig. 6. Position of source S_2 as determined from the 'best' intersection of 6 great circle segments. The box shows the region corresponding to the one standard deviation uncertainty in the source position and the cross indicates the center of the box. A radio contour map of SN1572 (Baldwin, 1967) is superimposed on the position of S_2.

or spectral index of the radio emission, do not correlate with X-ray fluxes, surface brightness in radio and diameter of the remnants (quantities which are not, of course, independent) seem to show a direct correspondence. Roughly the X-ray emission seems to be inversely proportional to some power of the diameter.

It follows that not all supernova remnants can be assumed to be X-ray emitters, as it has often been done on the basis of coarser surveys (Poveda and Woltjer, 1968). In addition, we note that several of the galactic X-ray sources do not coincide with known supernova remnants. It appears, therefore, unlikely that X-ray emission from supernova remnants can be assumed to be the source of the bulk of the known galactic X-ray objects.

(2) *Galactic Distribution of X-Ray Sources*

As a result of previous surveys, we had adopted as a working hypothesis in a previous paper (Gursky *et al.*, 1967) the point of view that X-ray sources were associated with the galactic spiral arm. Assuming all sources to be of the same intrinsic luminosity, it appeared possible to obtain a consistent picture by assuming that sources with l^{II} between $-15°$ to $+17°$ lie in the Sagittarius arm and the ones with l^{II} between $+36°$ to $+80°$ corresponded to tangential directions of several arms. Figure 7 shows the degree of correspondence which appeared to exist between X-ray source directions and galactic spiral arms structures in neutral hydrogen as determined from the 21 cm radio studies of Sharpless (1965). The Cep-Lac sources were the ones observed by Friedman *et al.* (1967).

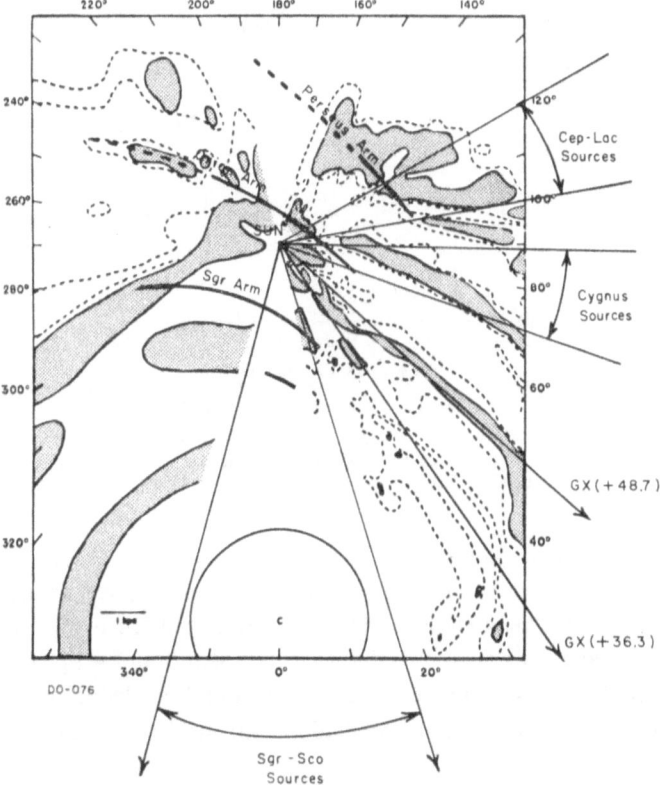

Fig. 7.   Direction of X-ray sources superimposed on a map of the galaxy prepared by Sharpless (1965). The map shows the isodensity contours of neutral hydrogen as derived from the 21 cm radio studies and the spiral arms (shown as curved segments of solid and dashed lines) as deduced from optical data. The names used for the arms (Sgr., Orion, and Perseus) are those given by Sharpless.

The fact that except fot the two supernova remnants described above, we observe in the most recent survey no additional sources in the Cep-Lac region makes the validity of our working hypothesis very doubtful. Sources in the Cep-Lac arm would have to be X-ray emitters with intrinsic luminosity less than 0.025 of Crab in order not to be observed.

Also, detailed examination of the spectra of several X-ray sources in the Cygnus and Scorpio-Sagittarius region shows evidence of detectable amounts of absorption at about 1–2 keV energies. The most noticeable effect was observed in Cyg X-3 by Gorenstein *et al.* (1967) and the data are shown in Figure 8. We note that also sources

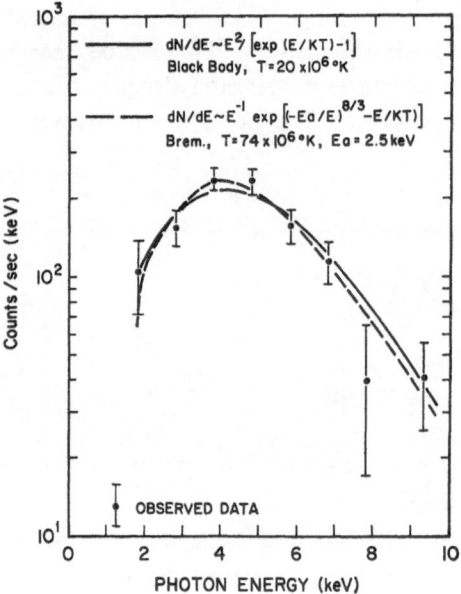

Fig. 8. For the source Cyg X-3 the observed histograms of counting rate vs. energy are compared to the calculated counter response to two assumed photon distributions: (1) $dn/dE \sim (Ea/E)^{8/3}$ exp $(-E/KT)/E$, (2) $dN/dE \sim E^2/(\exp(-E/KT)-1)$. The parameters Ea and T represent the best fit. Both spectra fit the data about equally well.

in the direction of Scorpio-Sagittarius show varying degrees of absorption. Assuming this absorption not to occur at the source, but in interstellar space, as indicated by the apparent correlation between visible light obscuration and X-ray absorption, we had noted that the distances we would derive were between 3 and 4 kpc. In view of the uncertainty both in the experimental data and in our knowledge of the absorption properties of the interstellar medium, we had felt this was not in contradiction to our simple model.

We now feel that a more consistent interpretation of the data might be that the sources in the Scorpio-Sagittarius region are, in fact, not associated with the nearby Sagittarius arm, but more distant from us and closer to the galactic center. I understand that Prof. Livio Gratton will discuss the distribution of galactic X-ray sources and compare it to population types of other objects in more detail as part of this symposium. I would only like to note here that the preceding considerations have important consequences both in the prediction of the total X-ray luminosity of our galaxy and in the age of the objects involved.

References

Baldwin, J. E.: 1967, in *I.A.U. Symposium No. 31* (ed. by H. Van Woerden), Academic Press, London and New York, Paper 56.
Friedman, H., Byram, E. T., and Chubb, T. A.: 1967, *Science* **156**, 374.
Giacconi, R., Gorenstein, P., Gursky, H., and Waters, J. R.: 1967a, *Astrophys. J.* **148**, L119.
Giacconi, R., Gorenstein, P., Gursky, H., Usher, P. D., Waters, J. R., Sandage, A., Osmer, P., and Peach, J. B.: 1967b, *Astrophys. J.* **148**, L129.
Gorenstein, P., Giacconi, R., and Gursky, H.: 1967, *Astrophys. J.* **150**, L85.
Gorenstein, P., Kellogg, E. M., and Gursky, H.: 1969a, *Astrophys. J.* **156**, 315.
Gorenstein, P., Kellogg, E. M., and Gursky, H.: 1969b, submitted to *Astrophys. J.*
Gursky, H., Gorenstein, P., and Giacconi, R.: 1967, *Astrophys. J.* **150**, L75.
Gursky, H., Kellogg, E. M., and Gorenstein, P.: 1968, *Astrophys. J.* **154**, L71
Poveda, A. and Woltjer, L.: 1968, *Astron. J.* **73**, 65.
Ryle, M., Elsmere, B., and Neville, A. C.: 1965, *Nature* **205**, 1259.
Sharpless, S.: 1965, in *Galactic Structure* (ed. by A. Blaauw and M. Schmidt), University of Chicago Press, Chicago.

THE FLUX OF HARD RADIATION FROM M87

G. J. FISHMAN, F. R. HARNDEN, Jr., and R. C. HAYMES

Dept. of Space Science, William Marsh Rice University, Houston, Texas, U.S.A.

This paper reports upon an observation of hard X-radiation from the direction of M87. Earlier (Haymes *et al.*, 1968a), we reported the apparent detection of hard radiation from the general direction of Virgo; the experiment discussed here, conducted 16 months later, recorded approximately the same flux from a smaller region of sky that tends to isolate M87 as the most likely candidate for the source.

Other groups (Adams *et al.*, 1969; Bradt *et al.*, 1967; Friedman and Byram, 1967) have detected soft X-radiation (photon energies between ~ 1 and ~ 10 keV) from M87. The present observation, conducted at energies greater than 30 keV, recorded additional evidence for a flux in agreement with an extrapolation of the power law that fits both the core radio spectrum and the soft X-ray flux from the radiogalaxy.

Most of the balloon-borne instrumentation used in the present work was the same as previously described (Haymes *et al.*, 1968b), with the exception of the detector itself. An additional active collimator was inserted in the beam in front of the actively collimated detector previously used and the assembly was flown as a narrow-angle device. The thickness of the additional collimator was nine inches of plastic scintillator (Nuclear Enterprises Corporation Type NE-102). It was viewed by four additional photomultipliers connected in anti-coincidence with the photomultiplier viewing the central NaI(Tl) crystal.

This thickness of plastic corresponds to at least 3.5 mean free paths of absorber at photon energies up to 100 keV. The additional collimator resulted in a half-flux angle that was 3° from the axis, at energies less than 100 keV. The purpose of this improvement in the angular resposne was to enable M87 to be isolated from 3C273, one of the other candidates for the source of hard radiation previously found from Virgo.

The balloon was launched from Palestine, Texas, at 0305 CST on December 11, 1968. It floated at a pressure altitude that averaged 3.2 mb for 5.2 hours, before termination occurred at 1110 CST. The pressures were measured with a Metrophysics Corporation transducer that senses the thermal conductivity of the atmosphere, as well as by aneroid devices.

Good telemetered data were received throughout the flight; none had to be discarded because of noise. We have rejected in the analysis the three slightly discordant background segments that were widely separated in time, along with their associated M87 segments. Hence the analysis is based on 133 min of Virgo data and approximately 118 min of background measurements; no known discrete sources of X-radiation were contained in the background data used in the analysis. It should be noted that the conclusions reached herein are essentially unaffected if all of the segments rejected are instead included in the analysis.

As in the previous experiment, the detector was continuously rotated about the polar axis at the sidereal rate, so that it would track M87 in hour angle. The azimuth of the polar shaft was automatically changed by 180° every 10.5 min after the azimuth servo system was turned on at 0440 CST. Background measurements were taken during each of the alternate segments when the detector was turned away from Virgo.

While the azimuth of the polar axis returns to the same angle (i.e., South) during each background segment, the detector azimuth does not do so, because of the continuous rotation. Hence the background was measured for a range of azimuths. The region of sky scanned in the background portions of the flight discussed here consisted of an arc of disconnected segments, each 10.5 min long and 6° wide (with equal exposures to the East and to the West because of the 0730 transit time of M87). The angular separation of the segments from M87 is twice the zenith angle of M87; the separation varied during the flight from 40° at meridian transit to approximately 90° near the beginning and end of the observing period.

As before, the average of the two background segments that were time-contiguous with an M87 segment was compared with the flux measured during that M87 segment. The results discussed below are inconsistent with a significant azimuthal variation of the photon flux. In addition, this technique permits corrections to be made for other background changes that may occur, such as those caused by changes in the zenith angle, altitude or latitude. The sensitivity of the method is shown by the upper limits found for Centaurus A (Haymes *et al.*, 1969). The upper limit set in that experiment at the 95% confidence level amounts to 0.7% of the background in the 40–100 keV interval; any background asymmetry is smaller than that.

The radiation from Virgo was found by averaging the background segments adjacent in time to a Virgo observation period and then subtracting this average from the data obtained while M87 was in the field of view. A residual was thus formed for each Virgo period.

Each residual was corrected for atmospheric absorption caused by the air along the line of sight between M87 and the detector. It was also corrected for pointing errors; the narrower beam-width employed in the present experiment increased the importance of the pointing error, which was mainly due to the variation in geomagnetic declination over the balloon's trajectory. Corrections for absolute efficiency and the escape of K X-radiation were applied as were the corrections for the absorption caused by the central photomultiplier, etc. (Haymes *et al.*, 1968b). These latter correction-factors were experimentally determined in our laboratory prior to the experiment. No gain change as great as ±1 channel was detected in the 128 channel pulse-height analysis conducted during the flight.

Figure 1 shows the results of the flight, corrected as outlined above. All pulse height channels are displayed, as well as the 13-channel sums that were formed in order to improve the statistics. No flux is observed above 100 keV; upper limits may be assigned to the higher energy radiation from M87. A small flux is seen in the energy interval 40–95 keV; it is 2.8 standard deviations (2.8σ) above the background at those energies.

The excess counting rate from M87 is constant to within statistics throughout the flight. This constancy, observed over a range of East-West and North–South azimuths, indicates that the observed flux is not due to an azimuthal asymmetry. Also, the increase is ~4 times any background asymmetry, as observed in the Centaurus A experiment, which makes it seem likely that the 40–100 keV enhanced flux is real and extraterrestrial.

This flux at the top of the earth's atmosphere from the direction of M87 is

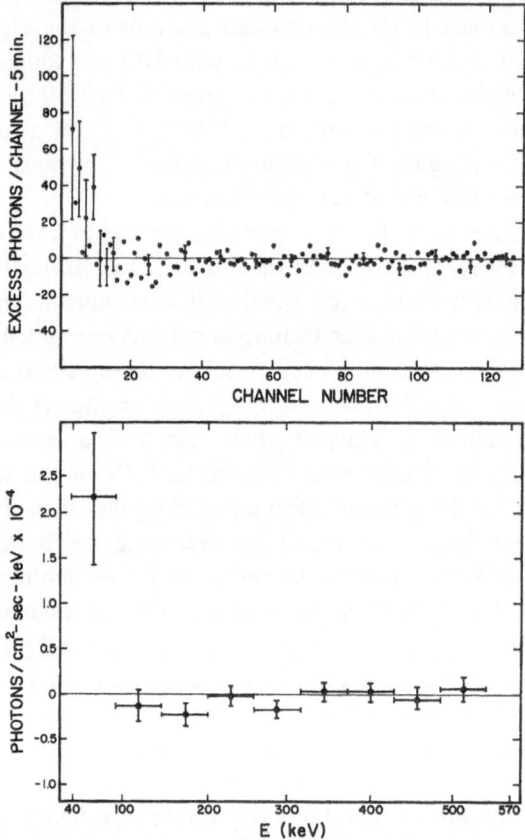

Fig. 1. Upper plot is the observed excess from the M87 region for each pulse height channel over the entire flight. The method of background subtraction is described in the text. Lower plot is the derived flux for 13-channel sums at the top of the earth's atmosphere. Error bars shown are ± one standard deviation for both plots.

$2.13 \pm 0.75 \times 10^{-4}$ photons/cm²-sec-keV. It is to be compared with the value of $1.22 \pm 0.52 \times 10^{-4}$ photons/cm²-sec-keV we previously reported for Virgo. The larger error in the present experiment is caused by the importance of the pointing error when using a narrow-beam device.

To within statistics, the flux agrees with that observed on August 10, 1967. If we assume M87 to be the source of hard radiation observed on both occasions and further assume that M87 has not varied its emission at these wavelengths during this period, we may combine the results of the two measurements. The weighted average of the two is 3.5σ above background; it is $1.51 \pm 0.43 \times 10^{-4}$ photons/cm²-sec-keV, in the energy interval 40–100 keV, at the top of the earth's atmosphere. It would appear from the agreement of the two measurements, that 3C 273 contributed little if any radiation to the flux found earlier.

These results on M87 do not agree with those reported by McClintock *et al.* (1969). Those investigators failed to detect radiation at energies up to 100 keV from M87 with a proportional-counter balloon-borne experiment conducted on October 25, 1968; they reported upper limits for the flux and concluded that a break exists in the M87 spectrum at energies between those of soft X-radiation and their threshhold at ~ 20 keV. While it is possible that M87 is a variable source, the difference most likely arises in the manner by which the background is determined. Specifically, we suggest that perhaps the constant-azimuth method (36° East and West from M87) and/or the

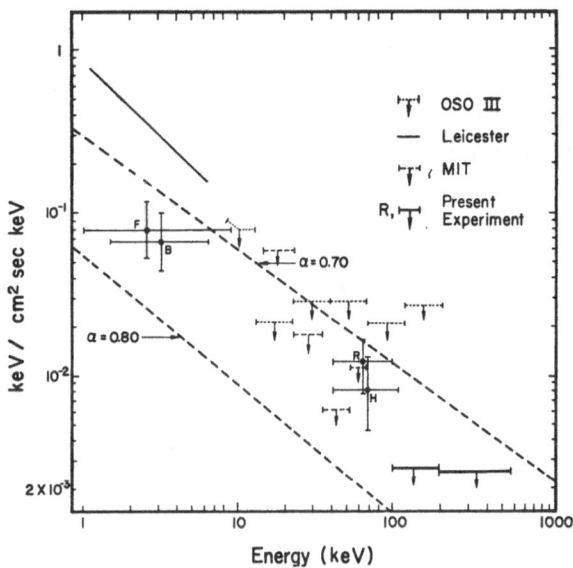

Fig. 2.   Measurements of M87 in the X-ray region. Data plotted are from the following references: 'F', Friedman and Byram (1967); 'B', Bradt *et al.* (1967); 'Leicester', Adams *et al.* (1969); 'MIT', McClintock *et al.* (1969); 'OSO III', Hudson *et al.* (1968), and 'H', Haymes *et al.* (1968). Also shown are two possible extrapolations of the radio core spectrum, indicated by values of their spectral index, α. Upper limits indicated are at the 2σ level; error bars associated with positive fluxes are $\pm 1\sigma$.

pulse-shape discrimination (PSD) technique used by McClintock *et al.* (1969) may be responsible for the difference. The sun, observed to exhibit two Importance One flares during their flight (ESSA, 1968), was only 34° from M87 and was closer than that to the field of view of their detector during their west background observations. Solar flares are known to have hard solar radiation associated with them; hard X-radiation was in fact detected from the sun during both of the flares in question by the Explorer 37 satellite (ESSA, 1968). Also, if the PSD technique is not 100%

efficient in rejecting charged particles, any slight East–West asymmetry in the charged-particle spectrum could give rise to the null result reported by those investigators. The experiment reported here was conducted when the sun was about 75° away from M87 and during a time of relative solar calm (ESSA, 1969).

Figure 2 shows, however, that there is no great inconsistency between the two measurements at the 2σ level. We also show in Figure 2 the 2σ upper limits we have deduced from our data at higher energies. At 506 keV, the upper limit on the red-shifted positron annihilation flux is 1×10^{-3} photons $cm^{-2}sec^{-1}$, and the 2σ upper limit on the 'overflow counting rate' (i.e., those pulses due to photons with energies greater than 570 keV) is 6×10^{-4} counts/cm²-sec.

Also shown in Figure 2 are two extrapolations into the X-ray region of the core radio power law spectrum. The upper limits established on the flux by the present experiment at high energies are consistent with a spectral steepening above 100 keV. However, a single power law, presumably synchrotron in origin, may be fit to the data on the radiation in the radio and soft X-ray regions of the spectrum, as well as to the 40–100 keV flux; such a curve is also consistent with the high-energy upper limits found in the present work. The several optical determinations that have been made are in disagreement.

Felten et al. (1969) have found that the spectral index of the best-fit power law is 0.75. Since the spectrum extends to at least 40 keV, it would appear that M87 is most luminous in the hard X-ray region of the spectrum.

Acknowledgments

We wish to express our thanks to Mr. A. C. Heath, who constructed much of the apparatus and helped during the flight. The research was supported in part by Project Themis under Contract N00014-68-A-0503 and in part by the Air Force Office of Scientific Research, United States Air Force, under Contract F44620-69-C-0083.

References

Adams, D. J., Cooke, B. A., Evans, K., and Pounds, K. A.: 1969, *Nature* **222**, 759.
Bradt, H., Mayer, W., Naranan, S., Rappaport, S., and Spads, G.: 1967, *Astrophys. J. (Letters)* **150**, L199.
ESSA, Solar-Geophysical Data, Reports No. IER-FB-291, 292: 1968, Boulder, Colo., Environmental Science Service Administration.
ESSA, Solar-Geophysical Data, Report No. IER-FB-293: 1969, Boulder, Colo., Environmental Science Service Administration.
Felten, J. E., Arp, H. C., and Lynds, C. R.: 1969, Contributions from the Kitt Peak National Observatory, No. 390.
Friedman, H. and Byram, E.: 1967, *Science* **158**, 257.
Haymes, R. C., Ellis, D. V., Fishman, G. J., Kurfess, J. D., and Tucker, W. H.: 1968a, *Astrophys. J. (Letters)* **151**, L9.
Haymes, R. C., Ellis, D. V., Fishman, G. J., Glenn, S. W., and Kurfess, J. D.: 1968b, *Astrophys. J. (Letters)* **151**, L131.
Hudson, H. S., Peterson, L. E., and Schwartz, D. A.: 1969, *Solar Phys.* **6**, 205.
McClintock, J. E., Lewin, W. H. G., Sullivan, R. J., and Clark, G. W.: 1969, Massachusetts Institute of Technology Preprint Number CSR-P-69-13. Submitted to *Nature*.

A ROCKET OBSERVATION OF COSMIC X-RAYS IN THE ENERGY RANGE BETWEEN 0.15 AND 20 keV

S. HAYAKAWA, T. KATO, F. MAKINO, H. OGAWA,
Y. TANAKA, and K. YAMASHITA

Dept. of Physics, Nagoya University, Nagoya, Japan

and

M. MATSUOKA, S. MIYAMOTO, M. ODA and Y. OGAWARA

Institute of Space and Aeronautical Science, University of Tokyo, Tokyo, Japan

Abstract. Cosmic X-rays were observed with three sets of proportional counters covering the energy range between 0.15 and 20 keV. The detector born on a spinning rocket scanned a celestial region in which the galactic latitude b^{II} changed from 30° to $-55°$ across the galactic plane in the Cygnus-Cassiopeia region. The spectrum of Cyg XR-2 thus obtained is represented by a thermal bremsstrahlung of temperature 3.4 keV modified by the interstellar absorption for the hydrogen column density of 3×10^{21} cm^{-2}. The diffuse component showed an interstellar absorption effect, which was however found much weaker than one would expect if the diffuse component were due entirely to be of extragalactic origin. The spectrum obtained in the highest latitude region is represented approximately by a power law $E^{-1.8}$ but shows a possible trough at about 1 keV.

1. Introduction

Observation of cosmic X-rays at low energies is known to be important in understanding the mechanism of X-ray emission and investigating properties of interstellar matter through the absorption of X-rays. For this purpose detectors sensitive to soft X-rays have to be used. This has been achieved by means of proportional counters with thin polypropylene windows.

In order to guarantee the reliability of experimental results, we used three kinds of counters whose sensitive regions partially overlap with each other. This also facilitated to obtain the energy spectrum in a wide energy range from 150 eV to 20 keV.

The X-ray detector was launched on board a Japanese sounding rocket (K-10-4) at 19:00 (JST) 14 Jan. 1969 from Kagoshima Space Center (31°N, 131°E), Institute of Space and Aeronautical Science, University of Tokyo. The rocket reached a maximum altitude of 226 km. The detector scanned a sky region nearly parallel to the galactic longitude, so that the effect of interstellar absorption could be obtained. The scanning path crossed the galactic plane in Cygnus-Cassiopeia region, so that Cyg XR-2 was clearly resolved. An X-ray source which probably coincided with Cas A was also resolved with a reasonable confidence level, although its separation from near-by sources and possible X-ray emission from the galactic disk was not yet unambiguous at the present stage of data analysis.

In the present paper, therefore, we restrict ourselves to results on the diffuse component and Cyg XR-2.

L. Gratton (ed.), Non-Solar X- and Gamma-Ray Astronomy, 121–129. All Rights Reserved

2. Instrumentation

The detector consisted of seven proportional counters, of which three counters had 120 micron Be windows filled with 250 mm Hg, $Xe + 10\%$ CH_4, the other four had 4 μ thick polypropylene windows with 200 Å aluminium coating. Two of them were filled with $Ar + 10\%$ CH_4, and the other two with $He + 10\%$ CH_4 of 1 atm, respectively. The effective area of each counter was 50 cm^2. Be counters covered the energy range 2–20 keV, whereas the polypropylene counters were also sensitive to soft X-rays of energies above 150 eV. The detection efficiency vs. energy is shown in Figure 1.

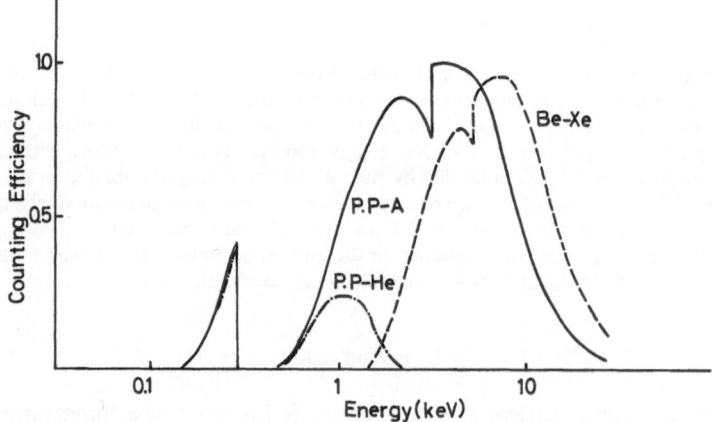

Fig. 1.   Efficiencies of the counters calculated for X-rays of normal incidence.

The counter axis was inclined by 76° with respect to the rocket axis and the field of view was defined by two sets of slats collimators (A and B), FWHM $4° \times 15°$. The collimator A was approximately parallel to the galactic plane in the Cygnus-Cassiopeia region and was placed in front of three Be counters and a pair of polypropylene counters, one Ar-filled and the other He-filled. The collimator B for another pair of Ar- and He-filled polypropylene counters was tilted by 70° with respect to the collimator A.

A door-type shutter that periodically opened and closed the field of view was incorporated in order to measure the environmental background counts separately. Inflight calibration was provided throughout the flight with ^{55}Fe sources mounted on the shutter. A weak C-K X-ray source served to monitor the He-filled polypropylene counter that is insensitive to K X-rays from ^{55}Fe.

Because of the high permeability of the polypropylene film for air, polypropylene counters were continuously flushed from an outside reservoir till the time of launch and were later closed by solenoid valves.

Pulse height of each count was telemetered to the ground. Inflight calibration data verified perfect functioning of the payload.

3. Performance of Experiment

X-ray counts were obtained during about 200 sec. According to the roll of the rocket at a rate of 4.2 c/sec with a small yaw angle, the counters scanned a fixed band on the celestial sphere, including Cyg XR-2 and Cas A, as shown in Figure 2. The results of added counts for every spin as a function of roll angle are shown for individual sets of counters and for given energy ranges in Figure 3. The background counting rates

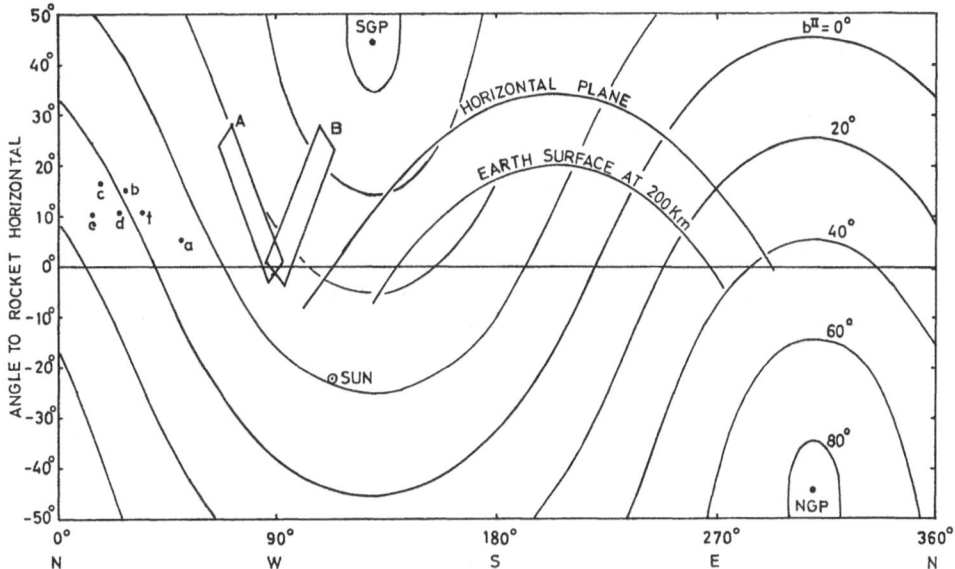

Fig. 2.   The scanned band on the celestial sphere in the rocket frame of reference. The rocket axis pointed $\alpha = 3^{h}19^{m}$ and $\delta = 0.5°$, whereby a small yaw motion is neglected. The fields of view of collimators A and B are indicated. Curves represent galactic latitudes b^{II} and the horizons at the ground surface and at an altitude of 200 km. The positions of the galactic north and south poles (N.G.P. and S.G.P., respectively), the sun and several known X-ray sources are indicated; (a) Cyg XR-2; (b) Cas A; (c) Cep XR-1; (d) Cep XR-2; (e) Cep XR-3; (f) Lac XR-1.

as deduced from the closed-shutter period are indicated by dotted lines. They agree with the counting rates when looking at the earth in the energy range above 1 keV, so that the earth can also be used as a shutter. As energy decreases, however, the difference between the counting rates when looking at the earth and when the shutter was closed increases. There is a significant difference in the lowest energy region.

In this energy region a strong peak was found in the direction of the western horizon and a weak peak near the eastern horizon. Since the sun was located by 35° below the counter axis when the axis passed by the direction indicated by an arrow in Figure 3, the western peak may be due either to scattered solar X-rays or to twilight glow. The counter was slightly sensitive to ultraviolet radiation of wavelengths below 3500 Å. The counters were subject to atmospheric background also in other directions.

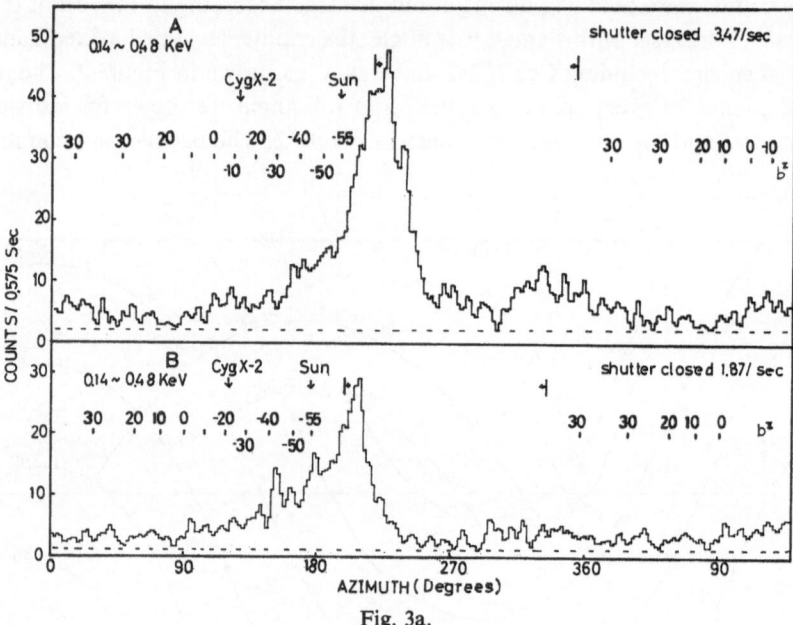

Fig. 3a.

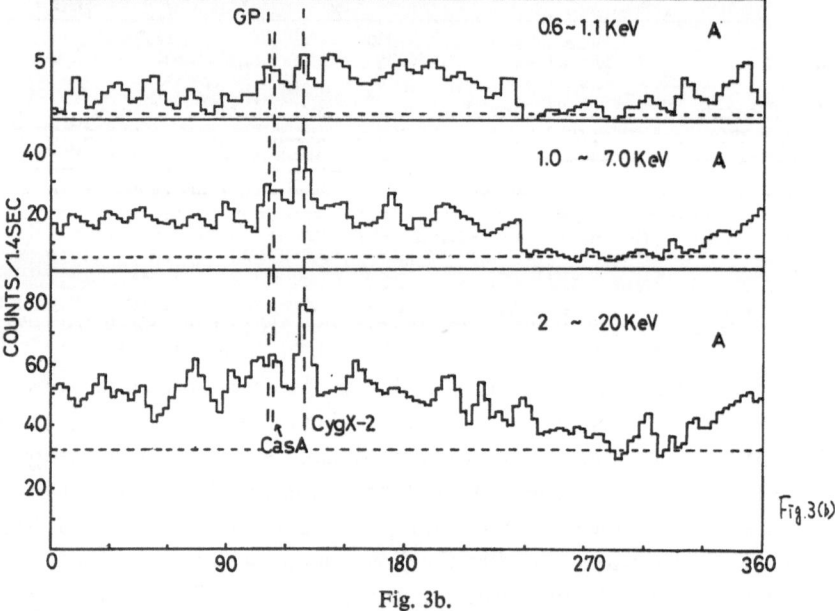

Fig. 3b.

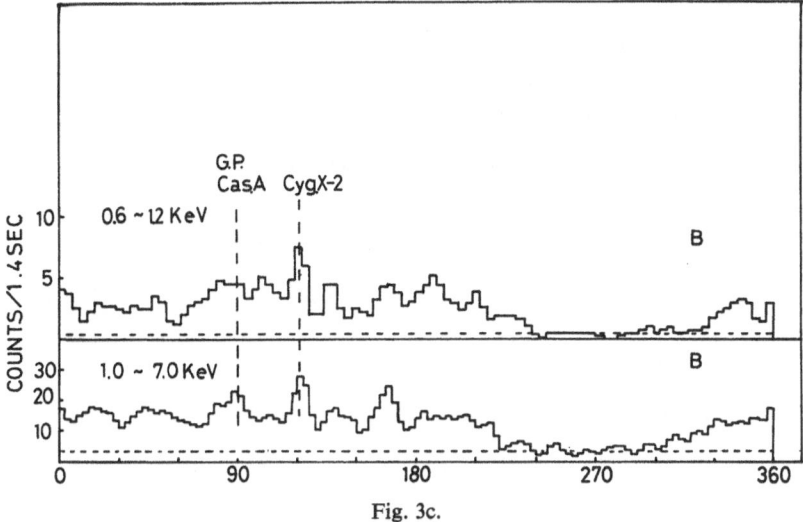

Fig. 3c.

Fig. 3. Counting rates vs. azimuth angle in the rocket frame of reference. – (a) Energy range 0.14–0.48 keV, collimators A and B, counts of He- and Ar-filled counters being added. – (b) Energy ranges 0.6–1.1 keV, 1.0–7.0 keV and 2–20 keV, collimator A. – (c) Energy ranges 0.6–1.2 keV and 1.0–7.0 keV, collimator B.

From the altitude dependence of the counting rate in the lowest energy region, the emission of the atmospheric background was limited to altitudes below 180 km. Thus only data free from the atmospheric background were used as cosmic X-rays.

Three kinds of counters gave consistent results in overlapping energy regions. The energy spectrum obtained with the xenon filled counters was slightly flatter than that with the argon filled counters. This seems to have been the case when Gorenstein *et al.* [1] observed Sco XR-1 on two different days, although they attributed the difference to the time variation of temperature. While we do not fully understand causes of the differences, we refer to results with the argon filled counters in the overlapping energy ranges, if any significant difference is found.

4. Cyg XR-2

Cyg XR-2 is significantly visible in the energy region above 0.6 keV, whereas it can not be seen in the lowest energy region. The energy spectrum of Cyg XR-2 is shown in Figure 4.

If this is represented by an empirical expression

$$j(E) = E^{-1} \exp(-E/kT) \exp(-E_a/E),$$ (1)

we obtain a dashed curve in Figure 4 for

$$kT \simeq 4 \text{ keV}, \quad E_a \simeq 0.5 \text{ keV}.$$ (2)

The value of kT is consistent with other results, whereas the value of E_a is much smaller than that given by Giacconi [2], $E_a \simeq 2$ keV.

If the last factor in the expression (1) represents the absorption effect, it may be more realistic to use the expression,

$$j(E) = E^{-1} \exp(-E/kT) \exp[-(E_a/E)^b] \qquad (3)$$

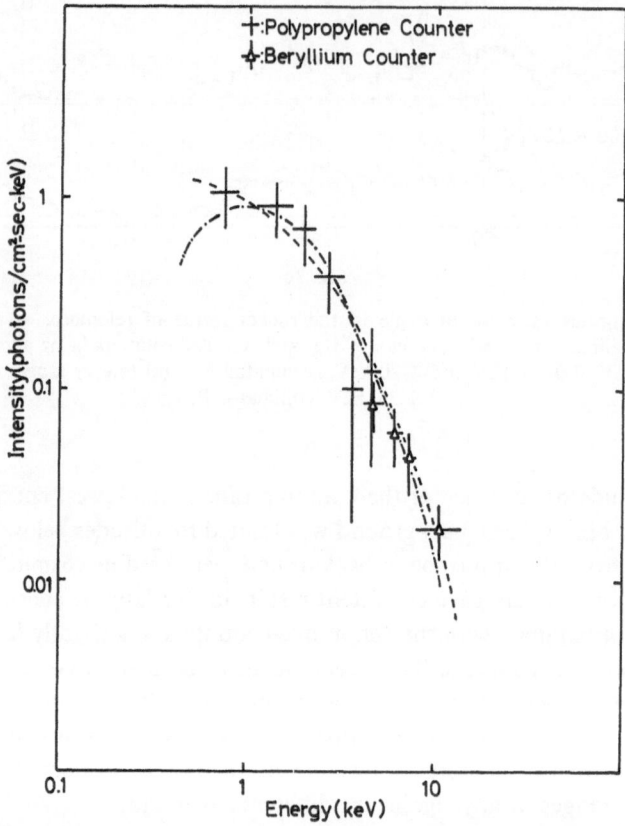

Fig. 4.   The energy spectrum of Cyg XR-2. Calculated curves for Equation (1) with (2) and Equation (3) with (4) are shown by dashed and dot-dashed curves, respectively.

with $b \approx 8/3$. The spectrum shown by a dot-dashed curve in Figure 4 represents the expression (3) for

$$kT = 3.4 \text{ keV}, \qquad E_a = 0.8 \text{ keV}. \qquad (4)$$

The latter value of E_a is related to the hydrogen column density by reference to the absorption cross section of interstellar gas of normal abundances [3],

$$\int n_H \, dl \approx 3 \times 10^{21} \text{ cm}^{-2}. \qquad (5)$$

For the average hydrogen density of $\bar{n}_H = 0.8$ cm^{-3} in the direction of Cyg XR-2, its distance is obtained as

$$l \approx 1.2 \text{ kpc.} \tag{6}$$

This is not inconsistent with the distance derived from the extinction of the possible optical counter part of Cyg XR-2.

5. Diffuse Component

The diffuse component of X-rays with energies smaller than 1 keV may show the effect of interstellar absorption. In order to see this effect, the intensities in the two lowest energy ranges are plotted against galactic latitude in Figure 5, in which the counts obtained with argon- and helium-filled counters with two different collimators are added up so as to increase statistics. The result is similar to the one given by Kraushaar [4].

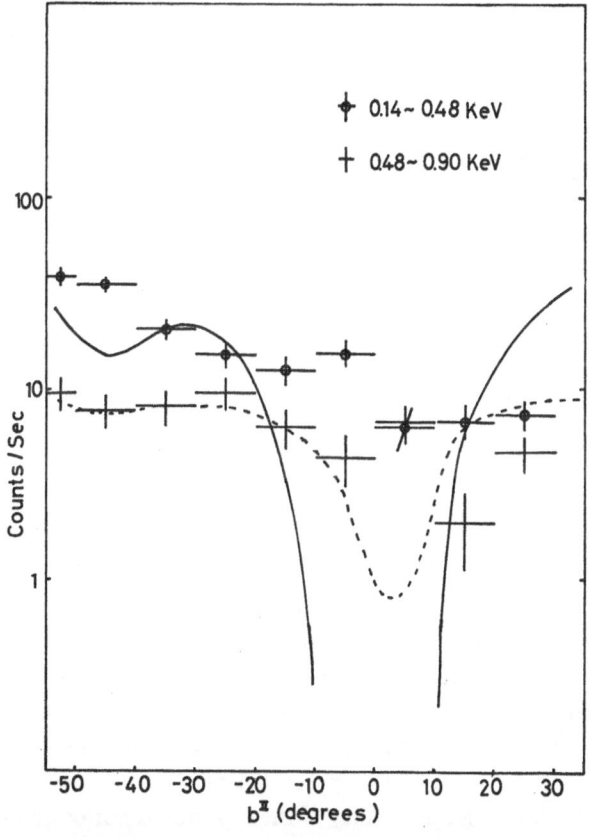

Fig. 5.   The latitude dependence of counting rates in the energy ranges 0.14–0.48 keV and 0.48–0.9 keV. The solid (0.14–0.48 keV) and dashed (0.48–0.9 keV) curves represent the ones expected from interstellar absorption based on 21-cm radio observation assuming zero helium abundance. The curves are normalized to the observed intensities at − 35°.

From Figure 5 one sees that the latitude dependence is not symmetric about the galactic plane. This cannot be accounted for by the latitude dependence of the column density of neutral hydrogen as observed by 21-cm radio emission. This is neither consistent with the one expected from the interstellar absorption based on 21-cm

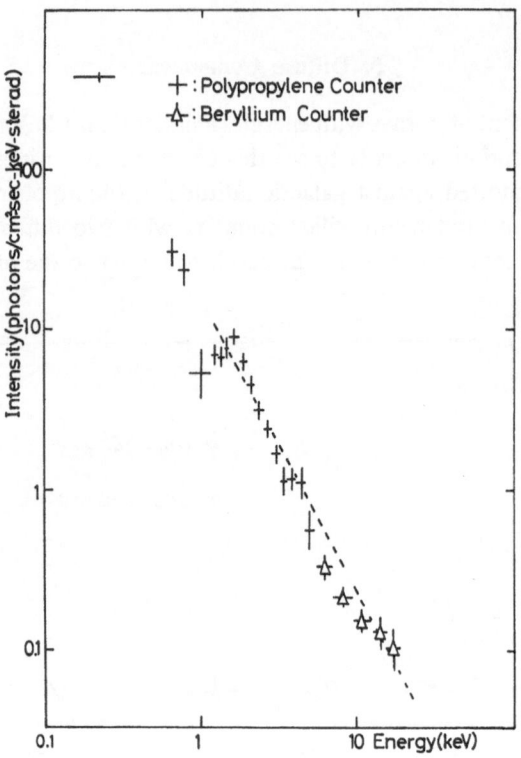

Fig. 6.   The energy spectrum of the diffuse component at galactic latitudes between $-40°$ and $-55°$. The dashed line represents the $E^{-1.8}$ spectrum.

radio observation and the normal abundances of interstellar gas. Such a feature was also observed by Bowyer *et al.* [5]. This could be due, at least in part, to the emission of soft X-rays from unresolved sources in our galaxy.

Since effects of the interstellar medium are not yet known, the spectrum of background X-rays cannot be unambiguously derived. Here we give the spectrum at the highest latitude region scanned, namely between $b^{II} = -40°$ and $-55°$. The spectrum shown in Figure 6 is approximately represented by a power law $E^{-1.8}$, except for a dip at about 1 keV. Although the reality of the dip should not be claimed too strongly at the present stage, this could again be due to a superposition of metagalactic X-rays subject to interstellar absorption and X-rays emitted in our galaxy. Except for the dip, the result is consistent with Kraushaar's.

References

[1] Gorenstein, P., Gursky, H., and Garmire, G.: 1968, *Astrophys. J.* **153**, 885.
[2] Giacconi, R.: this volume, p. 107.
[3] Bell, K. L. and Kingston, A. E.: 1967, *Monthly Notices Roy. Astron. Soc.* **136**, 241.
[4] Bunner, A. N., Coleman, P. L., Kraushaar, W. L., McCammon, D., Palmieri, T. M., Shilepsky, A., and Ulmer, M.: 1969, *Nature* **223**, 1222.
[5] Bowyer, C. S., Field, G. B., and Mack, J. E.: 1968, *Nature* **217**, 32.

ANGULAR SIZE AND POSITION OF THE
X-RAY SOURCE CYG-X-1

M. MATSUOKA, S. MIYAMOTO, J. NISHIMURA, M. ODA, Y. OGAWARA

Institute of Space and Aeronautical Science, University of Tokyo, Tokyo

and

M. WADA

Institute of Physical and Chemical Research, Tokyo

We performed a couple of balloon experiments to measure the size and the location of Cyg-X-1 using the techniques of the modulation collimator [1]. The angular periods of the modulation collimator were 26′ and 10′ for the respective flights corresponding to the approximate angular resolutions of the size determination and location, 3′ and 1′ respectively. Preliminary results of the experiment with 3′ resolution are reported here.*

The balloon was launched at Haranomachi, Japan at 0616 UT, September 23, 1968 and reached the ceiling of 8 g/cm² at 0820 UT. The passage of Cyg-X-1 at the meridian was expected at 1037 UT. The X-ray detector consists of two NaI (Tl) square scintillators, each 3 mm thick and 85 cm² area. The detector set underneath the modulation collimator detects the modulation of the X-ray flux while the X-ray source moves in the field of view and the orientation of the balloon gondola changes. We may determine the location of the source by means of the phase of the modulation and the size by the depth of the modulation.

A star sensor rigidly fixed to the structure which holds the modulation collimator was used to determine the aspect of the collimator with respect to the celestial sphere. The gondola was roughly pointed in such a way that one or two stars of Ursa Major

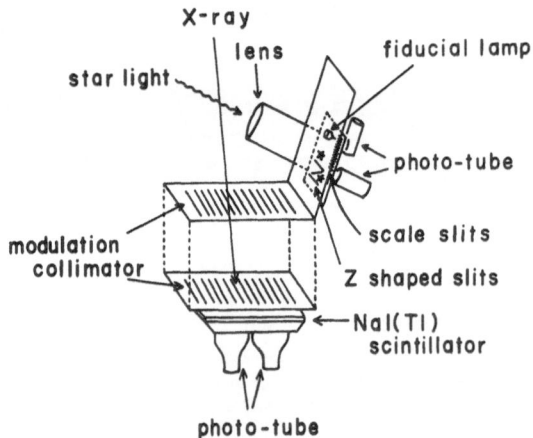

Fig. 1.   Schematical figure of the balloon-borne modulation collimator and star sensor.

* The details of the instrument and data analysis will be published separately.

L. Gratton (ed.), Non-Solar X- and Gamma-Ray Astronomy, 130–133. All Rights Reserved
Copyright © 1970 by the IAU

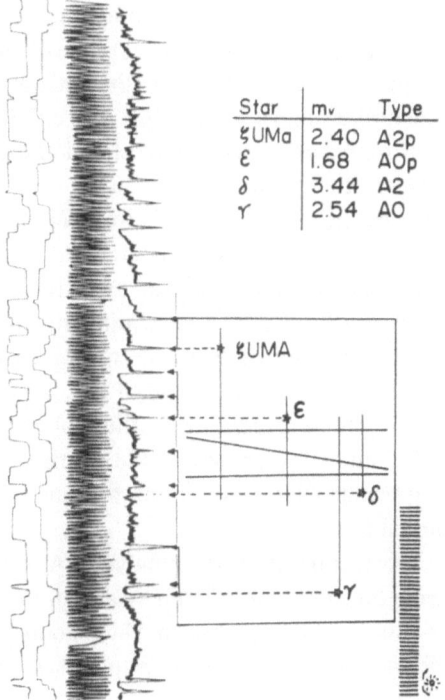

Star	m_v	Type
ʒUMa	2.40	A2p
ε	1.68	A0p
δ	3.44	A2
γ	2.54	A0

Fig. 2. Telemeter record showing the output pulses of star sensor and X-ray counters. The method to derive the direction of a star from the output pulses is illustrated.

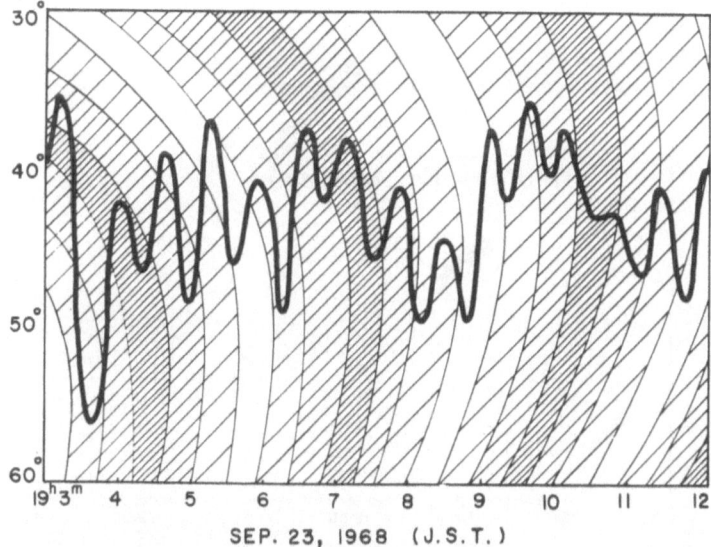

Fig. 3. The variation of the orientation of the instrument is obtained by means of the star sensor. The phase of expected modulation for a certain assumed position of an X-ray source is indicated

are always in the star sensor field of view. The star sensor is composed of a telephoto-lens and a moving diaphragm with slits at the focal plane of the lens which sweep star images and a fiducial lampmark. The Z-shaped slits cut on the diaphragm produce three light pulses for each star image. A row of slits on the diaphragm generates fiducial light pulses. Light pulses are converted to electric pulses by means of a photo multiplier and telemetered to the ground station.

An example of telemeter records is shown in Figure 2. The elevation and the azimuth of a star in the instrument frame of reference are obtained by the phase of appearance of star pulse with respect to the fiducial marks and the ratio of intervals between three pulses respectively. The precision of aspect determination using this star sensor has been proved to be less than an arc minute. The aspect was measured once every second.

A part of analysed data showing the variation of the orientation of the gondola is shown in Figure 3. If we assume a certain location of the source, we may draw contours as shown in the figure for phases of the transmission of the modulation collimator. Thus, we may synchronize X-ray counts and inspect the assumed location of the source for the modulation.

Data for one hour of observation have been analysed. A number of assumed points on the celestial sphere were thus tested for the phase and the amplitude of modulation. Figure 4 shows the candidate area of Cyg-X-1 in which good fits of data to the modulation obtained. The boundary of the area corresponds to one standard deviation error of the phase.

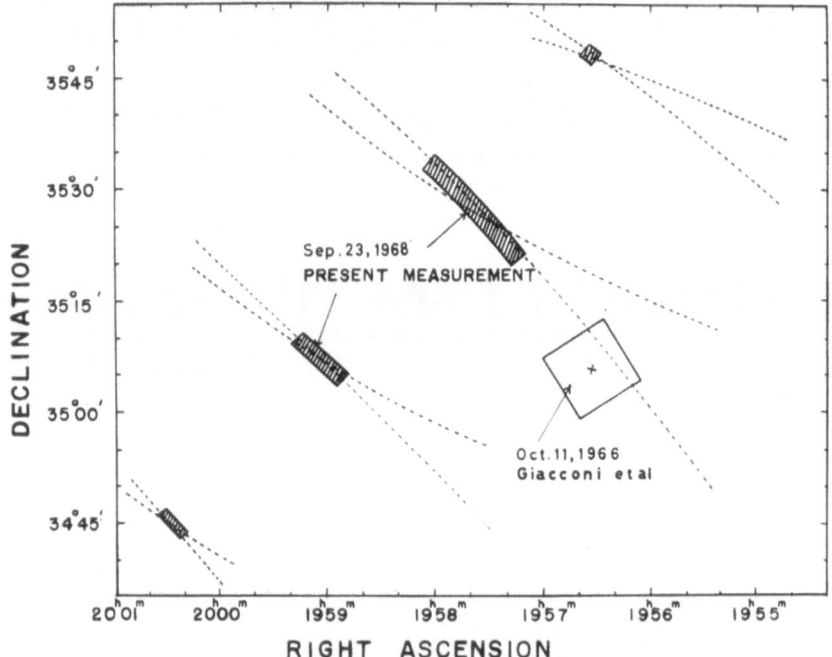

Fig. 4.   Location of Cyg-X-1. (1950).

Typical modulation is seen in Figure 5. The amplitude is consistent with the average observed intensity of Cyg-X-1, $0.14\pm0.01/\text{cm}^2$ sec ($25\sim50$ keV). From this modulation curve we conclude that the depth of modulation is not reduced more than 10% of the phase. It is, thus, concluded that the upper limit of the size is three arc minutes.*

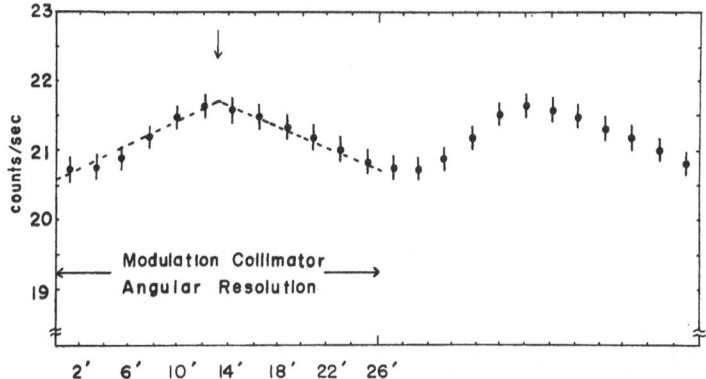

Fig. 5.   The observed modulation of X-ray flux presented for two cycles.

The location shown in Figure 4 is in disagreement with the previously determined location by the ASE group [2] beyond the statistical error. We do not yet understand the cause of this disagreement and we feel we should reserve this part of our conclusions until further analysis is completed of the fine measurement which is more free from the systematic error in the relative angle of the optical axis of the star sensor and the collimator. This reservation does not affect the present conclusion on the size of the source.

Acknowledgement

We wish to acknowledge valuable assistance by Mr. M. Fujii and Mr. S. Ohta.

References

[1] Bradt, H., Garmire, G., Oda, M., Spada, G., and Sreekantan, B. V.: 1968, *Space Sci. Rev.* **8**, 471.
[2] Giacconi, R., Gorenstein, P., Gursky, H., and Waters, J. R.: 1967, *Astrophys. J.* **148**, L119.

* F. Floyd concluded that the upper limit was one arc minute, also utilizing a modulation collimator flown on a balloon (1969, *Nature* **222**, 967).

INTERSTELLAR ABSORPTION OF X-RAYS EMITTED BY SUPERNOVA REMNANTS*

P. GORENSTEIN, E. M. KELLOGG and H. GURSKY

American Science and Engineering, Cambridge, Mass., U.S.A.

An X-ray observation of the Cassiopeia Region by the ASE group from a sounding rocket on December 5, 1968, has resulted in the determination of locations for two sources that are precise to about 0.1 of a square degree. The positions of two well-known radio sources Cas A and SN1572 (Tycho's Supernova), objects which are remnants of relatively recent galactic supernova, are consistent with these locations. Inasmuch as that region of the galaxy does not appear to contain nearly as large a concentration of objects as the galactic center, it is reasonable to make the identification between the X-ray sources and the supernova remnants on the basis of there being a small a priori probability of having an accidental coincidence within 0.1 square degrees. Cas A is almost certainly the same source as Cas XR-1 which the NRL group saw in an earlier survey [1]. During the December flight the Crab nebula was also observed for a short time interval.

The X-ray spectra of supernova remnants are of special interest. Many characteristics of these objects such as their age and distance are already known from previous observations in the radio and visible region so that it is possible to immediately give some physical significance to the X-ray results. Also in the radio and visible regions these objects have a finite angular width of a few arc minutes. If the X-ray emitting region is of comparable dimensions then considering the mass and distances involved, it must be rather transparent to its own emission up to rather long wavelengths. Consequently, observed photon deficiencies in the X-ray region are due rather unambiguously to the absorption effects taking place in the interstellar medium. In the case of the Crab, observations during a lunar occultation experiment [2] and with a modulation collimator [3] have indeed shown that most of the X-ray emission is from an extended region. In the case of Cas A and SN1572 there is as yet no information concerning the angular size in X-rays. Prior to the discoveries that a significant portion of the X-ray emission of the Crab is in a pulsar mode [4] it may have seemed safe to assume that the angular size of the X-ray source in Cas A and SN1572 was comparable to the radio source. We are examining the time profile of the distribution of counts from these two sources and so far have seen no obvious indication of a preferred frequency. In the absence of any contrary indication we assume these sources are large and hence transparent to their own X-ray emission.

On the order of 10^3 counts in the range 1.5–10 keV were detected from each of the three objects. Their spectral distribution was analyzed according to a method described previously [5] which makes a least squares comparison of the experimental

* Supported by the Office of Space Science Applications of National Aeronautics and Space Administration.

results to the expected instrumental response to functions of the following form:

$$dN/dE \sim e^{-(Ea/E)^{2.67}} E^{-\alpha} \tag{1}$$

$$dN/dE \sim (e^{-(Ea/E)^{2.67}} e^{-E/kT})/E. \tag{2}$$

In the case of Cas A and Tycho's Supernova it is not possible to determine from these data alone whether (1) or (2) provides the better fit. Results are given in Table I. With

TABLE I

X-ray characteristics of three supernova remnants (preliminary)

Source	Rel. Counts (1.5–11 keV)	Probable Dist. (pc)	Power Law Spectral Index	Ea (keV)	Exponential Temp (10^6K)	Ea (keV)
Crab Nebula (SN 1054)	1	1700	2.0 ± 0.1	<0.9	–	–
Cas A ($\sim$ 1700)	0.1	3400	3.3 ± 0.6	$1.35^{+.32}_{-.45}$	15 ± 6	$1.1^{+.2}_{-.5}$
Tycho's Supernova (SN 1572)	0.04	3500	2.3 ± 0.4	<1.6	27^{+17}_{-6}	1.1

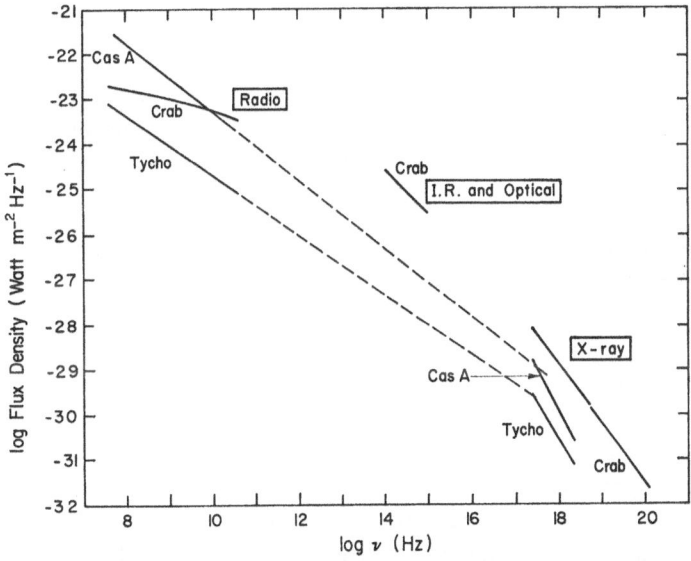

Fig. 1.   The electromagnetic spectra of three supernova remnants.

respect to (1) the Crab is in agreement with previous determinations and has a smaller value of α than the other two. The best values of the temperature T associated with (2) are decidedly lower than the value 10^8 K suggested by Shklovsky [6] in his model which attributes the X-ray emission to thermal bremsstrahlung at the outer boundary of an expanding shell that has been ejected by a supernova explosion.

The electromagnetic spectra of the three objects is shown in Figure 1. It is remarkable that the extrapolation of the radio spectrum of Cas A and Tycho's Supernova is in agreement with the observed X-ray intensity, although the X-ray results by themselves suggest that the spectral index has steepened in the X-ray region. Of course, this apparent agreement could be entirely fortuitous as we have no assurance that the same source mechanism is responsible for both the radio and X-ray emission. Indeed in Shklovsky's model they would not be related.

The parameter Ea represents the energy at which the interstellar opacity is one mean free path. In this energy range, ~ 1 keV, it is mainly determined by the concentration of O & Ne. Models of the interstellar medium relate the amount of O & Ne to the quantity of hydrogen along the line of sight [7, 8].

In principle the amount of neutral hydrogen along the line of sight to strong radio sources can be determined from the 21-cm absorption features that the interstellar medium imposes upon the spectra of the emitted continuum [9, 10]. However, such an estimate is dependent upon the value of an assumed temperature for the absorbing hydrogen clouds and may not be sensitive to all the diffuse hydrogen. The absorbing medium is largely confined to galactic spiral arms. For the Crab, it is the orion arm and for Cas A and Tycho's Supernova it is the orion plus perseus arm. Table II lists

TABLE II

Source	N_H [a]	Ea (theo.) [b]	Ea (obs.)
Crab	1.6×10^{21} H atoms/cm^2	0.73 keV	< 0.9 keV
Cas A	1.0×10^{22}	1.55 keV	$1.2_{-0.5}^{+0.3}$
Tycho's Supernova	1.0×10^{22} [c]	1.55 keV	< 1.6

[a] Refs. [9, 10].
[b] Calculated according to cross-sections of Bell and Kingston [8].
[c] N_H to Tycho is assumed to be the same as Cas A.

the amount of hydrogen as determined from the radio measurements, the value of Ea expected on the basis of Bell and Kingston's [8] cross-section and the preliminary value of Ea as determined in this measurement, assuming power law spectra.

The principal conclusion to be drawn from these data is that the interstellar medium is clearly not more opaque to X-rays of ~ 1 keV than had been expected and could possibly be less opaque. Consequently, the abundance of neon and oxygen relative to hydrogen over a significant portion of the galaxy as determined by line of sight X-ray absorption is consistent with and certainly not more than presently accepted values.

References

[1] Byram, E. T., Chubb, T. A., and Friedman, H.: 1966, *Science* **152**, 66.
[2] Bowyer, S., Byram, E. T., Chubb, T. A., and Friedman, H.: 1964, *Science* **146**, 912.
[3] Oda, M., Bradt, H., Garmire, G., Spada, G., Sreekantan, B. V., Gursky, H., Giacconi, R., Gorenstein, P., and Waters, J. R.: 1967, *Astrophys. J.* **148**, L5.
[4] Reports have been presented at this symposium of the observation of the pulsar in the Crab

Nebula in the X-ray regions by the NRL, MIT, Columbia U., and Rice U. groups.

[5] Gorenstein, P., Gursky, H., and Garmire, G.: 1968, *Astrophys. J.* **153**, 885.
[6] Shklovsky, I. S.: 1968, *Supernovae*, Chapt. II, Wiley-Interscience, New York.
[7] Felten, J. E. and Gould, R. J.: 1966, *Phys. Rev. Lett.* **17**, 401.
[8] Bell, K. L. and Kingston, A. E.: 1967, *Monthly Notices Roy. Astron. Soc.* **136**.
[9] Muller, C. A.: 1958, *Astrophys. J.* **125**, 830.
[10] Clark, B. G.: 1965, *Astrophys. J.* **142**, 1398.

THE POSSIBLE DETECTION OF IRON LINE EMISSION
FROM SCO X-1

S. S. HOLT, E. A. BOLDT and P. J. SERLEMITSOS

NASA, Goddard Space Flight Center, Greenbelt, Md., U.S.A.

Abstract. A rocket-borne measurement carried out on March 3, 1969 has yielded a net (dead-time-corrected) exposure of 10^4 cm^2 sec to the differential spectrum of X-radiation from Sco X-1. The data are fully consistent with radiation from an optically thin thermal source, even to the extent that K-emission from high ionization states of iron appears to be present. Such iron emission is consistent with cosmic abundance at the measured temperature.

1. Introduction

In a recent communication (Holt *et al.*, 1968) we discussed the practical possibility of searching for iron line emission with proportional counters as a positive indication of the thermal nature of discrete X-ray sources. In that communication, we announced our intention of performing an experiment to measure such possible emission from Sco X-1. We report, here, the results of that experiment.

2. Experiment

We observed Sco X-1 for 150 sec with essentially the same detector system used in our exposure to the Crab Nebula last year (Boldt, *et al.*, 1969). Almost all of the source exposure is obtained in two nominally identical argon-methane (10%) proportional counters with 2-mil beryllium windows. We have chosen to treat the two counters as two independent experiments, partly because the counter responses are not identical and partly because no additional information can be obtained from decreasing the number of degrees of freedom available in the data analysis. The data from one of these counters will be presented here, with the remark that the data from the other are completely consistent with any inferences which we have drawn from these data alone. Furthermore, we note that the data presented have *not* been folded back through the detector response; this procedure is necessarily non-unique, and can be particularly misleading if there is a suspicion of discontinuous structure in the input spectrum.

We have started our analysis by fitting the data (with background subtracted) to a single exponential in energy: the standard procedure for analyzing the emission from a suspected thermal source. Figure 1 exhibits this comparison with the best fit exponential normalized to the data between 4 and 15 keV (we pick 4 keV as the lower energy bound to ensure that we are above possible Si and S edges in the source, and to minimize possible systematic effects arising from minute contaminants in the Be window). The response of the detector has been exhaustively measured, with the effects of energy dependent efficiency, resolution, escape, internal anode fluorescence

L. Gratton (ed.), Non-Solar X- and Gamma-Ray Astronomy, 138–143. All Rights Reserved

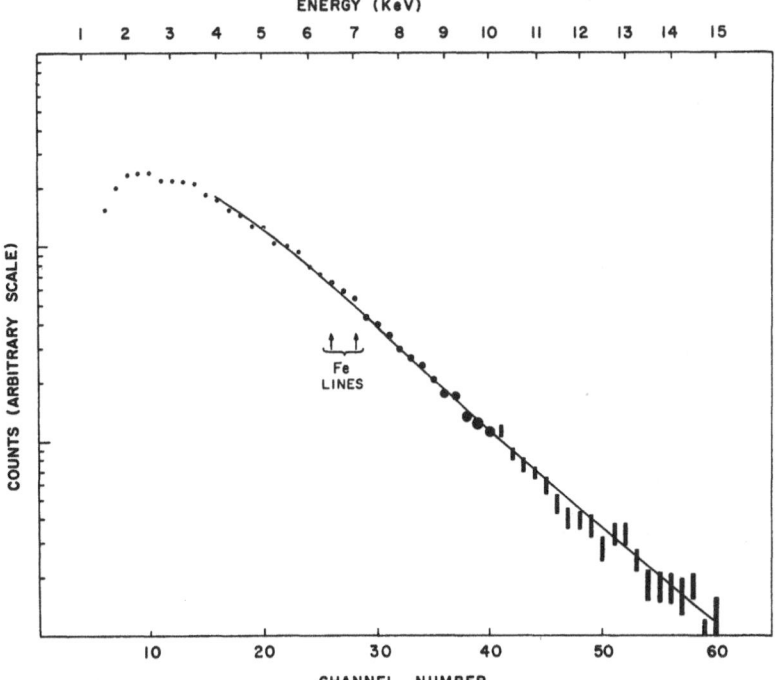

Fig. 1.   Comparison of the raw data from one of the two high-exposure counters (with dead-time-compensated background subtracted) with the best fit exponential in differential energy flux folded forward through the detector response. Each point represents $\frac{1}{4}$ keV channel width, and the statistical error is in the ordinate only. The points have been drawn as circles of radius 1σ for observational ease (above 10 keV it became impractical to draw circles). The best fit exponential is normalized to the data in the energy range indicated in the figure: 4–15 keV.

and incomplete energy conversion taken into account. The trace in Figure 1 corresponding to the trial spectrum folded forward through the detector response is, therefore, virtually free of systematic error. The fit looks quite good by eye, with the exceptions that there seems to be an excess near the energies corresponding to K-emission from the highest ionization states of iron (as given by Tucker, 1967), a small deviation below the comparison spectrum at lower energies, and an observable deviation below at high energies.

3. Data Analysis

Figure 2 indicates the standard quantitative measure of the goodness of fit in terms of χ^2. The best fit χ^2 is > 70 at 5.7 keV for 46 degrees of freedom for the simple exponential. χ^2 for a pure hydrogen plasma, also shown in the figure, is even worse, and is clearly excluded. The other counter exhibited a minimum χ^2 of about the same value at about the same temperature. Since the confidence level at which we obtain a value of χ^2 of 70 for 46 degrees of freedom is only $\sim .01$, and we have two such results which are statistically independent, the data obviously have too much statistical significance to be fittable to a unit gaunt factor optically-thin isothermal source with no structure.

The most straightforward way to arrive at a consistent input spectrum is to perturb

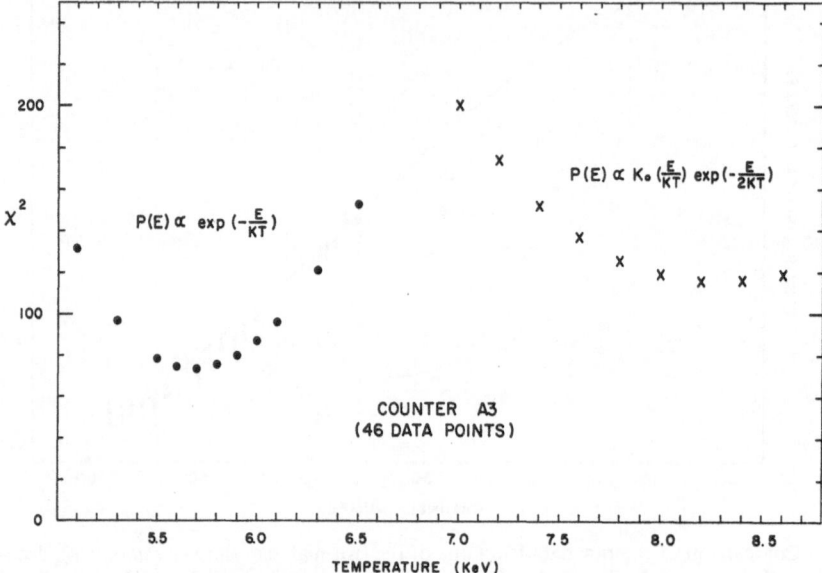

Fig. 2. χ^2 for trial exponentials (circles) and trial hydrogenic plasmas (crosses) as a function of temperature for the data shown in Figure 1.

the exponential in a physically reasonable manner. The use of an energy-independent gaunt factor implies the presence of heavy elements, so that the effect of line and recombination radiation on the output spectrum would seem to be the next logical consideration. We have been careful to consider the data above 4 keV only up to this point, because we need not consider the effect of elements with $Z < 26$ in a perturbation of the input spectrum if we restrict ourselves to this energy range, since, in fact, the first order effects above 4 keV will be due solely to iron. Figure 3 illustrates, schematically, what the effect of one line and its associated recombination edge would have on a comparison of real data and a pure exponential. The minimum χ^2 fit would be the dotted line, at a higher temperature than the actual source temperature, and if one were to plot the percentage deviation between the data points and the trial exponential folded through the detector response, one would expect to see a negative percentage deviation at energies below the line, and another dip at high energies. If the detector resolution is good enough, the intermediate energy structure may be resolvable; the effect of the line and edge emission would be observationally suppressed, however, by the fact that one would be comparing with respect to a higher continuum than actual.

Such a comparison is displayed in Figure 4. The data are the same data as in Figure 1, as is the best-fit exponential. At energies below the possible iron lines the deviation is negative, as it is at high energies (past 10 keV the statistical significance of the data points are too poor to plot on this scale). The possibility of iron line emission is

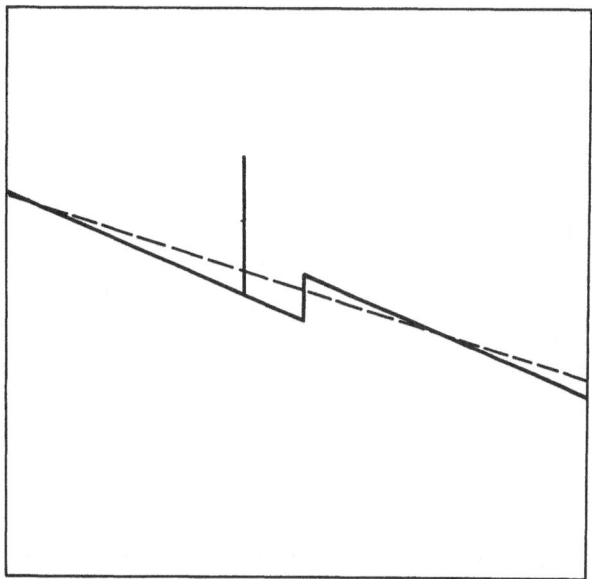

Fig. 3.   Schematic representation of the effect of elemental structure in an optically thin isothermal plasma (solid trace) on the determination of the minimum χ^2 exponential with which it can be fitted (dotted trace).

intriguing: the 1 keV bin at 6.9 keV is 2.5 σ above the best-fit exponential, and the $\frac{1}{4}$ keV points have the shape we would expect to see with our counter resolution. We must remember, also, that this indication of line emission can only be enhanced if we realize that the continuum with which it should be compared is that at lower energies.

4. Discussion

It is obvious from Figure 3 that the fit will be better than that obtained with a simple exponential if iron emission is included in the trial spectrum; it remains, however, to determine the amount which is needed for consistency. We have tried fitting Tucker's models II and III (Tucker, 1967), which should be representative of abundances in low and high mass supernovae, to our data. In both cases, the best χ^2 was much worse than that for a simple exponential because the iron abundance in these models is much higher than what we observe. The only way that we can get an acceptable value of χ^2 is with a smaller iron abundance: in fact, one which must be close to universal.

It is possible to test the hypothesis of universal abundance on the spectrum as a whole. Since, however, the contribution to the X-ray spectrum from each high ionization state of every element is small, it is necessary to include them all if one includes any for $Z < 26$. We intend to continue our analysis of the lower energy data (our sensitivity extends down to slightly below 2 keV), but this procedure, if done carefully, will necessarily take a considerable amount of time; this analysis will be complicated, also, by the possibility of optical thickness when we extend the procedure to energies below 4 keV. We reiterate, however, that the Sco X-1 spectrum above 4 keV is completely consistent with an isothermal, optically thin source of universal ele-

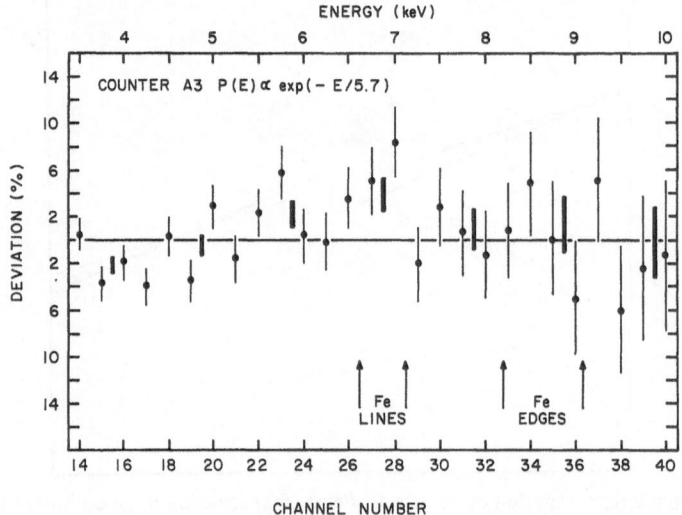

Fig. 4. Percentage deviation of the data from the best fit exponential shown in Figure 1. The points are the same ¼ keV points displayed in Figure 1, while the bars are 1 keV weighted averages of these points centered at the energies shown.

mental abundance. The temperature we measure is essentially identical to that measured by Fritz *et al.* (1969), and we can exclude distinct overabundances of iron such as demanded by Tucker's models II and III. We can also exclude a pure hydrogenic plasma, and on the basis of a χ^2 analysis, conclude that our data can be consistently fit with a source of universal elemental abundance, but not with a pure exponential.

We feel obliged to make a few qualifying comments on the interpretation of our experimental results. The first involves the possibility of a non-thermal (non-collisional) origin for the X-radiation with a spectral shape which masquerades as thermal. We cannot, of course, exclude such a possibility, but we note that the bump is observed at precisely the energy where a thermal source with at least universal abundance of iron would necessarily exhibit such a feature.

Another consideration which would appear to be in order would be the probability that a collisional source would be isothermal. There is no doubt that the celestial object from which the X-ray emission is observed is not isothermal when considered as a whole. The X-ray source, on the other hand, may be well represented by an isothermal distribution. Solar active regions (and even large solar X-ray flares) give every indication of being isothermal regions imbedded in a cooler solar corona. The emission from these regions (which have a temperature of only a few times the coronal temperature) completely dominates the observed solar X-ray spectrum. We believe that the objections which may be raised to an isothermal model for the X-ray emitting region on the basis of stellar models which demand temperature gradients are not necessarily valid; the X-ray emitting volume may be virtually isothermal (in analogy with solar active regions) without violating any precepts of stellar model building.

Finally, we address ourselves to the question of iron line emission, per se. Had the input spectrum been characteristic of thermal emission from a supernova remnant (Tucker's models II and III) we could have, unquestionably, made a positive and unambiguous identification of iron line emission (Holt *et al.*, 1968). Since, however, the data indicate that iron line emission, if present, is barely at the level of detectability, we cannot unambiguously conclude that it has been observed. The fact that the inclusion of iron, at approximately the level of universal abundance, is that perturbation of the simple exponential sufficient for complete consistency with our data, would seem to indicate that we have probably observed iron line emission from Sco X-1.

References

Boldt, E. A., Desai, U. D., and Holt, S. S.: 1969, *Astrophys. J.* **156**, 427.
Fritz, G., Meekins, J. F., Henry, R. C., and Friedman, H.: 1969, *Astrophys. J.* **156**, L33.
Holt, S. S., Boldt, E. A., and Serlemitsos, P. J.: 1968, *Astrophys. J.* **154**, L137.
Tucker, W.: 1967, *Astrophys. J.* **148**, 745.

HARD X-RAYS FROM THE SOUTHERN SKY

WALTER H. G. LEWIN, JEFFREY E. McCLINTOCK and WILLIAM B. SMITH

Massachusetts Institute of Technology, Cambridge, Mass., U.S.A.

Abstract. On March 20, 1969, we scanned the Southern sky for 8 hours in a manner similar to that during our October 15, 1967 observations.* A quick look at the data shows clearly that the brightness of Cen XR-2 in the energy range above 20 keV has significantly decreased since our first observation. At this stage of the data analysis we can put an upper limit on the intensity of Cen XR-2 which is $\frac{1}{3}$ the intensity observed by us on October 15, 1967. From further data analyses we will be able to determine the intensity of the source more precisely.**

During the same flight we observed Sco X-1, Nor XR-2, and the sources M-1, M-2, M-3, and M-4.‡ Results on these observations will become available by May 1970.

---

* Lewin, W. H. G., Clark, G. W., and Smith, W. B.: 1968, *Astrophys. J.* **152**, L49.
** The final results on Cen XR-2 will be available in Febr. 1970; they will be published shortly thereafter.
  ‡ Lewin, W. H. G., Clark, G. W., Gerassimenko, M., and Smith, W. B.: 1969, *Nature* **223**, 1142.

X-RAY POLARIZATION FROM SCO X-1 AND TAU X-1

R. NOVICK

Columbia University, New York, N.Y.

No text nor Summary was communicated by the Author.

IONOSPHERIC EFFECTS OF X-RAYS FROM
DISCRETE GALACTIC SOURCES*

S. ANANTHAKRISHNAN, S. C. CHAKRAVARTY and K. R. RAMANATHAN

Physical Research Laboratory, Ahmedabad, India

Abstract. Observation of the field strength of low frequency radio waves (164 kHz) transmitted during night hours from Tashkent and received at Ahmedabad show increased absorption around the sidereal times of the transit of the X ray stars Sco X-1 and Tau X-1. It is estimated that the ionization in the D region produced by these X-ray stars can explain the observed changes in field strength.

1. Introduction

Since the discovery of discrete X-ray sources in the galaxy about 7 years ago by Dr. Giacconi and his collaborators, numerous measurements have been carried out, particularly by the Naval Research Laboratory and the American Science and Engineering Co. of the U.S.A., to determine their exact location, angular diameter and spectral features. Of particular interest among the 40 or more sources discovered so far, are Sco X-1 located in the constellation Scorpio (Right accension $\alpha = 16$ h 15 m, Declination $\delta = -15°5$), Tau X-1 in the Crab nebula ($\alpha = 5$ h 31 m, $\delta = +22°1$), Cygnus X-1 ($\alpha = 19$ h 53 m, $\delta = +34°5$) and Centaurus X-2 ($\alpha = 13$ h 24 m, $\delta = -62°$). The present paper reports the results of an investigation to find out the effect on low frequency radio waves reflected from the upper atmosphere by X-rays from Sco X-1 and Tau X-1.

2. Low-frequency Recording Equipment at Ahmedabad

Since 1960, we have been recording at Ahmedabad the field strength of 164 kHz rádio waves transmitted from Tashkent (42 °N, 69 °E). The radiated power is in the neighbourhood of 150 kw and the transmission is on the air for approximately 20 hours a day. The receiving equipment is very simple and consists of a loop aerial, or a Wave Antenna terminated in the direction of the transmitter, feeding into a communication receiver, the detected output of which is amplified and fed to a strip chart recorder. Frequent calibrations are carried out with the aid of a standard signal generator. Care is taken to maintain the stable operation of the receiver. Basically, the instrument measures the variations in the amplitude of the radio waves caused by its passage through the atmosphere below the reflecting level. The normal height of reflection is in the D region round about 70–75 km, during the day and 85–90 km during the night. This region is ionised by X-rays from the sun, particularly during solar flares.

* Communicated by professor U. R. Rao.

L. Gratton (ed.), Non-Solar X- and Gamma-Ray Astronomy, 146–150. All Rights Reserved

3. Solar X-Rays and the D-region

It is well known that solar flares cause marked increases of ionization in the D-region resulting in short-wave fade-outs at broadcast frequencies and sudden anomalies in field strength of radio waves of long and very long wave lengths (30–300 kHz and 3–30 kHz). The work of Burnight (1948) and of Friedman's group at the Naval Research Laboratory clearly established that solar X-rays were primarily responsible for these effects. It is now known that ionization in the height range 65–90 km is very sensitive to X-radiation of wavelengths shorter than 10 Å (Labeyrie, 1968).

4. Effect of X-Ray Fluxes on the 164 kHz Field Strengths at Ahmedabad

Before describing the effects produced by the galactic X-ray sources, some details of the propagation path may be considered. The great circle distance between the transmitter at Tashkent and receiver at Ahmedabad is 2150 km, the single hop reflection point being located at geographic latitude 32°N, longitude 71°E. The radio waves

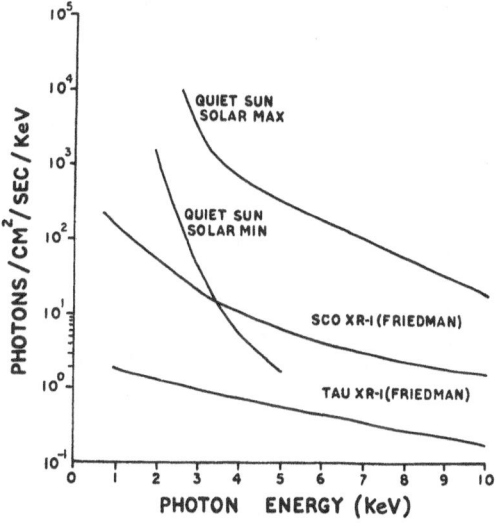

Fig. 1.   Comparison of non-flare solar X-ray spectra (sunspot maximum and minimum) with measured X-ray spectra of Sco X-1 and Tau X-1.

from Tashkent may arrive at Ahmedabad after a single oblique reflection in the ionosphere (single hop) or after two reflections in the ionosphere and one at the ground (two-hop) or even after more hops. In general the one-hop mode predominates up to distances of about 2500 km from the transmitter. The equivalent critical frequency for vertical incidence reflection corresponding to the oblique one-hop transmission frequency of 164 kHz is about 29 kHz.

An examination of the coordinates of the four sources, Sco X-1, Tau X-1, Cyg X-1 and Cen X-2, showed that the first three were favourably situated with respect to the reflection point to affect the field strengths at Ahmedabad. In the succeeding paragraphs the changes in the field strength produced by the transit of Sco X-1 and Tau X-1 across the reflection meridian will be discussed. A brief report of the ionospheric effects of X-rays from Sco X-1 is under publication (Ananthakrishnan and Ramanathan, 1969).

The records of nighttime field strengths (between 21 hr local time and 04 hr local time (were scrutinized. During daytime, it would be difficult to separate the effect of

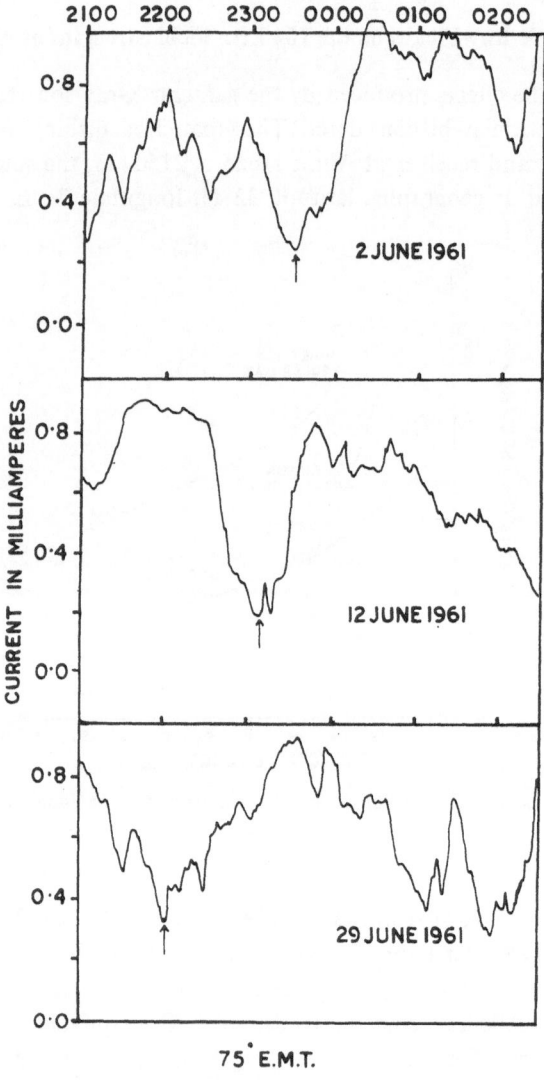

Fig. 2.   Night-time field strength of Tashkent (164 kHz) radio transmissions received at Ahmedabad.
Arrows indicate meridian transit of Sco X-1.

stellar X-rays from the effect of X-rays from the sun. The X-ray spectra of Sco X-1, Tau X-1 and of the quiet sun in sunspot maximum and minimum conditions are compared in Figure 1. The spectrum of the sun during sunspot minimum has been measured by Bordeau *et al.* and is taken from Whitten and Poppoff (1965). The spectrum for solar maximum is taken from an article by Labeyrie (1968). It may be noted that in sunspot minimum conditions, the flux of radiation from Sco X-1 can be greater than that from the sun in the high energy region above 4 keV. Calculations of the increase of electron density produced by the X-ray flux from Scorpio in the height range 70–90 km with currently available loss coefficients, show that the increase would be sufficient to produce a noticeable effect on reflected LF and VLF radio waves.

A simple calculation shows that Sco X-1 would transit over the reflection meridian at about midnight in June of each year. Examination of the field strength records for this period showed that there were significant decreases in the amplitude of the 164 kHz field strengths associated with the transmit. An example is shown in Figure 2. The arrows in the figure indicate the times at which the source reaches its maximum altitude at 32°N, 71°E.

The sidereal shift of about $1\frac{3}{4}$ hr between June 2 and June 29 can clearly be seen. In addition to the pronounced minimum associated with the transit of Sco X-1, indicated by the arrows, other minima of sporadic nature are occasionally observed on the records.

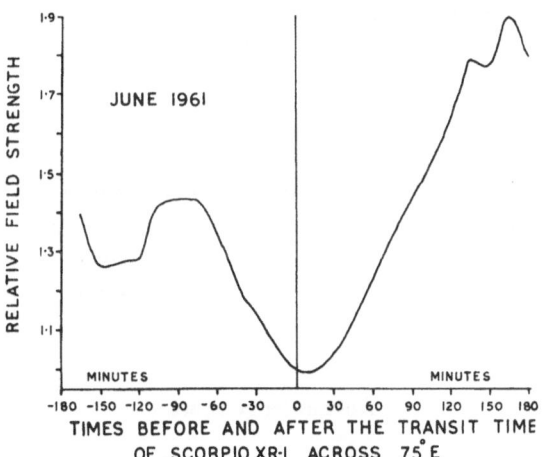

Fig. 3.   Decrease in amplitude produced by the X-radiations from Sco X-1 as brought out by superposed epoch analysis.

Figure 3 shows the result of a superposed epoch analysis carried out for the period May–June 1961. The diagram brings out clearly the decrease in amplitude of the signal strength produced by Sco X-1.

Figure 4 shows the result of a similar analysis carried out for Tau X-1. The reduction in amplitude, though clearly evident, is much weaker. The flux of radiation from this source is about an order of magnitude lower, in the energy range of interest.

5. Conclusion

The results discussed in Section 3 show that both Sco X-1 and Tau X-1 produce an increase in absorption, which means a change in ionization in the night-time lower ionosphere, of a magnitude large enough to affect the amplitude of low frequency radio waves traversing this region. A few remarks may be made by way of conclusion.

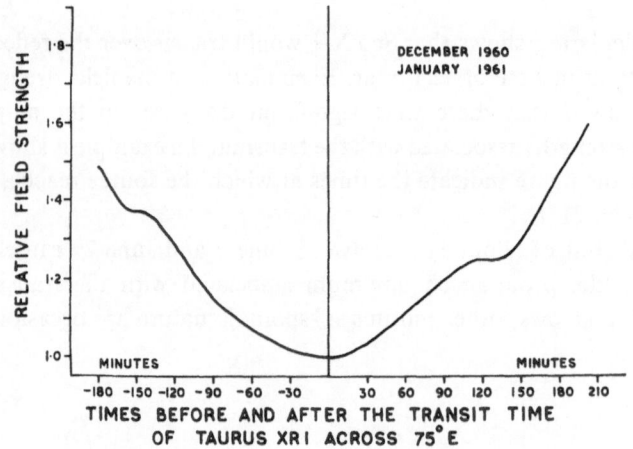

Fig. 4.   Decrease in amplitude produced by the X-radiations from Tau X-1 as brought out by superposed epoch analysis.

The inability of rockets to carry out observations for more than a few minutes and the inaccessibility to balloons of D region levels where the softer X-rays are absorbed, show the importance of ground-based observations which can show up the effect of X-ray fluxes from the sun, stars and galaxies. We can look back into the earlier low-frequency records to see if any time-variations in the fluxes can be found.

Acknowledgements

Our thanks are due to Dr. U. R. Rao and Dr. J. S. Shirke for helpful discussions.

References

Ananthakrishnan, S. and Ramanathan, K. R.: 1969, *Nature* (in press).
Burnight, T. R.: 1949, *Phys. Rev.* **76**, 165.
Labeyrie, J.: 1968, in *Electromagnetic Radiation in Space* (ed. by J. G. Emming), D. Reidel Publ. Co., Dordrecht, Holland, p. 125.
Whitten, R. C. and Poppoff, I. G.: 1965, *Physics of the Lower Ionosphere*, Prentice-Hall, Inc., p. 30.

NON-SOLAR GAMMA AND X-RAY ASTRONOMY:
OPTICAL OBSERVATIONS*

(Invited Discourse)

HUGH M. JOHNSON**

Lockheed Missiles and Space Company, Palo Alto, Calif., U.S.A.,
and Cerro Tololo Inter-American Observatory‡

An optical astronomer enters this field only by courtesy of those X-ray astronomers who pay some attention to accurate positional measurements of X-ray sources. So my first and last words are to ask X-ray observers to give more time to establishing positions of X-ray sources. It appears that in fact most effort has been spent on spectral measurements of X-rays, and this has led just to the classification of sources according to either of two mechanisms for the production of the continuum. In one or two early instances the extrapolated X-ray spectrum has been useful for predicting the brightness of the optical counterpart to be found. A typical uncertainty of making optical identification is that of Vel XR-1 for which Gursky *et al.* (1968) have given a position with an error box of 3 square degrees. One candidate I can suggest for this is CU Vel, the only variable star of the 1958 *General Catalogue of Variable Stars* inside the error box. It is interesting because it is assigned to the U Gem class with a range of photographic magnitudes from 10.7 to 15.5. The stellar spectrum has not been observed.

The optical light of identified sources may arise in another volume than the very energetic plasma which emits the X-rays. But we can hardly doubt that the various parts of an X-ray source interact. In this sense X-ray sources were first known to be certainly variable as soon as Sco X-1 was identified with a variable star. Much effort has been spent to observe the optical variability of Sco X-1 independently of X-ray observations. Periodicity has been especially sought, for it would lead to a great advance in understanding the source whether the periodic light curve arose from orbital or pulsational modes. Optical tests for periods in the range from 2 sec to many hours have been made on Sco X-1. Ephemeral or quasi-periodical effects have been suspected, e.g. at 0.5276 day (Van Genderen, 1969).

It is still impossible to conclude on the basis of photometric observations that either Sco X-1 or any other X-ray source is a binary or a pulsating star. However, the only known optical pulsar has been found in the Crab nebula, and the pulse is present in the X-radiation. It becomes interesting to look for short periods in other X-ray sources, although we have no guiding theory as to whether most X-ray sources can develop or maintain pulsar characteristics. J. C. Golson and I at Kitt Peak National Observatory have reached a preliminary negative result against periodicities down to

---

* Contributions from the Cerro Tololo Inter-American Observatory, No. 75.
** Visiting astronomer, 1969, Cerro Tololo Inter-American Observatory.
‡ Operated by the Association of Universities for Research in Astronomy, Inc., under contract with the National Science Foundation.

L. Gratton (ed.), Non-Solar X- and Gamma-Ray Astronomy, 151–161. All Rights Reserved
Copyright © 1970 by the IAU

0.002 sec for Sco X-1. To do this we have used a stroboscope which interrupts the beam of a star into a conventional d.c. photometer. The stroboscope was constructed with the aid of A. J. Gardiner at the Kitt Peak National Observatory in accordance with our rough plan. The stroboscopic frequencies can be increased steadily from zero to 500 sec^{-1} for several minutes. We have used u, b, v, y filters as well as no filters over the 1P21 photomultiplier. The recorder graph of the amplified current output on March 27 and April 12, 1969, shows about the same noise whether the stroboscope is running or the beam is passed without interruption (Figure 1). Laboratory tests of

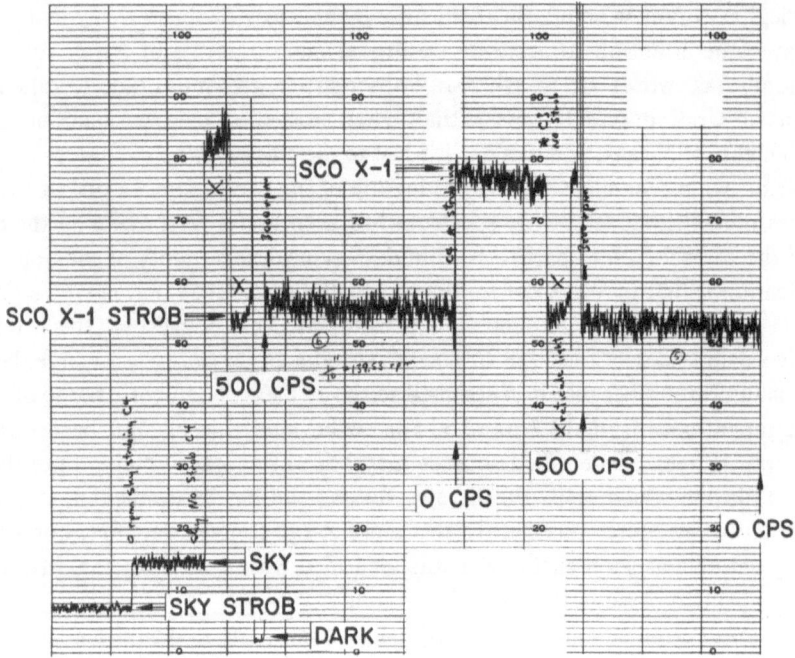

Fig. 1.   Recorder chart from April 12, 1969, of the d.c. output of an unfiltered 1P21 photomultiplier, showing response to light of Sco X-1 governed by a stroboscope running at pass-stop frequencies which rise steadily from zero to 500 sec^{-1}. No resonances are apparent.

the stroboscope on a.c. lamps confirm during similar frequency scans that, when the stroboscopic period or multiple of it approximates to the period of the lamp, the amplitude of the recorder graph becomes very great compared with random-noise fluctuations. Although we do not see such resonances in any of the records of Sco X-1, we can refine this first application and also extend the technique to some other objects.

Narrow-band photometry can concentrate on selected lines in the optical spectra of identified X-ray sources. Figures 2, 3, and 4 show narrow-band observations of WX Cen, the nucleus of NGC 5189, and the object called GX3+1. The first two of these three objects are rival candidates for the identification of Cen XR-2. The pass-bands are specified in Figure 5 and Table I. The 60-inch reflector of the Cerro Tololo Inter-American Observatory was used on April 8–12, 1969, UT, and the 36-inch on

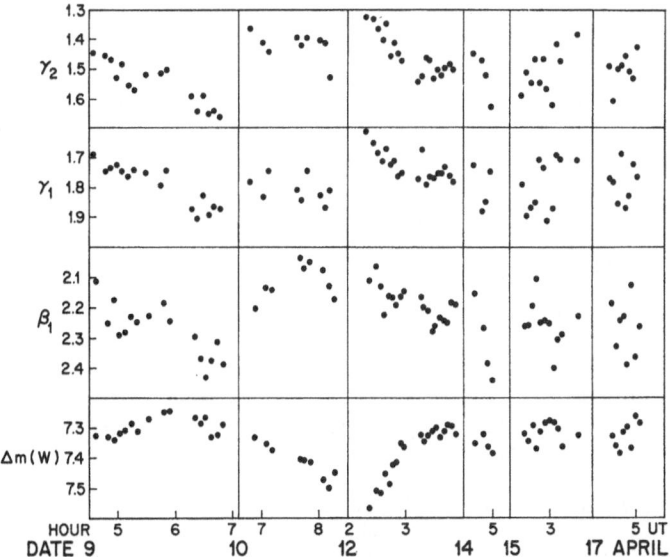

Fig. 2.   Light curves of WX Cen in $\Delta m(W) = m$(wide Hβ) (object) – m (wide Hβ) (comparison star), in $\beta_1 = m$ (narrow Hβ) – m (wide Hβ), in $\gamma_1 = m$ (narrow $\lambda 4648$) – m (wide $\lambda 4694$), and in $\gamma_2 = m$ (narrow $\lambda 4684$) – m (wide $\lambda 4694$). The comparison star is HD 114461, type F0, $m_{pg} = 6.9$.

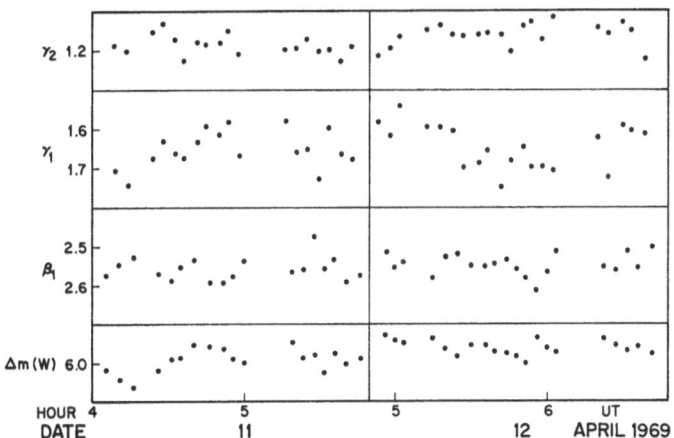

Fig. 3.   Light curves of the central star of NGC 5189 as in Figure 2. The comparison star is HD 117694, type B9, $m_{pg} = 8.8$.

April 13–17, for the work. The integration time per filter was always 20 sec except on April 9. The corresponding start-to-stop times were 20 sec for $m(W)$, 85 sec for β_1 and γ_2, and 130 sec for γ_1, where β_1, γ_1, and γ_2 are composed of pairs of integrations with narrower-band filters centered on single wider-band integrations. On April 9 the integration time per filter was 30 sec and start-to-stop times were 30 sec for $m(W)$, 110 sec for β_1 and γ_2, and 180 sec for γ_1. The reductions assume a linear change of the light between the first and last integrations of each plotted datum. In short, the

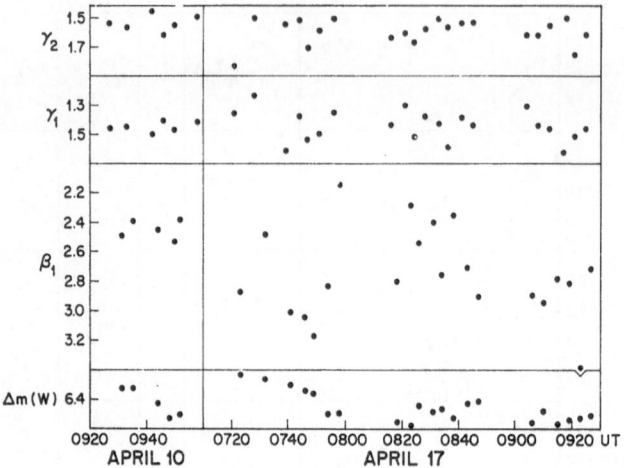

Fig. 4. Light curves of GX3 + 1 as in Figure 2. The comparison star is HD 160972, type A5, $m_{pg} = 8.6$.

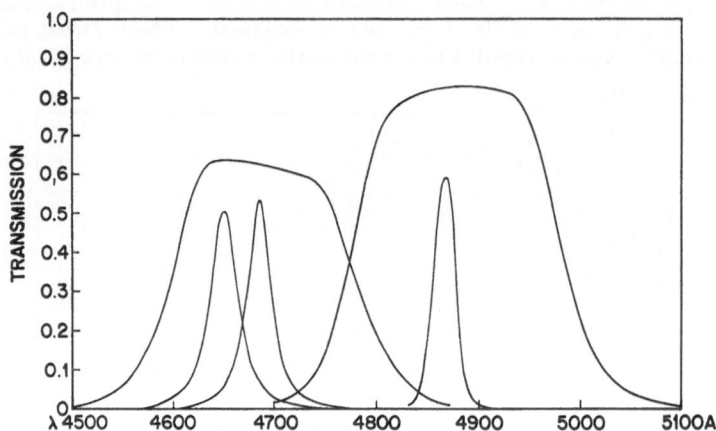

Fig. 5. Passbands used to observe X-ray sources at Cerro Tololo Inter-American Observatory in April, 1969. The wide and narrow bands around Hβ give, respectively, $m(W)$ and $m(N)$, while $\beta_1 = m(N) - m(W)$ is a measure of the strength of Hβ. According to the areas under these curves $\beta_1(0) = 2.531$ magnitude corresponds to null Hβ. The other narrow bands transmit CIII, NIII 4640 and HeII 4686, respectively, while the wide band over them serves with them to give an index γ_1 of the equivalent width of λ 4640 and an index γ_2 of the equivalent width of λ 4686. The laboratory value for null λ 4640 is $\gamma_1(0) = 1.812$ magnitude, and for null λ 4686 it is $\gamma_2(0) = 1.954$ magnitude.

methods are much the same as in the first of a series of narrow-band photometry by Johnson and Golson (1968a). The filter set used at Cerro Tololo is different from the set used in the beginning of the series, but the significance of $\Delta m(W)$ is very nearly the same as it was then. The Hβ index differs from the β' of previous reports simply in scale and zero-point; γ_1 and γ_2 distinguish narrow emissions around λ4640 and λ4686 which were passed together under a filter which we previously had used for them.

TABLE I

Photometric passband characteristics of equipment used at
the Cerro Tololo Inter-American Observatory

Filters				Spectral features
No.	λ_{eff}(A)	Equivalent width (A)	Transmission at lines	
387	4648	22.8	0.42	NIII 4634–41, CIII 4647–51
406	4684	20.0	0.51	HeII 4686
407	4694	121	0.63	NIII, CIII, HeII
217	4865	17.4	0.57	$H\beta$ (narrow)
222	4880	179	0.82	$H\beta$ (wide)

The separation is incomplete, of course, for very broad and blended $\lambda\lambda$4640–4686. The rms errors in this work may be estimated from the rms error of the β_1 measurements of the comparison stars of the three objects to be discussed. This error is ± 0.007 mag.

The principal features of the blue spectrum of WX Cen are $H\beta$, HeII 4686, and NIII, CIII 4640. The lines are broadened but the λ 4640 feature is partially resolved from the dominant HeII 4686 (Eggen *et al.*, 1968). According to the Cerro Tololo photometry there is on some, but not on all, days a correlation between the continuum as represented by $\Delta m(W)$ and the line indices in the sense that, as the continuum brightens, the equivalent width of any of the observed emission features decreases. Keeping all of the data there is apparently a positive correlation between γ_1 and γ_2, and between β_1 and γ_2. The index γ_1 is well correlated with γ_2 and β_1 on April 9 and April 12, but it has distinctly dropped in relative brightness on the 10th. The times of the β_1 data lag about 3 min behind the times of the γ_1, γ_2 data, but this may have little effect within the time scale of the major variability. WX Cen certainly varies within an hour and probably within a few minutes, but 'flaring' is not clearly present. These data could result from a volume of slightly variable line fluxes in the presence of a more highly variable continuum flux.

The blue spectrum of the central star of NGC 5189 apparently shows very strong, broad emissions of OVI 3811–34, and HeII probably partially blended with CIII, NIII 4640 (Blanco *et al.*, 1968). $H\beta$ is not contained on the published spectrogram. The present narrow-band photometry of the central star was done with a diaphragm of 10″ diameter. If the light of the nebula is distributed in the photographic region like that of the central star, and also uniformly over its surface, about 15% of the total light passed by the diaphragm is estimated to be nebular. However, the relative contribution of the nebular lines in the filters of Table I is probably much less than the estimate suggests, and the light curves of Figure 3 should pertain quite well to the central star. There is some correlation between $\Delta m(W)$ and β_1 of the same sense as in WX Cen, and the quality of variability in these light curves resembles that of WX Cen. However, $\langle \beta_1 \rangle > \beta_1(0) = 2.531$ magnitude (the value of computed null equivalent width of $H\beta$), and this implies net absorption rather than net emission close to Hβ. A similar effect in GX3+1 is mentioned below. Also, the variations of γ_1 and γ_2 are definitely not so well correlated in the central star of NGC 5189 as the variations of these indices are in WX Cen. The mean values of γ_1 and γ_2 show that the complex of emissions around $\lambda\lambda$ 4640–4686 has greater equivalent width in the spectrum of the central star of NGC 5189 than in the spectrum of WX Cen.

The spectrum of the ultraviolet object in Sagittarius, identified with the X-ray source GX3+1, is similar to that of the central star of NGC 5189 but with even greater broadening and blending of emission features (Blanco *et al.*, 1968; Freeman *et al.*, 1968). Figure 4 shows variability of GX3+1 in $\Delta m(W)$ much as in an earlier report (Johnson and Golson, 1968b), and β_1 again goes through a large and erratic dispersion as did β' in 1968. The mean of β_1 mimics an absorption-line at Hβ. As in 1968, this may represent the net gap between broad emission wings of He II 4686 and O v 4925–40. Emissions of probably C III, N III, and He II blend and spread over our narrow passbands (and indeed fill much of the wider passband of the system) which make up the indices γ_1 and γ_2. These gross effects on the passbands must alter the meaning of the absolute values of γ_1 and γ_2 from the meaning they would have for narrow lines. We can still say that their magnitudes show that the mean ratio of the equivalent widths of λ 4640/λ 4686 is greater in GX3+1 than it is in the central star of NGC 5189; again the variability of γ_1 and γ_2 is apparently uncorrelated.

If the emitting volumes of X-rays and optical radiation interact in X-ray sources, we ought to look for correlations in the variability by taking simultaneous data in two or more *wide-spaced* passbands. The most spectacular of the light variations of Sco X-1, the so-called 'flares', have escaped such inspection. But on four instants measurements of flux density in the range 2–20 keV have been made from rockets while a ground-based astronomer cooperatively observed Sco X-1 in *UBV* (Mark *et al.*, 1969). The four pairs of flux-density data have been analyzed by these authors to derive some physical properties of Sco X-1 which could not be approached by means of X-ray data alone, namely plasma volume radius (about 10^9 cm) and density (about 10^{16} cm^{-3}), but the correlation between variations of radius, density, and temperature is not clear from these data. Sco X-1 has been observed on May 22, 1968, with balloon-borne detectors working at 20–40 keV, and simultaneously with the 84-inch reflector at Kitt Peak National Observatory (Pelling *et al.*, 1969). The X-ray flux then appeared to vary less than a 25% slow rise and fall of the continuum around Hβ during a period of several hours. At the same time Hβ emission and the emissions around $\lambda\lambda$ 4640–4686 also varied little compared with variation in the continuum. A full report will be published elsewhere. Sco X-1 has also been observed by B. H. Andrew and C. R. Purton at the Algonquin Radio Observatory at 4.6 cm for several hours on April 14, 1969, and by me simultaneously at Cerro Tololo Inter-American Observatory in a first attempt to correlate radio and optical variability. The data are not reduced in time for this Symposium.

In short, simultaneous X-ray/optical/radio photometry of variable X-ray sources has been undertaken only for Sco X-1, and the data still cannot all be correlated in terms of simple models. The characteristics of *UBV* (broadband) variability of Sco X-1 are sufficient to suggest that the optical continuum of Sco X-1 is optically thick (Johnson, 1968a). The slope of the spectrum of Sco X-1 from the visible to 2.2 μ also suggests an infrared-dominant blackbody in Sco X-1 (Neugebauer *et al.*, 1969). These results agree with the models of Mark *et al.* (1969), the ones derived partly from X-ray data. But the radiofrequency data of Andrew and Purton (1968) and Ables (1969) cannot be accommodated in such pictures of Sco X-1. The radio data may require a basically two-component model, of which the one contributes soft X-rays and the other identifies perhaps with the optical line-emitting cool-gas component (Riegler and Ramaty, 1969).

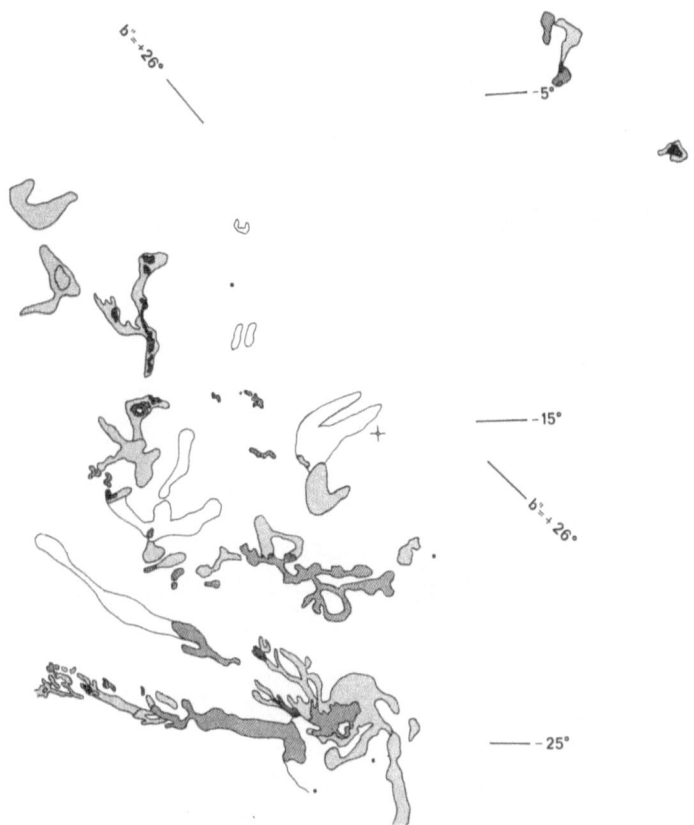

Fig. 6. Dark nebulae around Sco X-1 (cross mark). The darkest clouds are given the blackest shading. Compare with Lynds (1962) for a map of the larger area which shows the unusual aspect of these clouds in relation to the general obscuration in the Milky Way.

No *characteristic* X-rays have been found in cosmic sources other than the sun. Thus the optical line spectrum of X-ray sources is the only present means of making radial-velocity measurements or abundance analyses of them. Not much progress has actually been made because of the complex behavior of the few available radial velocities of Sco X-1 and Cyg X-2, the very great breadth of the lines in some other sources, and, in respect to standard abundance analyses, doubts about applicability to such anomalous objects. We can only reinforce the negative photometric conclusion on the subject of periodic binary or pulsational motions, and note the weakness or absence of Balmer lines as compared with the strength of helium in some sources such as GX3+1 which are like Wolf-Rayet stars.

All model-making starts with the distance of a source. Because the earliest proper-

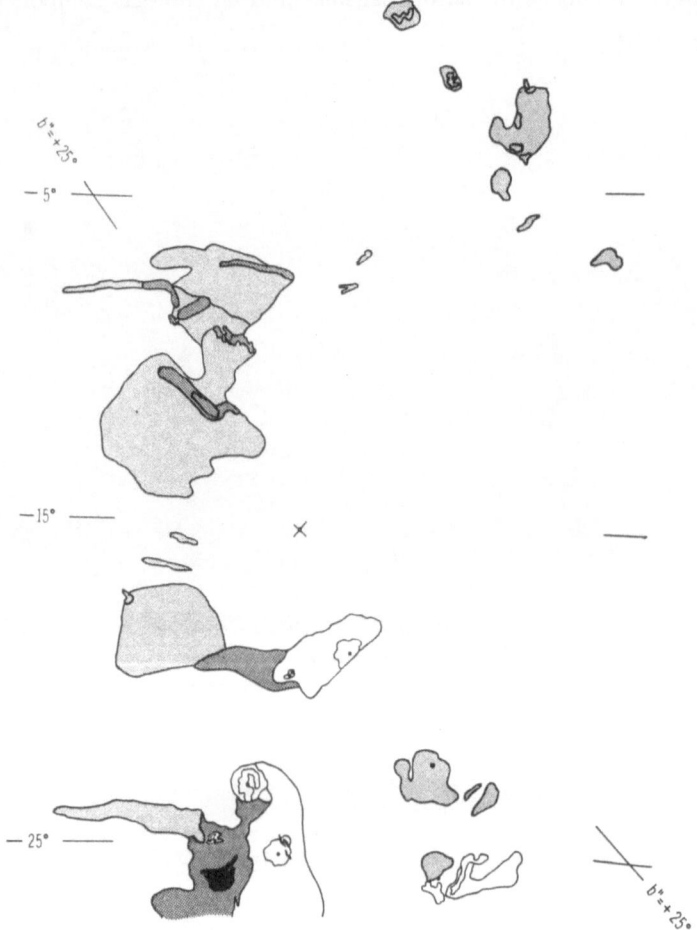

Fig. 7. Luminous nebulae around Sco X-1 (cross mark). The nebulae which are brighter on the blue than on the red Palomar Sky Survey are unshaded. Nebulae that appear red are shaded; the degree of shading is supposed to represent the brightness of the nebulae.

motion measurements of Sco X-1 (Johnson and Stephenson, 1966; Luyten, 1966) included mean errors of about $\pm 0\overset{''}{.}017$ – larger than the proper motions – it has been concluded by Luyten and later by others that Sco X-1 is very distant. Now Sofia *et al.* (1969) have determined the proper motion of Sco X-1 with greater accuracy, i.e. with mean errors of $\pm 0.''0022$, and the proper motion agrees with the proper motions of members of the Sco-Cen association. The conclusion is that Sco X-1 is most probably at the distance of the association, 170 psc according to Bertiau (1958), and that it belongs to a strong population I group. Several years ago I suggested (Johnson, 1966) that Sco X-1 is in the Sco-Cen association because it appeared to be centered in arc-like distortions of the interstellar dust clouds which Lynds (1962) had mapped before the discovery of Sco X-1. If Sco X-1 is immersed in interstellar clouds, we must ask whether the current radiation of Sco X-1 affects them. This inquiry may be applied to the strength of interstellar Ca II K, which Wallerstein (1967) has found to be unusually great in the spectra of Sco X-1 and in HD 146935, a star separated 10' from Sco X-1 on the sky, in comparison with the strength of Ca II K in stars somewhat farther from Sco X-1. However, here we ask whether any of the red interstellar clouds in the vicinity

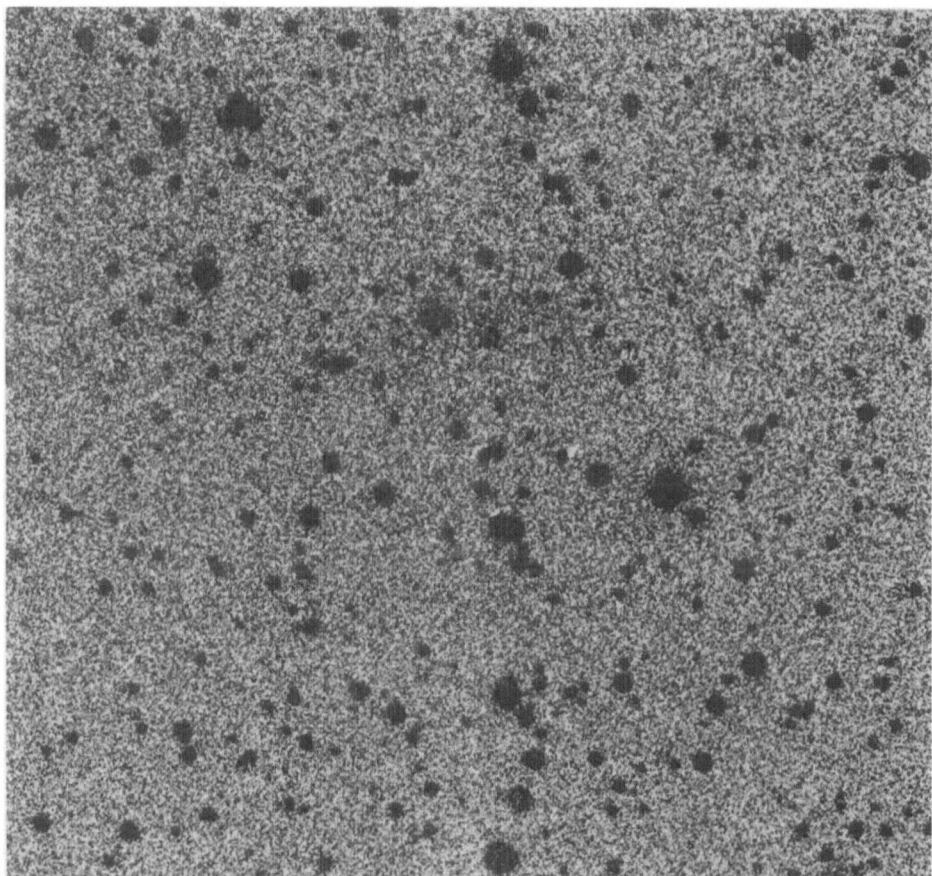

Fig. 8. An area of 8' × 9' centered on Cas A. A diffuse arc of radius 2' may be seen in the northern quadrant above center (East to the left). It is apparently the optical continuum of the supernova remnant in the passband $\lambda\lambda$ 5200–6300. The high-contrast print on Kodalith Ortho paper was made from a sandwich of Kodalith Ortho film copies of two plates exposed in the 48-inch Schmidt telescope on May 6 and 7, 1967.

of Sco X-1, found on the Palomar Sky Survey and catalogued by Mrs. Lynds (1965), are radiating in Hα or are scattering the light of a red star. This is a future observational problem. We must also discover whether the radiation of Sco X-1 contributes to the photoionization of the neighboring red clouds. Ultimately new information about the spectrum of Sco X-1 below 912 Å may be obtained starting from its site in interstellar matter. Figures 6 and 7, respectively, are maps of dark and luminous clouds near Sco X-1 prepared by Mrs. Lynds. Regarding the history of interstellar matter around Sco X-1, I note that the proper motion of ζ Oph is roughly equal in magnitude and opposite to that of Sco X-1. ζ Oph is a 'runaway' O star in the Sco-Cen association. The high speed may reflect a supernova event (Blaauw, 1961).

Optical observations of known supernova remnants may be as relevant to X-ray astronomy as to radio astronomy, but the material is too extensive for review here. The principal thing to note is that parts of supernova remnants may fluoresce in the optical band because they are supplied with photons below 912 Å. Optical variability may lead to information about the stability of XUV sources in supernova remnants. The optical continuum of the Crab nebula is observed to be much stronger than possible optical continua in other supernova remnants. Figure 8 illustrates an attempt to photograph the yellow continuum of the Cas A synchrotron continuum (Johnson, 1968b). A photoelectric observation by J. C. Golson and me at Kitt Peak National Observatory in November, 1968, apparently confirms this continuum. The observation is the first epoch in a series intended to monitor long-term variability of [O III] 5007 and the continuum in Cas A. The Crab nebula is similarly monitored. The whole subject of the effects of XUV-source radiation on the interstellar environment deserves theoretical and observational effort.

Acknowledgements

This work was supported by the Lockheed Independent Research Program and by the Office of Naval Research under contract No. N00014-69-C-0147.

References

Ables, J. G.: 1969, *Astrophys. J. (Letters)* **155**, L27.
Andrew, B. H. and Purton, C. R.: 1968, *Nature* **218**, 855.
Bertiau, F. C.: 1958, *Astrophys. J.* **128**, 533.
Blaauw, A.: 1961, *Bull. Astron. Inst. Netherl.* **15**, 265.
Blanco, V., Kunkel, W., Hiltner, W. A., Chodil, G., Mark, H., Rodrigues, R., Seward, F., and Swift, C. D.: 1968a, *Astrophys. J. (Letters)* **152**, L135.
Blanco, V., Kunkel, W., and Hiltner, W. A.: 1968b, *Astrophys. J. (Letters)* **152**, L137.
Eggen, O. J., Freeman, K. C., and Sandage, A.: 1968, *Astrophys. J. (Letters)* **154**, L27.
Freeman, K. C., Rodgers, A. W., and Lyngå, G.: 1968, *Nature* **219**, 251.
Genderen, A. M. van: 1969, *Astron. and Astrophys.* **2**, 6.
Gursky, H., Kellogg, E. M., and Gorenstein, P.: 1968, *Astrophys. J. (Letters)* **154**, L71.
Johnson, H. M.: 1966, *Astrophys. J.* **144**, 635.
Johnson, H. M.: 1968a, *Astrophys. J.* **154**, 1139.
Johnson, H. M.: 1968b, *Canadian J. Phys.* **46**, S481.
Johnson, H. M. and Golson, J. C.: 1968a, *Astrophys. J.* **153**, 307.

Johnson, H. M. and Golson, J. C.: 1968b, *Astrophys. J. (Letters)* **154**, L77.
Johnson, H. M. and Stephenson, C. B.: 1966, *Astrophys. J.* **146**, 602.
Luyten, W. J.: 1966, *I.A.U. Circ.* No. 1980.
Lynds, B. T.: 1962, *Astrophys. J. Suppl.* **7**, 1.
Lynds, B. T.: 1965, *Astrophys. J. Suppl.* **12**, 163.
Mark, H., Price, R. E., Rodrigues, R., Seward, F. D., Swift, C. D., and Hiltner, W. A.: 1969, *Astrophys. J. (Letters)* **156**, L67.
Neugebauer, G., Oke, J. B., Becklin, E., and Garmire, G.: 1969, *Astrophys. J.* **155**, 1.
Pelling, R. M., Matteson, J. L., Peterson, L. E., Johnson, H. M., and Golson, J. C.: 1969, 129th AAS meeting (Honolulu).
Riegler, G. R. and Ramaty, R.: 1969, *Astrophys. Letters* **4**, 27.
Sofia, S., Eichhorn, H., and Gatewood, G.: 1969, *Astron J.* **74**, 20.
Wallerstein, G.: 1967, *Astrophys. Letters* **1**, 31.

SPECTROSCOPIC AND STATISTICAL PROPERTIES OF
X-RAY SOURCES*

(Invited Discourse)

LIVIO GRATTON

University of Rome, Italy

1. Introduction

Among the little less than 50 known X-ray sources, only 7 had been identified with optical objects at the end of 1968 (Gratton, 1968) and a couple or so have been added during the present Symposium; some identifications are still uncertain.

It is remarkable that, apart from the case of Supernova remnants, one of which was the first X-ray source to be identified, all other objects which have been predicted as likely sources of X-rays have never been observed as such. This is by no means intended to discourage theoretical predictions like those contained, for instance, in a recent paper by Biermann (1969). Since X-rays from the sun have been observed, it is obvious to assume that many more stars will emit more or less intense X-ray fluxes, and it is quite reasonable to look first at those which are considered more likely. But it is clear that all predictions must be regarded as very uncertain until we do not understand better the nature of the most powerful sources. Hence the discussion must be necessarily limited to those sources for which direct observational evidence is available.

The present evidence shows quite clearly that these sources belong to (at least) three different categories.

(a) Galactic extended objects.

(b) Galactic star-like objects.

(c) Extragalactic objects.

Only one object of class (c) is at present known with a reasonable certainty (the radiogalaxy M87) and this class will not be discussed here.

Objects of class (a) usually have an X-ray spectrum with a strong tail towards the high energies; two of them have been known since some time and one more has been added during this Symposium by the ASE group (Gorenstein *et al.*, 1969); the position

---

* According to the original program a discussion of the spectral properties of the X-ray sources in the optical region was to be presented at the Symposium by E. M. Burbidge; the Organizing Committee invited also A. Sandage to discuss optical identifications, relationships with known optical objects and other astrophysical problems. Unfortunately Sandage was unable to attend the Symposium and a few weeks ago E. M. Burbidge wrote that she also could not come.

Therefore the present paper was prepared rather hurriedly to cover these points, mainly with the purpose of introducing the discussion on them. The author asks to be excused for its many deficiencies and hopes that the optical astronomers present at the meeting will make some compensation and perhaps supply more new material and subject for discussion.

L. Gratton (ed.), Non-Solar X- and Gamma-Ray Astronomy, 162–172. All Rights Reserved
Copyright © 1970 by the IAU

of the latter coincides convincingly with Tycho's supernova of 1572. There is therefore little doubt that objects of this class are remnants of supernova explosions. It is quite possible that all supernova remnants would show X-ray emission if observed with detectors of sufficient sensitivity, although of course this cannot be proved.

Objects of class (b) have a much steeper X-ray spectrum than those of class (a). In the high energy region Sco X-1 – the prototype of class (b) objects – is fainter than the Crab nebula, although the reverse is true in the 1–10 keV domain. The four objects for which an identification exists, Sco X-1, Cyg X-2, GX 3+1, Cen XR-2 (the last two objects are still doubtful), have optical counterparts of star-like appearance showing peculiar light variations and spectral features strongly resembling those of ordinary Novae at minimum light.

In the rest of this paper I will shortly summarize the published results on the spectra of the four objects assigned to class (b); a short discussion will follow concerning the evidence which can be obtained from the distribution of known X-ray sources on the celestial sphere.

2. Optical Spectra of the Star-like X-Ray Sources (Class b)

1. Scorpio X-1

The identification of this source (Sandage et al., 1966) on the basis of the accurate position obtained by the ASE and MIT groups (Gursky et al., 1966) is beyond doubt. Observations of the spectrum have been published by Sandage et al. (1966), Ichimura et al. (1966), Jugaku (Babcock, 1967), Wallerstein (1967), Westphal et al. (1968).

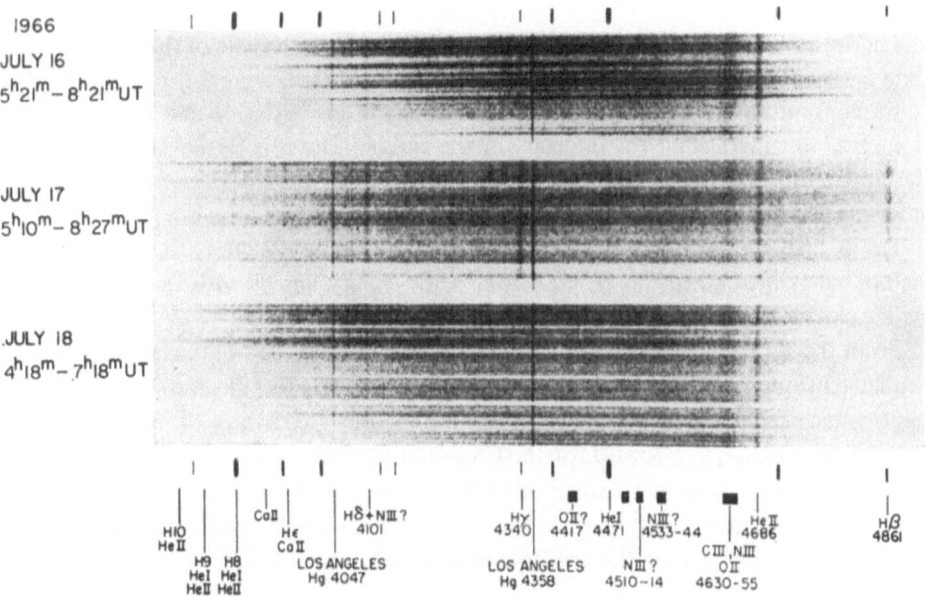

Fig. 1.   Spectra of Sco X-1 (from Sandage et al., 1966).

Figure 1 is from Sandage *et al.* paper of 1966 and Figure 2 from that by Westphal *et al.* (1968); on the latter various identified emission lines are marked. They correspond to the Balmer series up to H_{12}, to HeI, HeII (the line 4686 is the strongest in the whole spectrum), OII, NIII. The identification of the lines marked FeII is considered as uncertain by the authors. No forbidden lines were convincingly identified.

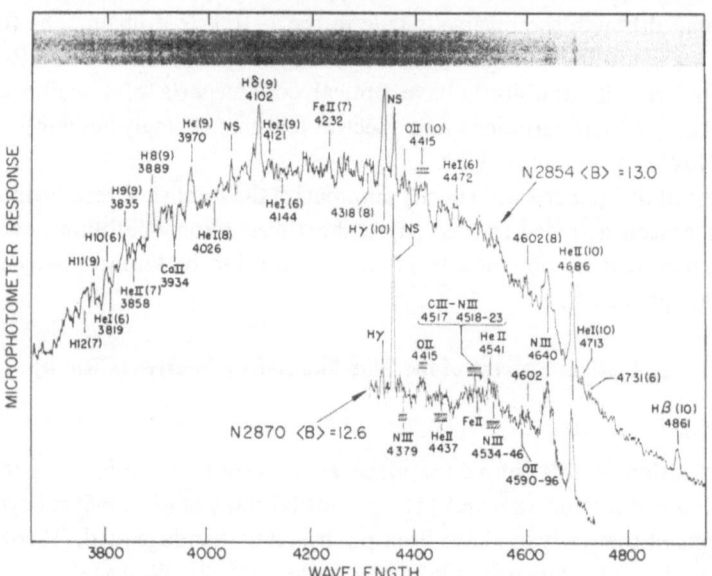

Fig. 2.   Microphotometer tracings of spectra of Sco X-1 (from Westphal *et al.*, 1968).

The authors note that "the most impressive single characteristic of the spectroscopic data is the large intensity variation of the hydrogen lines from night to night, relative to the continuum". This is correlated with the magnitude variation and the correlation is such as to suggest that the intensity of the emission lines is in reality constant and the apparent variation is due to the continuum; when the continuum is fainter the emission lines appear brighter by contrast.

The K line of CaII in absorption is almost certainly interstellar; the H line is also visible but faint, due to the H_ε emission. The velocity agrees with that of the interstellar clouds in the same region of the sky (Wallerstein, 1967).

From the K line a reliable estimate of the distance can be obtained; according to Wallerstein one obtains thus a lower limit of 270 pc for the distance of the source, but a value of 1000 pc or even more is not impossible. According to Westphal *et al.* (1968) from Wallerstein's ratio for the Ca doublet, K/H = 1.4, a distance of 240 pc is obtained, in fair agreement with Wallerstein's lower limit.

In view of the many uncertainties Westphal *et al.* favour a distance of 500 pc; with a B-absorption of 0.9 mag the absolute magnitude of Sco X-1 is

$$M_B = 3.6;$$

at this distance the total energy output in the X-ray spectrum is

$$P = 2 \times 10^{37} \text{ erg sec}^{-1},$$

or 10^3 or 10^4 times the power radiated in the optical spectrum.

This distance estimate agrees well enough with that from observations of the soft X-ray flux, which however gives contradictory results. I think that most astronomers would feel inclined to trust the interstellar line results and to agree that a distance $300 < D < 1000$ pc is quite realistic. Of course the ratio between the X-ray and optical fluxes is almost independent of distance.

Radial velocity measurements give very interesting and puzzling results. From the three strongest lines (He II 4686, H_γ and H_δ) it is found that on two nights the He II and H lines were changing the respective velocities in an opposite way: on July 17, 1967, He II changed from -155 to -215 km/sec, H_γ from -125 to -35 km/sec, H_δ from -192 to -112 km/sec. A similar variation occurred in the following night.

Thus not only the velocities from He II and from H were different but the former increased, while the latter decreased in absolute value. Also the H_δ velocities were systematically more negative than those of H_γ.

According to Westphal *et al.* (1968) simple binary motion cannot explain these variations; gas streams in which atoms of different excitation move in a widely different way are of course not impossible to imagine, but clearly more observations are needed.

It has been suggested by Braes and Hovenier (1966), Blaauw (1967), O'Dell (1967) that Sco X-1 might be a member of the Scorpio-Centaurus association. The best evidence should be based from the proper motion, which according to Gatewood and Sofia (1968) is in very good agreement with that corresponding to the Scorpio stream. But there is some disagreement between proper motions from different sources (Luyten, 1966; Johnson and Stephenson, 1966; Klare 1967) and the matter is still doubtful.

If Sco X-1 were a member of the Scorpio-Centaurus Association, its absolute magnitude – $M_V = +7$ – would be difficult to reconcile with the distance obtained from the interstellar lines (Wallerstein, 1967). This point is very important and should be further discussed.

I will not discuss here the discrepancy between X-ray and optical fluxes which results if both are explained as thermal bremsstrahlung from an optically thin source, nor the correlation between X-ray, optical and radio variations (Chodil *et al.*, 1968, Ables, 1969).

2. Cyg XR-2

The identification of Cyg XR-2 (Giacconi *et al.*, 1967) may be also considered fairly well established. Spectroscopically this is an even more complicated object than Sco X-1; published results are due to Lynds (1967), Burbidge *et al.* (1967), Kristian *et al.* (1967), Kraft and Demoulin (1967); see Figure 3.

The general impression is that of an object more or less similar to Sco X-1, on whose

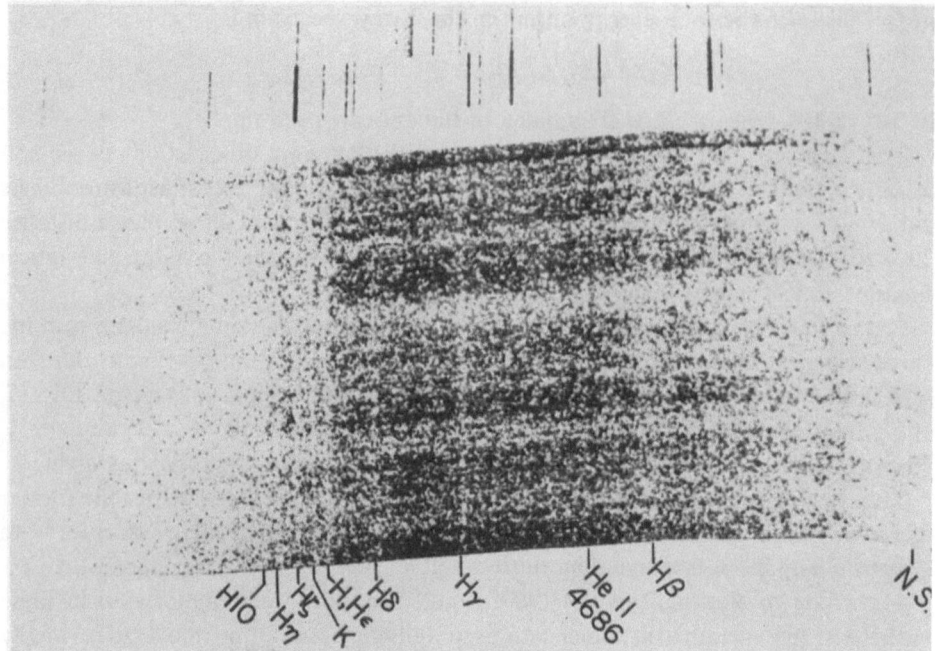

Fig. 3.   Spectrum of Cyg X-2 (from Lynds, 1967).

spectrum another spectrum closely resembling that of a G subdwarf is superimposed. This somewhat simplistic interpretation is supported by the photoelectric scanner photometry by Peimbert *et al.* (1968); they conclude that the observed continuum may be explained as due to at least two sources: one flat or of an early type (O9 V) component and one of a late type (later than F2 V). The flat spectrum is in fair agreement with the prediction from the observed X-ray flux, if one assumes that it is due to thermal bremsstrahlung. However the attenuation of the bremsstrahlung radiation is more than what might be expected from interstellar absorption alone.

The G-dwarf shows a large and variable negative velocity. By assuming that it is due to a 'normal' G or late F subdwarf (absolute magnitude $M = +5$) a distance of 700 pc may be obtained for the source; this reduces to little more than 500 if interstellar absorption is considered.

The absolute value of the X-ray flux of Cyg X-2 is difficult to evaluate, mainly because the interstellar attenuation is not well known; apparently it is somewhat fainter than that of Sco X-1; a realistic estimate is about 10^{36} erg sec^{-1}. Compared with the optical luminosity of a G-subdwarf (10^{33} erg sec^{-1}) we get a ratio of 10^{3} between the power radiated in the X-ray and in the optical domains. Like in the case of Sco X-1 this ratio is almost independent of the assumed distance.

The large variations of the radial velocity are not readily explained by orbital motion (Burbidge *et al.*, 1967; Kraft and Demoulin, 1967). If taken literally the average of the radial velocity would give a systemic velocity around -250 km/sec

(Kristian *et al.*, 1967). But this interpretation is doubted by Sofia and Wilson (1968) mainly on the basis that the total mass would come out too large for a G-subdwarf; they suggest that a systemic velocity of about − 120 km/sec would be more consistent with a reasonable value of the mass. The difference between the observed velocity and this value should then be due to motion inside the system, which is also not easy to understand.

3. *Cen XR-2 and GX 3 + 1*

The identifications of the last two sources is still open to doubt. Cen XR-2 was identified by Eggen *et al.* (1968) with a previously known peculiar variable, WX Cen, mainly on account of the character of the light variation and of the spectroscopic behaviour which follow closely the pattern corresponding to the optical counterpart of Sco X-1. Figure 4 shows the spectrum published by these authors, who comment upon its close resemblance to that of Sco X-1.

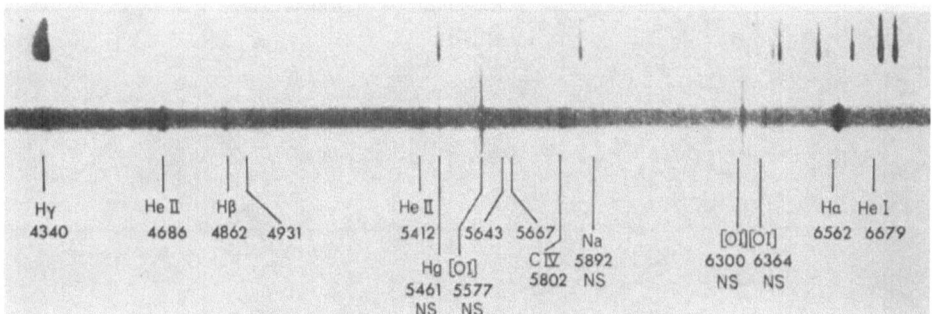

Fig. 4.   Spectrum of WX Cen = Cen XR-2 (from Eggen *et al.*, 1968).

The emission lines are somewhat broader than those of Sco X1; the H_α line has a faint narrow extension at both sides of the stellar spectrum, which however may be probably due to the general diffuse H_α emission in this galactic region. A distance of 500 pc is obtained if the absolute magnitude of the object is the same as that of Sco X-1. As it is known the X-ray flux underwent a very large decrease of intensity during 1967.

The identification of GX 3 + 1 with a faint star with a peculiar spectrum was suggested by Blanco *et al.* (1968a) and later Blanco *et al.* (1968b) reported a spectrum with broad emissions of He ɪɪ 4686 and O vɪ 3811 and 3843. Freeman *et al.* (1968) find however that the spectrum is not very similar to that of old Novae or Sco X-1, but is rather suggesting a Wolf-Rayet star or a 'supernova'. The star is situated inside a faint ring nebula.

3. Distribution of X-Ray Sources in Galactic Coordinates

It has been observed that the distribution of X-ray sources on the celestial sphere suggests some relationship with known objects and this might afford some clue to their nature.

The following classes of objects may conceivably be considered as reasonable candidates and have in fact been suggested:

(a) supernova remnants;
(b) old novae;
(c) Wolf-Rayet and related high temperature objects;
(d) pulsars.

Of course these objects do not exhaust the list, but I thought it better to confine myself to them; also all pulsars are at least potentially related to supernova remnants and viceversa.

The distribution in galactic coordinates of the 45 X-ray sources known at the end of 1968 is shown in Figure 5; like in other similar plots the concentration of these

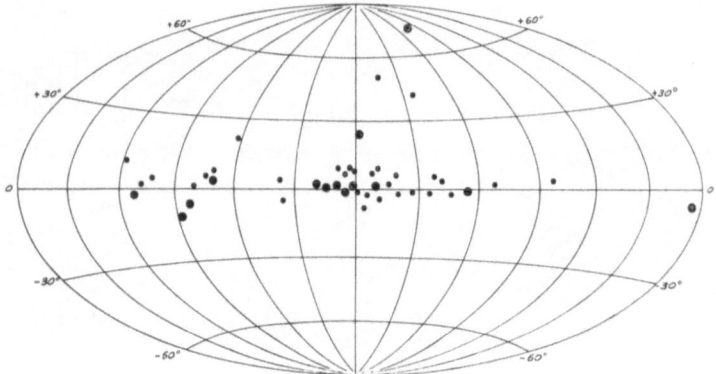

Fig. 5.   Distribution of X-ray sources in galactic coordinates.

sources towards the galactic plane and towards the galactic centre is very striking and may be taken as an established fact.

Nevertheless it may be observed that since the experimenters are – reasonably enough – anxious to obtain at least some positive results during a flight, by far the majority of the experiments were planned in order to cover the galactic belt or a part of it as soon as the galactic concentration of the sources was discovered. Therefore the chances of discovery of a faint source at high galactic latitude is much smaller than at low latitude, especially if one takes into account the variability of many sources. In Figure 6 I tried to give some idea of how densely the different parts of the sky have been covered by the experiments whose results have been published before 1969. The corresponding selection effect should be considered in an accurate study of the distribution of the sources.

The relationship between X-ray sources and supernova remnants has been thoroughly discussed by Poveda and Woltjer (1968) and is of course proved by some known individual cases. But among the 10 possible identifications suggested in Poveda and Woltjer's paper, it was found that Vel XR-1 does not coincide with Vel X nor

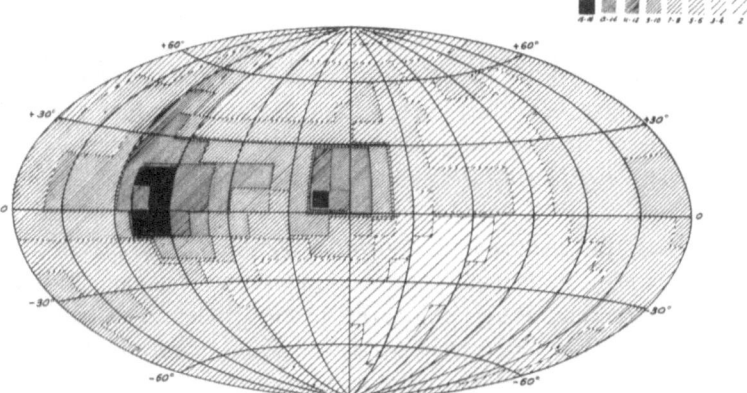

Fig. 6.   Covering of the sky by X-ray experiments prior to January 1969. The shading of the various
areas corresponds to the number of flights which covered the areas themselves.

with Vel Y (Gursky *et al.*, 1968) and Tycho's supernova is not coincident with the
source Cep XR-1 observed by Friedman and his coworkers (Friedman *et al.*, 1967) but
with another source which was missed by them and discovered later by the ASE group
(In this volume).

On the other side the distribution of supernova remnants does not show any remarkable clustering in the direction of the galactic centre and seems more connected
with the spiral arms. This fact and the examples of wrong identifications mentioned
above strongly suggest – in my opinion – that only a very small number of the known
X-ray sources will be found to belong to this class.

The distribution of old Novae is shown in Figure 7 as given by Payne-Gaposchkin

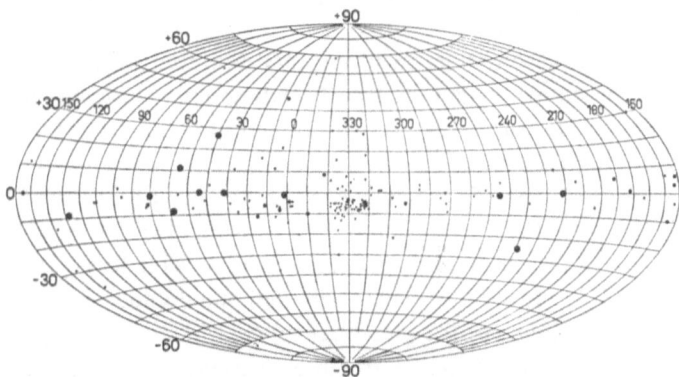

Fig. 7.   Distribution of novae in galactic coordinates (from Payne-Gaposchkin, 1957).

(1957). In this case the similarity to the distribution of the known X-ray sources is
indeed striking; the clustering in the Sagittarius region and also in the direction of
Cygnus and the scarcity of both classes of objects between galactic longitude 150° and
250° is remarkable.

On the other side, until present, not a single X-ray source could be associated with a known nearby old nova. Clearly there are for this fact only two possible explanations.

Either strong X-ray emission is actually a general feature of a certain stage of the development of a nova, but this hypothetical 'X-ray stage' does not come immediately after the main outburst, but much later. Although not impossible, this explanation seems rather unlikely; if true our present ideas on novae should be considerably reviewed, especially considering the enormous power radiated by X-ray sources in the X-ray spectrum.

Or, alternatively, the majority of known X-ray sources, although superficially resembling an old nova in their photometric and spectral behaviour form a different class of objects (Mumford, 1967). This seems to me a much more likely possibility, which, however, leaves us where we were as far as the nature of X-ray sources is concerned.

There are of course other known objects having the same galactic distribution as old novae and, hence, as the X-ray sources. For instance, planetary nebulae (Payne-Gaposchkin, 1957) are very well known to have almost the same galactic distribution; nevertheless, according to Minkovski (1948), there is no direct relationship between planetary nebulae and novae.

I believe that the only reasonable conclusion which can be drawn from this kind of evidence is that the known X-ray sources, with a few exceptions, belong as galactic objects to an intermediate system, like novae and white dwarfs, but the physical and evolutionary implications of this fact are completely obscure.

This, however, rules out the possibility that among them there be a large proportion of Population II objects (spherical subsystems) and carts some doubts upon the interpretation of the large negative velocities observed in Sco X-1 and Cyg XR-2 as being the actual velocities of the centers of gravity of these objects, in agreement with Sofia and Wilson (1968) conclusion.

But, for the same reason, it seems to me very doubtful that a large percentage of the known X-ray sources might be true Population I objects connected with OB Associations as it was suggested by various authors (Braes and Hovenier 1966; Gursky *et al.*, 1967; Sofia and Wilson, 1968).

I will only mention, in passing, that some suggested identifications of X-ray sources with Wolf-Rayet stars (Vel XR-1 with ζ Pup or γ Vel, Lac XR-1 with HD 211853) have been either disproved or are very uncertain. If Wolf-Rayet stars possess any X-ray flux – and I do not see any reason why they should not – it is fainter than that corresponding to the limits of the present surveys.

The connection of X-ray sources with pulsars deserves a few more words. A recent discussion of this problem was published by Friedman *et al.* (1969), who point out that the distribution of pulsars in galactic coordinates (Figure 8) is very dissimilar from that of the X-ray sources; also the distances commonly accepted for pulsars do not fit with those of the X-ray sources.

Observations at 234, 256 and 405 MHz (Friedman *et al.*, 1969) have shown – I think – conclusively that the radioemission from Sco X-1 is not pulsed with periods in the range 0.1–5 sec.

A direct relationship of pulsars with *known* X-ray sources is thus extremely unlikely. Another (indirect) relationship may of course exist, since some X-ray sources
are connected with supernova remnants and pulsars are now believed to be neutron
stars formed following a supernova explosion. It is not unlikely, therefore, that at

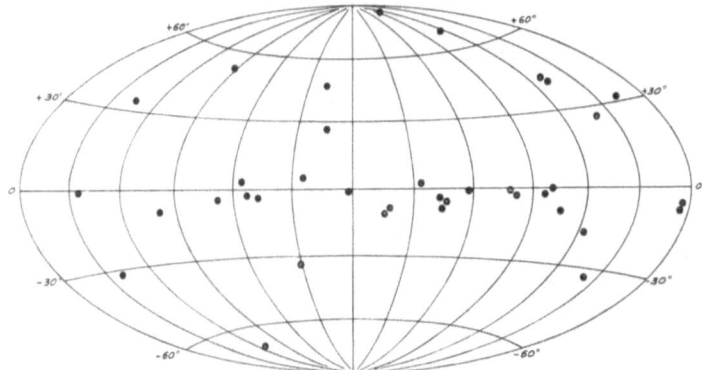

Fig. 8.   Distribution of known pulsars in galactic coordinates (April 1969).

least one pulsar is nested in all supernova remnants and, conversely at a certain
distance of all pulsars some remnants of the outburst are still lingering which might
be found to emit X-rays.

A more direct relationship was also communicated for the first time in this symposium, that is the emission of pulsed X-rays by the pulsar itself. This is of enormous
importance for the theory of pulsars, but I think that hardly anybody would maintain
that the X-ray flux from Sco X-1 and similar sources could be of this type.

To conclude, I believe that we can safely risk the following statement which
summarizes the situation: apart from a few objects which are clearly supernova
remnants, *the vast majority of the known X-ray sources form a rather homogeneous
class of objects which are members of an intermediate galactic subsystem, like novae
and planetary nebulae.* The main characteristic of these objects is a very large ratio
– from 10^3 to 10^4 – between the power emitted in the X-ray and optical spectra. In the
optical range they are very similar to old novae with regard to light variations and
spectroscopic behaviour; indeed were it not for the X-ray flux they would have been
called old novae. If they are also close binary systems – which I do not think at
present established – one of the components is a late type dwarf and the other a very
high temperature object; in this case one should explain how a star can survive very
close to an object from which it receives a quantity of energy of the same order or
even larger than that produced by its own nuclear sources.

References

Ables, J. G.: 1969, preprint.
Babcock, H.: 1967, Annual Report of the Director of Mt. Wilson and Palomar Obs. for 1966–67
   p. 288.

Biermann, L.: 1969, Max Planck Inst. München preprint.
Blaauw, A.: 1967, Discussion at the *I.A.U. Symposium No. 31*, p. 473.
Blanco, V., Kunkel, W., Hiltner, W. A., Lyngå, G., Bradt, H., Clark, G., Naranan, S., Rappaport, S., and Spada, G.: 1968a, *Astrophys. J.* **152**, 1015.
Blanco, V., Kunkel, W., and Hiltner, W. A.: 1968b, *Astrophys. J.* **152**, L137.
Braes, L. L. E. and Hovenier, J. W.: 1966, *Nature* **209**, 360.
Burbidge, E. M., Lynds, C. R., and Stockton, A. N.: 1967, *Astrophys. J.* **150**, L95.
Chodil, G., Mark, H., Rodrigues, R., Seward, F. D., Swift, C. D., Turiel, I., Hiltner, W. W., Wallerstein, G., and Mannery, E. J.: 1968, preprint.
Eggen, O., Freeman, K., and Sandage, A.: 1968, *Astrophys. J.* **154**, L27.
Freeman, K. C., Rodgers, A. W., and Lyngå, G.: 1968, *Nature* **219**, 251.
Friedman, H., Byram, E. T., and Chubb, T. A.: 1967, *Science* **156**, 374.
Friedman, H., Fritz, G., Henry, R. C., Hollinger, J. P., Meekins, J. F., and Sadeh, D.: 1969, *Nature* **221**, 345.
Gatewood, G. and Sofia, S.: 1968, *Astrophys. J.* **154**, L69.
Giacconi, R., Gorenstein, P., Gursky, H., Usher, P. D., Waters, J. R., Sandage, A., Osmer, P., and Peach, J. V.: 1967, *Astrophys. J.* **148**, L129.
Gorenstein, P., Kellogg, E. M., and Gursky, H.: 1969, this volume, p. 134.
Gratton, L.: 1968, Royal Soc. – Conference on X-Ray Astronomy, London (in press).
Gursky, H., Giacconi, R., Gorenstein, P., Waters, J. R., Oda, M., Bradt, H., Garmire, G., and Sreekantan, B. V.: 1966, *Astrophys. J.* **146**, 310.
Gursky, H., Kellogg, E. M., and Gorenstein, P.: 1968, preprint.
Ichimura, K., Ishida, G., Jugaku, J., Oda, M., Osawa, K., and Shimizu, M.: 1966, Tokyo Astron. Obs. Reprint, No. 301.
Johnson, H. M. and Stephenson, C. B.: 1966, *Astrophys. J.* **146**, 602.
Klare, G.: 1967, *Z. Astrophys.* **67**, 249.
Kraft, R. P. and Demoulin, M. H.: 1967, *Astrophys. J.* **150**, L183.
Kristian, J., Sandage, A., and Westphal, J. A.: 1967, *Astrophys. J.* **150**, L99.
Luyten, W. J.: 1966, I.A.U. Circular No. 1980.
Lynds, C. R.: 1967, *Astrophys. J.* **149**, L41.
Minkovski, R.: 1948, *Astrophys. J.* **107**, 106.
Mumford, G. S.: 1967, *Publ. Astron. Soc. Pacific* **79**, 283.
O'Dell, C. R.: 1967, *Astrophys. J.* **147**, 855.
Payne-Gaposchkin, C.: 1957, *The Galactic Novae*, North-Holland Pub. Co., Amsterdam.
Peimbert, R., Spinrad, H., Taylor, B. J., and Johnson, H. M.: 1968, *Astrophys. J.* **151**, L93.
Poveda, A. and Woltjer, L.: 1968, *Astron. J.* **73**, 65.
Sandage, A., Osmer, P., Giacconi, R., Gorenstein, P., Gursky, H., Waters, J., Bradt, H., Garmire, G., Sreekantan, B. V., Oda, M., Osawa, K., and Jugaku, J.: 1966, *Astrophys. J.* **146**, 316.
Sofia, S. and Wilson, R. E.: 1968, *Nature* **218**, 73.
Wallerstein, G.: 1967, *Astrophys. Letters* **1**, 31.
Westphal, J. A., Sandage, A., and Kristian, J.: 1968, *Astrophys. J.* **154**, 139.

IDENTIFICATION OF X-RAY SOURCES AT
CERRO TOLOLO INTER-AMERICAN OBSERVATORY

W. E. KUNKEL and V. M. BLANCO

The telescope used for identification of X-ray sources at the Cerro Tololo Inter-American Observatory is the Curtis Schmidt Telescope. This telescope is located at Tololo by agreement with the University of Michigan. The field of the telescope is $5° \times 5°$ and its correcting aspheric plate has negligible chromatic aberration over the range 3200 to 10 000 Å. This latter feature makes it an ideal instrument for multicolor survey work.

For optical identification of X-ray sources, the technique followed aims at the discovery of objects with strong ultraviolet excess. Eastman 103aO plates are used, and a given plate is exposed first through a Schott UG1 or UG2 filter, and then, after the telescope is moved about 30″, exposed again through a GG13 filter.

Experience shows that relative exposures of about 6 to 1 for the ultraviolet exposure as compared with the blue exposure, yield a pair of images that are comparable in size and density for earlier O-type stars. Objects whose ultraviolet image, in these plates, is stronger than the blue image are likely to have strong ultraviolet excess. To reach magnitude 18, exposures of 7 min for the blue and 42 min for the ultraviolet are required. To this limit, experience shows that there is an expectancy, depending on location in the Milky Way, of from one to four objects with strong ultraviolet excess per 4 square degrees in the neighborhood of the galactic center. Thus it is rather important that the smallest possible error box is available before an attempt at identification is made.

To follow up tentative candidate objects (1) objective spectra of about 500 Å/mm dispersion at Hγ are obtained, yielding information to about magnitude B=15. The principal telescope for follow-up photometry and spectrography is a 60-inch (1.5 m) reflector with modified Ritchie-Chretien optics. A Cassegrain image-tube spectrograph can be used with this telescope, yielding usable spectra to B=17.

Identification at Cerro Tololo has been attempted for about 12 sources and tentative candidates have been obtained for Cen XR-2, GX3+1, and Vela X-1.

Work on the first of these, Cen XR-2, was done late in 1967 and the central star in the irregular planetary nebula NGC 5189 was picked out as a possibility (Blanco, 1967). About the time that this identification survey was going on, Feast in South Africa also mentioned this planetary nebula as a possibility (Feast, 1967). Photometry of this star confirmed its marked ultraviolet color.

The main spectral features are exceedingly strong and broad emission of O VI 3811 and 3834 and of He II 4686. The spectrum is thus not at all like that of Sco XR-1. Prior to this work, this type of spectrum had been found by other investigators in a number of central stars in planetary nebulae. A sequence can indeed be established of a number of these objects, and the star in NGC 5189 appears to be, because of the

L. Gratton (ed.), Non-Solar X- and Gamma-Ray Astronomy, 173–175. All Rights Reserved
Copyright © 1970 by the IAU

strength of the O VI 3811 and 3834 emissions, an extreme example of the members of such a sequence.

The identification of Cen XR-2 with the star responsible for the excitation of the planetary nebulae NGC 5189 is no longer fashionable. The peculiar variable WX Cen, which shows spectral and variability features that are similar to the object identified as Sco XR-1, is currently a candidate favored by several investigators (Eggen and Lyngå, 1968).

Nevertheless, it is worth pointing out that another tentative identification made at Tololo for the source GX3 + 1 resulted in an object having photometric and spectro-scopic features that are practically identical with those of the star in NGC 5189 (Blanco *et al.*, 1968).

In the case of GX3 + 1, the error box was about 15 sec of arc on a side, and about $\frac{1}{16}$ of a square degree in area.

The ultraviolet object was found within the box. From frequency of ultraviolet objects in the surroundings of GX3 + 1, the expectancy of one such object in the kind of survey made, is less than one in fifty. This low expectancy is the main argument in favor of the identification of GX3 + 1, and in turn, the fact that this candidate and the central star in NGC 5189 have similar colors and spectral features gives support to the latter identification.

A search for an optical candidate for the X-ray source in Vela (Vela XR-1, or GX263 + 3) was conducted in a region of about ten square degrees centered on $8^h 57^m$ and $-41° 15'$ (Epoch 1950) using the double exposure method previously described. Ten candidates found this way, all fainter than B = 15.3, were examined photo-electrically with the 60-inch reflector. The hottest (or most ultraviolet) object at

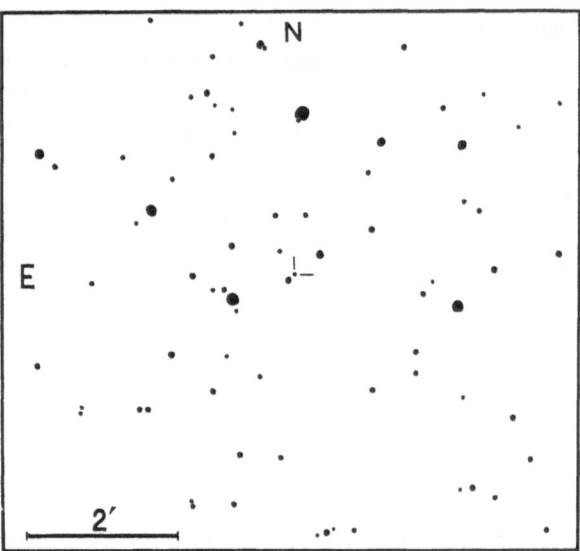

Fig. 1.   Identification Chart for the Optical Candidate of Vela XR-1. 30-minute exposure on 103a-E with RG-2 filter, Schmidt telescope.

position (Epoch 1950) $\alpha = 8^h\ 57^m\ 55^s$ and $\delta = -41°\ 13'.9$ is quite close to the center of the uncertainty box published by Gursky *et al.* (1968), well within its borders. A chart for identification is shown in Figure 1. The magnitude and colors, determined from a single night's observations, are $B = 17.3 \pm 0.1$, $B - V = +0.21 \pm 0.10$, and $U - B = -0.92 \pm 0.12$. The rather large mean errors result from a close optical companion about five seconds of arc distant, near p.a. $\simeq 135°$. Photometry for this star is $B = 15.92 \pm 0.07$, $B - V = +0.91 \pm 0.05$, and $U - B = +0.07 \pm 0.08$. Since some trouble was encountered with the equipment, these values should be considered provisional. Within these limits, the colors lie near the black body line for both stars, as though both were white dwarfs. Unwidened image-tube spectra at 100 Å/mm dispersion of the fainter, hotter star show a smooth continuum with no lines in either emission or absorption, and in one of two spectra of the brighter, cooler star, H and K of Ca II show faintly in absorption.

Since a number of conflicting interpretations of these preliminary results appear possible, further observations of this pair of stars are planned for the next season.

References

Blanco, V. M.: 1967, I.A.U. Circular No. 2035.
Blanco, V. M., Kunkel, W. E., and Hiltner, W. A.: 1968, *Astrophys. J.* **152**, L137. See also Freeman, K. C., Rodgers, A. W., and Lyngå, G.: 1968, *Nature* **219**, 251.
Eggen, O. J. and Lyngå, G.: 1968, *Astrophys. J.* **153**, L195.
Feast, M. W.: 1967, *Nature* **215**, 1158.
Gursky, H., Kellogg, E. M., and Gorenstein, P.: 1968, *Astrophys. J.* **154**, L71.

OAO OBSERVATIONS OF SCO X-1*

R. C. BLESS, A. D. CODE, T. E. HOUCK, C. F. LILLIE, and J. F. McNALL

Space Astronomy Laboratory, Washburn Observatory

On several occasions in the past month we attempted to observe the bright X-ray source in Scorpio using the OAO 8-inch photometers. These instruments isolate ultraviolet spectral bands by means of interference filters which have half-widths of about 300 Å. Digital and analog outputs both are available and integration times of 8 and 64 sec were used. The length of observing time was about 30 min per orbit. Sky background measurements about 30 arc min away from Sco X-1 source were interspersed with those of the X-ray source, enabling the large contribution of the background to be subtracted. These OAO observations are near the limit of our equipment and are therefore rather noisy. A sudden increase in the intensity of Sco X-1 less than 25% would be undetectable in the OAO photometers. During the observations described here, which were made on April 9, 11^h14^m UT to 11^h43^m UT, simultaneous observations were made by Dr. Jerome Kristian with the 200-inch telescope and multichannel photometer. Kristian observed a slow decrease in the flux from Sco X-1 during the first several minutes of simultaneous observations. We also observed such a decrease in intensity but did not see a 10% flare which the 200-inch saw at 5550 Å.

OAO filter photometry was obtained in 9 bands from 3300 Å to 1380 Å (Ly-α excluded). After correcting for interstellar extinction of 0.23 magnitudes in $(B-V)$ and using the interstellar extinction curve in the ultraviolet given in *Astrophys. J.* **153**, 561, we found that in frequency units the spectrum over this interval was essentially flat. The flux was about 1.5×10^{-24} ergs/cm^2/sec/cycle/sec. The 200-inch observation at 5550 Å is slightly more than one-half this value. These preliminary results are consistent with free-free radiation from a hot gas.

Additional OAO observations of Sco X-1 are underway.

---

* Communicated by Dr. W. L. Kraushaar.

L. Gratton (ed.), Non-Solar X- and Gamma-Ray Astronomy, 176. *All Rights Reserved*
Copyright © 1970 by the IAU

ON THE MAGNITUDE-COLOR RELATION FOR CYGNUS X-2
AND WX CENTAURI*

GEORGE S. MUMFORD

Tufts University, Kitt Peak National Observatory**, and Cerro Tololo Inter-American Observatory**

Abstract. While the short-term fluctuations in the brightness and color of the source Sco X-1 have been found to be correlated in the sense that the system is bluest when brightest, such a result does not appear in the observations of Cyg X-2 or WX Cen described here.

A connection between brightness and color index for the X-ray source Sco X-1 was found by Mook (1967), who showed that in the case of night-to-night variations this object was bluer when brighter. A similar correlation is suggested in the nightly changes for Cyg X-2 (Kristian *et al.*, 1967). On the other hand, a pronounced trend is not evident in the observations of WX Cen, optically identified with the source Cen X-2 (Eggen *et al.*, 1968), though there is a suggestion the trend may be in the opposite sense.

One wonders if the correlation is also evident in the short-time-scale brightness fluctuations that these objects are observed to undergo during a given night. Recent observations by Stępień (1968) indeed show the effect in the case of Sco X-1.

To my continuing program of photoelectric observations of novalike variables, I have added certain optically identified X-ray sources because the two classes of objects appear to be photometrically similar. The data for Cyg X-2 were obtained with the No. 2, 36-inch Cassegrain reflector at Kitt Peak National Observatory, as were those for U Geminorum. A similar instrument was used at Cerro Tololo Inter-American Observatory for observations of WX Cen. In all cases observing techniques and reduction methods are similar to those described earlier (Mumford, 1964).

On the night of October 5, 1967, Cyg X-2 was observed in blue and yellow light for about 4.5 hrs. During this interval Lynds (unpublished) also made observations with the image-tube spectrograph at the Kitt Peak 82-inch reflector. Neither set of data gives any suggestion of orbital motion.

The chief photometric result is shown in Figure 1. Over the interval of observation, the brightness of the object varied by some 0.3 magnitude and it appeared *redder* when *brighter* unlike Sco X-1.

A similar result derived from 2 hrs of observation on WX Cen the night of February 15, 1969, is shown in Figure 2. The scatter of points is less here than in the first diagram because of the smaller activity of WX Cen.

Correlations like these for novalike variables will be discussed in detail elsewhere. Suffice it to say for the present that U Gem, the prototype of a subclass of novalike stars, displays both effects at certain times. For example, immediately following prima-

* *Contributions from the Kitt Peak National Observatory*, No. 527.
** Operated by the Association of Universities for Research in Astronomy, Inc., under contract with the National Science Foundation.

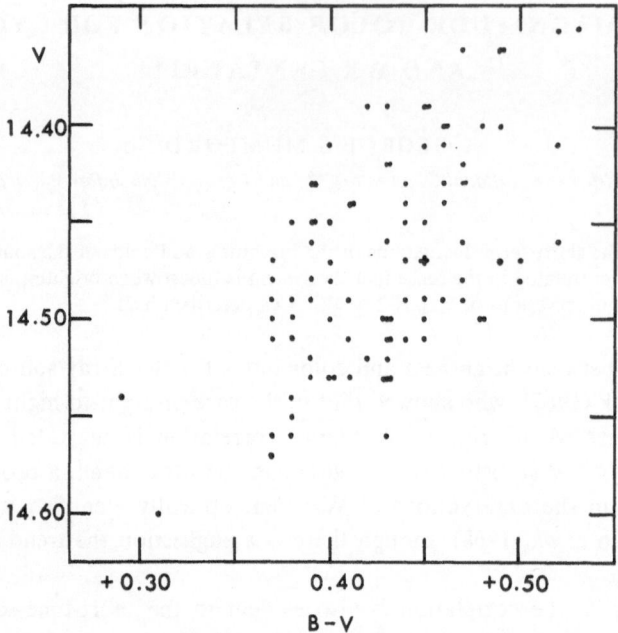

Fig. 1.   The color-brightness relation for Cyg X-2. Ordinate is visual magnitude on the standard system; abscissa is B–V color index. Notice the object appears redder when brighter.

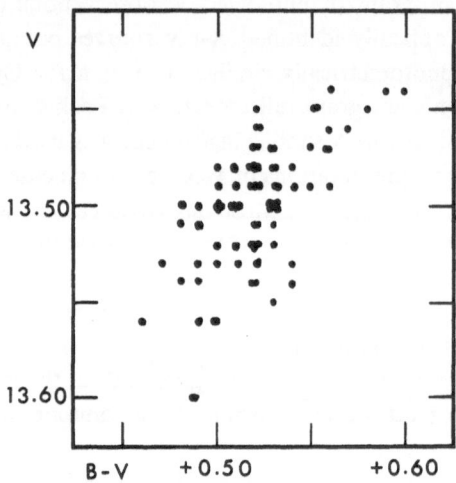

Fig. 2.   The color-brightness relation for WX Cen. Ordinate and abscissa as in Figure 1. Notice this object, too, appears redder when brighter.

ry eclipse, when the red component of the binary pair dominates, U Gem is redder when bright. Prior to primary eclipse, however, as the blue component and hypothesized ring or disk surrounding it become prominent, the system becomes bluer when brighter. These effects are displayed in Figure 3. Clearly this changing color-brightness relation results from orbital motion.

We do not know, at present, whether the correlations described above for Cyg X-2

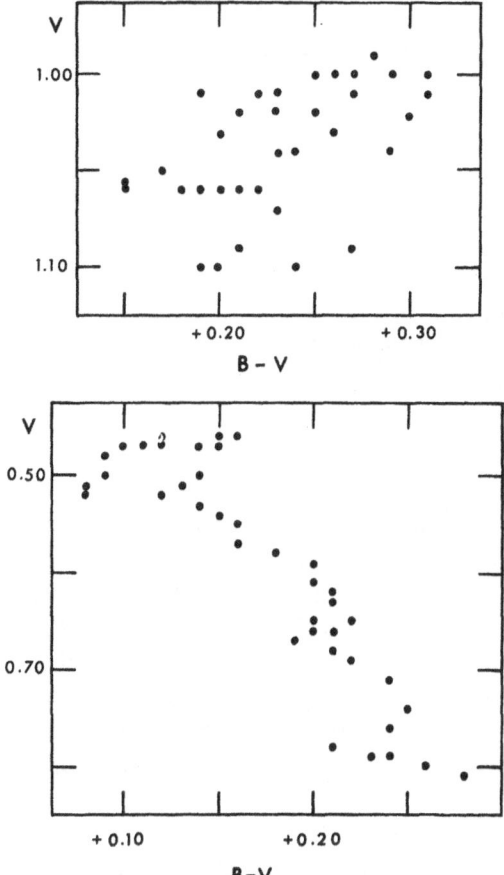

Fig. 3.   The color-brightness relation for U Gem. Here the visual magnitudes, ordinate, have been taken with arbitrary zero point; the B–V colors are on the standard system. In the upper diagram, when the red component predominates, the system is redder when bright. However, when the blue component predominates, lower diagram, the system is bluer when bright.

and WX Cen change in the course of time. If they do, we may have evidence for the binary nature of these two objects. If no such change occurs, it would appear that these two X-ray sources are optically unlike Sco X-1 in this respect.

Acknowledgement

This work has been supported by grants from the National Science Foundation.

References

Eggen, O. J., Freeman, K. C., and Sandage, A. R.: 1968, *Astrophys. J. Letters* **154**, L27.
Kristian, J., Sandage, A. R., and Westphal, J. A.: 1967, *Astrophys. J. Letters* **150**, L99.
Mook, D. E.: 1967, *Astrophys. J. Letters* **150**, L25.
Mumford, G. S.: 1964, *Astrophys. J.* **139**, 436.
Stępień, K.: 1968, *Astrophys. J. Letters* **151**, L15.

ON THE NATURE OF THE X-RAY SOURCES
SCORPIUS X-1 AND CENTAURUS X-2*

SABATINO SOFIA

Dept. of Astronomy, University of South Florida, Tampa, Fla., U.S.A.

Abstract. Preliminary evidence is presented which indicates that the star WX Cen, associated with the X-ray source Centaurus X-2 belongs to the Scorpio-Centaurus association. It is pointed out that this result greatly strengthens the similar results for the source Scorpius X-1, and lends support to the idea of the existence of a class of X-ray stars formed by young, subluminous objects, probably supernova remnants, thus neutron stars.

On carrying out an accurate proper motion study (by the plate overlap technique) of 83 stars in the region surrounding the optical object associated with the X-ray source Sco X-1, it was found that Sco X-1 together with at least three more of these stars have proper motions very similar to the nearby B stars which are well known members of the Scorpio-Centaurus association (Sofia *et al.*, 1969). Following the general practice in such cases, we concluded that Sco X-1 is probably a member of the association. Thus, the accurate distance to the source is about 200 parsecs, and the upper limit for its age is 20 million years (cf. Bertiau, 1958). We soon realized (Gatewood and Sofia, 1968) that the resulting source parameters impose some severe constraints on the nature of the X-ray star. In particular, since Sco X-1 appears substantially below the main sequence (see Figure 2 of Gatewood and Sofia's paper), according to well established results in stellar evolution, the object must have evolved off the main sequence in less than 20 million years. This means (Iben, 1967) that the main sequence mass of the object must have been above 10 $M_\odot$, and according to Colgate and White (1966), an object that massive would have undergone a supernova explosion and probably now be a neutron star.

The idea of a relationship between X-ray sources and neutron stars is not new. Soon after the first detection of X-ray sources, Chiu and Salpeter (1964) proposed that they were hot neutron stars. The observed characteristics of Sco X-1 are, however, very different from those expected according to the early picture of the X-ray emitting neutron stars. In particular, the spectrum (from the X-ray to the visual region) requires most of the radiation to originate in a cloud which surrounds the neutron star, and has a radius of at least a hundred times that of the star (cf. Shklovsky, 1967).

In view of the implications of these results, much effort was spent in checking the accuracy of the method. For example, it was easy to verify that stars #13 and #25 in the catalogue in the paper by Sofia *et al.*, for which spectral classification was available, perfectly fit the main sequence once corrected for a distance modulus of

---

* Astronomical Contributions from the University of South Florida at Tampa, No. 23.

L. Gratton (ed.), Non-Solar X- and Gamma-Ray Astronomy, 180–182. All Rights Reserved
Copyright © 1970 by the IAU

6.1 mag. and an absorption of 0.1 mag., corresponding to the values derived for the Scorpio-Centaurus association by Bertiau (1958). The same happened for the star #37 which was classified GOV from a spectrum taken by J. Hunter at the Yale 40 inch telescope. It is well known, however, that the main weakness of any proper motion result is the ever present possibility that, through a series of coincidences, a star not belonging to a certain group would have the exact tangential component of its peculiar velocity (both in direction and magnitude) to offset distance differences so as to appear to move within the group. I am now pleased to report that a preliminary study indicates that the star WX Cen, recently identified with the source Cen X-2 (Eggen *et al.*, 1968) is also a member of the Scorpio-Centaurus association. Its location in the H–R diagram very near Sco X-1 appears to indicate both that the identification of Cen X-2 is correct, and that the results for Sco X-1 are real, and not the consequence of possible freak coincidences pointed out above.

In a work performed at the University of South Florida in collaboration with H. Eichhorn and G. Gatewood, we determined the positions and proper motions of 124 stars in the region of Cen X-2, including the star WX Cen. The early epoch plate (1897) was kindly provided to us by Mr. Harley Wood, of the Sydney Observatory, while the recent epoch plates were taken for us by Mr. G. A. Harding at the Cape Observatory in South Africa. The position and proper motion of Cen X-2 in the FK_4 system are

α_{1950}	μ	δ_{1950}	μ'
$13^h9^m38\overset{s}{.}11$	$-0\overset{s}{.}0030$	$-63°7'51\overset{''}{.}3$	$-0\overset{''}{.}004$
±0.030	$\pm.00090$	$\pm.16$	$\pm.0056$

The errors are rms errors.

At least five other stars with photographic magnitudes ranging from 7.8 to 14 have been found to have proper motions similar to that of Cen X-2. The only one for which we have a spectrum, CPD $-63°2636$, fits the main sequence of the Scorpio-Centaurus association which in this region has a distance modulus of 6.3 mag. (Bertiau, 1958). The above proper motion may be compared with that of the nearest member in Bertiau's list, namely HD 116087, whose proper motion components (μ, μ') in the system of the FK_4 are $(-0\overset{s}{.}0034\pm.00057, -0\overset{''}{.}011\pm.0041)$ respectively, the errors being rms errors.

It is in summary concluded that, if additional checks confirm this preliminary result, substantial evidence exists which indicates the existence of a class of X-ray stars, of which Sco X-1 and Cen X-2 are members, consisting of young, subluminous objects which probably are supernova remnants, and thus according to current ideas, neutron stars.

Acknowledgement

The work reported has been supported by the NSF Grant No. GP 8121, for which the author is very grateful.

References

Bertiau, F. C.: 1958, *Astrophys. J.* **128**, 533.
Chiu, H. Y. and Salpeter, E. E.: 1964, *Phys. Rev. Letters* **15**, 599.
Colgate, S. A. and White, R. H.: 1966, *Astrophys. J.* **143**, 626.
Eggen, O. J., Freeman, K. C., and Sandage, A., 1968, *Astrophys. J.* **154**, L27.
Gatewood, G. and Sofia, S.: 1968, *Astrophys. J.* **154**, L69.
Iben, I.: 1967, *Ann. Rev. Astron. Astrophys.* **5**, 571.
Shklovsky, I. S.: 1967, *Astrophys. J.* **148**, L1.
Sofia, S., Eichhorn, H., and Gatewood, G.: 1969, *Astron. J.* **74**, 20.

AN X-RAY PULSAR IN THE CRAB NEBULA

BRUNO ROSSI

Massachusetts Institute of Technology, Cambridge, Mass., U.S.A.

I wish to report the preliminary results obtained by Hale Bradt, William Mayer and Saul Rappaport of MIT in a recent rocket flight devoted entirely to the observation of the X-ray emission from the Crab Nebula. The flight had been planned originally for April 3. Because of administrative difficulties, it was delayed until April 26.

The launching took place from the White Sands Missile Range. The instrumentation included a bank of proportional counters with 2 mil Be windows about 800 cm^2 in area, and a second bank of counters, with windows of $\frac{1}{3}$ mil Al, $\sim$90 cm^2 in area. The results reported here are based on the data provided by the Be window counters, whose spectral sensitivity extended from about 1.5 to about 10 keV.

The rocket was provided with an aspect control system, which kept the Crab Nebula within the field of view of the counters (11° × 11°) during most of the useful flight time (about 190 sec). In this time, about 300 000 X-ray pulses were recorded, with a timing accuracy of about 1 ms. Simultaneously, the WWV time standard was also recorded. This made it possible to correlate the X-ray data with optical observations of the pulsar in the Crab that were carried out within one-half hour of the flight by R. Edward Nather, Brian Warner, and Malcolm MacFarlane at the

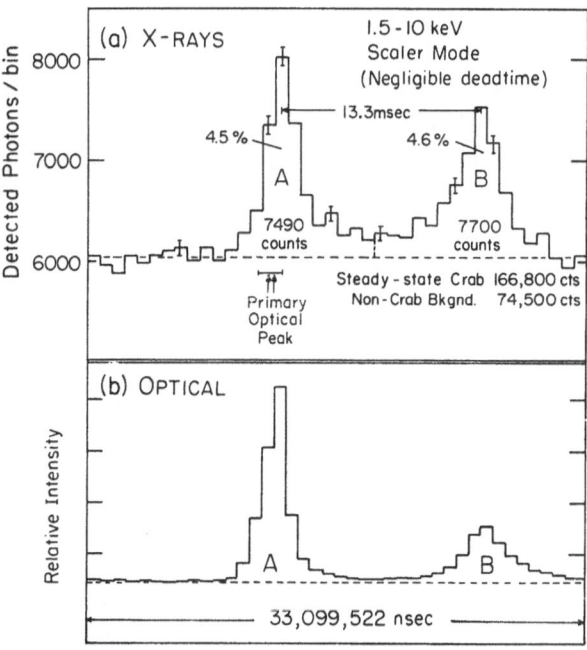

Fig. 1.   X-ray and optical data from the MIT and McDonald groups respectively.

L. Gratton (ed.), Non-Solar X- and Gamma-Ray Astronomy, 183–184. All Rights Reserved

McDonald Observatory, and by Jerome Kristian at the Mt. Palomar Observatory.

The occurrence times of the X-ray pulses were reduced to the module of the precise apparent period of the optical pulsar (33.0995 ms). This interval was then divided into 40 equal time 'bins' of about 0.83 ms each, and the pulses were distributed among these bins. The resulting distribution is shown by Figure 1a. For comparison, Figure 1b (adapted from a paper by Warner *et al.*, 1969) shows light curves of the optical pulsar, as obtained at the McDonald Observatory.

One sees that the X-ray emission undergoes strong periodic variations whose essential features match those of the optical emission. Within each period, the X-ray curve, like the optical, exhibits two peaks of different widths. The narrow X-ray peak is simultaneous with the principal narrow optical peak, within the 1 ms uncertainty of the correlation. The separation of the two peaks is about 13.5 ms both in the X-ray and in the optical range of the spectrum. The X-ray intensity between the two peaks remains appreciably above background, as does the optical intensity.

On the other hand, the strength of the pulsating component, relative to the background, is much greater for the X-ray than for the visible radiation. In the X-ray range, the total pulsating power amounts to about 9% of the background due to the continuous emission of the nebulosity, whereas in the optical range the pulsating power amounts to about 0.1% of the background. Moreover, the two X-ray peaks have about the same area, whereas the principal narrow optical peak has a considerably larger area than the secondary wide peak. Note that, in the pulsed mode, the total X-ray power is about 100 times the total optical power and about 10^5 times the radio power; thus the pulsar in the Crab emits by far the largest amount of its radiating energy in the form of X-rays.

The figures quoted above are still preliminary. However, it should be pointed out that the X-ray data on which Figure 1 is based were taken from a channel for which the dead-time corrections should be negligible.

The spectral data on the pulsating X-ray emission are still meager. At the present state of the analysis, there is no clear evidence for a difference in the spectra of the pulsating and of the continuous components in the X-ray region.

Reference

Warner, B., Nather, R. E., and MacFarlane, M.: 1969, *Nature* **222**, 233.

A SEARCH FOR PULSED HARD RADIATION
FROM THE CRAB NEBULA

G. J. FISHMAN, F. R. HARNDEN, Jr., and R. C. HAYMES

Dept. of Space Science, William Marsh Rice University, Houston, Tex., U.S.A.

This paper reports the detection of repetitive pulses of hard X-rays and gamma rays in data obtained in mid-1967 from the Crab Nebula. The pulses are believed to originate from NP0532 since the observed repetition period is what would be expected on the basis of an extrapolation of recent optical data obtained for NP0532 back to the time of the observation.

The experiment reported here was conducted at balloon altitudes on June 4, 1967 and has been previously described (Haymes *et al.*, 1968). Briefly, the actively-collimated γ-ray detector has a field of view characterized by a half-flux angle of 12° from the axis. Its associated pulse height analyzer was arranged to detect and analyze those photons having energies greater than 35 keV.

While the balloon floated at approximately 130000 feet altitude, data were collected for $3\frac{1}{2}$ hours; for 10-min intervals the detector was pointed at the Crab (to within $\pm 1°$) and alternately at a background region offset by 180° in azimuth. In the present analysis, we have searched 80 min of source data, and for reference and control purposes, 30 min of background data, for hard X-ray pulsations.

The detection of each photon was telemetered to ground receiving stations where it was recorded on analog magnetic tape along with time signals received from WWV.

The time at which each photon event occurred has been determined using WWV as a time base. Short-term speed fluctuations of the tape recorder were found to be less than $\frac{1}{10}$ of a millisec, peak to peak, in a 1-sec time interval. A delay time, introduced by the balloon-borne system, between the detection and recording of an event consists of a constant and a variable component. The constant component merely introduces a phase shift (~ 2 millisec) and does not affect the present analysis. The variable delay is due mainly to the photon-energy dependence of the pulse height analysis time. For approximately one percent of the events, this delay may be as large as 0.5 millisec; otherwise, it is less than 0.2 millisec.

In the present analysis, the apparent time of arrival of each photon was measured with 1.0 millisec resolution. The nearest millisec interpolation between the 1-sec pulses from WWV was accomplished with an Astrodata Corporation Model 6190 time-code generator, which has a time base stable to one part in 10^8. The times of photon arrivals and of the 1-sec WWV pulses, as measured by the time-code generator, were recorded on digital tape.

The slow-down rate for NP0532, determined optically by Warner *et al.* (1969), was used in a linear 575-day extrapolation from their measurements to the time of the flight. The extrapolation resulted in a predicted heliocentric period 'P_0' of 33.071 758 $\pm$ ± 0.000047 millisec at 0930 CDT on the morning of the flight. With the aid of a

L. Gratton (ed.), Non-Solar X- and Gamma-Ray Astronomy, 185–189. All Rights Reserved
Copyright © 1970 by the IAU

digital computer, searches for periodicity were made over 1.4×10^5 NP0532 periods, using a superposed epoch analysis. Over 200 searches were made at periods differing by 5 nanosec increments and covering a 1200 nanosec range which included the geocentric period corresponding to P_0. Each search determined the number of counts per millisec for each millisec of phase.

During the search at each assumed geocentric period (figured for an epoch at 0930 CDT), the value of the period was updated at 20-min intervals to account for the apparent change in the period of NP0532. The effects for which corrections were made throughout the flight, in order of their importance, are: the change in the diurnal radial velocity component; the change in the earth's orbital radial velocity component; the intrinsic slow-down of NP0532; and the balloon drift motion (change in propagation delay). The above terms had the net effect of increasing the apparent period of NP0532 by 18 nanosec over the 190-min interval during which the 80 min of data were obtained.

The diurnal and orbital radial velocity terms were calculated for the position of the south-preceding central star of the Crab Nebula as given by Minkowski (1968). This object has been identified as the source of the optical pulsations (Cocke et al., 1969; Lynds et al., 1969).

For a barycentric period of 33.071 783 millisec, one peak at the 3.9σ (σ = standard deviation) level and another, 14 msec later in phase, at the 3.0σ level were detected. This was the only period that resulted in a peak exceeding 3 standard deviations above the mean number of counts. Observation of the interpulse with the expected phase delay considerably enhances the statistical significance of the period determination. The period was then varied about this value in 1-nanosec increments. No further peak increases were observed, nor did any additional structural features emerge. It appears that the barycentric period has been established to ± 10 nanosec since shifting the period by this amount reduces the peaks to ~ 2 standard deviations.

In Figure 1 the resulting pulse profile is shown along with the optical data of Lynds et al. (1969). A double-pulse structure rather like that found at optical wavelengths is evident; 14 millisec separate the interpulse from the main pulse. The interpulse is somewhat wider than the main pulse, which is also true for the optical pulse profile. It may be significant that the power contained in the interpulse seems nearly equal to that in the main pulse in the energy region above 35 keV and therefore relatively greater than it is in the optical. If it is statistically significant, this may be evidence for either a temporal variation or a wavelength-dependence of the interpulse structure. In the soft X-ray region (1–10 keV), Fritz et al. (1969) have also observed the interpulse to be stronger, relative to the main pulse, than has been observed in the optical region. This seems to support the wavelength-dependence of the pulse profile, but additional observations are required in both the soft and hard X-ray regions.

The value found for the period is only 25 nanosec greater than P_0, a discrepancy that is well within the range of periods allowed by the uncertainties in the radio and optical slow-down rates. This suggests that there have been no discontinuous changes in the period of NP0532 between June 4, 1967 and Februari 22, 1969 of the type observed in the Vela pulsar, PSR 0833–45 (Radhakrishnan and Manchester, 1969; Reichley and Downs, 1969).

The present early detection of the pulsar permits an improved estimate of the slow-down rate to be made. Using recent period data (Warner *et al.*, 1969), we calculate a linear coefficient of 36.51 ± 0.02 nsec day^{-1} over the 575-day interval, which compares

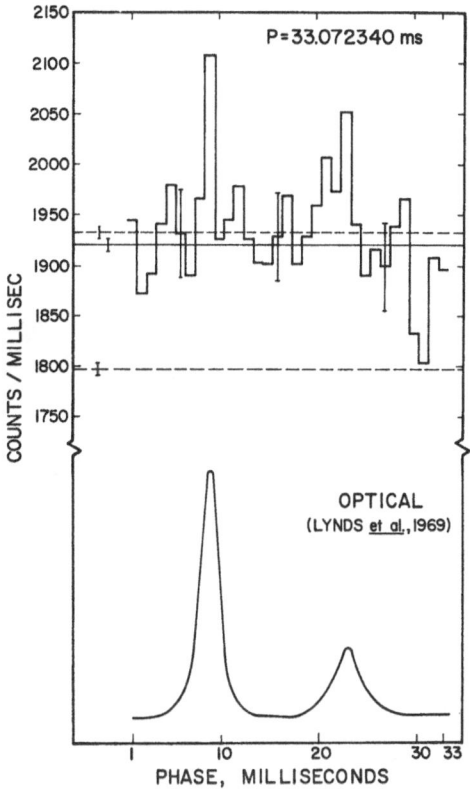

Fig. 1.   Pulse profile found for photon energies greater than 35 keV from pulsar NP0532 on June 4, 1967. The lower dashed line indicates the mean contribution from the background. The upper dashed line is the mean count rate; the solid horizontal line is the mean without peaks. The error bars shown are $\pm$ 1 standard deviation. The value shown for P is the geocentric period at 0930 CDT and the optical data are shown for comparison purposes.

with the McDonald Observatory value of 36.55 ± 0.08 nsec day^{-1} and with the Arecibo value (Richards and Comella, 1969) of 36.48 ± 0.04 nsec day^{-1}. The agreement of the above linear coefficients over the long time interval between measurements allows an upper limit of 3×10^{-4} nsec day^{-2} to be placed on the magnitude of the second order term, $|\mathrm{d}^2 P/\mathrm{d}t^2|$ at the 2σ level.

Figure 2 shows the spectrum of the entire Nebula and of the pulsed component in various energy regions. Since all 128 channels of the pulse height analyzer were combined for the present analysis, no spectral information is yet available. In plotting the point representing the present analysis at 120 keV, we have assumed that the pulsed spectrum is like that of the hard radiation from the Nebula. 120 keV is the geometric mean of the energies (35–560 keV) over which the Nebula's spectrum was previously determined (Haymes *et al.*, 1968).

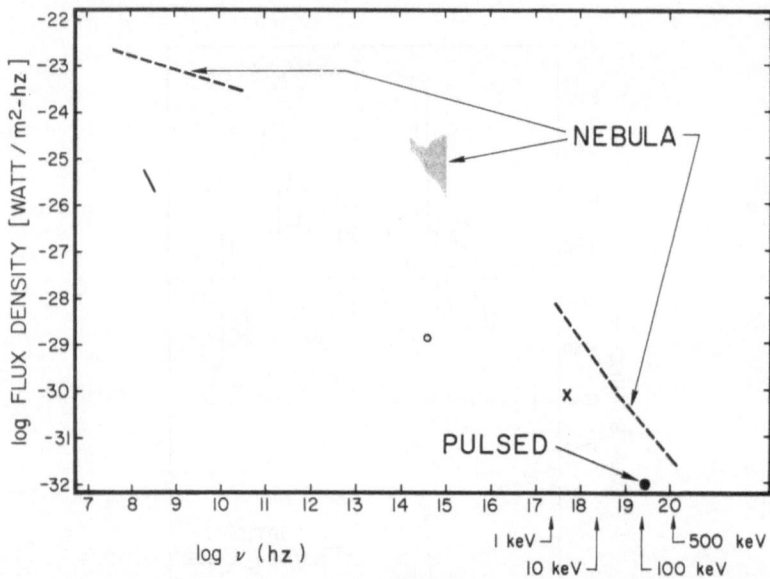

Fig. 2.   Tentative spectrum for NP0532. The Nebula spectra are from Haymes *et al.* (1968). The pulsed radio spectrum is from Richards and Comella (1969); the optical point, from Lynds *et al.* (1969). The soft X-ray point (×) is due to Fritz *et al.* (1969) and the present measurement is shown at 120 keV under the assumption that the spectral shape is like that of the Nebula above 35 keV.

With this assumption, the X- and γ-ray pulsed intensity is $7 \pm 2\%$ of the Crab Nebula. This is equivalent to an average pulsed power of $1.6 \pm 0.4 \times 10^{-9}$ ergs cm^{-2} sec^{-1} at the top of the earth's atmosphere; the hard X- and γ-ray regions contain a major component of the pulsed luminosity of NP0532 if the spectral index is like that of the Nebula as a whole. The percentage of pulsed luminosity is of the same order as that observed in the soft X-ray region (Fritz *et al.*, 1969), indicating a softening of the pulsed spectrum between the optical and the hard X-ray regions.

Acknowledgements

Thanks are due to Professors F. C. Michel and H. C. Goldwire for their advice, encouragement and aid in the calculations.

The research was supported in part by the Air Force Office of Scientific Research, Office of Aerospace Research, United States Air Force, under Contract No. F44620-69-C-0083.

References

Cocke, W. J., Disney, M. J., and Taylor, D. J.: 1969, *Nature* **221**, 525.
Comella, J. M., Craft, H. D., Jr., Lovelace, R. V. E., Sutton, J. M., and Tyler, G. L.: 1969, *Nature* **221**, 453.
Fritz, G., Henry, R. C., Meekins, J. F., Chubb, T. A., and Friedman, H.: 1969, *Science* **164**, 709.
Haymes, R. C., Ellis, D. V., Fishman, G. J., Kurfess, J. D., and Tucker, W. H.: 1968, *Astrophys. J.* **151**, L9.
Lynds, R., Maran, S. P., and Trumbo, D. E.: 1969, *Astrophys. J.* **155**, L121.
Minkowski, R.: 1968, in *Nebulae and Interstellar Matter* (ed. by B. M. Middlehurst and L. H. Aller), University of Chicago Press, Chicago, p. 639.
Radhakrishnan, V. and Manchester, R. N.: 1969, *Nature* **222**, 228.
Reichley, P. E. and Downs, G. S.: 1969, *Nature* **222**, 229.
Richards, D. and Comella, J. M.: 1969, *Nature* (preprint).
Warner, B., Nather, R. E., and MacFarlane, M.: 1969, *Nature* **222**, 233.

X-RAY PULSAR IN THE CRAB NEBULA

H. FRIEDMAN

Naval Research Laboratory, Washington, D.C., U.S.A.

Prof. Friedman did not communicate any text or Summary of his communication.

X-RAY PULSAR IN THE CRAB NEBULA

R. NOVICK

Columbia University, New York, N.Y.

No text nor Summary was communicated by the Author.

A SEARCH FOR HIGH-ENERGY γ-RAYS FROM PULSARS

G. G. FAZIO, D. R. HEARN, H. F. HELMKEN, G. H. RIEKE, and T. C. WEEKES

*Mt. Hopkins Observatory, Smithsonian Astrophysical Observatory, Amado, Ariz.,
and Harvard University, Cambridge, Mass., U.S.A.*

Abstract. The 10-m optical reflector at Mt. Hopkins, Ariz., was used to search for cosmic γ radiation from pulsars by detection of atmospheric Čerenkov light generated by energetic particle showers. In the energy region of 10^{11}–10^{12} eV, no evidence of pulsed γ-ray emission was found from either NP0532 (Crab Nebula) or CP1133.

1. Introduction

Two pulsating radio sources (pulsars), NP0532 (Crab Nebula) and CP1133, have been investigated for pulsed γ-ray emission in the energy region from 10^{11} to 10^{12} eV. Such radiation could be produced at the pulsar source by Compton scattering between high-energy electrons and their synchrotron-emitted photons or by high-energy proton collisions. In either case, detection of γ radiation from pulsars would support the theory that these objects are the origin of cosmic radiation in the Galaxy.

2. Observations

At energies above 10^{11} eV, cosmic γ-rays produce high-energy particle showers in the atmosphere; these showers can be detected with ground-based instruments by the Čerenkov light bursts they produce.

The 10-m light reflector at Mt. Hopkins, Ariz., has been used to search for such γ-ray emission from pulsars. The detector has been described in another paper in this symposium (Fazio *et al.*, 1969).

For each source, the reflector, operating in the coincidence mode, was moved in order to track the source for periods of 45–90 min. While the source was being tracked, each coincidence pulse was recorded on one track of a magnetic tape. On a parallel track, timing pulses (typically 10 kHz) from the clock at the satellite-tracking station on Mt. Hopkins were recorded. The frequency stability of the clock pulses was 1 part in 10^{10}. The magnetic tapes were analyzed on a 100-channel pulse-height analyzer operating in the multiscaler mode. The analyzer was recycled after two pulsar periods, with the channel advance and the recycle signal generated from a vernier scaler system driven by the clock pulses (Horowitz, 1969). Thus, the arrival time of each Čerenkov pulse was sorted into time intervals corresponding to the phase of one pulsar period. A monitor was constructed to detect any loss of clock pulses when the magnetic tape was replayed. The analyzer was stopped when a single missing pulse was detected. The pulsar periods were corrected for earth orbital and axial rotation.

L. Gratton (ed.), Non-Solar X- and Gamma-Ray Astronomy, 192–195. All Rights Reserved
Copyright © 1970 by the IAU

TABLE I

Summary of data and results

Source	Date (GMT)	Observation time (min)	Apparent period (sec)	Time interval (msec/channel)	Energy threshold (10^{11} eV)	Flux upper limit (10^{-10}/cm^2 sec)
NP0532	2/10/69	37.4	0.033097092	0.7	1.2	1.4
		37.8	0.033097104			
NP0532	2/10/69	37.4	0.033097092	2.1	1.2	2.6
		37.8	0.033097104			
CP1133	3/16/69	82.5	1.18795000	24	1.4	0.90
CP1133	2/10/69	30.4	1.18785700	24	10	0.12

3. Results

No evidence of periodic high-energy (10^{11}–10^{12} eV) γ-ray emission was detected from either NP0532 or CP1133. The results are summarized in Table I. The calculated upper limit to the flux was based on the largest positive fluctuation observed in the total number of pulses stored in each time interval. To the corresponding number of standard deviations in this channel, three standard deviations were added to compute

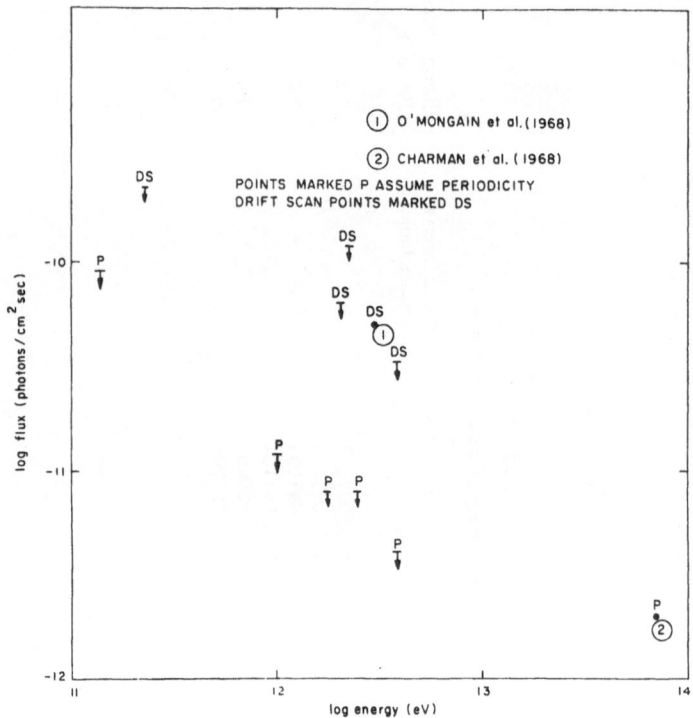

Fig. 1. Summary of results from CP1133. The drift-scan upper limits (DS) are from the results of Fazio *et al.* (1968), modified by the results of Rieke (1969). The periodicity (P) upper limits are from the results of this work, Rieke (1969), and Fazio *et al.* (1968).

the number of Čerenkov pulses per unit time above the background rate. When the results from two tapes were combined, e.g., NP0532, the upper limit was based on the sum of the maximum fluctuation in each case.

The results from several experiments are shown in Figure 1. This experiment does not confirm possible evidence for periodic γ-ray emission from CP1133 previously reported by Charman *et al.* (1968) at 7×10^{13} eV. O'Mongain *et al.* (1968) reported possible evidence for a continuous flux from CP1133 at 4.5×10^{12} eV. Our previous work (Rieke, 1969; Fazio *et al.*, 1968) did not support this result, and the present results do not indicate that the flux occurs in bursts at the pulsar frequency.

Acknowledgement

We wish to thank Roger Carson, Don Hogan, and Ed Horine for their assistance during this experiment.

References

Charman, W. N., Jelley, J. V., Orman, P. R., Drever, R. W. P., and McBreen, B.: 1968, *Nature* **220**, 565.

Fazio, G. G., Helmken, H. F., Rieke, G. H., and Weekes, T. C.: 1968, *Nature* **220**, 892.

Fazio, G. G., Helmken, H. F., Rieke, G. H., and Weekes, T. C.: 1970, this volume, p. 192.

Horowitz, P.: 1969, *Rev. Sci. Instr.* **40**, 369.

O'Mongain, E. P., Porter, N. A., White, J., Fegan, D. J., Jennings, D. M., and Lawless, B. G.: 1968, *Nature* **219**, 1348.

Rieke, G. H.: 1969, Smithson. Astrophys. Obs. Spec. Rep. No. 301.

RADIO, OPTICAL AND X-RAY EMISSION FROM PULSARS

B. BERTOTTI

European Space Research Institute (ESRIN), Frascati, Rome

A. CAVALIERE

Laboratorio Gas Ionizzati (EURATOM-CNEN), Frascati, Rome

and

F. PACINI

Laboratorio di Astrofisica, Frascati, Rome

Abstract. We examine the electromagnetic spectrum of the Crab Nebula pulsar NP 0532. The present observations are in rough agreement with the idea that the orbit of the radiating particles involves two different radii of curvature. One of the possible models requires electrons with a typical energy of the order of 100 MeV and a density $\approx 10^9$ cm^{-3}. A correlation length of about 10 cm could then give the coherence of the radio emission. Slow pulsars are unlikely to emit an appreciable amount of optical or X-ray radiation but a weak infrared emission is possible.

1. We can now be reasonably confident that the timing and the basic energy source of pulsars are both related to the rotation of a magnetic neutron star (Gold, 1968; Pacini, 1968). This confidence is based upon the observed repetition rate which is consistent with the neutron star idea but in most cases exceeds the Jeans frequency of a white dwarf. Furthermore, the period increase observed in the case of several pulsars can be explained as loss of rotational energy (Gold, 1968; Pacini, 1968; Gold, 1969; Finzi and Wolf, 1969; Bertotti *et al.*, 1969a; Gunn and Ostriker 1969a), but it cannot be easily interpreted in a vibrating model. In the particular case of the Crab Nebula pulsar NP 0532, the loss of rotational energy is of the same order as the total energy requirement of the Crab Nebula, about 10^{38} ergs sec^{-1} (largely under form of relativistic particles). It appears therefore that the rotation of a magnetic neutron star is connected not only to the pulsar phenomenon, but also to the overall high energy activity of Supernovae Remnants (Pacini, 1967, 1968; Gold, 1969; Bertotti *et al.*, 1969a; Gunn and Ostriker, 1969a, b; Finzi and Wolf, 1969).

An understanding of pulsars requires also a description of the interaction between the spinning star and the surrounding plasma, and furthermore an analysis of the radiation process.

Several suggestions have been put forward concerning the first of these points, namely how rotational energy can be released from the star. These involve:

(a) rigid corotation of the magnetosphere up to a critical distance, limited by $R = c/\Omega$, where Ω is the angular rotation frequency of the star (Gold, 1968, 1969; Goldreich and Julian, 1969; Michel and Tucker, 1969; Michel, 1969);

(b) emission of electromagnetic energy from an oblique rotator, which, for distances r greater than c/Ω, has the character of magnetic dipole radiation (Pacini, 1967, 1968; Bertotti *et al.*, 1969a; Gunn and Ostriker, 1969a, b).

The radiation processes which have been suggested are essentially induced scattering from relativistic particles (Ginzburg *et al.*, 1968, 1969; Bertotti *et al.*, 1969a; Cavaliere, 1969) and synchrotron-like radiation (Gold, 1968, 1969; Michel and Tucker, 1969). These processes seem to be good candidates to account for the high radio brightness, broad band and polarization of the pulsar radiation.

On the observational side, while most sources appear to emit only in the radio band, the recent discovery of optical (Cocke *et al.*, 1969) and X-ray (Bradt *et al.*, 1969; Fritz *et al.*, 1969; Haymes *et al.*, 1969) flashes from NP0532 extends, in this particular case, the spectral band to be accounted for by the emission mechanisms.

In this paper, we shall concentrate upon the electromagnetic spectrum of NP0532 and examine a simple synchrotron-like radiation process based indeed upon the vectorial acceleration of relativistic particles with $\gamma = E/mc^2 \gg 1$ (See also Bertotti *et al.*, 1969b).

Our main assumption will be that the overall kinematics of relativistic particles around the star entails two widely different radii of curvature ϱ and hence two fundamental frequencies c/ϱ: these qualitative features result in a number of detailed models. Here we shall concentrate mainly upon the emission which results, at $r \geqslant c/\Omega$, from electrons forced to rotate at the basic frequency Ω by the stellar electromagnetic field and also circling around the lines of force of the local magnetic field B. The first motion can be due to the well-known Gold's idea of corotation up to the speed of light cylinder $R = c/\Omega$; alternatively, outside R, it can be forced upon by the absorption of the angular momentum carried by the magnetic dipole radiation emitted from the star. For NP0532 we have $\Omega = 200 \text{ sec}^{-1}$. On the other hand, the magnetic field at distances $\approx c/\Omega$ must be about 10^6 G (corresponding to a surface field of about 10^{12} G) if the energetic stockpile of the Crab Nebula is to be drawn from the stellar rotation through the agency of the magnetic field. This would imply a gyrofrequency in the rest frame $\omega_L = eB/mc$ much larger than the frequency Ω.

We note explicitly that the same qualitative considerations would be relevant even if the motion has a torsion and a tangential acceleration. In particular, they are valid in the case of emission of particles from the polar regions (Goldreich and Julian, 1969; Radhakrishnan and Cooke, 1969), provided that the particles simultaneously move away from the poles along the field lines and also circle around them. (We thank Dr. M. Rees for calling our attention on this point.)

In the following, we shall minimize the number of free parameters by taking a small dispersion in the particles energy, i.e. $\Delta\gamma/\gamma \ll 1$. Fit to detailed, forthcoming spectra may require relaxing this constraint in a second approximation: for the time being, we consider powers emitted across large spectral regions and we recall that the radio spectrum of NP 0532 has a peak about 100 MHz (Comella *et al.*, 1969) while the optical spectrum is roughly flat and shows an ultraviolet excess (Cocke *et al.*, 1969).

2. We recall briefly the emission characteristics of the motion of relativistic particles (for a more detailed account, see, e.g. Ginzburg and Syrovatskii, 1965, 1969). For these particles, the high frequency features of the radiation depend essentially upon the radius of curvature ϱ (or the associated angular frequency $\bar{\Omega} = c/\varrho$) at those points of the orbit whose tangent lies within an angle $1/\gamma$ with the visual line of the observer (for single particles the peak emission is indeed concentrated within a cone of width $1/\gamma$).

The spectrum is composed of harmonics of the fundamental frequency $\bar{\Omega}$; this structure is irrelevant for $\omega \gg \bar{\Omega}$, and the corresponding spectral power $p(\omega)$ can be approximated as follows

$$p(\omega) \approx \frac{e^2}{c} \bar{\Omega}\gamma \left(\frac{\omega}{\omega_{\mathrm{cr.}}}\right)^{1/3} \quad \text{for} \quad \omega \lesssim \omega_{\mathrm{cr.}} \tag{1}$$

$$p(\omega) \approx \frac{e^2}{c} \bar{\Omega}\gamma \left(\frac{\omega}{\omega_{\mathrm{cr.}}}\right)^{1/2} \exp\left(-\frac{\omega}{\omega_{\mathrm{cr.}}}\right) \quad \text{for} \quad \omega > \omega_{\mathrm{cr.}}, \tag{2}$$

where the critical frequency is given by $\omega_{\mathrm{cr.}} = \frac{3}{2}\bar{\Omega}\gamma^3$.

For the case of a helicoidal path and for the density effects, see Ginzburg and Syrovatskii (1969).

For our purposes, coherence effects are very important, as indicated by the very high brightness temperature of the pulsars radio emission. In the following we shall assume that coherence stems from bunching of particles (Gold, 1968, 1969), even if other causes are possible (see Ginzburg et al., 1968, 1969). One particular, but not exclusive, cause of correlations in the electron density can be electrostatic instability feeding on relative streaming energy.

Coherent radiation by bunches of particles with correlation length L and density n would give an emission per unit volume

$$P(\omega) = p(\omega) n (1 + F). \tag{3}$$

The coherence factor F, in the case where turbulent fluctuations have frequencies smaller than the radiation angular frequency ω, is of the form

$$F = nL^3 f(c/\omega). \tag{4}$$

Here $f(c/\omega)$ is the dimensionless Fourier transform of the density correlation function and vanishes for $L \gg c/\omega$. For a gaussian bunching, for example, we obtain

$$f(c/\omega) = \exp\left[-\left(\frac{L\omega}{4\pi c}\right)^2\right] \tag{5}$$

but different and weaker dependences upon the frequency are conceivable.

3. We shall now discuss two different possibilities:

(a) all the observed spectrum is associated with the basic frequency Ω (the gyrofrequency could then give rise to a high-energy γ-rays emission);

(b) the radiospectrum is associated with the frequency Ω while the optical and the X-ray emission are related to the gyrofrequency.

We shall find that the first possibility implies much higher values for the particles energy and for their density than the second one.

Let us investigate first the case where the entire electromagnetic spectrum observed up to now is connected with the basic angular frequency Ω. It is then easy to show that the observed ratio of power in the optical and X-ray band (respectively indicated by $\Delta\omega_0 \approx 5 \times 10^{15}$ sec^{-1} and by $\Delta\omega_x \approx 10^{19}$ sec^{-1}), fixes the critical frequency $\omega_1 = \frac{3}{2}\Omega\gamma^3$. The relation

$$\frac{P_0}{P_x} \approx 10^{-2} \approx \left(\frac{\omega_0}{\omega_1}\right)^{1/3} \left(\frac{\omega_1}{\omega_x}\right)^{1/2} \frac{\Delta\omega_0}{\Delta\omega_x} \exp\left(-\frac{\omega_x}{\omega_1}\right) \tag{6}$$

gives then, roughly, $\omega_1 \approx 10^{18}$ sec^{-1} and implies $\gamma \approx 3 \times 10^5$. We stress, however, that the exponential in Equation (6) follows from $\Delta\gamma/\gamma \ll 1$.

We can now determine an average coherence factor F over the radio band $(\Delta\omega_R \approx 10^9$ sec$^{-1})$ from the observed ratio

$$\frac{P_R}{P_0} \approx 10^{-2} \approx F\left(\frac{\omega_R}{\omega_0}\right)^{1/3} \frac{\Delta\omega_R}{\Delta\omega_0}. \tag{7}$$

We obtain $F \approx 5 \times 10^6$.

The average radio power $P_R = \varepsilon 10^{31}$ ergs sec^{-1} tells us how many particles nV are emitting the observed radiation from a volume V

$$P_R \approx \frac{e^2}{c} \Omega\gamma \left(\frac{\omega_R}{\omega_1}\right)^{1/3} nVF\Delta\omega_R. \tag{8}$$

Here ε is a factor introduced to take into account the beaming of the radiation: if we assume that the pulsar can be detected from most directions, ε is of the order of the ratio between the pulse length τ and the period T, namely $\varepsilon \approx 10^{-1}$. We then obtain $nV \approx 5 \times 10^{38}$; assuming for the emitting volume a value ΔR^2 with $\Delta \approx 10^7$ cm (inferred from the existence of sub-millisecond pulse structures), the particle density turns out to be $n \approx 5 \times 10^{15}$ cm^{-3}.

We now examine, with similar arguments, the other case where the radio spectrum is determined by the basic frequency Ω, while the optical and the X-ray emissions are connected to the Larmor gyration and hence to the critical frequency $\omega_2 = \frac{3}{2}(\omega_L/\gamma)\gamma^3$. The relation

$$\frac{P_0}{P_x} \approx \left(\frac{\omega_0}{\omega_2}\right)^{1/3} \left(\frac{\omega_2}{\omega_x}\right)^{1/2} \frac{\Delta\omega_0}{\Delta\omega_x} \exp\left(-\frac{\omega_x}{\omega_2}\right) \tag{9}$$

gives now $\omega_2 \approx 10^{18}$ sec^{-1}. This corresponds to $B\gamma^2 \approx 5 \times 10^{12}$ G.

The radio spectrum can now be decreasing either because of loss of coherence or because we are dealing with frequencies $\omega > 0.3\omega_1$, where ω_1 is again the critical frequency associated with the basic rotation. In the latter case we must have $\gamma \approx 200$. The corresponding value for B, deduced from the previous expression $B\gamma^2 \approx 5 \times 10^{12}$, is about 10^6 G, in agreement with a previous independent estimate.

The observed ratio

$$\frac{P_R}{P_0} \approx \frac{\Omega\gamma}{\omega_L}\left(\frac{\omega_2}{\omega_0}\right)^{1/3}\frac{\Delta\omega_R}{\Delta\omega_0}F \tag{10}$$

gives the numerical value for the coherence factor $F \approx 10^{13}$.

Finally, the absolute radio power gives again the number density and we obtain $n_e \approx 10^9$ cm^{-3}, leading to a correlation length $L \approx 10$ cm.

4. We have not discussed here the pulsed nature of the source but only its electro-magnetic spectrum. We recall however that the radiation in a fixed direction comes from particles whose velocity vector lies within an angular distance of order $1/\gamma$ from this direction: the pulse width could therefore be a consequence of the periodical filling up (due to the rotation of the star) of such a region in phase space. For example, one could have a rotating velocity distribution of angular size $\Delta\theta \approx 2\pi\tau/T$. If $\Delta\theta > 1/\gamma$. one expects a larger linear polarization for spectral bands close or beyond the relevant critical frequency. In our case (b), we must therefore expect more primary polarization in the radio emission than in the optical emission of NP 0532; the X-rays should again be strongly polarized. For pulsars having a smaller duty cycle τ/T, we expect the eventual emergency of the polarization features of single emitters: this seems to be the case for the source CP 0328 (Clark and Smith, 1969; Cocconi, 1969).

Finally, we notice explicitly that the above emission scheme explains why optical and X-ray emissions show up for NP 0532 and not for other pulsars. Indeed, for slow pulsars, the local value of the magnetic field at distances $r \approx c/\Omega$ would be much smaller than for NP 0532, provided that the surface field on the neutron star is comparable. For slow pulsars, therefore, the peak of the gyroemission would lie in the infrared, possibly around 10^{11} or 10^{12} MHz. Observations in this spectral range are desirable, even if the power radiated (proportional to B^2) is likely to be quite low. Among the pulsars known at present, only the Vela pulsar can be expected to emit (weakly) in the optical range.

References

Bertotti, B., Cavaliere, A., and Pacini, F.: 1969a, *Nature* **221**, 624.
Bertotti, B., Cavaliere, A., and Pacini, F.: 1969b, *Nature* **223**, 1351.
Bradt, H., Rappaport, S., Mayer, W., Nather, F. H., Warner, B., Macfarlane, M., and Kristian, J.: 1969, *Nature* **222**, 728.
Cavaliere, A.: 1969, 'Emission Processes in Plasma', paper given at the Pisa Meeting on Pulsars, 14–16 April.
Clark, R. R. and Smith, F. G.: 1969, *Nature* **221**, 724.
Cocconi, G.: 1969, preprint.
Cocke, W. J., Disney, M. J., and Taylor, D. J.: 1969, *Nature* **221**, 525.
Comella, J. M., Craft, H. D., Lovelace, R. V., and Sutton, J. M.: 1969, *Nature* **221**, 453.
Finzi, A. and Wolf, R. A.: 1969, *Astrophys. J.* **155**, L107.
Fritz, G., Henry, R. C., Meekins, J. F., Chubb, T. A., and Friedman, H.: 1969, *Science* **164**, 709.
Ginzburg, V. L. and Syrovatskii, S. I.: 1965, *Ann. Rev. Astron. Astrophys.* **3**, 297.
Ginzburg, V. L. and Syrovatskii, S. I.: 1969, *Ann. Rev. Astron. Astrophys.* **7**, 375.
Ginzburg, V. L., Zhelezniakov, V. V., and Zaitsev, V. V.: 1968, *Nature* **220**, 535.

Ginzburg, V. L., Zhelezniakov, V. V., and Zaitsev, V. V.: 1969, *Astrophys. Space Sci.* **4**, 464.
Gold, T.: 1968, *Nature* **218**, 731.
Gold, T.: 1969, *Nature* **221**, 25.
Goldreich, P. and Julian, W. H.: 1969, *Astrophys. J.* **157**, 869.
Gunn, J. F. and Ostriker, J. P.: 1969a, *Nature* **221**, 454.
Gunn, J. F. and Ostriker, J. P.: 1969b, *Phys. Rev. Letters* **22**, 728.
Haymes, R. C., Freeman, K. C., and Harnden Jr., F. R.: 1969, *Astrophys. J.* **156**, L107.
Michel, F. C.: 1969, *Phys. Rev. Letters* **23**, 247.
Michel, F. C. and Tucker, W. H.: 1969, *Nature* **223**, 277.
Pacini, F.: 1967, *Nature* **216**, 567.
Pacini, F.: 1968, *Nature* **219**, 145.
Radhakrishnan, V. and Cooke, D. J.: 1969, *Astrophys. Letters* **3**, 225.

ROTATING NEUTRON STARS, PULSARS,
AND COSMIC X-RAY SOURCES

WALLACE H. TUCKER

Rice University, Houston, Texas, U.S.A.

The purpose of this paper is to discuss the relationship between rotating neutron stars, pulsars, and cosmic X-ray sources. The latter may be divided into at least two classes: the sources with large angular diameters, such as the Crab Nebula, and those with small angular diameter, such as Sco X-1. I submit that a basic model, consisting of a rotating neutron star losing mass in the presence of a large magnetic field, can account for both types of X-ray source. The extended sources represent the case where the energy in the 'neutron-star wind' is greater than the magnetic energy. The streaming protons and electrons deposit their energy far out into the nebula in a shock transition region. The relativistic electrons responsible for the extended sources of radio, optical and X-ray emission are produced in the transfer of energy between the protons and electrons in the shock wave, and by magnetic pumping in hydromagnetic waves which are generated by fluctuations in the mass loss rate. The compact sources, such as Sco X-1, represent the other extreme where the magnetic energy dominates, so that no mass loss occurs. The particles are then accelerated and radiate in radiation belts around the neutron star, resulting in a source with a small angular diameter.

Consider first the X-ray source in the Crab Nebula: I am referring only to the extended, steady source; the pulsed emission, amounting to 7% of the total [1], will not be discussed in what follows.

Of fundamental importance for the theory of the Crab Nebula is the observation that the period of the pulsar located there is increasing at a rate of one part in 2400/ year [2]. This slowing down has been interpreted as being due to a torque on a rotating neutron star [3–6] and implies a rotational energy-loss rate comparable to the luminosity of the entire nebula [7]. This agreement has led to the general view that the two phenomena, the rotational energy-loss rate of the neutron star, and the luminosity of the Crab Nebula, are related. For this to be so, energy must be transferred from the rotating neutron star to relativistic particles and then from relativistic particles to radiation.

The second part of this chain is fairly well understood; relativistic electrons in a magnetic field can radiate very efficiently by means of the synchrotron mechanism, and the observations [8, 9] are consistent with this interpretation.

If we accept the synchrotron mechanism, then the observed size of the X-ray source places severe restrictions on the way in which the relativistic electrons are produced, as stressed recently by Burbidge and Hoyle [10]. The observations of Oda *et al.* [11] show that the projected diameter of the source of low energy X-rays is about 2 light years, for an assumed distance of 5000 light years.

The maximum energy that an electron can have after traversing a distance from

L. Gratton (ed.), Non-Solar X- and Gamma-Ray Astronomy, 202–207. *All Rights Reserved*
Copyright © 1970 by the IAU

r_0 to r in a magnetic field varying as

$$B_\perp = B_{\perp_0}(r_0/r)^n \tag{1}$$

is given approximately by

$$\gamma_{max} = E_{max}/mc^2 = 10^{19}(2n-1)/B_{\perp_0}r_0[1-(r_0/r)^{2n-1}], \tag{2}$$

where $B_\perp$ = component of magnetic field perpendicular to the electron velocity. For the case of a uniform field, $r_0 = 0$, and

$$\gamma_{max} = 10^{19}/B_{\perp_0}^2 r \tag{3}$$

For $n \geqslant 1$ and $r \gg r_0$.

$$\gamma_{max} = 2n \times 10^{19}/B_{\perp_0}^2 r_0, \tag{4}$$

which shows that the electron loses most of its energy in the region where the magnetic field is large. On the other hand, in order to generate X-rays with energies of 4 keV, electron energies corresponding to

$$\gamma = 10^6/2B_\perp^{1/2} \tag{5}$$

are required. Since the magnetic field is practically uniform and equal to about 5×10^{-4} gauss over the region occupied by the X-ray source, energies corresponding to $\gamma = 2 \times 10^7$ are needed. In addition, Equation (3) shows that the maximum radius for the soft X-ray source is $r_{max} \cong 2 \times 10^{18}$ cm, just barely compatible with the observations. Of course, this result is sensitive to the assumed magnetic field; for $B_\perp = 2 \times 10^{-4}$ gauss, $r_{max} \approx 7 \times 10^{18}$ cm. If the electrons are produced in a high field region and injected into the nebula, Equation (4) shows that the magnetic field $B_{\perp_0}$ and the size r_0 characterizing the accelerating region must satisfy $B_{\perp_0}r_0 \leqslant 10^{12}n$.

Consider now some of the proposed mechanisms transferring angular momentum away from the neutron star in the light of this requirement. The basic model [3–6] consists of a rotating neutron star losing mass in a large magnetic field. The various theories differ on important questions such as the strength of the magnetic field and the amount of mass loss, etc.

In one class of model, the energy in the magnetic field is much greater than the energy of the mass flow away from the star, and the torque on the neutron star is produced by the radiation of very low frequency magnetic dipole radiation [4, 5]. In this model the acceleration of charged particles occurs in a region having a size of the order of 10^8 cm, and a magnetic field strength $B_0 \sim 10^6$ gauss, so that $B_0 r_0 \sim 10^{20}$. Hence, the electrons lose almost all their energy before escaping the region, regardless of the energies attained in the accelerating region. This model is a very efficient generator of synchrotron radiation but can not inject energy into the nebula in the form of relativistic electrons at anything like the required rate.

In the other type of model, the energy in the mass flow is much greater than the magnetic energy density, and the torque is produced by much the same process that slows down the sun and other stars. In this case, the flow of the plasma carries out the magnetic field lines to form a predominantly radial magnetic field structure which takes the form of Archimedean spirals as a result of the rotation of the neutron star. The radial extension of the magnetic field transfers angular momentum to the out-flowing plasma and thereby results in a significant angular momentum loss. The energy density of the flow, which resides almost entirely in the protons, is fixed by the observed torque at about 10^{38} erg/sec ([6]; for more details, see [12]). This energy will be deposited in the nebula as a whole when the energy density of the flow drops to a value equal to the energy density of the ambient medium. For the Crab Nebula, the energy density of the ambient medium is of the order 10^{-8} erg/cm^3 [13], so that the energy deposition must occur at a distance $r \sim 10^{17}$ cm from the neutron star. At this point, we might expect that a standing shock wave or some kind of transition region would exist. Fluctuations in the mass loss rate would result in fluctuations of the shock position and the generation of hydromagnetic waves.

Using this general picture, we can interpret the features which are observed in the active region near the center of the nebula. The most recent and exhaustive study of these features is given by Scargle [14]. He finds a permanent wisp which moves about but is confined to a definite region near the center of the nebula, about 10^{17} cm from the central star. It is proposed here that this wisp be identified with the transition region, or shock wave. In fact, Scargle states that the structure of the central wisp "is just the configuration which might be produced by a small object emitting ionized gas in all directions in an initially uniform magnetic field". A number of moving wisps that gradually damp out are also observed. These features may be interpreted as hydromagnetic waves generated by fluctuations in the mass loss rate.

Melrose [15] has shown that magnetic pumping in the moving wisps, considered to be hydromagnetic waves, can account for the large electron energies necessary to explain the observed X-ray emission. Note that $B_0^2 r_0 \sim 3 \times 10^{10}$, so that synchrotron losses are not a problem. In this way the extended source of X-ray emission can be related to the observed rate of increase of the period of the pulsar. On the basis of this model there should be some correlation between a strong train of pulses and the appearance of a new wisp in the nebula, with an appropriate time lag.

The shock wave may also be a source of relativistic electrons. The rate at which relativistic particles are generated at the shock wave depends on the mass loss rate. For large mass loss rates ($> 10^{17}$ gm/sec) the flow velocities are non-relativistic, whereas for the minimum mass loss consistent with Maxwell's equations ($\sim 2 \times 10^7$ gm/sec) energies of the order of 10^3 BeV are attained [12]. If the electrons and protons share the energy equally in the shocked plasma, then 10^3 BeV electrons are produced. These electrons would have a lifetime of the order of 100 years and would emit optical radiation. Due to their long lifetime, they would diffuse throughout the nebula before emitting most of their energy. The observed optical luminosity of the nebula is un-certain because of the effects of interstellar absorption. Estimates range from 10^{36} erg/sec to a value somewhat greater than 10^{38} erg/sec [16]. Of course, the electron en-ergies might be considerably lower, and the radiation consequently shifted to lower frequencies. Observations in the far infrared should soon decide whether the Crab is anomalously bright there.

Concerning the X-ray emission from other pulsars, I have listed in Table I the rotational energy loss rate for the seven pulsars for which slowing down times have

TABLE I

Expected fluxes from pulsars

Pulsar	(dE/dt) n.s.	Estimated distance	Flux at earth
NP 0532	7×10^{37} erg/sec	2 kpc	2×10^{-7} erg/cm² sec
PS 0833	8×10^{35}	0.468	3×10^{-8}
CP 0950	1×10^{32}	0.048	4×10^{-10}
CP 1133	1×10^{31}	0.026	2×10^{-10}
CP 0834	2×10^{31}	0.253	3×10^{-12}
CP 0808	$<6 \times 10^{29}$	0.097	$<6 \times 10^{-13}$
CP 1919	3×10^{30}	0.093	3×10^{-12}

been measured. The computations assume that the moment of inertia is the same for all pulsars, $= 10^{45}$ gm cm². Using the slowing down rates and distances as given in [18] and [19], the total energy flux at earth is easily computed and is given in the fourth column of Table I. It would appear that only NP 0532 and PS 0833 are capable of producing detectable X-ray fluxes at earth. The recent observations of Gursky et al. [20] set a limit of 10^{-9} erg/cm² sec on the X-ray flux from Vela X and rule out any identification of Vela XR-1 with PS 0833. However, the intensity of line emission from Vela X is $\sim 3 \times 10^{-8}$ erg/cm² sec [21], so maybe all the energy goes into the excitation of line emission rather than the production of relativistic electrons.

Finally, consider briefly the case where the magnetic energy density is much greater than the particle energy density. It was shown above that if relativistic electrons were produced around such an object, they would immediately radiate all their energy by the synchrotron process. While such a model is not adequate to explain the X-ray source in the Crab Nebula, it is an attractive possibility for the compact X-ray sources such as Sco X-1.

A major difficulty for any model of Sco X-1 is the shape of the optical spectrum, which shows that the spectrum must turn over around 10^{15} Hz. Requiring that this be due to synchrotron self-absorption fixes the angular diameter of the source in terms of the flux density $S_\nu(\text{Wm}^{-2}\ \text{Hz})$, and the magnetic field $B_\perp$:

$$\theta \cong 4 \times 10^{16} S_\nu^{1/2} B_\perp^{1/4} \nu^{-5/4}. \tag{6}$$

Setting $\nu = 10^{15}$ Hz, and $S_\nu = 10^{-26}$ [22], yields

$$R \cong 2 \times 10^6 B^{1/4} d_{\text{kpc}}\ \text{cm}, \tag{7}$$

where d_{kpc} is the distance to the source in kiloparsecs.

In order for the synchrotron process to produce the required luminosity we must have, for a uniform spherical source

$$L_{syn} = 2 \times 10^{-15} B^2 \gamma^2 N \tfrac{4}{3} \pi R^3 = L_x \simeq 10^{37} d_{kpc}^2 \text{ erg/sec}, \qquad (8)$$

where N is the relativistic electron number density. It must satisfy the inequality

$$N < B^2/8 \, \pi mc^2 \gamma \qquad (9)$$

if the magnetic field energy is to be dominant. Equations (5)–(9) imply that we must have $R > 4 \times 10^7 d_{kpc}^{16/17}$ cm. On the other hand, if we require that the surface field on the neutron star B_{ns} be less than 10^{14} gauss, then $R < 7 \times 10^7 \, d_{kpc}$ cm. Thus, R must be about 10^7 cm. Choosing $d_{kpc} = 0.2$, we arrive at a model for Sco X-1 described by the following parameters:

$$
\begin{aligned}
R &\approx 10^7 \text{ cm} \\
B &= 5 \times 10^7 \text{ gauss}, \quad B_{n.s} = 5 \times 10^{10} \text{ gauss} \\
\gamma &= 5 \times 10^3 \qquad\qquad N = 6 \times 10^5 \text{ cm}^{-3}
\end{aligned}
\qquad (10)
$$

This discussion relates only to the emission mechanism and is independent of the assumed energy source. However, note that a rotating magnetic dipole [4, 5] can produce 4×10^{35} erg/sec if the frequency of rotation $\Omega = 10^3$ rad/sec, and the magnetic moment $M = 5 \times 10^{28}$ gauss cm^3. For this rotation frequency, the acceleration of electrons should occur around $R = c/\Omega \approx 3 \times 10^7$ cm, in agreement with the above estimate for the size of the emitting region. Pulsed emission, if it occurs, should have a period of about 6 m sec.

Acknowledgements

This research has been supported by the National Aeronautics and Space Administration. It is a pleasure to acknowledge helpful discussions with A. J. Dessler and F. C. Michel.

References

[1] Haymes, R., Fishman, G., and Harnden, R.: 1969, *Astrophys. J.* (in press).
[2] Comella, J., Craft, jr., H. D., Lovelace, R. V. E., Sutton, J. E., and Tyler, G.: 1969, *Nature* **221**, 453.
[3] Gold, T.: 1969, *Nature* **221**, 25.
[4] Gunn, J. and Ostriker, J.: 1969, *Nature* **221**, 454.
[5] Pacini, F.: 1968, *Nature* **219**, 145.
[6] Michel, F. and Tucker, W.: 1969, *Nature* **223**, 277.
[7] Finzi, A. and Wolf, R.: 1969, *Astrophys. J. (Letters)* **155**, L107.
[8] Ginzburg, V. and Syrovatskii, S.: 1965, *Ann. Rev. Astron. Astrophys.* **3**, 300 (Annual Rev., Palo Alto), and references cited therein.
[9] Haymes, R. C., Ellis, D. V., Fishman, G. J., Kurfess, J. D., and Tucker, W. H.: 1968, *Astrophys. J. (Letters)* **151**, L9. – Peterson, L., Jacobson, A., Pelling, R., and Schwartz, D.: 1968, *Canadian J. Phys.* **47**, 437. – Gorenstein, P., Kellogg, E., and Gursky, H.: 1969, *Astrophys. J.* **156**, 315. – Gould, R. J.: 1967, *Amer. J. Phys.* **35**, 5.
[10] Burbidge, G. and Hoyle, F.: 1969, *Nature* **221**, 847.
[11] Oda, M., Bradt, H., Garmire, G., Spada, G., Sreenkantan, B. V., Gursky, H., Giacconi, R., Gorenstein, P., and Waters, J. R.: 1967, *Astrophys. J. (Letters)* **148**, L5.

[12] Michel, F. C.: 1969, *Astrophys. J.* **158**, 727.
[13] Woltjer, L.: 1958, *Bull. Astron. Inst. Netherl.* **14**, 39.
[14] Scargle, J.: 1969, *Astrophys. J.* **156**, 401.
[15] Melrose, D.: 1969, *Astrophys. Space Sci.* **4**, 246.
[16] Geisel, S.: 1969, Rice University Preprint.
[17] Cameron, A. G. W.: 1969, 'Supernovae, Neutron Stars, and Pulsars'. Yeshiva University Preprint.
[18] Rohlfs, K., Grewing, M., and Mebold, U.: 1969, 'Improved Pulsar Distances'. Astron. Inst. Bonn Preprint.
[19] Grewing, M. and Priester, W.: 1969, 'Pulsar Periods as Age Indicators'. Astron. Inst. Bonn Preprint.
[20] Gursky, H., Kellogg, E., and Gorenstein, P.: 1968, *Astrophys. J. (Letters)* **154**, L71.
[21] Milne, D.: 1968, *Australian J. Phys.* **21**, 201.
[22] Neugebauer, G., Oke, J., Becklin, E., and Garmire, G.: 1969, *Astrophys. J.* **155**, 1.

EMISSION MECHANISMS IN X-RAY SOURCES

(Invited Discourse)

L. WOLTJER

Dept. of Astronomy, Columbia University, New York, N.Y., U.S.A.

1. Introduction

A large body of spectral information on X-ray sources has now become available, but the interpretation remains ambiguous. If temperature variations and finite optical depth effects are taken into account, almost any spectrum can be fitted to a model of thermal bremsstrahlung. If a suitable energy spectrum is adopted for the relativistic electrons, a wide variety of synchrotron spectra becomes possible. Although one or the other interpretation may seem artificial in some cases, it nevertheless should be pointed out that a strictly isothermal source would be a miracle and that power-law type energy spectra of the relativistic particles can apply over only a limited range of energies. More satisfactory progress can be made when the spectral data are augmented with structural information, when emission lines can be studied and polarization can be measured. Not only do the intensities of emission lines give much more detailed information on the temperature and density in a hot gas than can be derived from the continuum, but at sufficient resolution velocity fields can also be studied. Useful structural information probably can be obtained only if spatial resolution of 1 arc-min or better is achieved. But one has only to look at the situation in radio astronomy to see how essential this information is for the building of quantitative models.

For the near future polarization data are vital for separating thermal and nonthermal radiation, although the situation is not as unambiguous as might have been hoped on account of the possible effects of electron scattering in asymmetrical thermal sources (Angel, 1969). But again, a combination with high spatial resolution is required to profit in full from the polarization information.

In the following we shall review the various processes that have been discussed as possible mechanisms for the X-ray emission in galactic sources.

2. Black-Body Emission

Immediately after the discovery of the first X-ray sources it was realized that a neutron star with a surface temperature of the order of 2×10^7 K would radiate 10^{38} ergs/sec, essentially with a black-body spectrum, mainly at X-ray wavelengths. Subsequent studies of the cooling of neutron stars indicated that such an object with a surface temperature of 2×10^7 K would have an internal temperature near 10^9 K, and would quickly lose most of its thermal energy through neutrino processes. As a consequence, it would cool too rapidly to be of interest. Some doubts remain, however, concerning

L. Gratton (ed.), Non-Solar X- and Gamma-Ray Astronomy, 208–215. All Rights Reserved
Copyright © 1970 by the IAU

the correctness of these cooling calculations. As has been pointed out by several authors it is not unlikely that a large part of the body of a neutron star may be made of superconductive and superfluid matter. This could have important effects, both on the specific heat of the matter and on the rates of the neutrino processes (Ginzburg, 1969). In addition, the cooling calculations presuppose that the internal thermal energy is the only source of energy that need be considered. However, it appears to be likely that the magnetic and pulsational energy may be comparable to or larger than $10^{50 \pm 1}$ ergs. The slow dissipation of magnetic energy would occur mostly in the outer shell of the star, especially if the interior were superconductive and the fraction of the energy going into neutrinos might be kept comparatively low because the interior would not be very hot. Detailed calculations to check this point quantitatively are in progress. Thus, although at the moment no X-ray sources are known in which there is reason to believe that black-body radiation is dominant, it may be premature to exclude this process altogether.

3. Supernova Shells

The next three mechanisms all relate to supernova remnants. Five such remnants with an age less than 1000 years are known (SN 1006, Crab Nebula SN 1054, Tycho SN 1572, Kepler SN 1604 and the strong radio source Cas A which appears to have originated in an unobserved supernova event around 1700). Not much is known about SN 1006, although a radio source identification has been suggested. The Crab Nebula (at 1.5 kpc), Tycho's supernova (at 3–5 kpc) and Cas A (at 3–4 kpc) have all been identified as X-ray sources, while Kepler's supernova probably is so distant (5–10 kpc) that it would be undetectable if it radiated at the same power level as the others. It thus appears likely that most young supernova remnants (there may be 20 such objects with ages below 1000 years in the Galaxy) are X-ray sources. So far, no convincing X-ray source identifications with older supernova remnants like the Cygnus Loop have been made. Most supernova remnants appear to have initial expansion velocities of the general order of 10000 km/sec (except for the Crab Nebula, with 1500 km/sec), but there is controversy as to the total initial masses and initial energies involved. Estimates have been made ranging between 0.1 and 10 solar masses and between a few times 10^{49} and 10^{52} ergs.

The most direct way in which X-rays may be generated in a supernova remnant is by the interaction between the expanding shell and the interstellar (or circumstellar) medium. A strong shock will separate the compressed gas from the undisturbed gas; a density jump of a factor 4 and a temperature $T \sim \frac{1}{8} m_p V^2 / k$ (in the case of hydrogen, with m_p the proton mass, k the Boltzmann constant and V the shock velocity) are expected at the shock. Cooling effects are slow at 10^9 K (corresponding to $V \simeq 10^9$ cm/sec) and much of the radiation would come from gas at a temperature not much below this value. The medium would be optically thin to X-rays and consequently a flat spectrum is expected. This is at variance with the spectral information available for Cas A and Tycho's object (Gorenstein *et al.*, this volume p. 134). Also for the Crab

Nebula, where the temperature would be lower, a spectral fit with this model seems impossible. Moreover, in all cases, the interstellar density that is needed to account for the observed intensities is quite high; in the case of the Crab Nebula 70 cm^{-3} (Heiles, 1964), a value almost a factor 100 larger than appears probable on other grounds.

4. Synchrotron Radiation

The prime candidate for this mechanism is the Crab Nebula. A discussion of the motions of the filaments which were measured by Trimble (1968) leads to a most probable distance of 1500 pc (see Appendix). Much is known about the spectrum of the object (Figure 1) from radio to X-ray wavelengths, the main uncertainty being whether the radio spectrum turns up beyond 10000 Mc/sec or not. For the moment we consider the measurements that do not show this rise as the more reliable. In

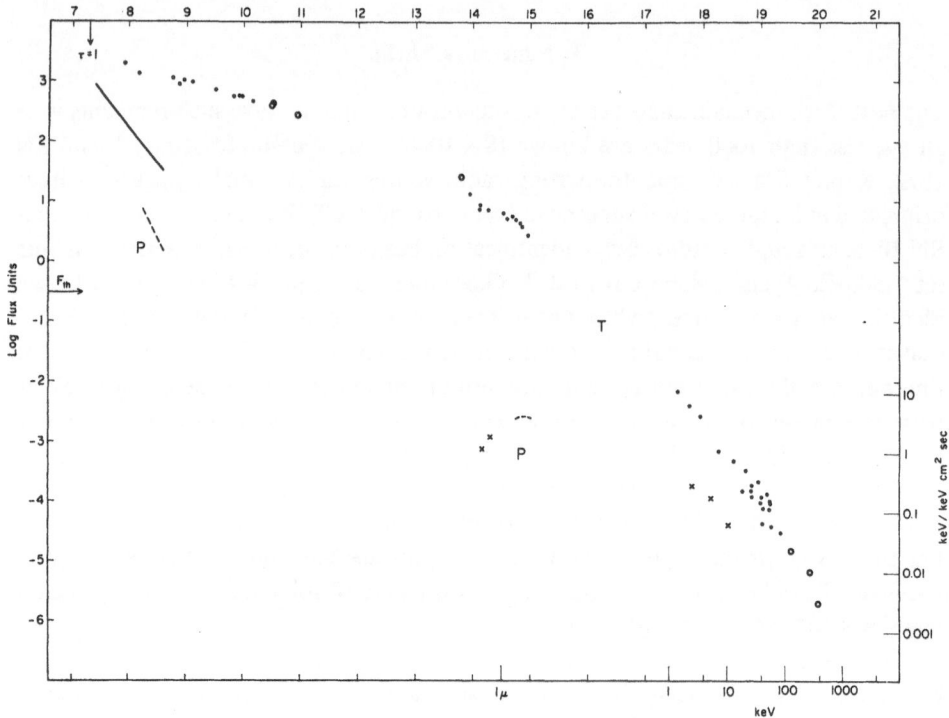

Fig. 1. The Spectrum of the Crab Nebula. Filled and open (for less reliable data) circles represent intensities from the Nebula as a whole. An upper limit to the far ultraviolet radiation inferred from the ionization equilibrium in the filaments is indicated by T. The filamentary shell becomes optically thick at the frequency labelled $\tau = 1$, while the thermal flux from the shell at higher frequencies (but with $h\nu \ll kT$) is indicated by F_{th}. The solid line represents the spectrum of the small-diameter component discovered by Hewish; and the dashed lines and crosses represent the time averaged spectrum of the pulsar. The scale on the left is (logarithmically) in flux units (1 f.u. $= 10^{-26}$ W m^{-2} Hz^{-1}), while the scale at the top gives the logν, with the frequency ν in Hz. For the X-ray region the intensities are also shown in keV/keV cm^2 sec. and the photon energy in keV. All optical and infrared data have been corrected for $1^{\mathrm{m}}5$ interstellar absorption in the visual.

plotting the optical and infrared points a correction for interstellar absorption and reddening must be made. The spectrum shown in Figure 1 has been corrected for a visual absorption of $1^m.5$, based on the available measurements of colors of nearby stars. The optical spectrum shows a distinct wiggle and it is clear that a simple smooth synchrotron spectrum does not provide a good fit in detail. If the interstellar absorption correction is arbitrarily reduced to $0^m.75$, the optical spectrum becomes smoother, but also steeper, and a fit to the X-ray data is no longer possible. Thus, although most of the spectrum can be represented by a smoothly varying function, some discrepancies remain which require further study. Because different parts of the nebula have somewhat different spectra, it is important that measurements be done at all wavelengths with good spatial resolution. On integrating the spectrum of Figure 1, we obtain a total luminosity of 9×10^{37} ergs/sec for the whole nebula, most of the total arising from the spectral region between the optical and X-ray wavelengths. It can be shown (Woltjer, 1958) that the electrons responsible for the optical radiation must have lost too much energy to be a possible relic from the original supernova event, but must have been accelerated more recently. More direct evidence for continuing activity in the nebula comes from the 'light ripples' discovered by Baade and discussed recently by Scargle (1969). These are ridges of light moving in the central region of the Nebula at speeds of up to 0.1 or 0.2 c. At least in some cases the light appears to be polarized in such a way as to indicate propagation transverse to the local magnetic field. Most probably these ripples are hydromagnetic waves produced in the neighborhood of the pulsar. Unless the amplitudes of the waves are very large, the propagation velocities indicate that the density of non-relativistic gas in the central parts of the nebula must be quite low (~ 0.1 cm^{-3}).

The discovery of a pulsar in the Crab Nebula has greatly contributed to the development of a consistent picture, in that it almost certainly represents the basic source of energy for the Nebula. Although the radiation mechanism has not yet been determined, it appears probable that pulsars are rotating neutron stars with a strong magnetic field. If the rotation period is identified with the period of the pulsar, then the observed increase of the period corresponds to a slowing down of the rotation and a decrease of the rotational energy. If we consider a 1-solar-mass neutron star of 10 km radius rotating as a solid body with a 30 ms period and with the period increasing at the observed rate, the energy loss is 2×10^{38} ergs/sec.

As we pointed out at the Fourth Texas Symposium, the assumption of solid body rotation may be doubtful. The braking effects act on the outside of the object and the inside could be expected to rotate faster than the outside, especially if superconductive effects exclude the magnetic field from part of the interior. It is interesting to note that differential rotation tends to be unstable in many cases. These instabilities will cause a variable coupling between the core and the mantle of the object and can be expected to cause decreases in period like the one observed in Vela pulsar. Because reliable viscosity and conductivity coefficients are not available, it is difficult to explore the situation more quantitatively.

If we assume that at least a good part of the neutron star rotates with more or less the same angular velocity, the figure of 2×10^{38} ergs/sec should not be changed too much. Of course, different neutron star models and masses can be considered, and thus this figure is uncertain by a factor of 3 at least. Nevertheless, it is remarkably close to the input needed to maintain the nebular radiation.

Can the rotational energy of a neutron star be efficiently converted into energetic particles? The answer seems to be positive. Pacini (1967) and Gunn and Ostriker (1969) argue that a rotating neutron star with an inclined dipolar field will radiate low frequency electromagnetic waves and that these waves will accelerate particles with high efficiency. Goldreich and Julian (1969) have shown that the electric fields present around a rotating magnetic neutron star – even in the axisymmetric case – will accelerate particles and set up current systems which may result in substantial magnetic fields. They make the very interesting suggestion that the magnetic field of the Crab Nebula may be understood this way. In both approaches a surface field of a few times 10^{12} Gauss is implied by the slowing down of the pulsar, while electrons may attain energies of 10^{13} eV or more, which is adequate to account for the hard X-rays observed in the Crab Nebula.

In a qualitative fashion, a combination of synchrotron losses, diffusion of particles and field inhomogeneities (as evidenced by the light ripples) acting on a power law injection spectrum, can explain the nebular spectrum. We know that different parts of the nebula have somewhat different spectra. To make reliable models it would be of great value to have not only total intensities, but also intensity distributions at different wavelengths. In particular, one would like to know the diameter of the nebula in hard X-rays, which would give direct information on the relative rates of particle diffusion and radiative losses rather close to the pulsar.

The radio spectra and X-ray spectra of Cas A and Tycho's supernova remnant may each be fitted with one power-law type spectrum. The absence of an easily noticeable optical continuum is understandable in view of the smallness of the expected brightness and (at least in Cas A) the very large interstellar absorption in the galactic plane. The main problem with a synchrotron interpretation is that the spectra should steepen at higher energies because of radiative losses. A possible way to avoid this would be a very rapid diffusion of particles out of these remnants.

5. Hot Gas Inside the Supernova Remnant

Sartori and Morrison (1967) have proposed that the X-rays of the Crab Nebula (and the same could apply to the other supernova remnants) arise in a gas heated by the activity in the remnant. Typical parameters would be a solar mass of gas with a density of 100 particles per cubic centimeter at temperatures ranging from 10^7 K to 10^9 K. The strongest argument in favor of this idea was that it seemed difficult to accelerate relativistic particles without producing a large amount of heat. This argument, however, has lost much of its force by the discovery of the pulsar and of possible mechanisms by which a pulsar can accelerate relativistic particles with high efficiency. The high density of the gas would not allow the fast propagation of the light ripples in the Crab Nebula and consequently a rather complex distribution of the gas would be needed. At the same time the hotter parts of the gas would have a considerable pressure and their containment would pose serious problems.

6. Hot Gas Surrounding Neutron Star / 7. Hot Gas Surrounding White Dwarfs

Generally these two possibilities, which have been considered by Shklovsky (1967) and Burbidge and Prendergast (1968), among others, have been considered in the framework of double star evolution, with one component of the double star evolving and transferring mass to the degenerate component. Mass falling onto a neutron star or white dwarf may become quite hot. If a small piece of matter without angular momentum falls freely onto a white dwarf and thermalizes its kinetic energy, a temperature of 10^9 K could be reached. The real situation is much more complex, however. Angular momentum is quite important, while in a steady flow the pressure effects also need be considered. As a consequence the infalling gas may lose a good part of its energy by radiating at temperatures well below the maximum value. Nevertheless the results of Burbidge and Prendergast in particular are quite promising in that they show that in plausible flows, temperatures of the order of 10^7 K may be obtained even with white dwarfs.

The main difficulty for models of this type is that despite extensive spectroscopic studies no specific evidence for binary motion has been found in Sco X. However, it is not clear at the moment how effectively one may 'hide' the binary characteristics by the gas streams in the system. Perhaps one will have to consider an alternative mode of producing a hot corona around a white dwarf or neutron star.

From the observational side some support is available for the idea that a hot gas cloud with white-dwarf type dimensions is responsible for the X-rays of Sco X. Such a gas cloud would be optically thin at X-ray wavelengths. The 1–10 keV spectral observations indicate a temperature of 5×10^7 K on the average, although the harder X-rays would probably require a higher temperature. If we adopt 500 pc for the distance we find from the X-ray intensities $N_e^2 V = 1.4 \times 10^{60}$. Neugebauer et al. (1969) find that the infrared data follow about a Rayleigh-Jeans type law which indicates the source to be optically thick. For a spherical source of radius R they obtain from the infrared intensities and the Rayleigh-Jeans law, $R^2 T = 3 \times 10^{25}$. Neugebauer et al. subsequently make the assumption that the cloud is spherical and uniform in its physical properties. From these results it then follows that $R = 8 \times 10^8$ cm and $N_e = 2.7 \times 10^{16}$ cm^{-3}. Neugebauer et al. observe that the optical depth for electron scattering is large (~ 14 from center to front) while the total optical depth for absorption (taking into account the path length increase caused by electron scattering) is about 6 at 10000 Å and 2.5 at 3000 Å, which possibly is a bit high in view of the flattening of the optical spectrum, but of the right order of magnitude.

It is very striking that the dimensions of the X-ray source turn out to be comparable to those of white dwarfs. A more probable model then is a shell of hot gas around a white dwarf. Taking, for example, a white dwarf radius R^* of 7×10^8 cm we would have, with the outer radius of the shell again equal to 8×10^8 cm, $N_e = 4.8 \times 10^{16}$ cm^{-3}. The optical depth of the shell in electron scattering now is only about 4 and the absorption optical depth more nearly unity or less at optical wavelengths.

Clearly models of this type are overly simplistic. We hardly can expect the gas to be isothermal either near the interface with the white dwarf atmosphere or towards the outside. The main effect of this might be to somewhat increase the outer radius of the shell.

In the optical spectra of Sco X emission lines are seen (N II, O III, H, etc.). The spectroscopic evidence indicates that the density is rather high. The same conclusion is inferred from the low degree of ionization in the strong ultraviolet radiation field of Sco X. If we take the temperature of the gas to be 30000 K the density has to exceed 10^{11} cm^{-3} for gas at 10^{10} cm from the center. Of course at greater distances lower densities become possible.

Radio emission has been detected at 6 cm with a flux per unit frequency interval about equal to that at optical wavelengths. If the radiation were thermal it would have to come from a fairly large region ($R \approx 10^{14}$ cm for $T = 10^4$ K if the medium were optically thick, and larger otherwise). In the optically thin case both the radio emission and the Hβ emission are proportional to N_e^2 integrated over the source, and it is easily verified that the Hβ emission which corresponds to the observed radio emission exceeds the observed Hβ emission by a factor of 20. Optical depth effects will increase the discrepancy. This appears to imply that the radio source is non-thermal. The Parkes data at 11 cm, reported at this symposium, appear to confirm this conclusion. If the radio source is indeed Sco X, the interesting result would be that a thermal X-ray source would be associated with a nonthermal radio source, and consequently with the acceleration of fast particles.

Appendix

THE DISTANCE OF THE CRAB NEBULA

The extensive material on proper motions and radial velocities obtained by Trimble (1968) provides a basis for determining the distance of the Crab Nebula. The basic method of equating the tangential velocities derived from proper motions at the edge of the Nebula with the radial velocities observed near the center meets the following difficulties:

(1) The systematic motion of the Nebula as a whole should be corrected for.

(2) The velocity field does not resemble that of a smoothly expanding shell very well: considerable irregularities are present.

(3) The Nebula is not axisymmetric and some assumption as to the three-dimensional shape must be made.

If we disregard these three difficulties and consider the Nebula in the mean as an expanding spherical shell, it appears most reasonable to simply equate the 'largest' proper motion observed with the 'largest' radial velocity. By arranging the motions in order of absolute magnitude and taking the 'largest' motion equal to the average

TABLE I

Distance determinations for the Crab Nebula

Velocities used	'Mean largest' proper motion	'Mean largest' radial velocities	Distance
All	0″.203/yr	1284 km/sec	1340 pc
Major axis	0″.189/yr	1247 km/sec	1390 pc
Minor axis	0″.157/yr	1247 km/sec	1680 pc

Adopted distance 1500 parsec

of the third through tenth largest values, we obtain the first entry of Table I.

Next we consider only filaments whose position angle (as seen from the center) differs less than 30° from that of the major axis. Near each end of the major axis we take the 2d–5th largest proper motions and average the eight values. Similarly we take the average of the 2d-5th largest positive – and negative – radial velocities. The resulting average largest motions should be independent of the systematic velocity of the Nebula. By equating the radial and tangential motions, the second entry in Table I is obtained; and by applying the same method of filaments near the minor axis, the third entry. On the basis of the results of Table I, we adopt a value of 1500 parsec for the distance of the Crab Nebula.

Acknowledgement

This research was supported in part under Contract AF 49 (638) 1358.

References

Angel, J. R. P.: 1969, to be published.
Burbidge, G. R. and Prendergast, K. H.: 1968, *Astrophys. J. (Letters)* **151**, L83.
Ginzburg, V. L.: 1969, *Nature*, to be published.
Goldreich, P. and Julian, W. H.: 1969, *Astrophys. J.*, to be published.
Gunn, J. E. and Ostriker, J. P.: 1969, *Nature* **221**, 454.
Heiles, C.: 1964, *Astrophys. J.* **140**, 470.
Neugebauer, G., Oke, J. B., Becklin, E., and Garmire, G.: 1969, *Astrophys. J.* **155**, 1.
Pacini, F.: 1967, *Nature* **216**, 567.
Sartori, L. and Morrison, P.: 1967, *Astrophys. J.* **150**, 385.
Scargle, J. D.: 1969, *Astrophys. J.* **156**, 401.
Shklovsky, I. S.: 1967, *Astrophys. J. (Letters)* **148**, L1.
Trimble, V. L.: 1968, *Astron. J.* **73**, 535.
Woltjer, L.: 1958, *Bull. Astron. Inst. Neth.* **14**, 39.

THEORIES OF DISCRETE X-RAY AND γ-RAY SOURCES

(Invited Discourse)

JAMES E. FELTEN

Institute of Theoretical Astronomy, University of Cambridge, England

Abstract. This is a critical review of theories of known discrete X-ray sources. The Crab is omitted, having been dealt with in Woltjer's review. Two of the identified sources, Sco X-1 and Cyg X-2, seem to be of the same sort. A binary or gas-stream model like that of Prendergast and Burbidge, with dimension $R \sim 10^9$ cm and density $n \sim 10^{16}$ cm^{-3}, appears reconcilable with the observed features of these sources, though much detailed work remains to be done. Neither object is yet known to be binary. Theoretical work becomes more difficult if, as appears to be the case at least for Sco X-1, the objects are optically thick due to electron scattering; this may affect the optical and X-ray spectra.

The recent searches for iron lines in the X-ray spectrum of Sco X-1 are reviewed briefly. The calculations and the energy resolution are not yet good enough to make this a dependable test of models.

Several possibilities are offered for explaining the excess radio flux from Sco X-1.

Other theories of Sco X-1-type sources are discussed briefly. The theory of Manley and Olbert seems a little superfluous when the gas-stream theory is still in a strong position.

There are serious discrepancies between X-ray and optical estimates of the distance to Sco X-1. 21-cm measurements must also be considered. The situation is reviewed, and ways out of the difficulty are discussed.

Cen X-2 seems to be like Sco X-1, but several other unidentified sources have hard spectra like the Crab. It is tempting to speculate that most of the galactic sources are supernova remnants.

The extended γ-ray source in the galactic plane may be the extrapolated unresolved sum of galactic X-ray sources, as suggested by Ogelman. There are several other possibilities.

M87 is the only established extragalactic source. Radio, optical and X-ray observations are summarized and graphed. A power-law extrapolation to the X-ray band is far from mandatory; nevertheless the optical flux from the jet is known to be synchrotron radiation. The time-scale difficulties in the jet are described, and several theories of the survival of the optical electrons are reviewed.

Processes for producing X-rays other than thermal bremsstrahlung and synchrotron radiation are listed. These other processes are characterized by low efficiency, and are likely to be unimportant in discrete sources, though several have attracted attention with reference to the diffuse background.

1. Introduction

The first thing to be said is that I am essentially a novice at this subject. Most of my own work in X-ray astronomy has been concerned with the cosmic background rather than with discrete sources. Consequently I have had to work pretty hard to prepare this review, but let us hope that I have brought to it the characteristic virtues of the amateur. Those who find my conclusions naive may take comfort; they may well be right!

Let me try to indicate briefly what I shall do and not do. I *will not* try to develop 'theories' of each object by just collecting and repeating the data we have heard reported. Nor will I deal with the newly found X-ray pulsar in the Crab, the theory of which is too young to admit useful review, and really belongs in a pulsar conference anyway. Instead I shall confine myself to discussing the prototype objects, starting

L. Gratton (ed.), Non-Solar X- and Gamma-Ray Astronomy, 216–237. All Rights Reserved
Copyright © 1970 by the IAU

with Sco X-1 and bringing in data from Cyg X-2, which seems to be similar in most ways. The other prototype of galactic sources is the Crab, but I am leaving discussion of this entirely to Professor Woltjer, to avoid needless duplication. I shall speculate a little on the nature of the many unidentified galactic sources and add some remarks about γ-ray sources and particularly possible sources in the galactic plane. Then I shall move on to the extragalactic sources, M87 being the only one of these which can bear much comment, and, judging by reports here, the only one which is well established. Finally I shall discuss some of the 'exotic' processes of photon production which have been suggested but are not known to be important in any observed source.

2. Scorpius X-1

This is of course the prototype of the thermal objects. Its identification by Giacconi, Sandage and coworkers (Sandage *et al.*, 1966) was a milestone in astronomy, made possible by the accurate X-ray position; we need more such positions. Its optical properties (Johnson *et al.*, 1967; Westphal *et al.*, 1968; Johnson, 1968; Neugebauer *et al.*, 1969) are by now familiar: a blue continuum, only slightly polarized (Hiltner *et al.*, 1967), variable in time, accompanied by emission lines of both low and high excitation which suggest a wide range of temperature within the source. The emission lines also seem to vary in intensity. The line wavelengths vary irregularly, suggesting gas streams of some sort; sometimes different lines seem to move in antiphase, but no regular period has been established, so the object is not, as of now, a spectroscopic binary. The same is true of Cyg X-2 (Kraft and Demoulin, 1967; Kraft and Miller, 1969) where the emission lines are accompanied by many stellar absorption lines. Although no period has been established for either object, the spectra are complicated, and one still suspects that binary motions or even multiple systems might be present.

It was realized early that the X-ray spectrum of Sco X-1 is nicely fitted by an exponential curve for optically thin thermal bremsstrahlung, with $T \approx 5 \times 10^7$ K. More recently it has been confirmed that this best-fit T varies (Gorenstein *et al.*, 1968; Chodil *et al.*, 1968), and also that there is at high photon energies a nonthermal tail (Peterson and Jacobson, 1966; Buselli *et al.*, 1968), apparently also variable (Lewin *et al.*, 1968a; Overbeck and Tananbaum, 1968; Riegler, 1969). There is a correlation between the optical continuum brightness and the temperature of the thermal X-ray curve, but its nature is in doubt. I shall return briefly to these variations later.

3. A Binary Model

The distance D to Sco X-1 was first estimated as 100–1000 pc, by analogy with the old novae which it resembled. Even this rough estimate was useful because, with the bremsstrahlung nature of the source established, it made possible an estimate of $\langle n_e^2 \rangle V$, and the small (starlike) image gave an upper limit on V, so that a lower limit on $\langle n_e^2 \rangle$, and hence an upper limit on the cooling time, could be given. This was done by Johnson (1966) and others, who showed that the cooling time was at most a few years, and shorter than the known history of the optical object, so that a continuing source of energy seemed necessary.

This fitted in well with the notion, first discussed at the Noordwijk symposium (Burbidge, 1967b) but based on an earlier suggestion by Hayakawa and Matsuoka (1964), that the Sco source is a binary system in which gas is escaping from a more extended component and falling into the gravitational potential well of a compact component, which Shklovsky (1967) suggested was a neutron star. Cameron and Mock (1967) put forth a white dwarf instead. They also made the interesting point that the energy release by such infall is self-limited by the pressure of the radiation produced, which, when it balances the gravity, will prevent further matter from falling in. For a central object of 1 $M_\odot$ the limit is $\approx 2 \times 10^{38}$ erg/sec, which is comfortably above the putative luminosity of the Sco source.

It is easy enough to show that the highest temperature you can get by completely converting the kinetic energy of infall to thermal energy is

$$T_k \sim 10^7 \text{ K} \left(\frac{M}{M_\odot} \frac{R_\odot}{R} \right), \tag{1}$$

where M and R are the mass and radius of the condensed central object. But this is only an upper limit; in the real system it is doubtful that such a high T could be attained. The only real investigation is that by Prendergast and Burbidge (1968), with a computer. They assumed gas to be released from one component with angular momentum typical of a binary orbit. This angular momentum prevents it from falling straight in upon the other component. Instead it forms a swirling disk, and a steady state is reached, with a temperature gradient established in the disk. With $R = 1.5 \times 10^{10}$ cm $\approx$ $\approx \frac{1}{3} R_\odot$ and $M = 10^{33}$ gm $\approx \frac{1}{2} M_\odot$, the highest temperature they achieved (at the surface of the 'primary' star, where the infall terminates) was only

$$T_{\text{surf}} \sim 10^{-2} T_k, \tag{2}$$

amounting to $\sim 2 \times 10^5$ K and clearly not adequate for a source like Sco X-1. The reason why we have (2) instead of (1) is that when the gas reaches the stellar surface it is still rotating. The angular momentum prevents all the kinetic energy from being thermalized. It is possible that the interaction and connection of the swirling gas with the stellar surface would improve the thermalization and raise T, but Prendergast and Burbidge found this aspect of the problem too difficult for adequate treatment.

4. Complications in an Optically Thick System

Of course T can also be raised by taking R smaller, and there is now evidence that R is quite small. Figure 1 is from the recent paper by Neugebauer et al. (1969). Their observations of Sco X-1 in the infrared and visible show quite clearly that the source is optically thick there, and even perhaps in the ultraviolet. So instead of just getting a lower limit on n_e, they introduce another relation by assuming that the infrared spectrum is on the Rayleigh-Jeans part of a black body curve at the temperature T indicated by the X-rays. This leads to a source dimension $R \sim 10^9$ cm (roughly the radius of a white dwarf) and to $n_e \sim 10^{16}$ cm^{-3}. It is now possible to check the model for consistency, by calculating optical depths. The optical depth due to free-free re-absorption is only $\tau_{ff} \approx 1.3$ in the infrared and 0.14 in the ultraviolet. This is not really adequate to produce a blackbody spectrum. But the optical depth due to electron scattering *is* quite large in this model, $\tau_{es} \approx 10\text{-}20$ depending on the distance D assumed.

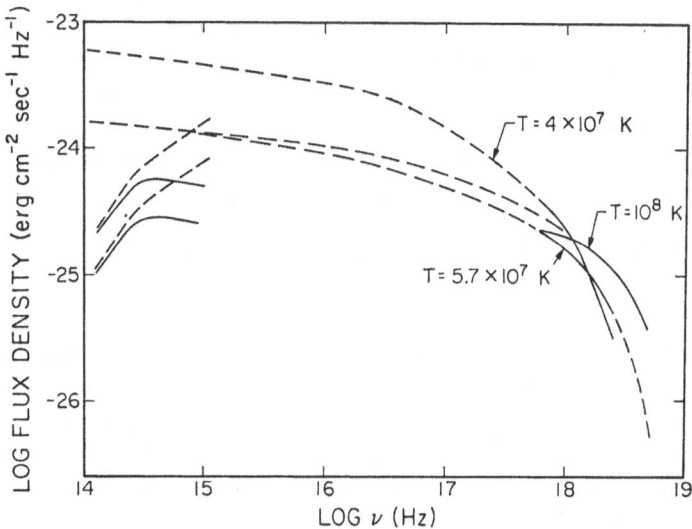

Fig. 1.   The spectrum of Sco X-1, from Neugebauer *et al.* (1969). The solid curves near $\log \nu = 18$ are the X-ray fluxes observed at different times. These have been extrapolated (broken curves) to the visible region, assuming free-free radiation by hydrogen at the appropriate temperatures. Between $\log \nu = 14$ and 15 are the infrared and visible fluxes observed at different times (solid curves), and (broken curves) the same fluxes corrected appropriately for interstellar reddening. The failure of these curves to fit the X-ray extrapolations shows that the source is optically thick in the visible.

And, say these authors, the 'true' optical depth is then essentially the harmonic mean, $\tau = \sqrt{(3\tau_{es}\tau_{ff})}$, and is $\gg 1$ in the infrared, justifying the assumption of a blackbody curve. Now I do not think this is quite sufficient; I think that to get the blackbody curve, instead of a graybody, you really need $\tau_{ff} \gg 1$ and not merely $\sqrt{(3\tau_{es}\tau_{ff})} \gg 1$. This needs to be more carefully looked at. But considering the ill-determined distance, unknown geometry, and nonuniform temperature of the source, such a simplifying assumption may not be out of place, and the results do suggest a small source, with $R \sim 10^9$ cm. Chodil *et al.* (1968) got about the same R by just assuming $\sqrt{(3\tau_{es}\tau_{ff})} \sim 1$ in the blue, without the blackbody assumption.

I wish to emphasize that a value for τ_{es} as large as 10–20 will have important effects. A photon produced in this medium will be scattered $\sim \tau_{es}^2 \sim 100\text{-}400$ times before escaping. If this is a photon with initial energy $h\nu \gg kT$, these are classical Compton scatterings. The mean energy loss per collision by the photon is then

$$\delta(h\nu) \sim \left(\frac{h\nu}{m_e c^2}\right) h\nu; \tag{3}$$

even for $h\nu$ as low as 10 keV this will become important after ~ 50 scatterings, and the effect is clearly to degrade high-energy photons and turn the photon spectrum above $h\nu \sim kT$ into a graybody spectrum at temperature T – which of course is not very different from the bremsstrahlung exponential shape in this energy range anyway. X-ray lines will be smeared too, as already mentioned here by Professor Novick in discussion. Manley and Olbert (1969) called attention to this degradation by Compton scattering, though they seem even to have overstated it by forgetting that with the electrons 'hot' ($kT \sim$ few keV) instead of at rest, the energy loss for the hard photons is self-limiting. Softer photons, of course, are *gaining* energy in the Compton collisions. Here, for $h\nu \ll kT$, the mean energy gain per collision is

$$\frac{\delta(h\nu)}{h\nu} \sim \left\langle \frac{\nu}{c} \cdot \frac{1}{2}\left(\frac{c+\nu}{c} - \frac{c-\nu}{c}\right)\right\rangle \sim \frac{\langle \nu^2 \rangle}{c^2} \sim \frac{kT}{m_e c^2}, \tag{4}$$

which is $\sim 10^{-2}$, and in 100–400 collisions this too can become important, turning the spectrum into a graybody – which in fact is what the optical spectrum looks like. The good fit to a ν^2 law in the optical band may indicate that we are in a regime where $\tau_{es}^2 kT/(m_e c^2) > 1$ but $\tau_{ff} \ll 1$; then we expect a 'pure graybody'. At lower ν, where $\tau_{ff} \gtrsim 1$, the photon emission is adequate to make the spectrum flatten and move up onto the true blackbody curve.

Of course it is not sufficient to have in this source only one homogeneous region, with a high T. Cooler regions are needed too, to produce the emission lines. We do not know the density in these regions, but it is probably high. If it is similar to that in the X-ray regions, then the volume of cool gas needed is $\sim 10^{-4}$ that of the hot gas (Johnson *et al.*, 1967). This is not much, and it could be just filaments. The requirement is

$$n_{cool}^2 \, V_{cool} \approx 2 \times 10^{54} \left(\frac{D\text{pc}}{100}\right)^2 \tag{5}$$

In models like this we have to be very careful. If the cool gas is imbedded in the hot gas, the optical lines may be smeared beyond recognition by scattering in the cloud of hot electrons; they would in the model of Neugebauer *et al.*, discussed above. On the other hand, if the hot gas is surrounded by the cool gas, the photoelectric absorption of X-rays will become large around 1 keV (which contradicts observations, e.g. Rappaport *et al.*, 1969) if the column depth of cool gas is as large as $nl \sim 10^{22}$ atoms/cm^2. If heavy elements are present in normal abundance, this remains true even if the temperature of the cool gas is high enough to ionize H and He (Bell and Kingston, 1967). From (5), setting $l \sim V^{\frac{1}{3}}$, we can keep $nl < 10^{22}$ if $n_{cool} \gtrsim 10^{10}$ cm^{-3}. Comparing this with $n_{hot} \sim 10^{16}$ in the Neugebauer model, we see that the cool gas can be spread out quite a bit around the hot source. You will easily recognize that estimates like these are based on a quasispherical geometry; a pancake configuration, as suggested by the binary hypothesis, relaxes such difficulties quite a bit.

Another interesting feature of these high-density models was pointed out by Manley and Olbert (1969). The blackbody temperature corresponding to the X-ray energy density inside the hot cloud of Neugebauer *et al.* is $T \approx 4 \times 10^5$ K! This means that the appearance of the central star, assuming there is one, must be significantly affected by the impinging X-ray flux, since the star must heat its surface up to 4×10^5 K just to shed the energy it acquires from the X-rays. This could explain some of the difficulty in finding conclusive evidence of a binary. It also shows that the optical object must be abnormal in every respect, and that arguments based on analogies with familiar objects are likely to mislead.

Yet another set of constraints on models of this sort is provided by the theoretical and experimental time scales for variations. A few years ago we only knew that the object was small enough to be starlike, and that the cooling time had to be shorter than a year or so. This was already felt to be a possible embarrassment, since the optical object had appeared on plates for 70 years. Now, with R down to 10^9 cm and n up to 10^{16} cm^{-3}, the theoretical cooling time is in the millisecond range. We know there *are* variations in the observable X-ray temperature, but observations are not extensive enough to give a good idea of them. It is pretty clear, however, that the variations in T are not as dramatic as the theoretical cooling time would suggest. There must be a continuing energy source, which may be sporadic, as suggested by the apparent flaring behavior (Lewin *et al.*, 1968a). But then there must also be an analog of capacitance in the system, such as a surrounding reservoir of low-temperature gas, which gradually feeds energy into the hot gas and prevents it from cooling suddenly. The energy source may be variable in time average over a week or a month; then slow variations in T may be explained. This is only speculation, and no one seems to have looked very carefully at the problem. It appears that such behavior might be achievable in a model of the Prendergast-Burbidge type.

Simultaneous optical and X-ray observations have not been numerous. What data there are (Chodil *et al.*, 1968) suggest that the optical continuum is brightest when the X-ray temperature is *lowest*. This is the reverse of what you would expect from a straightforward interpretation of the high-density model. I do not want to spend more time on this, so I shall just refer you to the discussion in the paper cited.

5. Iron Lines

Tucker (1967) pointed out that there is a sensible way to test a thermal model when high spectral resolution is available in the X-ray band: namely, to look for characteristic X-ray emission lines of heavy elements, particularly the lines of H-like and He-like iron ions near 7 keV. Holt *et al.* (1968) tried this on observations of the Crab to see if they could already eliminate Sartori and Morrison's (1967) thermal model for that object, and even found that they could, at least for certain choices of element abundance. However, it seems that this test is not so easy. The NRL group (Fritz *et al.*, 1969) had observations of even higher resolution on Sco X-1, which we think *is* thermal, and they could not see any iron lines either! Further improvements in spectral resolution are needed, and obviously we will not have faith in this test of models until we demonstrate that the lines *can* be seen in at least one thermal object. I think also that there is some inconsistency between the calculations performed by the NASA group and by the NRL group*, and between them and the work of Wally Tucker on which they both relied. Wally is the *eminence grise* of this subject, but unfortunately he is not here to give us an opinion. I do not like to take sides in a struggle between two arms of the U.S. government, but it appears to me that the analysis of Holt *et al.* is more nearly correct. Possibly there are also some misprints in the Tucker paper. This all needs reworking.

6. Sco X-1 as a Radio Source

It is necessary to say something about the radio observations of Sco X-1 (Ables, 1969). It is a weak but detectable source at 6 cm, and varies by roughly a factor of 10 in a few hours; the average strength is $\approx 3 \times 10^{-28}$ w m^{-2} Hz^{-1}. This is well *below* the extrapolation of the optically thin X-ray bremsstrahlung spectrum to radio wavelengths. But now we know that the bremsstrahlung spectrum already becomes optically thick in the visible, and so the radio flux now represents an *excess*. There is also a weak positive radio result on Cyg X-2 (Moffet and Berge, 1968)**; it lies about a factor of 10 *above* the bremsstrahlung extrapolation. But this may be a similar story. There are many ways of explaining an excess. We could have some large region at low T producing thermal radiation at the longer wavelengths, or we could have some non-thermal process occurring, such as synchrotron radiation. Riegler and Ramaty (1969) and Feldman and Silk (1970) both have proposed this. I have not studied these proposals hard enough to understand them very well, but it seems clear that, with observations at only one radio frequency and no spectral information, there is a good deal of elbow room for theorists. In both of these models the fields contemplated are rather strong, 1–100 gauss, but this may be appropriate for a circumstellar region. The radio phenomenon may be something like a solar flare. Feldman and Silk suggest that nonthermal electrons in a power-law spectrum produce the radio emission, and also, through nonthermal bremsstrahlung, the variable X-ray tail at $h\nu > 30$ keV. This would be much like what we see on the sun (Holt and Ramaty, 1969).

7. Other Models

I am coming to the end of what I, as a novice, can say about the theories of sources of the Sco X-1 type. The most elaborate work has been done by Prendergast and

* *Inter alia*, the 'correction' proposed by the NRL group to the earlier NASA paper appears to be in error.
** Note added June 30, 1969: Apparently this result was spurious (Purton and Andrew, 1969).

Burbidge, but it does not begin to cover the complexities of the problem. It seems to me that no dramatic new theoretical ideas are in the offing, and that we simply have on our hands a rather dirty problem in 'applied maths'. I should mention a modification of the binary idea, proposed by Cameron (1969); he suggests that a single star has formed a kind of planetary nebula *manque*; it gave the planetary shell too little energy to escape, so it stopped expanding, and now dribbles slowly back onto the star, producing the X-rays. This model, of course, is even less worked out than the other.

Little has been said here about the contrasting theory of Sco X-1 due to Manley and Olbert (1969), and I might be accused of neglecting it. Indeed this is a ponderous preprint they have sent out, and I have to confess that I have not studied it carefully enough to say anything very intelligent. The model involves a pre-stellar cloud, containing gas at $\approx 10^5$ K which produces the emission lines, and also hydromagnetic turbulence which accelerates electrons by a Fermi-type process, forming a power-law electron spectrum with a high-energy cutoff. The synchrotron emission can then match the observed X-ray spectrum pretty well. There are two adjustable parameters describing the turbulence, plus the customary n and T for the gas. A possible objection to this theory is that there does not seem to be any place in it for a *star*, so that we cannot apply it to Cyg X-2, where sharp stellar absorption lines are seen.

At La Jolla I shared an office with Wayne Stein, who sat there poring over his observations of infrared stars, trying to make some sense of them. One day I glanced at his notes and saw that he had scribbled in the margin, 'What is this little star saying? What is he trying to tell us?' Since then I like to think of the X-ray data in the same terms. It seems to me that Sco X-1 and Cyg X-2 are trying as hard as they can to 'tell us' (a) that they are essentially thermal objects, and (b) that the energy is coming from gas streams and/or binary motion. A complicated theory like that of Manley and Olbert which starts off in a different direction, when the need for such a departure has not been demonstrated, may be ingenious, but it has a low a priori probability of being right.

8. Absorption in the Sco X-1 Spectrum

Now I want to introduce one final topic related to Sco X-1, namely the question of its distance, and the interstellar absorption in its direction. I have left this until last because it may have more intrinsically to do with theories of the interstellar medium than with theories of X-ray sources; nevertheless I should discuss it, if only because I know rather more about it than about some of the source theories. There are now two contradictory determinations by optical astronomers of the distance to Sco X-1. Sofia *et al.* (1969) have measured a proper motion and identified Sco X-1 as a member of a subgroup within the Scorpio-Centaurus association. This puts it at $D = 170$–200 pc. Wallerstein (1967), and later Westphal *et al.* (1968), measured the interstellar H and K absorption lines of Ca^+ and found them stronger than in any nearby stars (in particular, stronger than in any stars of the Sco-Cen association); they concluded $D > 300$ pc. Whichever D is correct, it is large enough to pose difficulties with the observed *lack* of X-ray absorption. Sco X-1 has galactic latitude $b^{II} \approx 24°$, so that by the time we reach $D = 170$ pc we are 70 pc above the galactic plane. Thus almost half the total column density of atomic hydrogen in this direction should lie between us and Sco X-1, or *more* if $D > 170$ pc. How much hydrogen is there in this direction? I have not been able to find a high-resolution 21-cm radio map for this part of the sky, but on a 5° grid with 2° resolution (McGee *et al.*, 1966) the closest points give

$$N_H^{(\infty)} \equiv \int_0^{\infty} n_H \, dl \approx 1.8 \times 10^{21} \text{ atoms/cm}^2 ,$$ (6)

and the variation seems rather smooth.

With what should we compare this? There is an X-ray measurement at 0.25 keV by the NRL group (Fritz *et al.*, 1968). The flux reported has decreased by a factor of 6 from a previous measurement by the same group, and so we might have a little residual skepticism about the latest result. But for sake of argument let us accept it. It lies quite nicely on the bremsstrahlung curve extrapolated from higher X-ray energies and is therefore consistent with *zero* photoelectric absorption at 0.25 keV; apparently $\tau = 0.5$ is an upper limit for the optical depth. The absorption at this energy is due mainly to H and He, and if we know the abundance ratio we can use the theoretical cross sections (Bell and Kingston, 1967) to derive from the X-ray result an upper limit to $N_H^{(D)} \equiv \int_0^D n_H dl$ out to Sco X-1. Table I shows some results, for several assumed values of the ratio $n_H : n_{He}$. Note particularly the last column. If we reject the NRL datum, we must fall back upon measurements by the MIT and Livermore groups (Rappaport *et al.*, 1969; Hill *et al.*, 1968) at 0.6 keV, which indicate $\tau < 0.5$ at this energy, and the last column gives the corresponding limits.*

We must compare these numbers, particularly the NRL numbers, with $N_H^{(\infty)}$ in (6). You can see that the result is rather sensitive to the unknown helium abundance. Case (a) might possibly be consistent with (6), but a zero abundance of He in interstellar matter cannot seriously be entertained. Case (c) is for the 'cosmic abundances' of Aller (1961) and gives the biggest discrepancy; in any case there is some observational evidence, summarized by Biermann (1969), for an He abundance lower than this in interstellar matter. Case (b) is for 25% He by mass, as produced by a big bang without any further processing of elements, and it seems that we must assume at least this much He; also it corresponds to the smallest observational estimate for interstellar matter. So we have to explain the discrepancy with (6).

It is possible that there is a little 'hole' in the direction of Sco X-1, so that the amount of H out to 170 pc, or even out to infinity, is smaller than we think. A high-resolution 21-cm map would clarify this. But other evidence suggests that the phenomenon is more general. Observations of the Lyman-α absorption line in spectra of nearby stars

---

* These numbers are a little larger than those derived by Rappaport *et al.* This is mainly because I have included only absorption by H and He. 0.6 keV is above the K-edges of C, N and O, and if these elements are present the numbers in the last column have to be decreased by a factor which can be as large as 2 or 3.

(Jenkins and Morton, 1967) have also indicated less H than would be expected from the 21-cm observations. Kerr (1969), in a review, has discussed the idea that the sun is in a local region of low gas density, perhaps 200 pc in radius. The 21-cm observations give no information about regions so close, and the H seen in 21-cm radiation could all be farther away. More observations will be needed to test this notion, and X-rays may play a significant part.

TABLE I

Some limits, derived from X-ray measurements, on the column density of hydrogen out to Sco X-1

		$N_H{}^{(D)} \equiv \int_0^D n_H \, dl$	
	$n_H : n_{He}$	NRL (0.25 keV)	MIT (0.6 keV)
(a)	$\infty : 1$	$< 5 \ \times 10^{20}$ cm^{-2}	$< 7 \ \times 10^{21}$
(b)	12:1	$< 1.4 \times 10^{20}$	$< 2.3 \times 10^{21}$
(c)	6:1	$< 8 \ \times 10^{19}$	$< 1.4 \times 10^{21}$

A small value of $N_H^{(D)}$ will solve the problem of Table I, but it will not explain the *strength* of the Ca$^+$ lines in the optical spectrum. Westphal *et al.* (1968) assert that there is no star within 300 pc having such strong Ca$^+$ lines. From these lines they derive by a classical method

$$\int_0^D n_H^2 \, dl \approx 2.7 \times 10^{20} \text{ cm}^{-5}. \tag{7}$$

From this and $N_H^{(D)} < 1.4 \times 10^{20}$ we have

$$[n_H] \equiv \frac{\int_0^D n_H^2 \, dl}{\int_0^D n_H \, dl} > 2 \text{ cm}^{-3}, \tag{8}$$

which is not unreasonably high if the medium is cloudy. The 'classical method', however, is known generally to give incorrect results. There is a well-known calcium discrepancy, which goes in the wrong way for us and is probably connected with the tendency of the interstellar calcium to get locked up in grains. The effect is counteracted to some extent by extra ionization due to low-energy cosmic rays, but typically (G. B. Field, private communication; I cannot take time to go into more detail here) the number (8) would become $2 \times 40 = 80$ instead of 2, and this is a pretty extreme requirement for the interstellar clouds. Perhaps the lines originate instead in a local abnormal region near Sco X-1. This possibility was discounted by Wallerstein (1967) and by Westphal *et al.* (1968) because the velocity of the K-line is only 7 km/sec, perfectly typical of an interstellar cloud. Still, this whole question clearly needs re-examination.

Of course the easiest ways out of these difficulties involve disbelieving the X-ray data and/or their interpretation. If we merely reject the NRL result (i.e. switch from column 2 to column 3 in Table I) it appears that no real inconsistencies would remain, except that in case (c) we might have some difficulty with strong X-ray absorption by C, N, and O. Alternatively, we might just assume that the intrinsic spectrum of Sco X-1 from 1 to 0.25 keV has an additional, steeply rising, low-temperature component, not observed at higher energies. This then allows the observed flux at 0.25 keV to reflect a fair amount of absorption. It is not clear how much of this low-temperature (10^6–10^7 K) flux could be allowed without betraying itself in the 0.6 keV measurements. Spectral data in the soft band should clear this up quite soon.

X-ray absorption is a valuable tool for learning about the interstellar medium. A soft X-ray observation of the Crab would be especially valuable; an attempt was made by the Livermore group (Grader *et al.*, 1969), but there were atmospheric background problems. Absorption has been seen at a few keV for two sources in Sagittarius, presumably buried deep in the gas of the disk (Gorenstein *et al.*, 1967; Rappaport *et al.*, 1969). The K-lines of interstellar O and Ne at 0.53 and 0.87 keV should particularly be looked for (Felten and Gould, 1966; Bell and Kingston, 1967), and the degree of energy resolution already achieved (Hill *et al.*, 1968) is not far from that required to show them.

9. Other Galactic Sources

This is all I want to say about the Sco X-1 and Cyg X-2 type sources. Centaurus X-2, identified as the star WX Cen by Eggen *et al.* (1968), seems to belong to the same category. It is known to be a variable X-ray source, possibly of thermal character (Harries *et al.*, 1967), but it has a nonthermal tail at high energies, like Sco X-1 (Lewin *et al.*, 1968b). As for the unidentified galactic sources, many of the X-ray spectra are ill-known and still amenable to either a thermal or a power-law interpretation. It has long been known, though, that the Sagittarius sources as a group are distinctly harder than Sco X-1 (Giacconi *et al.*, 1965), and lately we see that the individual spectra of Cyg X-1, Cyg X-3, Cyg X-4, Lup X-1, GX 3 + 1, and GX 354 − 5 all seem to prefer a power-law fit (Peterson *et al.*, 1968; Buselli *et al.*, 1968; Hudson *et al.*, 1969; Rocchia *et al.*, 1969). This tempts me, for one, to the hypothesis that most of the unidentified sources are supernova remnants, like the Crab, or related objects in the disk population. Dr. Gratton has included in his review the very diagram I wished to show, comparing the distribution of X-ray sources in galactic coordinates with that of known novae. Curiously, I would have drawn conclusions different from his. I would have suggested that the mean latitude of X-ray sources is significantly smaller than that of novae, and that the former are true spiral-arm Population I objects rather than intermediate objects like the latter. This was also the point of view of Gursky *et al.* (1967), and I will refer you to the attractive figures in their paper rather than reproducing them here. If the Sagittarius and Cygnus sources are in the spiral arms, at distances of several kpc, then their absolute luminosities are comparable to the Crab. But if the Sagittarius sources are clustered around the galactic center, like novae, then they must be much brighter than Sco X-1. This, it seems to me, suggests the former as prototype rather than the latter. I should also refer you to the coincidences between supernova remnants and X-ray sources found by Poveda and Woltjer (1968). But this question must be settled by observation, and I should not waste more of your time speculating on it.

10. γ-Ray Sources

This seems the moment to say something about γ-ray sources. No point sources have been established, though there is continuing suspicion, with observations at the limit of sensitivity, that the Crab and possibly the pulsar CP 1133 are sources at 10^{11}–10^{13} eV (Fegan *et al.*, 1968; O'Mongain *et al.*, 1968; Charman *et al.*, 1968). Throughout the γ-ray band the sensitivities of observations need to be increased before much can be expected. Observations with poor angular resolution near 100 MeV have, however, shown quite clearly that there is a band of emission along the galactic plane (Clark *et al.*, 1968). It is possible to interpret this as π-γ secondaries from cosmic rays colliding with gas in the galactic plane, but the intensity is 20–50 times higher than would have been expected. Raising the cosmic-ray density or gas density in the inner parts of the Galaxy causes a variety of more or less severe difficulties for cosmic-ray theory; the matter is too complicated to discuss at length here. Alternatively, the intense far-infrared radiation reported by Shivanandan *et al.* (1968), if real, may permeate the galactic disk, and then Compton scattering on cosmic-ray electrons can give the γ-ray flux. This point of view has been argued by Cowsik and Pal (1969) and by Shen (1969).

There is a third explanation, however, which seems reasonable and has been discussed by Ogelman (1969): The γ-ray flux may be merely the unresolved sum of the galactic X-ray sources, extrapolated to the 100-MeV band. If most of the unidentified X-ray sources have power-law spectra $n(v)\,dv \propto v^{-2}dv$, as observed for the Crab, Cyg X-1, and the Sagittarius sources as a group, then the numbers work out about right. Indeed there is even a peak at the location of the Crab in the data of Clark *et al.*, and it may be that this represents a real observation of the Crab as a γ-ray source. Eliding details, if we have synchrotron X-rays from these objects, it does not transgress the bounds of the possible to have synchrotron γ-rays as well. If these supernova remnants are the sources of galactic cosmic rays, making this hypothesis is really equivalent to saying that the γ-rays are the tail of the synchrotron radio spectrum of cosmic-ray electrons in the Galaxy! This possibility was mentioned by Verma (1968) and much earlier by Friedlander (*c.* 1962). (At this early date, synchrotron γ-radiation was thought preposterous, so Friedlander suffered heavy criticism and withdrew the idea.) Of course at these high frequencies the emission must all occur near the cosmic-

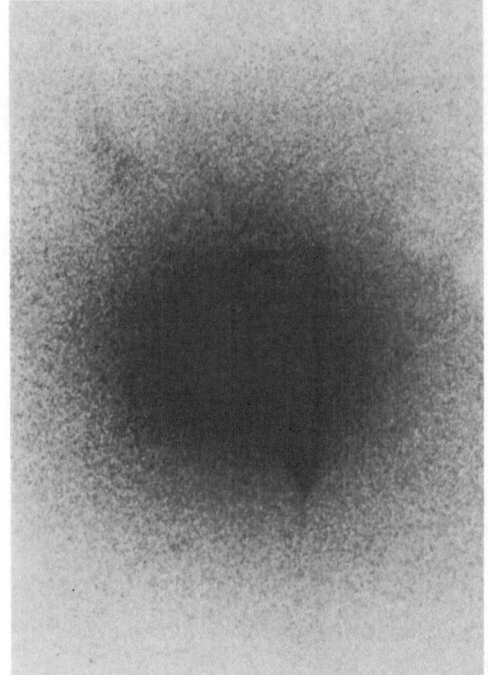

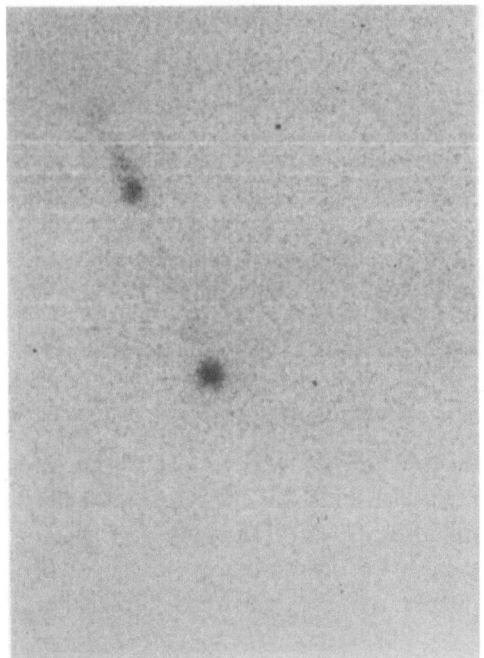

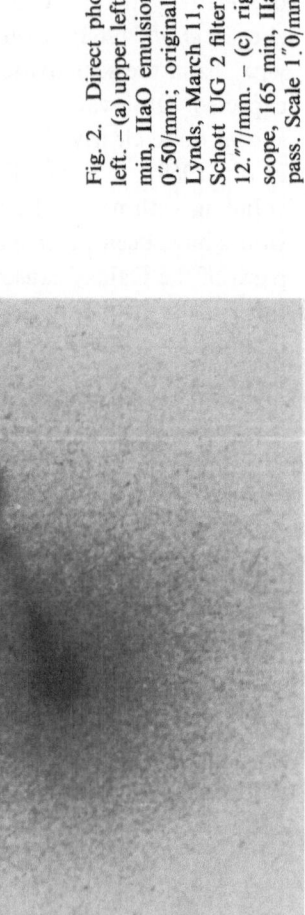

Fig. 2. Direct photographs of the M87 jet. North is at top, East at left. – (a) upper left: by H. C. Arp, Feb. 3, 1965; 200-inch telescope, 15 min, IIaO emulsion, λ3727 interference filter, 100-Å bandpass. Scale 0″.50/mm; original plate scale 11″/mm. – (b) lower left: by C. R. Lynds, March 11, 1964; 84-inch telescope, 60 min, IIaO emulsion, Schott UG 2 filter 2 mm thick. Scale 0″.50/mm; original plate scale 12″.7/mm. – (c) right: by H. C. Arp, June 2, 1967; 200-inch telescope, 165 min, IIaO emulsion, λ3727 interference filter, 100-Å bandpass. Scale 1″.0/mm; original plate scale 11″/mm.

ray sources (rather than diffusely throughout the Galaxy as at radio wavelengths) – but this is precisely Ogelman's suggestion.

11. Extragalactic Sources; M87

Among extragalactic objects one certainly expects the Magellanic Clouds to be sources, but reports at this conference leave us in doubt whether they have been detected. If the reported measurement of the Large Magellanic Cloud by the Livermore group (Mark *et al.*, 1969) *is* correct, the flux is just about what one would expect if the Cloud contains sources similar to those known in the Galaxy.

There is a weak high-latitude source in Leo, but I have heard nothing more of this since the first report by the NRL group (Byram *et al.*, 1966). I believe three other high-latitude sources have been announced at this conference. Of course any of these may well be extragalactic, but we need identifications.

This leaves us with M87 (Byram *et al.*, 1966; Friedman and Byram, 1967; Bradt *et al.*, 1967; Haymes *et al.*, 1968). The position of the X-ray source is still, I think, known only to a few degrees accuracy in right ascension, but in declination it coincides with the galaxy to within 0.1°, and at its high latitude the identification seems sure. Figures 2a, b, c (Felten *et al.*, 1970) show several recent photographs of the famous jet in this galaxy. Figures 2a and 2c are taken through an [OII] $\lambda 3727$ interference filter, but the jet radiation is probably continuum coming through the filter. (In Figure 2c, however, which is printed at half-scale to show the outer parts of the galaxy, the counter-jets (Arp, 1967) can be seen, and these apparently *are* line radiation.) The distance* from the nucleus of the galaxy to the tip of the jet is at least 1500 pc (depending on projection angle), and a filamentary extension can be seen. The bright part of the jet is at least 400 pc long, but it is a matter of dispute whether the bright condensations within it are resolved (de Vaucouleurs *et al.*, 1968; Felten *et al.*, 1970); their diameters are certainly $\leqslant 70$ pc. The bluish continuum of these knots is strongly polarized, making it quite certain that we are seeing optical synchrotron radiation.

Now there is also a strong radio source in the central region of this galaxy. Figure 3 is a radio map from the Cambridge interferometer (Macdonald *et al.*, 1968). The source is some 50″ long East-West, longer than the optical jet, and overlaps to the opposite side of the galactic nucleus, though it seems to be aligned parallel to the jet. Other observations (Lequeux, 1962) suggest that there are two peaks at the ends of this area. The source has not been resolved perpendicular to the jet. It is known, however, that finer structure is present; a few percent of the flux arises in hot spots <0″01 in diameter. Clearly this radio source is related to the optical jet but not identical in form; the exact relation is debatable.

Figure 4 (after Felten, 1968) collects flux densities for this object over a wide frequency range. The radio 'halo' spectrum is subtracted away to yield that shown for the 'core' source. A power-law extrapolation $v^{-0.75}$ is shown, but we see that the

---

* Distances here are based on $1'' = 72$ pc, for $D = 14.8$ Mpc, given recently by Sandage (1968).

JAMES E. FELTEN

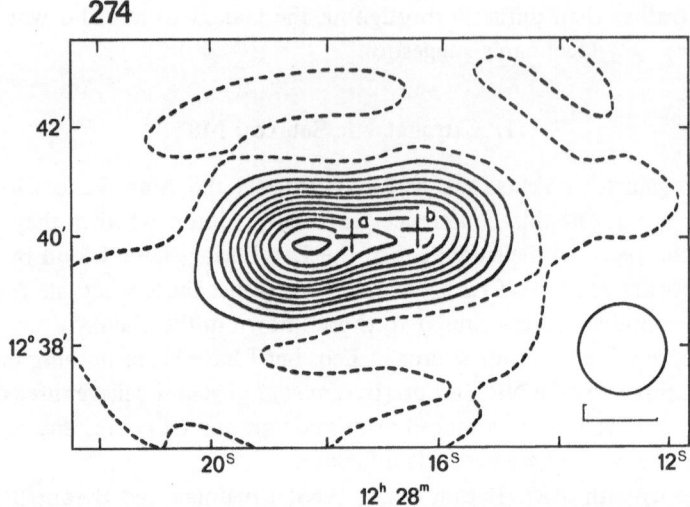

Fig. 3.   A radio map of the core source in M87 at 1407 MHz (Macdonald *et al.*, 1968). 'a' is the position of the galaxy nucleus and 'b' the tip of the optical jet. The vertical scale is compressed, for coordinates have been chosen to make the beam shape circular; the unequal orthogonal arms show the length of 20″ of arc in each coordinate. Beam size is shown. The source is unresolved perpendicular to the jet axis.

evidence for this is by no means conclusive. Though the scatter in optical measurements of the jet is large, recent optical and infrared results suggest a possible falloff in the optical band. In soft (1–10 keV) X-rays, there is now a considerable discrepancy between the Leicester results reported here (Adams *et al.*, 1969) and the earlier NRL and MIT measurements ('F' and 'Br' in Figure 4). Above 10 keV we have only a rash of upper limits, and one positive result (Haymes *et al.*, 1968), which is now much in doubt because, as shown, it conflicts with a 2σ upper limit from McClintock *et al.* (1969). Lacking positive observations in the hard band*, we are free to draw in almost any kind of fit to the soft X-ray data; a thermal bremsstrahlung curve at $T = 10^8$ K is shown as an example (Sartori and Morrison, 1967).

Nevertheless we know that the optical flux *is* synchrotron radiation. Regardless of the X-ray situation, the optical radiation poses big problems in a system as large as this. Suppose the fast electrons are injected at the nucleus. If equipartition prevails in the jet (and this assumption leads to $B \approx 3 \times 10^{-4}$ gauss), then the lifetimes of the electrons radiating at optical frequencies are only ≈ 100 yr. They cannot reach the distant outer parts of the jet! If B is weaker than the equipartition value, the lifetimes become longer, but still the electrons have to stream outward along the jet in order to make it, and the hose instability would then be expected to 'break off' the outer jet from the particle source at the nucleus (Felten, 1968). Another possibility is that the

* The published upper limit at 50 MeV (Frye and Wang, 1968) is a factor of 5 below the power-law extrapolation. Greisen (1968) has an unpublished measurement at 1 GeV which is said to lie on the extrapolation, but the statistics are poor.

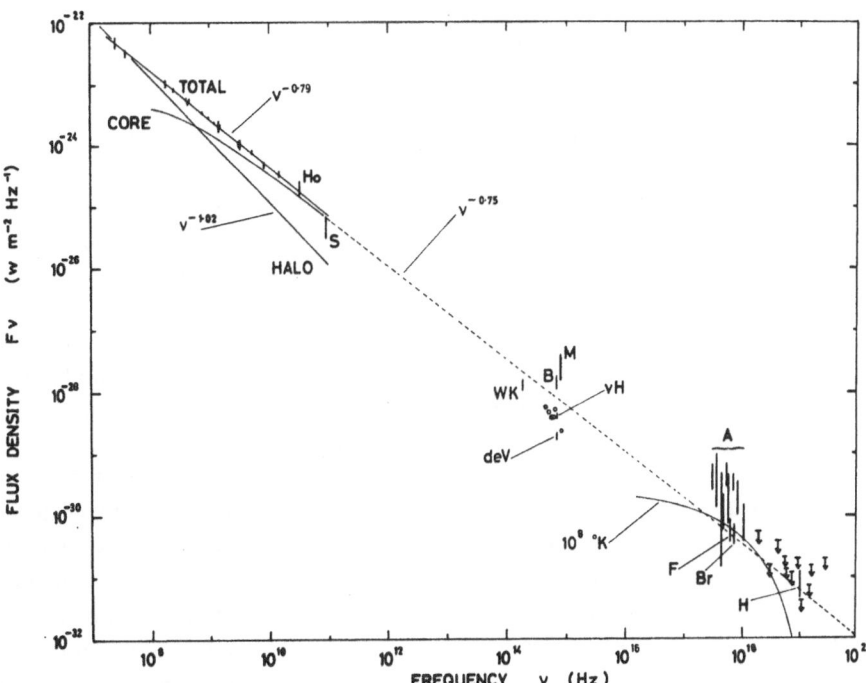

Fig. 4.   The electromagnetic spectrum of the M87 jet. References to many of the data may be found
in the paper by Felten (1968). Recent results include: Ho: Hobbs *et al.* (1968). S: Schorn *et al.* (1968).
WK: Wisniewski and Kleinmann (1968). Open circles: Pronik *et al.* (1967). deV: de Vaucouleurs
et al. (1968). A: Adams *et al.* (1969). The upper limits shown above 10^{18} Hz are at 2σ and are due to
Hudson *et al.* (1969), Overbeck and Tananbaum (1968), and McClintock *et al.* (1969). The positive
result 'H' is that of Haymes *et al.* (1968). The power-law extrapolation shown has equation
$F_\nu = 10.6 \times 10^{-18}\nu^{-0.75}$. A sample thermal bremsstrahlung curve at 10^8 K is also drawn through the
X-ray measurements.

ambient plasma is also moving outward, carrying the fast particles and field within it.
In this case there need not be any streaming of fast particles relative to field – or if
there is, and the field as a result becomes unstable and jumbled, the moving cloud still
carries fast particles and field along. Therefore in this model we can make B quite
small, and the synchrotron lifetimes long. But an upper limit on the lifetime of an
optical electron is always set by Compton loss to the high-density optical and infrared
photons (the Compton-synchrotron process), and the lifetime this allows the optical
electrons is only about 2×10^4 yr.* If the optical electrons are to survive the journey
of > 5000 light years from the nucleus, then, the plasma cloud must move at $v \gtrsim 0.2c$.
This is possible, though perhaps not very likely. Such a model can be tested better

---

* Anyway, Rieke and Weekes (1969) have recently pointed out that the field in the jet cannot be
arbitrarily weak; as we decrease it, the required flux of fast electrons goes up, and eventually the
Compton-synchrotron γ-rays at $\sim 10^{13}$ eV produced by these electrons would exceed observational
limits. This happens for $B \sim 10^{-5}$ gauss, a value which again gives a maximum lifetime $\sim 2 \times 10^4$ yr
for optical electrons in the jet.

when we have knowledge of the spectrum and, particularly, the spatial distribution of the X-ray source.

Time-scale difficulties of this sort can be overcome if one is free to postulate large inhomogeneities in the field. Then a fast electron can enjoy a period of 'rest' in a weak field, move briefly into a small volume of strong field where it can radiate optical or even X-ray photons, and then 'rest' again (Apparao, 1967). It is difficult, however, to maintain such inhomogeneities. A shock wave, e.g., will not usually propagate a field ratio greater than about 4:1, but we need a ratio more like 100:1 if it is to be of much use. Burbidge and Hoyle (1969) have suggested that the Crab contains massive condensed objects, which can retain strong fields at their surfaces, and Burbidge (1967a) proposed that condensed bodies of 10^6–$10^8 M_\odot$ were actually ejected from the nuclei of galaxies to form objects like the M87 jet. Such massive bodies could of course be injecting the cosmic rays as well as providing the strong fields in which they radiate.

Another model for the M87 jet, suggested when the time-scale problem first became apparent (Burbidge, 1956), is the 'secondary-production' model, in which a large reservoir of cosmic-ray primaries is injected initially, and later provides continuing injection of fast electrons throughout the confinement volume by pion production in collisions with ambient gas nuclei, followed by π-μ-e decay. I have recently investigated several variants of this (Felten, 1968; Felten et al., 1970). The fundamental difficulty which arises is the large cosmic-ray pressure introduced by the primaries, equivalent to that exerted by a field of $\sim 10^{-2}$ gauss. It is not likely that this can be balanced by a general field in the galaxy or by any other external agency. If the cosmic rays are confined to the volume of the optical jet, the energy content is $\sim 10^{56}$ erg, and an ambient gas density $n \sim 400$ cm^{-3} is needed within the jet to maintain it against the internal pressure for $\sim 10^5$ yr. It is possible to imagine conditions under which this large mass of gas ($3 \times 10^7 M_\odot$) would not yet have been detected spectroscopically, but it is not easy. If, on the other hand, the cosmic rays occupy a much larger volume (as suggested by the size of the core radio cloud), and the optical knots are visible simply because they are regions of dense gas where the p-p collisions occur, then the gas density required is somewhat lower, but of course the total energy involved in the cosmic-ray primaries becomes much larger. In none of these models is a time-scale much greater than 10^5 yr feasible. For further details, see the papers cited.

Of course we can always suppose that these optical electrons are being obtained through some continuous hydromagnetic acceleration process occurring in the outer part of the jet. I should mention, e.g., the 'galactic flare' model of Sturrock (1969, 1967), shown in Figure 5. The primeval intergalactic field, having been gathered in at the waist in the collapse of the protogalaxy, is amplified in the finished galaxy and forms a neutral sheet all around the equatorial plane. Tearing-mode instabilities can then cause field annihilation at any point in this neutral sheet. Sturrock's rough calculations indicate that electric fields of a few volts/meter can result and can accelerate particles to energies of 10^{18} eV. The observed luminosity of the jet can be supplied if the amplified field in the galaxy is as high as 10^{-3} gauss, and Sturrock finds this value appropriate to his model of galaxy formation if $M_{gal} \gtrsim 10^{10} M_\odot$. It should be noted that Sturrock's calculations assume essentially 100% efficiency for acceleration of fast particles in the field-annihilation process. Surely the true efficiency depends on the gas density, and one would think that most of the field energy would go into heat. The idea deserves a more careful investigation.

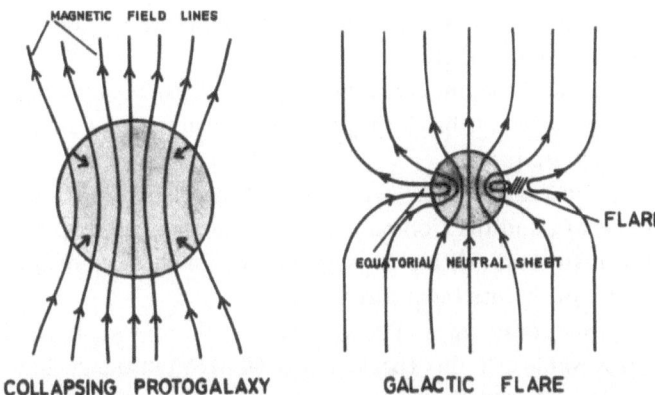

Fig. 5. Sturrock's (1969) model of the M87 jet as a 'galactic flare'. An unstable neutral sheet is formed in the equatorial plane of the galaxy by the primeval field trapped in the gravitational collapse of the protogalaxy. At any subsequent time, the tearing-mode instability may initiate field annihilation and reconnection anywhere on this neutral sheet.

I hope I have not bored you with this recital of theoretical possibilities. Perhaps it can fairly be said that all the models of M87 which are at all successful at the moment are faintly cranky – which only shows that here we are venturing into the unsolved problems of energy supply in extragalactic astrophysics. Let me mention that Haymes *et al.* (1969) have looked for X-rays above 34 keV from Centaurus A and not found any, which means that the X-ray flux from this object, if any, must lie well below the extrapolation of the power-law radio spectrum. This is not surprising, since we do not know of any optical synchrotron radiation from Cen A. For the moment M87 seems unique.

12. Other Processes

The processes of thermal bremsstrahlung and synchrotron emission, with which this review has been almost entirely concerned, have the advantage of high efficiency; that is, once you have your energy in hot gas or in fast electrons respectively, a good fraction of it tends to be given up through the specified process. Other mechanisms of X-ray production which theorists have suggested, but which are not known to be important in any of the discrete sources, are generally *inefficient* in competition with other loss processes acting. I shall discuss some of these briefly.

(a) *Compton (= 'inverse Compton') loss.* This depends on the ambient radiation density, which is quite low, at least by comparison with other processes acting in discrete sources. Its efficiency can, however, reach appreciable values under certain circumstances:

(i) In sources with high densities of synchrotron radiation, Compton scattering of the synchrotron photons on the fast electrons may occur (the Compton-synchrotron process). Gould (1965) pointed out that for the Crab, the expected flux of scattered photons around 10^{13} eV (where high-sensitivity detectors have most easily been achieved) was not far below the observational limit. This flux may have been observed by now, or may be shortly (Rieke and Weekes, 1969). Note also the relevance of this process to M87, discussed earlier.

(ii) In the intergalactic medium all other relevant densities are probably low, and Compton loss to the primeval blackbody photons becomes the dominant process for fast electrons, if any are present.

(iii) In discrete sources at an early epoch ($z \gg 1$) the blackbody radiation density is large, being $\propto (1+z)^4$, and the Compton process can again be dominant (Felten and Rees, 1969). Cases (ii) and (iii) are of great interest in theories of the diffuse background, but I will not discuss them further here.

(b) *π-γ γ-rays from cosmic rays.* The galactic cosmic rays probably make enough of these to be observable as a flux (peaked near 60 MeV) in the galactic plane before too long. The process is not likely to be important in discrete sources, but its occurrence does put some limits on models of these sources; e.g., in a secondary-production model of the M87 jet, the failure to observe π-γ photons around 10^{13} eV implies that $B < 7 \times 10^{-5}$ gauss (Felten, 1968).

(c) *Nonthermal bremsstrahlung, by either protons or electrons,* is inefficient, in that only $\sim 10^{-4}$ of the energy loss goes into photons; the bulk goes instead into ionization loss (or elastic Coulomb collisions if the ambient gas is ionized). Therefore these processes are rather extravagant in the particle energy required to explain a given photon flux. Again, however, they seem to be of interest in dealing with knotty problems of the diffuse background spectrum. Hayakawa (this conference), Silk and McCray (1969), and Boldt and Serlemitsos (1969) have all discussed models of this type. The strong heating of the gas implied by the required fluxes of fast particles should never be forgotten in these calculations.

(d) *Proton synchrotron radiation* has been considered by Rees (1968). It can hardly be important except in regions of very strong field, $B \gtrsim 100$ gauss.

13. Conclusion

Let me close this review by confirming what some of you may have gathered from my skeptical, even sardonic, tone: I have become something of a theoretical philistine in recent years. Big ideas do not thrill me much any more, because there are too many people having these big ideas, combining one particle flux with another like the ingredients of a cake, and not enough who are careful to check out the full consequences of their ideas. Also there is too much rationalizing and too little predicting – and in saying this I realize that my own papers are as vulnerable as anybody else's. Here in Rome, in a market square called Campo dei Fiori, a simple monument and eloquent inscription, set up after Rome was freed from the Vatican in the 19th century, mark the spot where Giordano Bruno was burned in 1600. Franco Pacini pointed out to me that Bruno was in fact burned for his heretical cosmological ideas, and suggested that if this punishment were reinstated, the present flood of astrophysical speculation would be much reduced! Well, no one is going to go that far, and if I am being too gloomy in these remarks, forgive me, and wait until the next symposium, when I am sure that the reviewer will take a more cheerful line.

Acknowledgements

I wish to thank Bob Gould, who passed on to me the opportunity to give this talk. Many colleagues gave me advice, and I should name especially George Field, Martin Rees, and Joe Silk, for without their help my amateur's task of digesting this material would have been even more painful than it was. I have also benefited from Geoffrey Burbidge's recent review (November 1968).

(Because of time pressure, parts of this paper had to be omitted in the oral presentation.)

Note added in proof (February 11, 1970): In this rapidly developing field it seems necessary to mention several recent papers pertinent to matters discussed above. Monte Carlo calculations for the broadening of X-ray lines by Thomson scattering in a thermal plasma have been presented by Angel (1969). There is evidence that iron lines are in fact present in the X-ray spectrum of Sco X-1 (Holt *et al.*, 1969).

Absorption of soft X-rays has been detected in the Crab spectrum, and also in Sco X-1 (Grader *et al.*, 1970). In the latter the absorption is apparently time-variable, which suggests that it is mainly circumstellar; then the optical Ca^+ lines are not indicators of the distance, which must be regarded as highly uncertain. Perhaps the value suggested by proper motions, $D \approx 200$ pc, is the best guess.

A discrete γ-ray source in Sagittarius is reported (Frye *et al.*, 1969). The diffuse γ-ray flux in the galactic plane may need some revision.

References

Ables, J. G.: 1969, *Astrophys. J. (Lett.)* **155**, L27.
Adams, D. J., Cooke, B. A., Evans, K., and Pounds, K. A.: 1969, *Nature* **222**, 757.
Aller, L. H.: 1961, *Abundance of the Elements*, Interscience Publishers, New York, pp. 179ff.
Angel, J. R. P.: 1969, *Nature* **224**, 160.
Apparao, M. V. K.: 1967, *Proc. Ind. Acad. Sci.* **A66**, 55.
Arp, H. C.: 1967, *Astrophys. Lett.* **1**, 1.
Bell, K. L. and Kingston, A. E.: 1967, *Monthly Notices Roy. Astron. Soc.* **136**, 241.
Biermann, L.: 1969, *Proc. Roy. Soc.* **A 313**, 357.
Boldt, E. A. and Serlemitsos, P. A.: 1969, *Astrophys. J.* **157**, 557.
Bradt, H., Mayer, W., Naranan, S., Rappaport, S., and Spada, G.: 1967, *Astrophys. J. (Lett.)* **150**, L199.
Burbidge, G. R.: 1956, *Astrophys. J.* **124**, 416.
Burbidge, G.: 1967a, *Nature* **216**, 1287.
Burbidge, G. R.: 1967b, in IAU Symposium No. 31, *Radio Astronomy and the Galactic System* (ed. by H. van Woerden), Academic Press, London, p. 463.

Burbidge, G.: 1968, paper read at Royal Society. *Proc. Roy. Soc.* **A313**, 331 (1969).
Burbidge, G. and Hoyle, F.: 1969, *Nature* **221**, 847.
Buselli, G., Clancy, M. C., Davison, P. J. N., Edwards, P. J., McCracken, K. G., and Thomas, R. M.: 1968, *Nature* **219**, 1124.
Byram, E. T., Chubb, T. A., and Friedman, H.: 1966, *Science* **152**, 66.
Cameron, A. G. W.: 1969, *Astrophys. Lett.* **3**, 171.
Cameron, A. G. W. and Mock, M.: 1967, *Nature* **215**, 464.
Charman, W. N., Jelley, J. V., Orman, P. R., Drever, R. W. P., and McBreen, B.: 1968, *Nature* **220**, 565.
Chodil, G., Mark, H., Rodrigues, R., Seward, F. D., Swift, C. D., Turiel, I., Hiltner, W. A., Wallerstein, G., and Mannery, E. J.: 1968, *Astrophys. J.* **154**, 645.
Clark, G. W., Garmire, G. P., and Kraushaar, W. L.: 1968, *Astrophys. J. (Lett.)* **153**, L203.
Cowsik, R. and Pal, Y.: 1969, *Phys. Rev. Lett.* **22**, 550.
De Vaucouleurs, G., Angione, R., and Fraser, C. W.: 1968, *Astrophys. Lett.* **2**, 141.
Eggen, O. J., Freeman, K. C., and Sandage, A.: 1968, *Astrophys. J. (Lett.)* **154**, L27.
Fegan, D. J., McBreen, B., O'Mongain, E. P., Porter, N. A., and Slevin, P. J.: 1968, *Canadian J. Phys.* **46**, S433.
Feldman, P. A. and Silk, J. I.: 1970, this volume, p. 257.
Felten, J. E.: 1968, *Astrophys. J.* **151**, 861.
Felten, J. E. and Gould, R. J.: 1966, *Phys. Rev. Lett.* **17**, 401.
Felten, J. E. and Rees, M. J.: 1969, *Nature* **221**, 924.
Felten, J. E., Arp, H. C., and Lynds, C. R.: 1970, *Astrophys. J.* (in press).
Friedlander, M. W.: *c.* 1962, unpublished work.
Friedman, H. and Byram, E. T.: 1967, *Science* **158**, 257.
Fritz, G., Meekins, J. F., Henry, R. C., Byram, E. T., and Friedman, H.: 1968, *Astrophys. J. (Lett.)* **153**, L199.
Fritz, G., Meekins, J. F., Henry, R. C., and Friedman, H.: 1969, *Astrophys. J. (Lett.)* **156**, L33.
Frye, G. M. and Wang, C. P.: 1968, *Canadian J. Phys.* **46**, S448.
Frye, G. M., Staib, J. A., Zych, A. D., Hopper, V. D., Rawlinson, W. R., and Thomas, J. A.: 1969, *Nature* **223**, 1320.
Giacconi, R., Gursky, H., and Waters, J. R.: 1965, *Nature* **207**, 572.
Gorenstein, P., Giacconi, R., and Gursky, H.: 1967, *Astrophys. J. (Lett.)* **150**, L85.
Gorenstein, P., Gursky, H., and Garmire, G.: 1968, *Astrophys. J.* **153**, 885.
Gould, R. J.: 1965, *Phys. Rev. Lett.* **15**, 577.
Grader, R. J., Hill, R. W., and Seward, F. D.: 1969, *Sky and Telescope* **37**, 79.
Grader, R. J., Hill, R. W., Seward, F. D., and Hiltner, W. A.: 1970, *Astrophys. J.* **159**, 201.
Greisen, K. I.: 1968, Brandeis Summer Institute lectures.
Gursky, H., Gorenstein, P., and Giacconi, R.: 1967, *Astrophys. J. (Lett.)* **150**, L75.
Harries, J. R., McCracken, K. G., Francey, R. J., and Fenton, A. G.: 1967, *Nature* **215**, 38.
Hayakawa, S. and Matsuoka, M.: 1964, *Prog. Theor. Phys.* Suppl. 30, 204.
Haymes, R. C., Ellis, D. V., Fishman, G. J., Glenn, S. W., and Kurfess, J. D.: 1968, *Astrophys. J. (Lett.)* **151**, L131.
Haymes, R. C., Ellis, D. V., Fishman, G. J., Glenn, S. W., and Kurfess, J. D.: 1969, *Astrophys. J. (Lett.)* **155**, L31.
Hill, R. W., Grader, R. J., and Seward, F. D.: 1968, *Astrophys. J.* **154**, 655.
Hiltner, W. A., Mook, D. E., Ludden, D. J., and Graham, D.: 1967, *Astrophys. J. (Lett.)* **148**, L47.
Hobbs, R. W., Corbett, H. H., and Santini, N. J.: 1968, *Astrophys. J.* **152**, 43.
Holt, S. S. and Ramaty, R.: 1969, *Solar Phys.* **8**, 119.
Holt, S. S., Boldt, E. A., and Serlemitsos, P. J.: 1968, *Astrophys. J. (Lett.)* **154**, L137.
Holt, S. S., Boldt, E. A., and Serlemitsos, P. J.: 1969, *Astrophys. J. (Lett.)* **158**, L155.
Hudson, H. S., Peterson, L. E., and Schwartz, D. A.: 1969, *Solar Phys.* **6**, 205.
Jenkins, E. B. and Morton, D. C.: 1967, *Nature* **215**, 1257.
Johnson, H. M.: 1966, *Astrophys. J.* **146**, 960.
Johnson, H. M.: 1968, *Astrophys. J.* **154**, 1139.
Johnson, H. M., Spinrad, H., Taylor, B. J., and Peimbert, M.: 1967, *Astrophys. J. (Lett.)* **149**, L45.
Kerr, F. J.: 1969, *Ann. Rev. Astron. Astrophys.* **7**, 39.
Kraft, R. P. and Demoulin, M.H.: 1967, *Astrophys. J. (Lett.)* **150**, L183.

Kraft, R. P. and Miller, J. S.: 1969, *Astrophys. J. (Lett.)* **155**, L159.

Lequeux, J.: 1962, *Ann. Astrophys.* **25**, 221.

Lewin, W. H. G., Clark, G. W., and Smith, W. B.: 1968a, *Astrophys. J. (Lett.)* **152**, L55.

Lewin, W. H. G., Clark, G. W., and Smith, W. B.: 1968b, *Nature* **219**, 1235.

Macdonald, G. H., Kenderdine, S., and Neville, A. C.: 1968, *Monthly Notices Roy. Astron. Soc.* **138**, 259.

Manley, O. P. and Olbert, S.: 1969, *Astrophys. J.* **157**, 223.

Mark, H., Price, R., Rodrigues, R., Seward, F. D., and Swift, C. D.: 1969, *Astrophys. J. (Lett.)* **155**, L143.

McClintock, J. E., Lewin, W. H. G., Sullivan, R. J., and Clark, G. W.: 1969, *Nature* **223**, 162.

McGee, R. X., Milton, J. A., and Wolfe, W.: 1966, *Australian J. Phys.* Suppl. 1.

Moffet, A. T. and Berge, G. L.: 1968, *Astrophys. J.* **153**, 997.

Neugebauer, G., Oke, J. B., Becklin, E., and Garmire, G.: 1969, *Astrophys. J.* **155**, 1.

Ogelman, H.: 1969, *Nature* **221**, 753.

O'Mongain, E. P., Porter, N. A., White, J., Fegan, D. J., Jennings, D. M., and Lawless, B. G.: 1968, *Nature* **219**, 1348.

Overbeck, J. W. and Tananbaum, H. D.: 1968, *Astrophys. J.* **153**, 899.

Peterson, L. E. and Jacobson, A. S.: 1966, *Astrophys. J.* **145**, 962.

Peterson, L. E., Jacobson, A. S., Pelling, R. M., and Schwartz, D. A.: 1968, *Canadian J. Phys.* **46**, S437.

Poveda, A. and Woltjer, L.: 1968, *Astron. J.* **73**, 65.

Prendergast, K. H. and Burbidge, G. R.: 1968, *Astrophys. J. (Lett.)* **151**, L83.

Pronik, V. I., Pronik, I. I., and Chuvaev, K. K.: 1967, *Astr. Zu.* **44**, 965; trans. *Soviet Astr.* **11**, 777 (1968).

Purton, C. R. and Andrew, B. H.: 1969, *Nature* **222**, 863.

Rappaport, S., Bradt, H. V., Naranan, S., and Spada, G.: 1969, *Nature* **221**, 428.

Rees, M. J.: 1968, *Astrophys. Lett.* **2**, 1.

Riegler, G. R.: 1969, dissertation, University of Maryland; Goddard Space Flight Center preprint no. X-611-69-1.

Riegler, G. R. and Ramaty, R.: 1969, *Astrophys. Lett.* **4**, 27.

Rieke, G. H. and Weekes, T. C.: 1969, *Astrophys. J.* **155**, 429.

Rocchia, R., Rothenflug, R., Boclet, D., and Durouchoux, P.: 1969, *Astron. Astrophys.* **1**, 48.

Sandage, A.: 1968, *Astrophys. J. (Lett.)* **152**, L149.

Sandage, A., Osmer, P., Giacconi, R., Gorenstein, P., Gursky, H., Waters, J., Bradt, H., Garmire, G., Sreekantan, B. V., Oda, M., Osawa, K., and Jugaku, J.: 1966, *Astrophys. J.* **146**, 316.

Sartori, L. and Morrison, P.: 1967, *Astrophys. J.* **150**, 385.

Schorn, R. A., Epstein, E. E., Oliver, J. P., Soter, S. L., and Wilson W. J.: 1968, *Astrophys. J. (Lett.)* **151**, L27.

Shen, C. S.: 1969, *Phys. Rev. Lett.* **22**, 568.

Shivanandan, K., Houck, J. R., and Harwit, M. O.: 1968, *Phys. Rev. Lett.* **21**, 1460.

Shklovsky, I. S.: 1967, *Astrophys. J. (Lett.)* **148**, L1.

Silk, J. and McCray, R.: 1969, *Astrophys. Lett.* **3**, 59.

Sofia, S., Eichhorn, H., and Gatewood, G.: 1969, *Astron. J.* **74**, 20.

Sturrock, P. A.: 1967, in *Plasma Astrophysics, Varenna lectures* (ed. by P. A. Sturrock), Academic Press, New York, p. 338.

Sturrock, P. A.: 1969, in *Plasma Instabilities in Astrophysics* (ed. by D. G. Wentzel and D. A. Tidman), Gordon and Breach, New York.

Tucker, W.: 1967, *Astrophys. J.* **148**, 745.

Verma, S. D.: 1968, *Astrophys. J.* **152**, 537.

Wallerstein, G.: 1967, *Astrophys. Lett.* **1**, 31.

Westphal, J. A., Sandage, A., and Kristian, J.: 1968, *Astrophys. J.* **154**, 139.

Wisniewski, W. and Kleinmann, D. E.: 1968, *Astron. J.* **73**, 866.

PREDICTED X-RAY FLUXES OF STELLAR CORONAS

C. DE LOORE and C. DE JAGER

Astronomical Institute Brussels, Belgium, and
Astronomical Institute Utrecht, The Netherlands

Abstract. Models of convection zones and corresponding mechanical energy fluxes were computed for 90 stellar photospheres with effective temperatures ranging from 2500 K to 41 600 K and acceleration of gravity values between 1 and 10^5. The most intense X-ray fluxes may be expected from stars with $T \approx 7200$ K and $\log g \approx 4$. Detectable X-ray fluxes could be expected from Procyon, α Cen and β Cas.

1. Computed Convective Models of Stellar Photospheres

Models of convective layers in stars were computed by one of us (De Loore, 1970), for 90 models of stellar photospheres for effective temperatures ranging from 2500 to 41600 K and values for the acceleration of gravity, g, between 1 and 10^5. The computations were based on Böhm-Vitense's (1958) mixing length theory. Mechanical fluxes were computed on the basis of Lighthill's considerations.

The computations show that there are two temperature ranges for which the largest values of the mechanical fluxes may be expected. These are first around 7200 K and secondly also around 3700 K. The variation of the mechanical flux with effective temperature and g-values is shown in Figures 1 and 2.

2. Expected Stellar Coronas

With the hypothesis that the mechanical energy flux is responsible for the heating of the corona, coronal models were constructed for the sun and for a 10 stars with effective temperatures between 5000 K and 8320 K for $\log g$-values of 4 or 5.

For main sequence stars the largest fluxes are generated in F-stars; stars with $T_{\mathrm{eff}} = 7130$ K and $\log g = 4$ possess also the hottest and almost dense coronas, with a computed temperature of 3.7×10^6 K and $\log N_e = 10.5$.

The solar corona computed in this way, on the basis of a photospheric mechanical flux of 0.14×10^8 erg cm^{-2} sec^{-1}, has a temperature of 1.3×10^6 K and $\log N_e = 9.8$. This density is apparently too high, but even when including in the computations all theoretical refinements proposed in the last few years by various authors it does not appear possible to obtain a solar coronal model with a smaller density.

3. Expected Stellar X-Ray fluxes

In estimating the expected stellar X-ray fluxes we make our computations in a differential way, by comparing them with the solar values. In that way possibly occurring small imperfections in the computations are hopefully greatly eliminated.

L. Gratton (ed.), Non-Solar X- and Gamma-Ray Astronomy, 238–241. All Rights Reserved

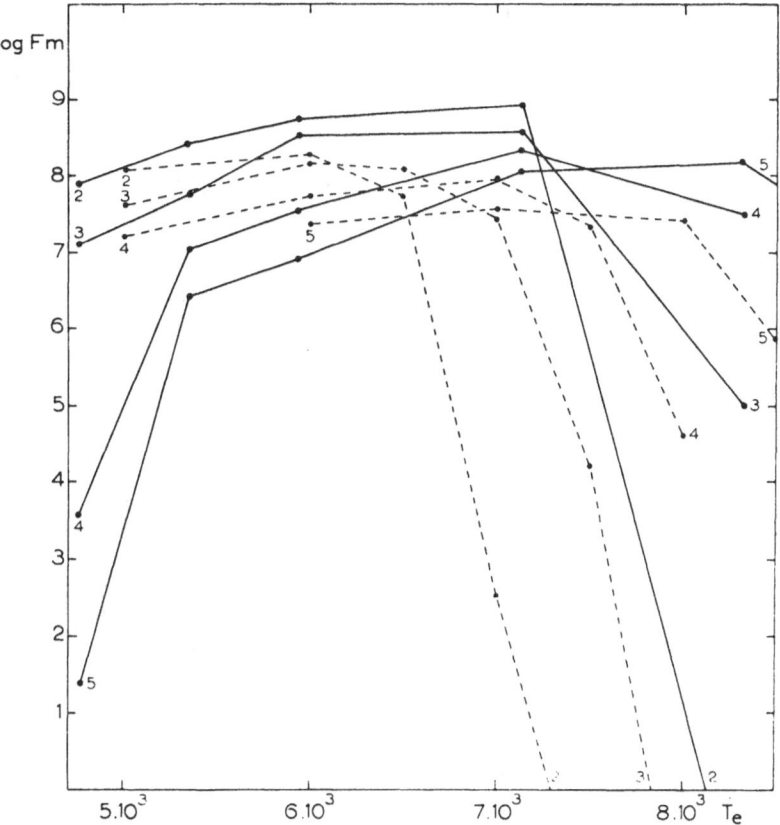

Fig. 1. Computed mechanical fluxes for stars. The curves are labelled with their $\log g$-values. Dashed are recent computational results of Nariai (1969).

The ratio between the expected X-ray flux of a star and that of the sun is given by:

$$\frac{F_X(\text{star})}{F_X(\text{sun})} = \left(\frac{R_{\text{star}}}{R_{\text{sun}}}\right)^2 \left(\frac{N_{e\,\text{star}}}{N_{e\,\text{sun}}}\right)^2 \left(\frac{H_{\text{star}}}{H_{\text{sun}}}\right) \left(\frac{d_{\text{sun}}}{d_{\text{star}}}\right)^2,$$

where R and d are the star's radius and distance; N_e is the electron density at the basis of the stellar corona; H is the coronal scale height.

From Allen (1962), we extracted the spectral types, the absolute magnitudes and the distances of a number of favourable stars. From the statistical relations, given in Allen, between the spectral type, the absolute magnitude, the radii, and the acceleration of gravity, we obtained values of the stellar radii and their g-values.

From the effective temperatures and g-values the temperatures and electron densities of the stellar coronas could be derived on the basis of the computations described in Section 2 of this paper.

Since the coronas are expected to be optically thin, the predicted X-ray flux may be proportional to the square of the electron densities. The precise value of the coronal temperature is less important for the predicted X-ray flux.

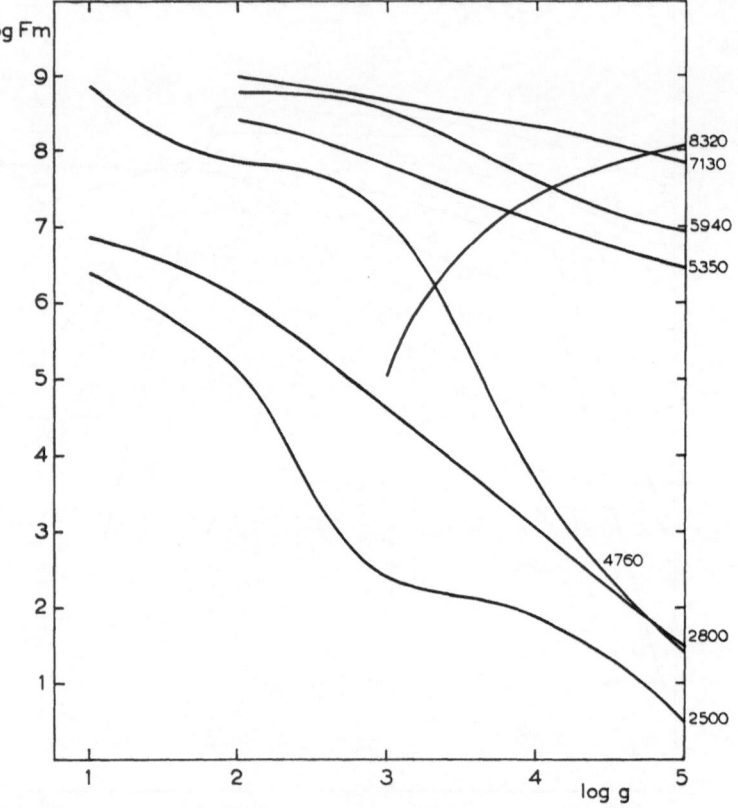

Fig. 2.   The same results as those of Figure 1, now plotted against log g, with the effective temperature as a parameter.

In comparing the predicted X-ray fluxes with those of the sun we used Friedman's (1959) observations, yielding a total solar X-ray flux during sunspot minimum of $0.13\ \mathrm{erg\ cm^{-2}\ sec^{-1}}$ and during the maximum he found values ranging up to $1\ \mathrm{erg\ cm^{-2}\ sec^{-1}}$. These are values integrated over the whole X-range. The corresponding numbers of photons are between 0.2 and 1.5×10^9 photons $\mathrm{cm^{-2}\ sec^{-1}}$ at earth's distance.

The results are found in Table I.

TABLE I

Star	Spectrum	T_{eff}	log g	log(R/R_{sun})	Corona log(N/N_{sun})	log distance (parsec)	Expected integral photon fluxes ($\mathrm{cm^{-2}\ sec^{-1}}$ earth)	
Procyon	F5	6500	4.3	0.2	+0.86	0.55	0.016	0.12
α Cen	G2	5800	4.4	0.1	0	0.11	0.008	0.06
β Cas	F2	6900	4.5	0.1	+1.20	1.15	0.0007	0.005
Sun	G2	5800	4.4	0	0	−5.3	0.2×10^9	1.5×10^9

In interpreting these expected integral photon fluxes it should be remembered that these values are the integral ones. With a given detector, which mostly only observes in a restricted wavelength band, the flux may easily be reduced by a factor of 100. Nevertheless, even when taking this into account, it does not look impossible that within a few years observations of the X-ray fluxes of Procyon, α Cen and perhaps also of β Cas, could be obtained.

References

Allen, C. W.: 1962, *Astrophysical Quantities*, Athlone Press, London, 2nd ed.
Böhm-Vitense, E.: 1958, *Z. Astrophys.* **46**, 108.
De Loore, C.: 1970, *Astrophys. Space Sci.* **6**, 60.
Friedman, H.: 1959, *Proc. Inst. Res. Eng.* **47**, 272.
Nariai, K.: 1969, *Astrophys. Space Sci.* **3**, 150, 160.

PHYSICAL PROCESSES IN CYGNUS X-2

ROBERT E. WILSON

Dept. of Astronomy, University of South Florida, Tampa, Fla., U.S.A.

In developing models for the point X-ray sources the general aim has been to find a common explanation for all such objects. However, important observational differences have already become apparent among the systems having optical identifications. A close examination of the evidence shows that those with the most certain optical identifications, Scorpio X-1 and Cygnus X-2, have very few features in common and, in fact, the model proposed here for Cyg X-2 can probably be excluded for Sco X-1.

Except for having generally similar colors, Sco X-1 and Cyg X-2 are decidedly different in almost all observational characteristics. The only lines in the spectrum of Sco X-1 are emission lines, while Cyg X-2 shows many lines in absorption but only one in emission. Cyg X-2 is known to be a spectroscopic binary*, whereas there is no evidence for binary motion in Sco X-1. Furthermore, the proper motion study by Sofia *et al.* [2, 3] points to probable membership for Sco X-1 in the Scorpio-Centaurus association, thus indicating a visual luminosity too low for even a single main sequence star and making it still less probable that the object is binary. Sco X-1 should also be much the younger of the two since the Scorpio-Centaurus association is only about 20 million years old, while the radial velocities of Cyg X-2 are difficult to reconcile with pure Population I kinematics [4], but are quite consistent with the older disk population. In addition Cyg X-2 has many spectroscopic peculiarities which make it not only different from Sco X-1 but unique among known stellar objects. We shall see, however, that it is possible to account for all of these peculiarities with one physically reasonable model. In fact, we shall even find this model similar to a kind of system previously identified from observations of a quite different sort.

Cyg X-2 is the only point X-ray source for which optical spectroscopy has provided really significant information, but it has not been possible to make straightforward interpretations of this information in terms of previously known objects. We briefly state here the basic well established observational facts. The system is a spectroscopic binary [5] with one component a G-star [6]. The distance is probably 500 to 700 parsecs [6], based on an assumed M_v of $+5^m.5$ for the G-star, which determines the X-ray luminosity to be about 25 $L_\odot$. Balmer absorption lines are seen [5, 6, 7], but these could scarcely be associated with the G-star since both their equivalent widths and their half-widths are far too great for any conceivable physical conditions that might be found on a G-star. That is, the equivalent widths correspond approximately to spectral type A8 and the half-widths to a temperature much higher than that needed

---

* In a recent paper, Kraft and Miller [1] have given radial velocity observations which, in their opinion, weaken the case for Cyg X-2 being a spectroscopic binary. However, we shall see that if the present model is correct, the lines measured by Kraft and Miller will be very poor indicators of binary motion.

L. Gratton (ed.), Non-Solar X- and Gamma-Ray Astronomy, 242–246. All Rights Reserved
Copyright © 1970 by the IAU

to completely ionize hydrogen. Sofia [8] has shown that all broadening mechanisms except mass motions can be eliminated, and since these mass-motions must be of the order of 10^3 km/sec, their association with any kind of recognizable G-star becomes still more unlikely. We also have the following additional observations:

(1) The Balmer line ratios are peculiar, with the *high* number lines relatively strong, Hβ weak and Hα absent.

(2) The radial velocity measures, although indicating binary motion, show much greater scatter than can reasonably be attributed to measuring errors.

(3) The Ca II and Balmer line velocities are in general agreement [5, 6, 7] although the Ca II lines are far too strong to be formed in the same region as the Balmer lines.

(4) Peimbert *et al.* [9], have shown that the absolute spectral energy distribution can be fitted by the sum of a normal G-star spectrum and an additional source approximately constant per unit frequency interval. This source makes the U, B, V colors bluer than those of a G-star alone.

(5) There are minute-to-minute fluctuations in optical brightness.

Although we have a diversity of clues into the nature of Cyg X-2, attention must repeatedly revert to one truly remarkable fact, which is that the Balmer lines appear in absorption. We have already seen that the broadening of the Balmer lines can only be due to mass motions of the order of 10^3 km/sec, and we also know that, since Balmer lines arise from an excited level, we should see them in emission rather than absorption unless they are formed immediately above a source of continuum radiation. Thinking first that this source of continuum radiation will be a stellar photosphere, we find ourselves trying to imagine vertical motions of 10^3 km/sec in the outer layers of a star – a situation for which no physical mechanism seems to exist. Convective motions should not approach this magnitude and, since the profiles are symmetrical, we can rule out motions which go only away from the observer, such as might be expected for matter falling onto a star. Falling matter would, of course, produce strongly assymetric profiles. Even if we relax all but the most fundamental physical constraints, there exist very few reasonable models involving mass motions of gas both toward and away from the observer at 10^3 km/sec while interposed between the observer and a star. In an accretion model, we could imagine approximately symmetrical motions being produced by infalling matter if the X-ray luminosity were sufficiently great so that radiation pressure could drive cells of the material back to considerable heights. However, this fails on many counts, such as the impossibility of maintaining hydrogen neutral in such temperature conditions, damping of motions by collisions of gas cells, and several other reasons. We might even picture optically thin material falling onto a compact object, but producing absorption lines because it is seen *projected* against the more distant G-component. This accounts for the symmetrical profiles but must be abandoned because it could only apply during a small phase interval, while the Balmer lines have been seen on all spectrograms of Cyg X-2. Strange as these ideas seem, they are worth mentioning to illustrate the extremes to which one must go to account for formation of the Balmer lines against a stellar continuum.

Of course, there does have to be a source of continuum radiation against which we see the Balmer lines, but from the preceding paragraph it appears that it cannot be a star. We may now pause to recall the various identifiable regions which must exist in Cyg X-2. We know there is a G-star because its spectrum has been observed, and we know there is at least one additional source of continuum radiation because one is required for the Balmer lines and because a source of excess blue radiation is required by the spectrophotometry. There seems to be no reason to doubt that these sources are the same. Finally we expect a region in which the X-rays are formed, which should be a very high temperature region and thus distinct from the other two.

All of these features will be present if the object is a semi-detached binary in which the primary component is a white dwarf accreting material which flows from a G-star secondary through the inner Lagrangian point. The X-rays will be produced at the white dwarf by the accretion mechanism as in the neutron star accretion model proposed by Sklovsky [10] for Sco X-1. We exclude a neutron star for Cyg X-2 because the explosion accompanying its formation would probably disrupt any close binary system [11]. The continuum source against which the Balmer lines appear will be the region of enhanced density in the general vicinity of the white dwarf where the infalling matter has been gravitationally focused. A general picture of how such material will move can be gained by inspection of trajectories computed by Kopal [12]. The trajectories show that the material will fall almost directly toward the white dwarf but, since the star presents a very small target, will not, in general, strike it directly. Once having entered this region of relatively high density, orbital energy will be lost through collisions and eventually the matter must fall onto the white dwarf. At any given time we expect a cloud of relatively high density which will show large Doppler motions due to the complicated gas streaming present. Since the Balmer line widths correspond to velocities of about 10^3 km/sec, the radius of the cloud should be about 10^5 km, for this is the distance from a 1-solar-mass white dwarf at which velocities of 10^3 km/sec are reached in free fall. The continuum radiation is due to heating of the cloud by the large X-ray flux, which should be about 10^{14} erg/cm^2 sec at 10^5 km from the white dwarf. The direct absorption of X-rays will be due to the heavy element constituents, with oxygen and neon the principal contributors. The cross section for their absorption depends on the abundances, but using solar abundances we find about 6 cm^2/gm for 3 keV X-rays. Due to the outward decrease in temperature the Ca II lines will naturally be formed at a greater distance from the center of the cloud than will the Balmer lines, thus explaining their relative sharpness since the mass motions will be smaller at this distance.

Three previously unexplained observational facts are natural consequences of the present model. The large scatter in the Balmer line radial velocities is expected from slight asymmetries in the gas streaming. These asymmetries need not be very great because the mass motions are of the order of 1000 km/sec while the binary orbital velocities are of the order of 200 km/sec. The reason for the weakness of the first two Balmer lines is clear when we realize that they are not formed on a stellar component and filled in by a bremsstrahlung spectrum, but are formed in the plasma cloud surrounding the white dwarf and filled in by light from the G-star. Since the G-star continuum becomes stronger relative to the cloud continuum toward longer wavelengths, the low number Balmer lines will be those most weakened. The minute-to-minute fluctuations in optical brightness are expected because the time scale for significant changes in the system is very short. It should take only about twenty minutes for material to fall all the way from the L_1 point to the cloud, so rapid brightness changes should definitely occur.

At the beginning we said that this model for Cyg X-2 would be found similar to systems recognized earlier from different observational evidence. These are certain old novae*, specifically the close binaries WZ Sge and DQ Her. For WZ Sge, which is actually a recurrent nova, Krzeminski and Kraft [13] have directly observed the lines of a white dwarf component, and for DQ Her one is indicated by the color and luminosity. WZ Sge shows gas streaming from the L_1 point toward the white dwarf, but it is not an X-ray source at presently detectable flux levels. In general, old novae which can be studied in detail spectroscopically often consist of a 'blue' primary component, which can sometimes be shown to be a white dwarf, and a 'red' secondary, of middle or late type [14]. Notice that the corresponding features of Cyg X-2 and the old novae have been established from entirely different evidence. In Cyg X-2 we must have a white dwarf because accretion onto a compact object is the only known X-ray generating mechanism sufficiently efficient to yield several tens of solar luminosities, while in WZ Sge we actually see the spectral lines of a white dwarf component. In Cyg X-2 we must have transfer of material from the secondary to the primary components because it is required by the accretion mechanism and by the Balmer line profiles, while in WZ Sge the phase dependence of the radial velocities shows that we are seeing matter flowing from the secondary component toward the white dwarf.

It seems highly probable that Cyg X-2 is a nova in some stage of development and, if we compare it to WZ Sge, we should probably favor a pre-nova interpretation. In WZ Sge the secondary component is spectroscopically invisible and has a mass, according to Krzeminski and Kraft, of only 0.03 $m_\odot$. If this is the actual mass of the object, it is somewhat argumentative whether it should be called a star. We seem to be viewing the remains of a star left over after a long interval of mass transfer. In Cyg X-2, however, most of the visible light comes from the secondary component, indicating that it still has a respectable mass. The identification of Cyg X-2 as a (probably pre-) nova is strengthened when we recall that its radial velocities indicate kinematics similar to those of old novae [4].

Cyg X-2 is an interesting object even when considered by itself, but it now appears much more so in view of its remarkable similarity to known post-novae. Further observations might solve one of the long standing problems of astrophysics, the search for the mechanism of the nova phenomenon.

---

* I am indebted to Dr. A. Mammano of Asiago for calling my attention to the similarity between Kraft's old novae and the present model for Cyg X-2.

Acknowledgement

I wish to thank Dr. Sabatino Sofia for extensive and very stimulating discussions of this model.

References

[1] Kraft, R. P. and Miller, J. S.: 1969, *Astrophys. J.* **155**, L159.
[2] Sofia, S., Eichhorn, H. K., and Gatewood, G.: 1969, *Astron. J.* **74**, 20.
[3] Gatewood, G. and Sofia, S.: 1968, *Astrophys. J.* **154**, L69.
[4] Sofia, S. and Wilson, R. E.: 1968, *Nature* **218**, 73.
[5] Burbidge, E. M., Lynds, C. R., and Stockton, A. N.: 1967, *Astrophys. J.* **150**, L95.
[6] Kraft, R. P. and Demoulin, M. H.: 1967, *Astrophys. J.* **150**, L183.
[7] Kristian, J., Sandage, A., and Westphal, J. A.: 1967, *Astrophys. J.* **150**, L99.
[8] Sofia, S.: 1968, *Astrophys. Letters* **2**, 173.
[9] Peimbert, M., Spinrad, H., Taylor, B. J., and Johnson, H. M.: 1968, *Astrophys. J.* **151**, L93.
[10] Sklovsky, I. S.: 1967, *Astrophys. J.* **148**, L1.
[11] Sofia, S.: 1967, *Astrophys. J.* **149**, L59.
[12] Kopal, Z.: 1959, *Close Binary Systems*, New York, p. 516.
[13] Krzeminski, W. and Kraft, R. P.: 1964, *Astrophys. J.* **140**, 921.
[14] Kraft, R. P.: 1964, *Astrophys. J.* **139**, 457.

THE ELECTROMAGNETIC SPECTRUM OF THE CRAB NEBULA

KRISHNA M. V. APPARAO*†

*Smithsonian Astrophysical Observatory and Harvard College Observatory,
Cambridge, Mass., U.S.A.*

The electromagnetic spectrum of the Crab Nebula has been determined experimentally in the radio, optical, and X-ray regions [1], in which it follows a power law of the type $S(v) = Av^{-\alpha}$, where $S(v)$ is the power (in watts/m^2 sec Hz), A and α are constants, and v is the frequency in Hz. Recent measurements [2–5], however, show a deviation from a power law in the microwave region (see Figure 1). In this paper, we

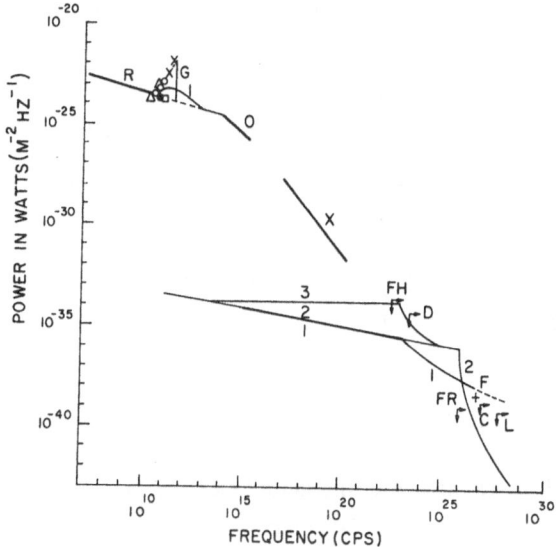

Fig. 1.   The electromagnetic spectrum of the Crab Nebula. The thick lines R, O, and X are the observed radio, optical, and X-ray spectra. The symbols around 10^{10} Hz indicate the observations of authors given in references [2–7]. The points with double arrows are upper limits and are integral values as given by Fazio *et al.* [10], Delvaille *et al.* [11], Fazio *et al.* [12], Chudakov *et al.* [13], Long *et al.* [14], and Fegan *et al.* [15]. The various curves are explained in the text. Curve 3 is the present calculation.

investigate the origin of this deviation and calculate the γ-ray spectrum due to this increase in the microwave photons via the Compton scattering from high-energy electrons.

The radio spectrum between the frequencies 2×10^7 and 2×10^{10} Hz follows a power law of the form $v^{-\alpha}$, with $\alpha = 0.28 \pm 0.05$. Between the frequencies 3×10^{10} and

* On leave from the Tata Institute of Fundamental Research, Bombay, India.
† This work was performed while the author held a U.S. National Academy of Sciences Senior Post-doctoral fellowship supported by the Smithsonian Institution.

L. Gratton (ed.), Non-Solar X- and Gamma-Ray Astronomy, 247–249. All Rights Reserved

3×10^{11}, the new observations indicate $\alpha \simeq -2$. (This change in the exponent is being disputed by Hobbs et al. [6] and Oliver et al. [7].) In the near-infrared and optical regions ($v > 10^{14}$ Hz), $\alpha \sim 0.8$. In the X-ray region, $\alpha \simeq 1$.

The difference in the exponent α between the radio spectrum and the optical spectrum is usually understood in terms of a model where there is continuous injection of electrons and where the electrons lose their energy predominantly by synchrotron radiation. Shklovsky [1] suggests that the change in slope at $v \sim 3 \times 10^{10}$ Hz could be due to injection of energetic electrons at the time of the explosion of the supernova; again due to synchrotron radiation, the electrons bunch together at a certain energy. It turns out that this hypothesis does not explain the increase of microwave radiation.

Let us consider the solution of the kinetic equation [8] for the differential energy spectrum of electrons $N(E, t)$, with a burst-injection spectrum at the time of the explosion of the form $KE^{-\gamma}$ (K and γ are constants) and a continuous injection of the term $qE^{-\gamma_1}$ (q and γ_1 are constants). With only the synchrotron radiation as the main energy-loss process, the solution is

$$
\begin{aligned}
N(E, t) = {} & KE^{-\gamma}(1 - \beta tE)^{\gamma - 2} \\
& + [qE^{-(\gamma_1 + 1)}/\beta(\gamma_1 - 1)] [1 - (1 - \beta tE)^{\gamma_1 - 1}].
\end{aligned}
\tag{1}
$$

Here t is the age of the nebula, and β is a function of the magnetic field H and is given by the equation for synchrotron loss, $(dE/dt) = -\beta E^2$. Equation (1) leads to the asymptotic forms

$$
N(E, t) = \begin{cases} KE^{-\gamma} + qtE^{-\gamma_1} & \text{for } E \ll 1/\beta t \\[2mm] \dfrac{qE^{-(\gamma_1 + 1)}}{\beta(\gamma_1 - 1)} & \text{for } E \gg 1/\beta t. \end{cases}
\tag{2}
$$

Comparing the asymptotic forms of the synchrotron spectra obtained from Equation (2) with the observed optical and radio spectra and using $H = 3 \times 10^{-4}$ gauss, we obtain K, q, γ, and γ_1. Using these, the actual electron spectrum, we derive $N(E, T)$, where T is the present age of the Crab Nebula. The synchrotron spectrum from this $N(E, T)$ is obtained between 10^{10} Hz and 10^{14} Hz (plotted as curve 1 in Figure 1). We see that such a burst injection is not sufficient to account for the observations. Beckman et al. [5] suggest that excess radiation in the microwave region may be due to a cool gas. We fit a Planck spectrum to the observations with the maximum at 1.2 mm (the observation with the highest frequency in the microwave region); the curve is shown as G in Figure 1.

We have calculated the γ-radiation due to the Compton-synchroton process by using the model given by Gould [9] and bearing in mind that the excess microwave photons are not of synchrotron origin. The γ-ray spectrum is plotted as curve 3 in Figure 1, along with the predictions from the calculation of Gould (curve 1) and that by Apparao (curve 2), who calculated the γ-radiation resulting from the Compton scattering of universal microwave photons from the electrons in the Crab Nebula. Selected observations of γ-radiation are also shown in Figure 1; these are the best upper limits. At energies $E_\gamma \geqslant 100$ MeV, the flux predicted is 2×10^{-5} photons/cm^2

sec, while the observed limit of Fazio *et al.* [10] is 3.5×10^{-5} photons/cm^2 sec. At energies $E_\gamma \geqslant 1$ GeV, the flux predicted is in 7×10^{-6} photons/cm^2 sec, which is to be compared with the limit 1.2×10^{-5} photons/cm^2 sec given by Delvaille *et al.* [11].

We conclude that the deviation from a power law of the spectrum of the Crab Nebula in the microwave region cannot be accounted for by the synchrotron process under the usual models. It could be of a thermal origin. If the excess of the microwave photons is due to a thermal gas, an upper limit to the temperature of ~ 5K is obtained by comparing the experimental upper limits of γ-radiation and the predicted fluxes. Further observations of the microwave spectrum and the γ-ray spectrum in the 100- to 1000-MeV region will throw light on this question.

References

[1] For references see Shklovsky, I. S.: 1968, *Supernovae*, Wiley, London.
[2] Tolbert, C. W.: 1965, *Nature* **206**, 1304.
[3] Tolbert, C. W. and Straiton, A. W.: 1964, *Nature* **204**, 1242.
[4] Kislyakov, A. E. and Na'umov, A. I.: 1968, *Soviet Astron.-AJ* **11**, 6.
[5] Beckman, J. E., Bastin, J. A., and Clegg, P. E.: 1969, *Nature* **221**, 944.
[6] Hobbs, R. W., Corbett, H. H., and Santini, N. J.: 1969, *Astrophys. J. Letters* **155**, L87.
[7] Oliver, J. P., Epstein, E. E., Schorn, R. A., and Soter, S. L.: 1967, *Astron. J.* **72**, 314.
[8] Kardashev, N. S.: 1962, *Soviet Astron.-AJ* **6**, 317.
[9] Gould, R. J.: 1965, *Phys. Rev. Lett.* **15**, 577.
[10] Fazio, G. G., Helmken, H. F., Cavrak, S. J., Jr., and Hearn, D. R.: 1968, *Canadian J. Phys.* **46**, Part 3 (Cosmic Ray Conf. Issue), 427.
[11] Delvaille, J. P., Albats, P., Greisen, K. I., and Ögelman, H. B.: 1968, *Canadian J. Phys.* **46**, Part 3 (Cosmic Ray Conf. Issue), 425.
[12] Fazio, G. G., Helmken, H. F., Rieke, G., and Weekes, T. C.: this volume, p. 192.
[13] Chudakov, A. E., Dadykin, V. L., Zatsepin, V. I., and Nesterova, N. M.: 1964, in *Proc. Intern. Conf. Cosmic Rays*, Jaipur, India, **4**, 199.
[14] Long, C. D., McBreen, B., Porter, N. A., and Weekes, T. C.: 1965, in *Proc. Intern. Conf. Cosmic Rays*, London, **1**, 318.
[15] Fegan, D. J., McBreen, B., O'Mongain, E. P., Porter, N. A., and Slevin, P. J.: 1968, *Canadian J. Phys.* **46**, Part 3 (Cosmic Ray Conf. Issue), 433.

A LOWER LIMIT TO THE MAGNETIC FIELD IN THE CRAB NEBULA FROM COSMIC γ-RAY EXPERIMENTS AT 10^{11} eV

G. G. FAZIO and H. F. HELMKEN

*Smithsonian Astrophysical Observatory and Harvard College Observatory,
Cambridge, Mass., U.S.A.*

and

G. H. RIEKE and T. C. WEEKES

*Mt. Hopkins Observatory, Smithsonian Astrophysical Observatory,
Amado, Arizona, and Harvard University, Cambridge, Mass., U.S.A.*

Abstract. The 10-m optical reflector at Mt. Hopkins, Arizona, has been used to search for cosmic γ-rays by the detection of atmospheric Čerenkov radiation from energetic particle showers. Approximately 100 drift scans of the Crab Nebula during 1968–69 have yielded no positive evidence of a γ-ray flux. The upper limit to the flux at 1.7×10^{11} eV is 2.0×10^{-10} photons/cm² sec. Assuming γ-rays of this energy are produced by Compton scattering, a lower limit on the average magnetic field in the Crab Nebula is 1.5×10^{-4} gauss. This experiment also verifies previous evidence that the high-energy electrons in the Crab Nebula are not the secondary products of high-energy proton interactions but must have been accelerated from lower energies.

1. Introduction

At energies above 10^{11} eV, cosmic γ-rays can interact with air molecules to generate a cascade of high-energy electrons, positrons, and secondary γ-rays. This 'air shower' is highly directional and proceeds with relativistic velocity along the direction of the initial γ-ray photon. The particles in the shower generate a cone of Čerenkov light about 2° wide along the shower axis. This cone of light will penetrate to sea level with a lateral spread of about 5×10^4 m². The duration of the light burst is approximately 10^{-8} sec. The detection of this light from γ-ray showers with energies as low as 10^{11} eV requires a dark site, a cloudless, moonless night, and a large, sensitive light collector. Unfortunately, high-energy cosmic-ray particles also generate air showers and bursts of Čerenkov light in the atmosphere. γ-rays can be distinguished only by a directional anisotropy in the distribution of Čerenkov light bursts.

2. Experimental System

In June 1968, a fully steerable 10-m light reflector was placed in operation at the 2320-m level of Mt. Hopkins, Arizona. The reflector, shown in Figure 1, consists of 248 individually adjustable hexagonal glass mirrors, each with a spherical figure and an aluminized front surface. The focal length of the reflector is 24 ft ($f/0.7$), which gives a field of view of about 1° diameter with a 5-inch photomultiplier tube at the focus. Additional properties of the detector are given in Table I.

Two operating modes have been used with the 10-m reflector. In one, a single photomultiplier was placed at the focus; the pulses from it were amplified and taken

L. Gratton (ed.), Non-Solar X- and Gamma-Ray Astronomy, 250–256. All Rights Reserved
Copyright © 1970 by the IAU

TABLE I

Experimental characteristics of the detector

Detector	Phototube	Effective area (m^2)	Effective solid angle (sterad)	Electronic bandwidth (MHz)	Photon density threshold (photon/m^2)	Altitude (m)	Zenith count rate (min^{-1})	Energy threshold (eV)	Collecting area (m^2)
Mt. Hopkins 10-m reflector	RCA 4522	60	1.5×10^{-4}	100	4–6	2320	300	9×10^{10}	1.3×10^4

Fig. 1.   The 10-m light reflector at Mt. Hopkins, Arizona.

to a discriminator. The discriminator threshold was set high enough so the random contribution to the counting rate was relatively small. The discriminator pulses were counted and printed out automatically. In the second mode, two photomultipliers were mounted on the focus ring of the reflector, with their centers separated by 1 ft. For a source on the reflector axis, the outer mirrors were aligned to form an image on one of the photomultipliers, and the inner ones to form an image on the other. The two photomultipliers were then used with a coincidence counting system.

Compared with a single-channel system, a coincidence system has advantages in stability and the ability to work into the noise level of the signal. In the coincidence system, the light-pulse broadening introduced by the reflector geometry is reduced from 6 nsec to 3 nsec. However, in this case, it has the disadvantage that each photomultiplier sees the whole reflector, including the mirrors aligned for the other phototube. Thus, the background light level is doubled.

A small lamp, driven by a servo-amplifier system, was placed in front of each phototube to keep the photomultiplier output current constant.

The 'drift-scan' technique was used to search for an anisotropy in the shower count rate. This technique consists of positioning the reflector at a point in the sky ahead in right ascension and allowing the earth's rotation to bring the source through the field of view. Depending on the declination, the duration of each scan is about 40 min, centered on the transit of the suspected source.

3. Observations

More than 14 possible sources of cosmic γ-rays have been investigated to date with the 10-m reflector. However, this paper will be concerned only with the data collected on the Crab Nebula. During the moonless periods from September 1968 to December 1969, 102 drifts scans were made on the Crab Nebula under excellent sky conditions.

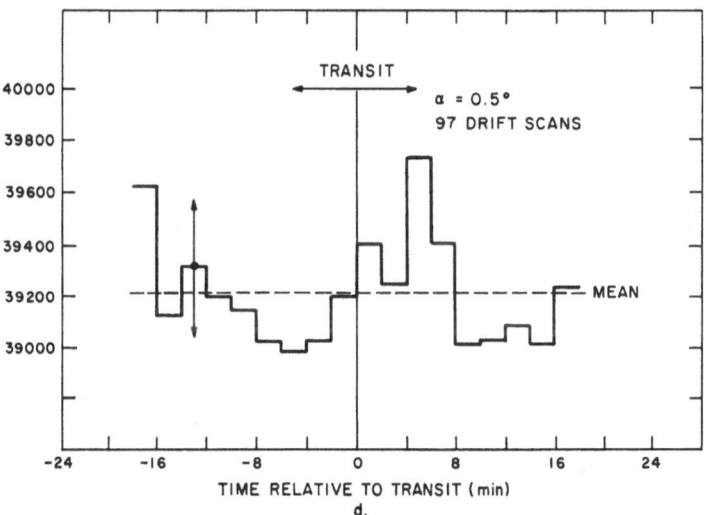

Fig. 2.   Histogram showing the cumulative counts from drift scans on the Crab Nebula.

A histogram summarizing the results of these drift scans is shown in Figure 2. In order to have a consistent rule for the rejection of scans in which nonrandom shifts in the counting rate occurred, any drift scan is rejected in which there is a statistical probability of less than 1% that the counting rate observed before the source entered the field of view is the same as the rate after it left the field. Five drift scans were rejected due to this criterion. A crude analysis of the accepted scans indicates that the observed fluctuations are 15% greater than those expected from statistical arguments. Therefore, the error quoted for the data is 1.15 times the standard deviation, and the flux limit is based on this error.

4. Results

As shown in Figure 2, there is no apparent anisotropy in the shower count rate greater than that expected from detection of a random angular distribution of showers. The upper limit (three standard deviations) to the cosmic γ-ray flux at energies greater than 1.7×10^{11} eV from the Crab Nebula is 2×10^{-10} photons/cm² sec. The observations were taken at an average zenith angle of 25°, and the excess counting rate from the source, subtracting background, was $+0.16 \pm 0.68$ min^{-1}. The source was in the field of view for a total of 776 min. In Figure 3, this upper limit is plotted along with the observations of other groups. The γ-ray threshold energy for this experiment is an order of magnitude lower than that of previous experiments.

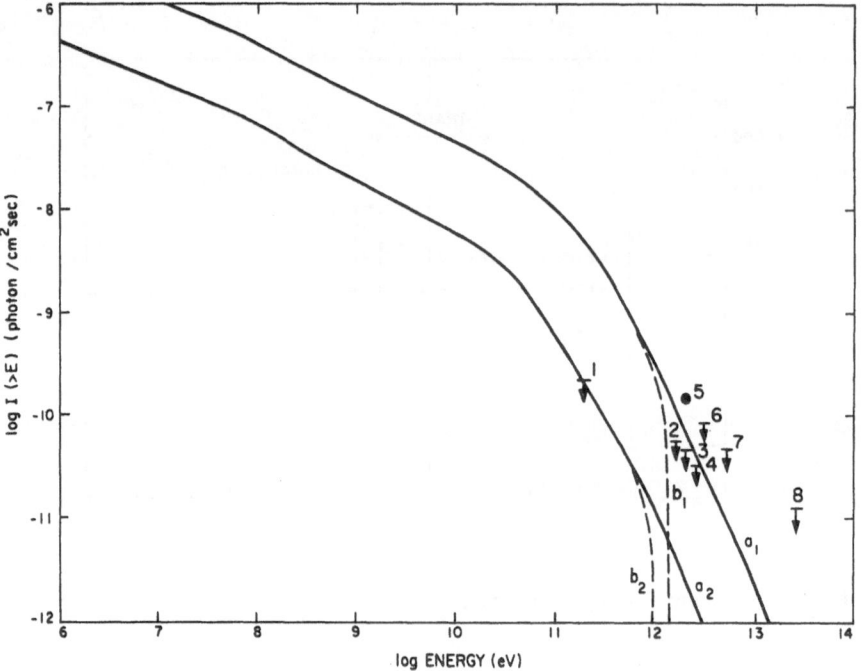

Fig. 3. Inverse Compton γ-ray flux from the Crab Nebula. Curves a_1 and b_1: $H = 10^{-4}$ gauss. Curves a_2 and b_2: $H = 3 \times 10^{-4}$ gauss. Curves a_1 and a_2 were computed assuming that the X-ray flux is synchrotron radiation; for b_1 and b_2, it was assumed that the synchrotron spectrum cuts off at 10^{16} Hz. Experimental point: 5, Fegan *et al.* (1968). Experimental upper limits: 1, this paper; 2–4, Rieke (1969), Fazio *et al.* (1968); 6, Tornabene and Cusimano (1968); 7, Chudakov *et al.* (1965); 8, Long *et al.* (1965).

5. Interpretation of the Results

Gould (1965) and Rieke and Weekes (1969) have calculated the γ-ray energy spectrum from the Crab Nebula, taking into account Compton scattering in the source between the synchrotron electrons and the observed synchrotron-emitted photons. The γ-ray flux predicted is a sensitive function of the magnetic field:

$$F(E_\gamma) \propto H^{-(1+m)/2},$$

where m is the spectral index of the relevant electron power-law spectrum. For the Crab Nebula, in the energy range of interest, the value of m is about 3.2; hence,

$$F(E_\gamma) \propto 1/H^{2 \cdot 1}.$$

Apparao (1967) has extended these calculations to include the collision of synchrotron electrons within the Crab Nebula with the photons of the 3 K universal blackbody radiation. These three calculations are the most accurate predictions of a γ-ray flux from the Crab Nebula that exist. Figure 3 shows the results of Rieke and Weekes (1969) for two values of the magnetic field. From this figure, it appears that the upper limit set by this experiment is sufficient to place a lower limit of 3×10^{-4} gauss on the strength of the magnetic field in the Crab Nebula. However, in view of the uncertainties in the experimental flux, the energy-threshold estimates, and the approximations in the theoretical calculations, a more reasonable lower limit is about 1.5×10^{-4} gauss. It should be emphasized that this limit does not depend on the extrapolation of the optical synchrotron spectrum to the X-ray region. A commonly accepted value for the field strength is 10^{-4} gauss.

One of the principal results from less sensitive experiments is evidence that the high-energy electrons in the Crab Nebula are not the product of high-energy proton interactions and therefore must have been accelerated from lower energies (Cocconi, 1960; Gould and Burbidge, 1967; Chudakov et al., 1965). The greater sensitivity of the 10-m reflector allows this result to be verified even more conclusively.

This experiment does not confirm possible evidence for a γ-ray flux from the Crab Nebula at 2×10^{12} eV reported by Fegan et al. (1968).

Several groups are attempting to detect γ-rays with an energy of 10^8 eV from the Crab Nebula. The most realistic predictions of this flux of photons are also given by the Compton-scattering calculations of Gould (1965), Apparao (1967), and Rieke and Weekes (1969). Assuming a magnetic-field intensity of 3×10^{-4} gauss in the Crab Nebula, the predicted flux is now about 10^{-7} photons/cm^2 sec, an order of magnitude lower than earlier predictions.

The pulsating radio source (NP0532) in the center of the Crab Nebula was also investigated for high-energy γ-ray emission, and the results are reported in another paper in this symposium.

Acknowledgements

We wish also to acknowledge the assistance of Roger Carson, Don Hogan, and Ed Horine during this experiment.

References

Apparao, M. V. K.: 1967, *Proc. Indian Acad. Sci.* **65A**, 349.
Chudakov, A. E., Dadykin, V. L., Zatsepin, V. I., and Nesterova, N. M.: 1965, in *Proc. (Trudy)*, *Lebedev Physics Inst.* (ed. by D. V. Skobel'tsyn), Consultants Bureau, New York, **26**, 99.
Cocconi, G.: 1960, *Proc. Intl. Conf. Cosmic Rays*, Moscow, **2**, 309.
Fazio, G. G., Helmken, H. F., Rieke, G. H., and Weekes, T. C.: 1968, *Astrophys. J. (Letters)* **154**, L83.
Fegan, D. J., McBreen, B., O'Mongain, E. P., Porter, N. A., and Slevin, P. J.: 1968, *Canadian J. Phys.* **46**, S433.
Gould, R. J.: 1965, *Phys. Rev. Letters* **15**, 577.
Gould, R. J. and Burbidge, G. R.: 1967, *Handbuch der Physik* **46/II**, 265.
Long, C. D., McBreen, B., Porter, N. A., and Weekes, T. C.: 1965, *Proc. Intl. Conf. Cosmic Rays*, London **1**, 318.
Rieke, G. H.: 1969, Smithson. Astrophys. Obs. Spec. Rep. No. 301.
Rieke, G. H. and Weekes, T. C.: 1969, *Astrophys. J.* **155**, 429.
Tornabene, H. S. and Cusimano, F. J.: 1968, *Canadian J. Phys.* **46**, S81.

SOME POSSIBLE IMPLICATIONS OF THE
RADIO EMISSION OF SCO X-1

P. A. FELDMAN and J. I. SILK

Institute of Theoretical Astronomy, Cambridge, England

Recent observations of radio emission from Sco X-1 at centimeter wavelengths by Andrew and Purton (1968) and by Ables (1969) have added an important new dimension to the study of galactic X-ray sources. We have considered two alternative interpretations of the radio emission from this object. First, we discuss whether it is possible to attribute the radio emission of Sco X-1 to thermal bremsstrahlung from an optically thin plasma at a temperature of 15 to 20 keV. This interpretation would have the merit of relating the radio properties to the existence of a hot, flaring plasma component whose presence has been invoked to explain the observations of hard X-rays ($\gtrsim 35$ keV) from this object (Peterson and Jacobson, 1966, Buselli *et al.*, 1968, Riegler *et al.*, 1968, Riegler, 1969, Agrawal *et al.*, 1969a, McCracken, 1969). That the 6-cm radio emission and the hard X-radiation should be found to vary by similar amounts over similar time scales lends this hypothesis a certain appeal. However, the measurement at 11 cm of flux density amounting to ≈ 0.07 flux units (1 flux unit $\equiv$ $\equiv 10^{-23}$ erg cm^{-2} sec^{-1} Hz^{-1}) reported at this Symposium by McCracken makes it virtually untenable to regard the radio emission of Sco X-1 as arising from thermal origins. Excessive optical and infrared continuum radiation would be produced; moreover, the shape of the radio spectrum is inconsistent with any conceivable thermal model. An explanation for the radio emission of Sco X-1 must therefore be sought in terms of some non-thermal radiation mechanism.

It is realistic to expect that a successful non-thermal model for the radio properties would be only one aspect of a comprehensive treatment of other physical phenomena observed in Sco X-1, such as the hard X-ray flux that is not satisfactorily explained by thermal models. It is also necessary to show how the presence of non-thermal components would relate to and interact with the 5 keV isothermal plasma that is widely believed to produce the observed photon spectra of Sco X-1 from the low-energy X-ray region to the near infrared by thermal bremsstrahlung.

We have considered a non-thermal model in which the radio flux is attributed to synchrotron-emitting electrons in the relativistic tail of a power-law energy distribution that produces hard X-ray quanta by non-thermal bremsstrahlung. However, the evidence presented earlier at this Symposium by McCracken (1969) and by Agrawal *et al.* (1969b) for sudden irregular changes in both the intensity and the spectral shape of the hard X-ray component makes it seem presumptive to describe detailed quantitative models at this time. Instead, we should like to restrict our discussion to some of the general properties that might arise from a non-thermal model for the radio emission of Sco X-1.

L. Gratton (ed.), Non-Solar X- and Gamma-Ray Astronomy, 257–259. All Rights Reserved
Copyright © 1970 by the IAU

The simplest non-thermal model for Sco X-1, in which the radio emission is produced by the electron synchrotron mechanism and the time scale for the radio variability is identified with the characteristic energy-loss time, requires the existence of rather strong magnetic fields *. These might be as intense as ~ 100 G, corresponding to a time scale of $\sim 10^4$ sec for the variations observed at 6 cm by Ables (1969). There are other possible ways to get significant radio flux changes over a few hours without necessarily invoking the presence of strong magnetic fields, but these alternatives do not have such interesting implications and hence we shall not attempt to consider them at this time.

An intriguing and possibly significant consequence of such a non-thermal model for the radio emission of Sco X-1 is the amount of circular polarization that might be produced by the synchrotron-radiating electrons. For relativistic electrons with a power-law differential energy spectrum and an isotropic velocity distribution emitting synchrotron radiation in a uniform magnetic field, the degree of circular polarization appropriate to the frequency of observation v is given by (Legg and Westfold 1967, 1968)

$$\Pi_c = C(\Gamma) \left[\frac{v_g \sin \theta}{v} \right]^{1/2} \cot \theta,$$

which is valid for $\theta \gg \gamma^{-1} \equiv (1 - \beta^2)^{1/2} \cdot C(\Gamma)$ is a slowly varying function of the electron energy spectral index Γ, v_g is the non-relativistic electron gyrofrequency, and θ is the angle between the magnetic field and the direction of observation. For $C \approx 1$, $v = 2727$ MHz (11 cm), and $B \sim 10^2$ G, the degree of circular polarization in the radio emission of Sco X-1 could be as high as 20 or 30%. Assuming that we are not situated at a pathological angle with respect to the magnetic field configuration of this source, it seems reasonable to expect $\sim 10\%$ circular polarization at decimeter wavelengths.

It is worth emphasising that both the amount and the sense of the circular polarization that an observer sees at any particular frequency are mainly determined by the configuration and strength of the magnetic field. The pitch-angle distribution of electron trajectories is relatively ineffective in this regard. Of course, Faraday rotation can have no depolarizing influence on the net degree of circular polarization that the radiation possesses. Only field reversals in the observer's line of sight can reduce the amount of circular polarization estimated for a uniform field with specified orientation.

The possible binary nature of Sco X-1 has been much discussed in recent years following the identification of the X-ray source with a variable optical object that seems to possess some of the characteristics of an ex-nova (Sandage et al., 1966). Attempts have been made by several authors – notably Shklovsky (1967), Cameron and Mock (1967), and Prendergast and Burbidge (1968) – to account for the observed X-ray emission in terms of physical processes appropriate to highly evolved, close binary systems. Although there is as yet no convincing evidence for *simple* binary motion in Sco X-1, the spectroscopic data are suggestive of complex 'binary' motions

---

* Riegler and Ramaty (1969) seem to have reached a similar conclusion from rather different premises.

possibly combined with gas streams (Westphal *et al.*, 1968). Several speakers at this Symposium have expressed the view that binary models are basically unsupported by observational evidence, but this is far from being a uniformly agreed conclusion. Wilson (1969) has even proposed a new binary model for Cyg X-2, an X-ray object similar to Sco X-1 with an optical counterpart that seems to be a short-period spectroscopic binary (Burbidge *et al.*, 1967; Kristian *et al.*, 1967; but see Kraft and Demoulin, 1967).

The existence of a measurable degree of circular polarization in the radio emission of Sco X-1 might provide a possibility of confirming the essential correctness of such models. A close binary system composed of two stars of approximately one solar mass each, separated by a distance of $\sim 2 \times 10^{12}$ cm* would complete an orbit about its centre of mass in about two weeks. If the magnetic field configuration is thus made to change its effective orientation towards the earth, one might expect the degree, and possibly also the sense, of the circularly polarized radio emission to vary regularly with this period. Observation of such an effect would be an important confirmation of the binary-star nature of the Scorpius type of galactic X-ray source.

References

Ables, J. G.: 1969, *Astrophys. J. Letters* **155**, L27.

Agrawal, P. C., Biswas, S., Gokhale, G. S., Iyengar, V. S., Kunte, P. K., Manchanda, R. K., and Sreekantan, B. V.: 1969a, this volume, pp. 94–106.

Agrawal, P. C., Biswas, S., Gokhale, G. S., Iyengar, V. S., Kunte, P. K., Manchanda, R. K., and Sreekantan, B. V.: 1969b, this volume pp. 94–106.

Andrew, B. H. and Purton, C. R.: 1968, *Nature* **218**, 855.

Burbidge, E. M., Lynds, C. R., and Stockton, A. N.: 1967, *Astrophys. J. Letters* **150**, L95.

Buselli, G., Clancy, M. C., Davison, P. J. N., Edwards, P. J., McCracken, K. G., and Thomas, R. M.: 1968, *Nature* **219**, 1124.

Cameron, A. G. W. and Mock, M.: 1967, *Nature* **215**, 464.

Field, G. B.: 1969, private communication.

Kraft, R. P. and Demoulin, M.-H.: 1967, *Astrophys. J. Letters* **150**, L183.

Kristian, J., Sandage, A. R., and Westphal, J.: 1967, *Astrophys. J. Letters* **150**, L99.

Legg, M. P. C. and Westfold, K. C.: 1967, *Proc. Astron. Soc. Australia* **1**, 27.

Legg, M. P. C. and Westfold, K. C.: 1968, *Astrophys. J.* **154**, 499.

McCracken, K. G.: 1969, this volume, p. 81.

Peterson, L. E. and Jacobson, A. S.: 1966, *Astrophys. J.* **145**, 962.

Prendergast, K. H. and Burbidge, G. R.: 1968, *Astrophys. J. Letters* **151**, L83.

Riegler, G. R.: 1969, Ph.D. thesis, Univ. of Maryland, published as NASA/Goddard Space Flight Center Report X-611-69-1.

Riegler, G. R. and Ramaty, R.: 1969, NASA/Goddard Space Flight Center Report X-611-69-123; 1969, *Astrophys. Lett.* **4**, 27.

Riegler, G. R., Boldt, E. A., and Serlemitsos, P.: 1968, *Bull. Am. Phys. Soc.* **13**, 1434.

Sandage, A. R., Osmer, P., Giacconi, R., Gorenstein, P., Gursky, H., Waters, J., Bradt, H., Garmire, G., Sreekantan, B. V., Oda, M., Osawa, K., and Jugaku, J.: 1966, *Astrophys. J.* **146**, 316.

Shklovsky, I. S.: 1967, *Astrophys. J. Letters* **148**, L1.

Westphal, J. A., Sandage, A., and Kristian, J.: 1968, *Astrophys. J.* **154**, 139.

Wilson, R. E.: 1969, this volume, p. 242.

* This linear dimension is about the minimum that would be consistent with the radio emission becoming optically thick to synchrotron self-absorption at wavelengths longer than 11 cm, assuming $B \sim 100$ G, $D \lesssim 100$ pc (Field, 1969), and $S_{11\,cm} \approx 0.07$ flux units.

OBSERVATIONAL RESULTS ON DIFFUSE COSMIC X-RAYS

(Invited Discourse)

MINORU ODA

Institute of Space and Aeronautical Science, University of Tokyo, Tokyo, Japan

Abstract. The present status of observations of the diffuse cosmic X-rays is discussed. The energy spectrum in the energy range 1–100 keV has been well established. The flux around 0.25 keV appears to be rather high. The basis of the classical argument that the integration of normal galaxies in the universe is not sufficient to explain the diffuse X-ray flux is re-examined. Recent observations around 0.25 keV are discussed and results are compiled.

1. Introduction

The purpose of this article is to discuss the present status of the observational investigations of the diffuse cosmic X-rays. The diffuse component of cosmic X-rays has been detected since the discovery of the cosmic X-rays. Experimental results have not been as convincing as those for point sources due to experimental difficulties in distinguishing the true cosmic component from the spurious component of terrestrial or local origin. Recently however more careful non-X-ray background discriminations have become common by means of the earth-shadowing effect, the shutter and the techniques of the rise time discrimination of counter pulses [1].

First, the energy spectrum will be discussed. The spectrum in 1–100 keV range has been established. The fact that the flux at 0.25 keV is consistent with the extension of a power law spectrum from 1–100 keV range or even higher is very important. Secondly, the argument to explain the diffuse component in terms of the integration of extragalactic sources will be discussed. Several observations in 0.25 keV band will be discussed. From the observations, in spite of some disagreement among them, interstellar absorption of very soft X-rays is concluded. Some details of the interstellar absorption are also discussed.

2. Energy Spectrum

The spectrum in the energy range of 1–100 keV has been well established. Figure 1 shows the results of various observations [2] which may be well represented by a power law $E^{-1.7\pm0.2}$ in the energy range 1–100 keV. It appears that the spectrum tends to steepen for higher energy and the exponent, 2.0, shows a better fit for 100–1000 keV. Indication of a *break* or *knee* of the spectrum, though not yet conclusive, is interesting from the point of theoretical interpretation of the diffuse cosmic X-rays. The energy range 100–1000 keV and beyond 1000 keV will have to be explored more thoroughly.

Several measurements have been carried out recently at energies close to 0.25 keV [3, 4, 5, 6, 7, 8]. A compiled value of the flux outside of the galaxy is represented by a double circle in the figure. The flux at 0.25 keV is on or above the extrapolation of the

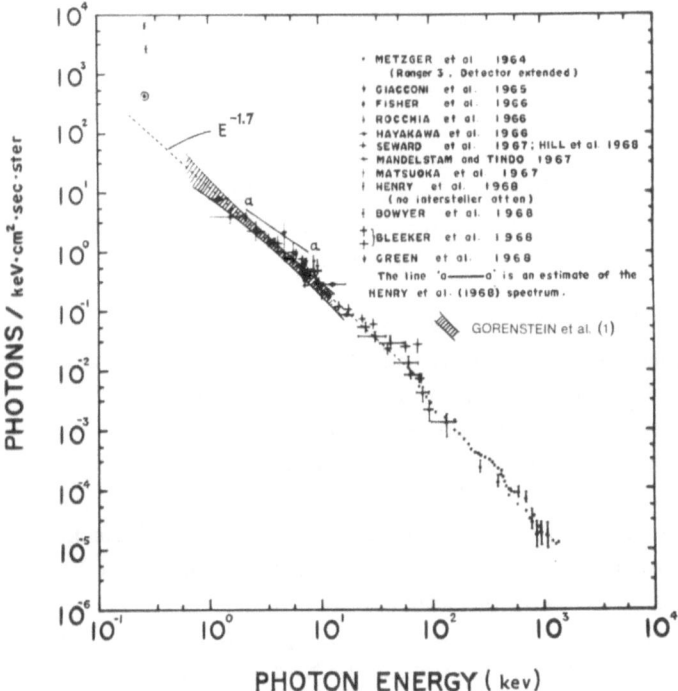

Fig. 1. Energy spectrum of the diffuse cosmic X-rays. The data are represented by a power law spectrum $E^{-1.7\pm0.2}$, in the energy range 1–100 keV. Measurements at 0.25 keV are compiled and the flux corrected for interstellar absorption is represented by a double circle.

spectrum at high energy range. The apparent absence of the absorption of the diffuse X-rays outside the Galaxy suggests that they originate in the intergalactic space and not in the galaxies in the universe. [9].

3. Extragalactic Objects

There has been a classical argument [10] that the contribution by galaxies in the universe is insufficient by a factor of 10–100 to explain the observed flux of the diffuse X-ray background. The argument was simply as follows: If we take the intensity of Tau-X-1 (Crab Nebula) as a standard we have approximately 20 sources of the intensity $\frac{1}{3}$ Tau-X-1. The distance to sources are 1.5 kpc for Tau-X-1 and 10 kpc or 3 kpc for other sources depending upon whether we consider other sources to be as far as the galactic center or to be in the nearby galactic arm.

If we take the absolute X-ray luminosity of Tau-X-1 to be 1, the luminosity of other sources for each assigned distance respectively is 10 and one approximately. The total number of sources, may be 100 or 1000 for these respective cases. Thus, the luminosity (X-ray) of our own galaxy may be equivalent to 1000 Tau-X-1. Assuming that all galaxies in the universe are similar to our galaxy, we may estimate the contribution of galaxies up to the Hubble distance. The fact that this estimate is less than the observed flux by a factor of 10–100 has led to a variety of theoretical considerations on the diffuse cosmic X-rays.

It is still worthwhile to re-examine the above classical arguments and ask the following questions.

(1) Is not the X-ray luminosity of our galaxy brighter than estimated on the basis of observed galactic point sources by a factor of ten or more?

The question (1), in other words, is whether there is a possibility that numerous unresolved and yet unobserved galactic sources increase the X-ray luminosity of our galaxy by some factor.

The observation of M31 (Andromeda Nebula) has been suggested in this respect. The Andromeda Nebula is known to be similar to our galaxy and its observation will provide an almost straightforward measurement of the absolute X-ray brightness of the galaxy. Its expected intensity is approximately $\frac{1}{300}$ of Tau-X-1 or more and is now within the possibility of observation but it has not been measured.

(2) Do some extragalactic objects much brighter than our galaxy contribute to the diffuse component more than normal galaxies? For example, cannot one percent of galaxies be brighter than a normal (our) galaxy by a factor of 1000 and altogether could the universe be ten times brighter?

Regarding question (2), if the diffuse component is mainly contributed by very bright objects, the intensity distribution of X-rays in the celestial sphere may not be as smooth as is expected due to the finite number of objects included in the field of view of the detector. We may derive a relation of the size of the field of view and the expected roughness (or granulation) of the diffuse component for an assumed brightness of each bright object.*

For example, if the density of the objects is one thousandth and their brightness is ten thousand times that of normal galaxies, with the field of view $4° \times 4°$ we may detect a fluctuation of the diffuse component of the order of 10%. Up to the present no evidence for the granulation has been reported.

Notice that in the Virgo cluster at a distance of 10 000 kpc three thousand galaxies crowd together and they may represent all kinds of galaxies (except for cosmological effect). If they are all like our galaxy, the total brightness of the cluster may be of the order of $\frac{1}{10}$ of Tau-X-1. If the *average* brightness is ten times the estimate of our galaxy, we may expect its brightness to be similar to Tau-X-1. Since the cluster spreads in a broad area in the sky it is not obvious that it should have been detected. One may design an experiment to detect the cluster and study if extraordinarily bright objects exist in it.

4. Interstellar Absorption of Soft X-Rays

The apparent isotropy of the diffuse component suggests the extragalactic origin. Most convincing evidence for the extragalactic origin may be acquired by observing the galactic or interstellar absorption of the diffuse component.

---

* The details of the argument will be published separately.

Figure 2 shows the expected interstellar transmission for various energies based upon Bell and Kingston's calculation [11]. The abscissae cover the range of the amount of hydrogen atoms along the line of sight corresponding to various galactic latitudes. It is seen that in the energy range >1 keV the galaxy is essentially transparent, for 0.5 keV a galactic latitude effect at low latitude is expected and for <0.25 keV, due to a strong interstellar absorption, a galactic latitude effect is expected only at high galactic latitudes.

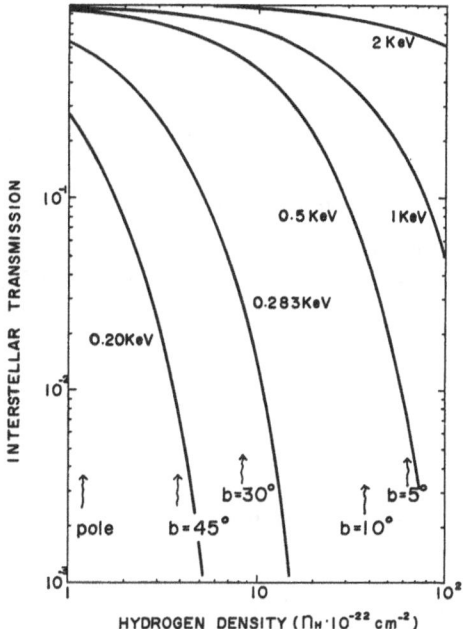

Fig. 2.   Interstellar transmission of soft X-rays computed based upon Bell and Kingston's calculation. n_H is the amount of hydrogen atoms along the line of sight. Corresponding galactic latitudes (b^{II}) are indicated.

5. Observation of 0.25 keV X-Rays

Techniques of detecting <0.5 keV X-rays quantitatively and with a reliability are not easy. Several groups have managed to fabricate proportional counters with extremely thin plastic windows so that the transmission band near the carbon K-absorption edge (0.28 keV) may be used. Spectral responses for few typical proportional counters of this type are shown in Figure 3.

Table I tabulates the features of the rocket experiments performed so far. Figure 4 summarizes rough sketches of observed regions of the sky in these experiments reproduced from the papers of these authors.

In what follows we shall summarize the experimental results and examine whether all the results may be combined. In the Berkeley experiment [3] if some observed patchiness in the sky is ascribed to local sources and can be ignored, there remains a

COUNTER EFFICIENCY

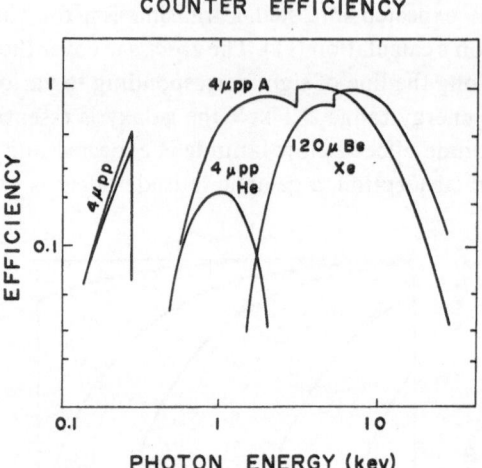

Fig. 3. Spectral response of proportional counter with thin plastic window, polypropylene (pp), and Be window. Transmission band at carbon K-absorption edge (0.28 keV) is clear.

TABLE I

Observations of soft X-rays (> 44 Å)

	Counter window	Gas	η (0.28 keV)	Calibration	A, Ω	Flight
Berkeley	$\frac{1}{8}$ mil mylar 200 Å nichrome	A	0.16	parallel counter Fe^{55} ground	500 cm² 0.021 str	Dec. 13, 66 0630 UT Brazil
N.R.L.	0.15 mil mylar 0.5 mil mylar	A A	0.12		200 cm² 200 cm² 0.024 str	Sep. 7, 67 2021 MST
L.R.L.	60 μg/cm² formvar	Ne-meth A-meth	0.65	nose cone isotope		May 14, 68 2038 Hawaii
Wisconsin	290 μg/cm² kimfol 585 μg/cm² mylar		0.42 0.14		250 cm² 250 cm² 0.019 str	Sep. 21, 68 0345 GMT
Calgary	0.25 mil mylar + 1350 Å Al		0.02	ground	30 cm² 0.117 str	Oct. 8, 68 0505 UT Canada
Nagoya- Tokyo	4 μ polypropylene 5 mil Be	He A Xe	0.45 – 	CK Fe^{55} shutter	200 cm² 190 cm² 0.015 str	Jan. 14, 69 1900 JST K.S.C.

general tendency showing the galactic latitude effect or interstellar absorption in the range of $n_H \approx 2$–15×10^{20}/cm². The attenuation appeared less than the estimate based upon Bell and Kingston's calculation.

In the NRL-experiment [4] the flight of the rocket was controlled to cover several

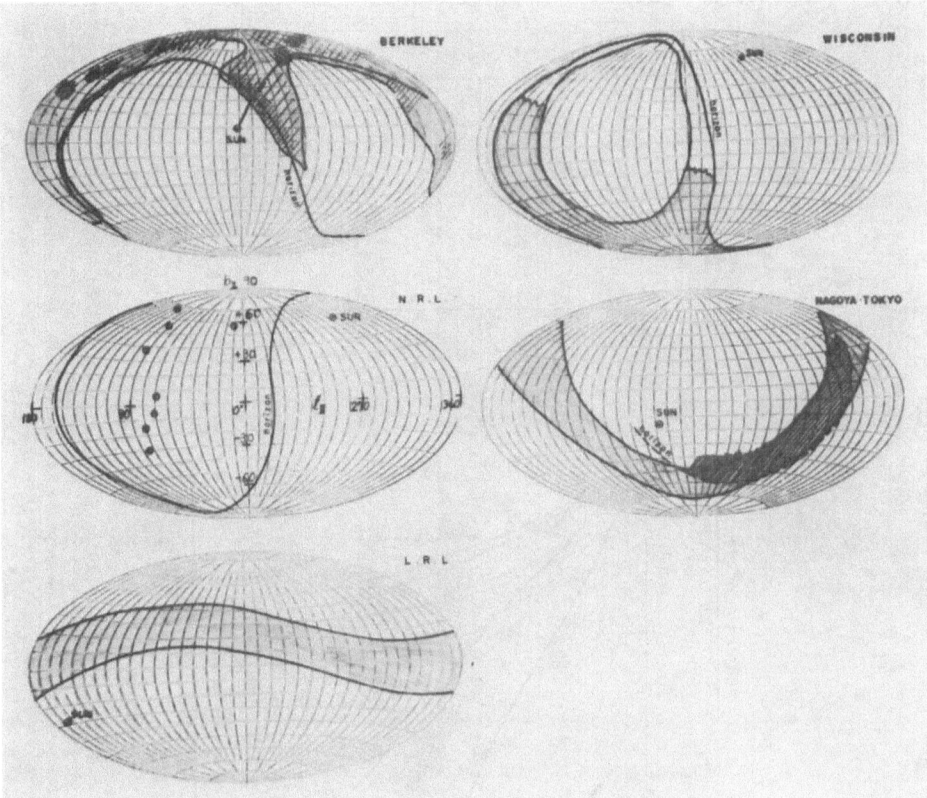

Fig. 4.   Region of the sky (galactic coordinate) scanned with 0.25 keV counters by various investigators (shadowed area).

spots in the sky including the low galactic latitude region and a region where the n_H is smallest (near the pole). Some points at low galactic latitude which showed strong intensities were attributed to cool (compared to normal X-ray sources) and nearby X-ray sources. This may imply the existence of numerous sources of this kind.

The region of the sky scanned by LRL group [5] covered the range of $n_H > 4 \times 10^{20}/$ cm^2. Except for some irregularity the intensity distribution was flat and no apparent galactic latitude effect was observed.

In the Nagoya-Tokyo experiment [6] the field of view of one of the detectors was designed in such a way that its long axis is precisely parallel to the galactic equator and hence the correspondence between the rocket azimuth and the galactic latitude is simple. In the observed range of $n_H > 4 \times 10^{20}/$cm^2 no obvious attenuation was detected. Yet an apparent difference was observed in counting rates when the shutter in front of the detector was open and closed.

Kraushaar and the Wisconsin group [7] covered a region extending from low galactic latitudes to high galactic latitude and observed a clear galactic latitude

266   MINORU ODA

TABLE II

Observer	Flux	n_H 10^{-20}	Counter window
Berkeley	1.9 ± 0.2	~ 2	$\frac{1}{8}$ mil mylar
NRL	2.6 ± 0.2	~ 2	$\frac{1}{8}$ mil mular
Wisconsin	5.5 ± 0.5	~ 2	290 μg/cm² kimfor
Nagoya-Tokyo	2.0 ± 0.2	~ 5	4 polypropylene

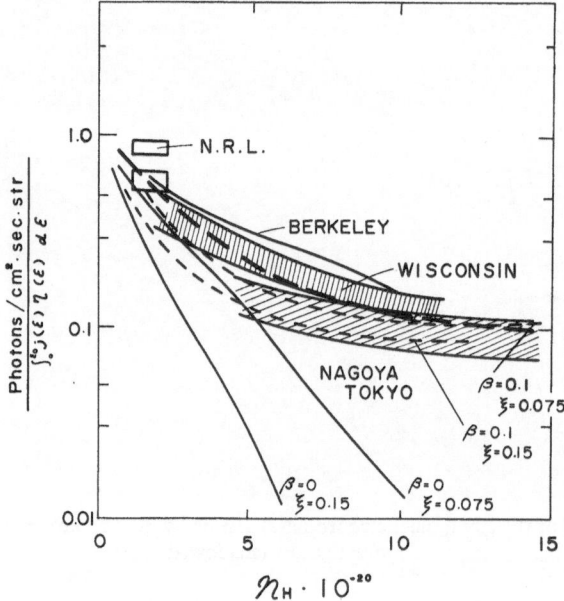

Fig. 5. Observed fluxes in 0.25 keV band of X-rays. The results are consistent with each other within a factor of 1.5–2 indicating the interstellar attenuation.

effect. The effect was also observed at a higher energy range 0.48–0.96 keV.

We summarize the *observed* fluxes in 0.25 keV band in Table II.

Results are summarized in Figure 5. If we ignore discrepancies by a factor of 1.5 or 2 considering experimental difficulties and some ambiguities in the estimate of the background counts to be subtracted to obtain the flux for low galactic latitudes, the combined results may be represented by the dashed curve.

If we accept this compilation, we find that the resulting attenuation curve is not compatible with the absorption by homogeneous interstellar gas of any composition which is represented by curves indicated as $\beta = 0$ in Figure 5. We have to explain the apparent flattening of the attenuation curve for higher values of n_H.

Curves in Figure 5 represent examples of calculated attenuation curves on an assumption that in addition to the component of extragalactic origin an isotropic flux contributes to the diffuse component. The flux may be represented as

$$\int_0^{\varepsilon_0} \frac{e^{-\tau} + \beta}{1 + \beta} j(\varepsilon)\, \eta(\varepsilon)\, d\varepsilon,$$

where $\tau = n_H \cdot (\varepsilon/\varepsilon_0)^{-3.2} (0.84 + 21.4\xi) \cdot 10^{-2}$; ξ: the number ratio of He/H in the inter-stellar gas; β: the fraction of the isotropic component; $j(\varepsilon)$: the counter efficiency as a function of photon energy ε; $\eta(\varepsilon)$: the flux at photon energy ε; $\varepsilon_0 = 0.28$ keV; $\eta\,(\varepsilon = \varepsilon_0)$ is taken as 400 photons/cm^2 sec str keV.

It is seen that the observational results are consistent with the existence of an *un-explained* isotropic component amounting to 10%, which may be of nearby galactic origin, or solar or even terrestrial origin.

Kraushaar discussed their experimental result according to the hypothetical clum-piness of H I cloud in the interstellar space which has also been discussed by Bowyer and Field [12].

The extension of the compiled curve to outside of our galaxy results in a flux at 0.28 keV of about 300–500 photons/cm^2 sec str keV which is barely within a reason-able limit of the extension of the spectrum for > 1 keV in Figure 1.*

6. Summary

(1) The energy spectrum expressed by a power law of the exponent 1.7 ± 0.2 in 1–100 keV range may be extended to 0.25 keV. There is a possibility that the flux at 0.25 keV is even higher than this extrapolation.

(2) A break of the spectrum near 100 keV is suspected though its existence is not conclusive yet.

(3) Several experiments have been suggested in relation to the model of superposed extragalactic sources for the explanation of the diffuse X-rays.

(4) Measurements of the very soft X-ray (0.25 keV) flux by several groups are compiled. Evidence for the interstellar attenuation indicating the extragalactic origin was obtained. The results, however, are in some disagreement with a theoretical prediction and this disagreement has been discussed.

Acknowledgements

I wish to acknowledge discussions of Prof. S. Hayakawa and Dr. M. Matsuoka and the permission of Prof. W. Kraushaar to include his results and discussions in this paper before his talk.

* An experiment by Wilson *et al.* of Calgary [8] is in disagreement with other experiments. They obtained a high flux at 0.25 keV, about 3000 photons/cm^2 sec str keV. The difference between this experiment and other experiments is that the counter window of Calgary experiment is relatively thick and the transmission at 0.25 keV is only a few percent whereas for other experiments the transmission was in the range of 15–40%. It may be dangerous to correct the data for such small transmission, but, on the other hand, the technical difficulties increase for a thinner window. This disagreement has not been thus far understood.

268 MINORU ODA

References

[1] Gorenstein, P., Kellogg, E. M., and Gursky, H.: 1969, *Astrophys. J.* **156**, 315.
[2] For references see [1] and Wilson, B. G.: 1968, ISAS Report, Institute of Space and Aeronautical Science, University of Tokyo, No. 428.
[3] Bowyer, C. S., Field, G. B., and Mack, J. E.: 1968, *Nature* **217**, 32.
[4] Henry, R. C., Fritz, G., Meekins, J. F., Friedman, H., and Byram, E. T.: 1968, *Astrophys. J.* **153**, L199.
[5] Grader, R. J., Hill, R. W., and Seward, F. D.: 1969, *Sky and Telescope*, Feb. issue, p. 79.
[6] Hayakawa, S., Kato, T., Makino, F., Ogawa, H., Tanaka, Y., Yamashita, K., Matsuoka, M., Miyamoto, S., Oda, M., and Ogawara, Y.: 1970, this volume, p. 121.
[7] Kraushaar, W.: this volume.
[8] Baxter, A. J., Wilson, B. G., and Green, D. W.: 1969, *Astrophys. J.* **155**, L143.
[9] Hayakawa, S.: preprint.
[10] Oda, M.: 1965, *Proc. Intern. Conf. in Cosmic Rays* Vol. 1, p. 68.
Hayakawa, S., Matsuoka, M., and Sugimoto, D.: 1966, *Space Sci. Rev.* **5**, 109.
Gould, R. J., and Burbidge, G. R.: *Handbuch der Physik*, Vol. 46, p. 281.
[11] Bell, K. L. and Kingston, A. E.: 1967, *Monthly Notices Roy. Astron. Soc.* **136**, 241.
[12] Bowyer, C. S. and Field, G. B.: preprint.

REVIEW OF OBSERVATIONAL RESULTS ON
γ-RAY BACKGROUND*

(Invited Discourse)

G. W. CLARK

Dept. of Physics and Center for Space Research,
Massachusetts Institute of Technology, Cambridge, Mass., U.S.A.

G. P. GARMIRE

Dept. of Physics, California Institute of Technology, Pasadena, Cal., U.S.A.

and

W. L. KRAUSHAAR

Department of Physics, University of Wisconsin, Madison, Wis., U.S.A.

Abstract. Recent observations in the X- and γ-ray region of the electromagnetic spectrum have given strong evidence for the existence of an extragalactic intensity with a slowly steepening power law spectrum in the region 10^3 to 10^8 eV. Further data from the OSO-III high energy γ-ray detector are in agreement with earlier published reports, and suggest that the γ-rays from high galactic latitudes have a softer spectrum than those from the galactic plane.

The previous paper by Dr. Oda of the University of Tokyo has reviewed the status of measurements of the diffuse radiation in the region below 100 keV. We shall be concerned here with the region of the electromagnetic spectrum above that energy.

Measurements of the diffuse radiation are difficult in this energy region. γ-rays are produced in collimators, in nearby pieces of apparatus, and in the earth's atmosphere by the ever-present charged particle cosmic radiation. In the region of a few MeV, in fact, Peterson (1967, 1969) has shown that the albedo from the earth is just equal to the apparent diffuse radiation. At higher energies, as will be discussed presently, the albedo is enormously greater than the diffuse radiation. Because γ-ray production in matter is such an important phenomena, the use of shutters, inactive collimators and background evaluation by viewing the earth – all important and useful devices in the lower energy region – are quite impossible in the energy region under discussion.

Figure 1, taken in part from a similar figure prepared by Gorenstein *et al.* (1969), summarizes representative measurements of the diffuse γ-radiation. Up to 1 MeV, at least, all measurements above 20 keV fall with reasonable consistency on a straight line of slope -2, indicating a photon number spectrum of the form dE/E^2. In the region 1–10 MeV, there are only measurements of Vette *et al.* (1970) indicated by 'Peterson *et al.* (1969-ERS)' on Figure 1. As with the measurement of Metzger *et al.* (1964) the observations were carried out far from the earth where albedo effects are small. The apparent deviation from a power law, if real, has possible cosmological indications as discussed by Stecker (1970).

* Supported in part by the National Aeronautics and Space Administration under contract NAS5-3205, grant NsG-386 (MIT), grant NSG-426 (CIT), and NGR 50-002-044 (University of Wisconsin).

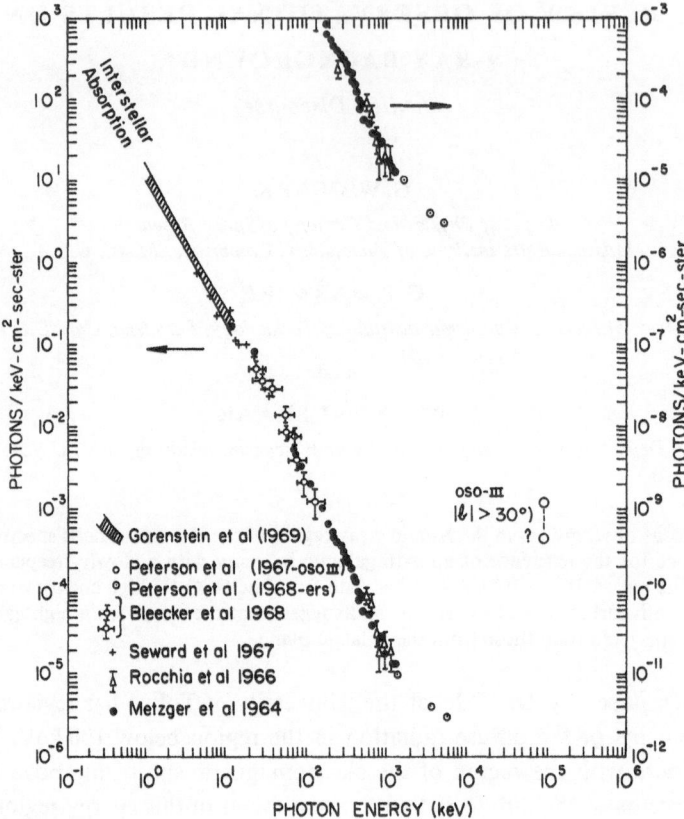

Fig. 1. Representative measurements of the apparently diffuse cosmic X- and γ-ray spectrum. Interstellar absorption is an important effect below 1 keV and the meaning of measurements in this range is unclear at present.

The highest energy measurement labeled 'OSO-III' at 100 MeV refers to the published results of Clark *et al.* (1968). Since that initial report, more observations have been reduced and while the earlier conclusions are unchanged, the statistical evidence is now appreciably improved.

Figure 2 shows the detected rate of γ-rays referred to a satellite-centered coordinate system with polar axis at the instantaneous zenith. The data have been separated into two parts; one in which the satellite was within 20° of the geomagnetic equator, the other in which the satellite was more than 20° from the geomagnetic equator. The horizon of the earth is brighter when the satellite is far from the equator because the earth's magnetic field permits a larger portion of the galactic cosmic ray flux to enter there. The counting rate for angles more than 40° above the horizon is statistically the same for both parts of the data. This is to be expected, of course, if these γ-rays are of celestial not terrestial origin.

The next several figures describe in various ways the anisotropic character of the detected high energy γ-radiation. Each point on the upper map of Figure 3 corre-

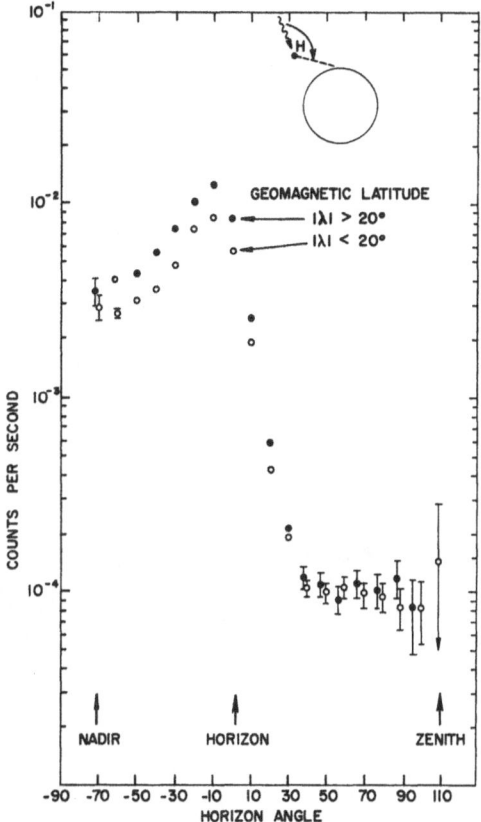

Fig. 2.   Distribution of high energy γ-rays relative to the earth.

sponds to the arrival direction in galactic coordinates of a γ-ray. In itself this map has little significance because the exposure of the instrument to various parts of the sky was not uniform. Correspondingly, each point on the lower map of Figure 3 corresponds to a certain time that the instrument spent viewing in the indicated direction. In other words, the density of points in a given region on the upper map divided by the density of points in the same region on the lower map is proportional to the directional gamma ray intensity. Once the data are available in the form described by Figure 3, variation of the intensity with galactic latitude, galactic longitude, etc. can be investigated conveniently.

Figure 4 shows the variation with galactic latitude, data from all galactic longitudes having been summed. We see a pronounced intensity peak at the galactic equator, and a definite non-zero intensity at all galactic latitudes. The shape of the pronounced rise near $b=0$ essentially reproduces the response of the instrument to a line source. The 'line' could be several degrees wide, of course. The data are sufficient to allow division into six regions of galactic longitude, as shown in Figure 5. The most pronounced peak at the galactic equator occurs near the galactic center, although significant peaks towards the equator but of lesser intensity are apparent elsewhere.

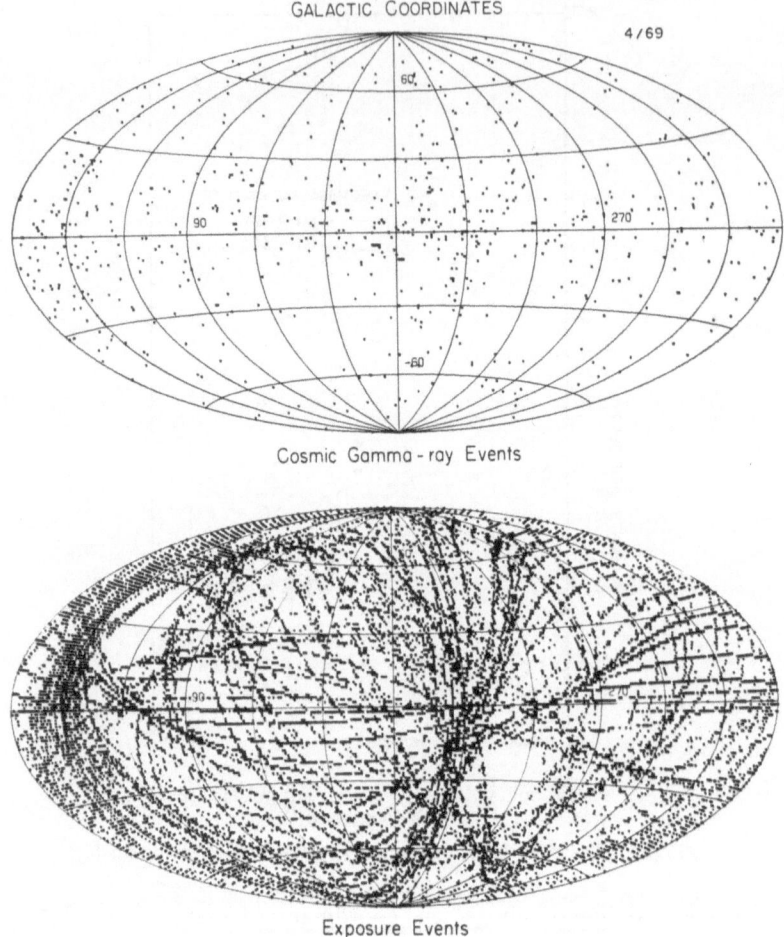

Fig. 3. Distribution of detected γ-rays in galactic coordinates (upper map). Each point on the lower map is proportional to a fixed amount of time that the instrument viewed in the indicated direction.

Figure 6 shows the galactic longitude distribution for all those γ-rays that arrived within 15° of the galactic equator. The strongest emission, as was evident from Figure 5, is from regions near the galactic center. The distribution in l, however, is much broader than the distribution to be expected from a point source at the galactic center.

One of the frequently discussed mechanisms for high energy γ-ray production is the collision of cosmic ray protons with nuclei of the interstellar gas. If the cosmic ray proton flux is the same everywhere in the galactic disc, the γ-ray intensity should be proportional to the columnar hydrogen density. In Figure 7 is shown the columnar hydrogen density averaged over the 5°, 10°, and 15° closest to the galactic equator plotted *versus* l. The dependence on l is surprisingly weak. This is because when one averages over several degrees in galactic latitude, much of the gas included is, in fact, relatively local. We conclude on these grounds alone that our data are not consistent

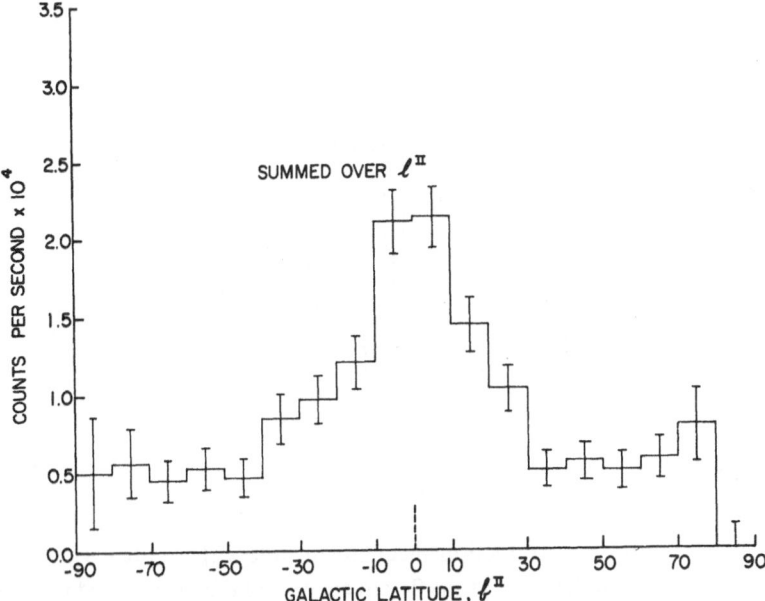

Fig. 4. Dependence of γ-ray intensity on galactic latitude. Here the data have been summed over all galactic latitudes.

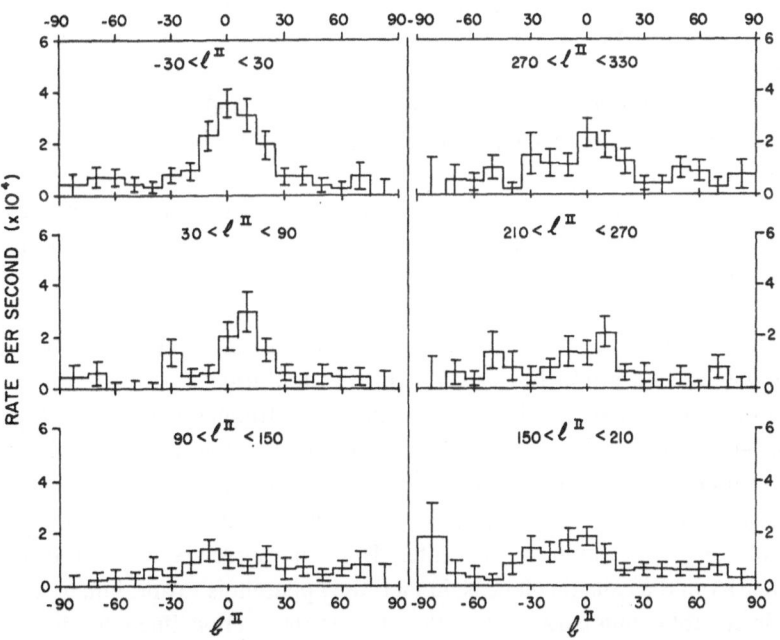

Fig. 5. Galactic latitude distribution for six regions of galactic longitude.

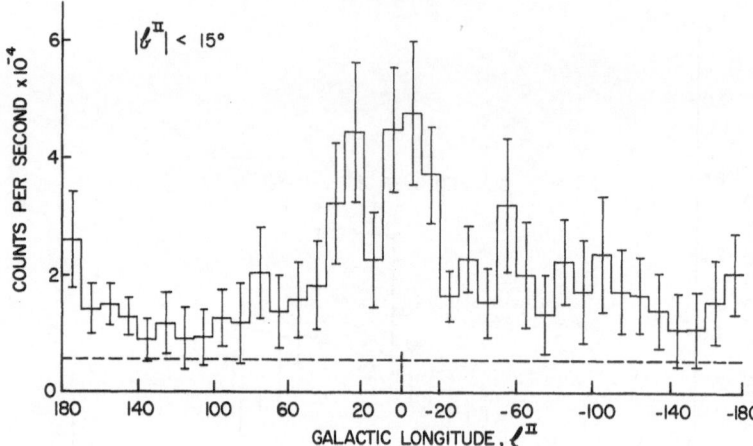

Fig. 6.   Dependence of γ-ray intensity near the galactic disc on galactic longitude. The dotted line
shows the average rate at high galactic latitudes.

with the nuclear collision production mechanism unless there are large amounts of
molecular or cool gas undetected in the 21 cm surveys and concentrated in the galactic
plane near the galactic center. In addition, as was pointed out in our initial paper
announcing the OSO-III results, the observed intensity is more than 10 times that
expected from the nuclear collision mechanism.

It is possible, of course, that cosmic rays are themselves concentrated towards the
galactic center. The non-thermal radio noise distribution in galactic longitude, as
indicated in Figure 7, may in fact be taken to indicate that this is likely. The radio
noise and high energy γ-ray intensities are distributed rather similarly in galactic
longitude.

The cumulative flux from discrete X-ray sources located within 15° of the galactic
plane has a distribution in galactic longitude similar to that of the high energy γ-rays.
This has also been pointed out by Ogelman (1969), who in addition has suggested
that when a power law spectrum of index 2 is assumed, the extrapolated X-ray
intensity falls near the measured γ-ray intensity. It is interesting to point out that when
extrapolating over 3 decades, an uncertainty of 20% in the index results in a dynamic
range of 16 to 1 within which 'agreement' may be claimed. Further, many X-ray
sources have energy spectra indicative of free-free not power law emission so that the
appropriateness of a power law extrapolation is doubtful.

Table I summarizes the predictions of some of the frequently discussed high energy
galactic γ-ray production mechanisms relative to our measured intensity near the
galactic center. The galactic center region is unique in many respects and it is likely
that at least a partial understanding of γ-ray emission can be more easily realized in
regions 60° or more away from the center. Here an appreciable fraction of the measured intensity in the galactic plane is, in fact, the apparent isotropic intensity discussed
further in the following paragraphs. We estimate the average line intensity in regions
more than 60° from the center to be about $\frac{1}{4}$ of the line intensity near the center.

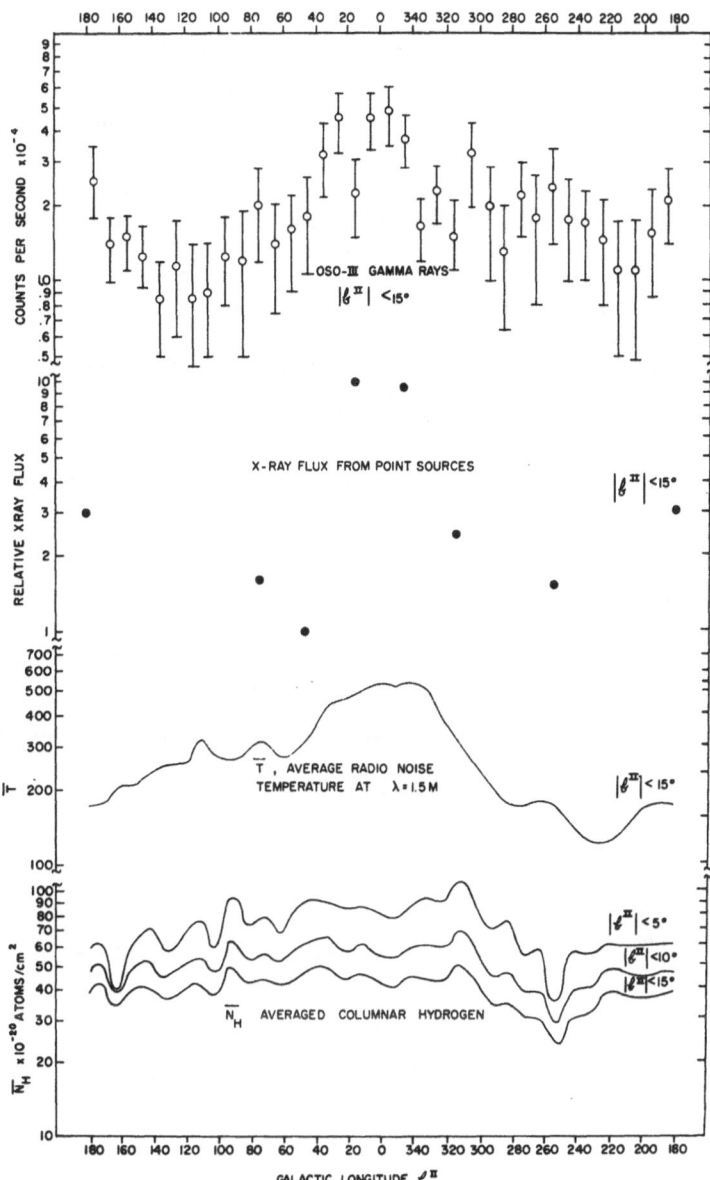

Fig. 7.   Galactic longitude dependence of γ-rays, cumulative X-ray flux, 1.5 meter non-thermal radio noise and columnar hydrogen density.

Because the longitude distribution of N_H (see Figure 7) is so nearly flat, the predicted line intensity from π° production or bremsstrahlung near the galactic center is about the same as elsewhere in the galactic plane. While this factor of 4 decreases the apparent discrepancy between predictions and observations appreciably, we do not wish to minimize the significance of the remaining difference. Indeed, we now have

TABLE I

γ-rays from galactic center region observed intensity $\approx 3 \times 10^{-4}$ cm^{-2}sec^{-1}rad^{-1}

Mechanism	Responsible momentum	Predicted Observed
π° production by nominal[a] CR protons on known gas	$P_p > 2$ GeV/c	0.07
Bremsstrahlung by nominal[a] CR electrons on known gas	$P_e > 0.1$ GeV/c	0.01
Inverse Compton by nominal CR electrons on known stellar photons	$P_e > 5$ GeV/c	0.02
Inverse Compton by nominal CR electrons on enhanced Becklin and Neugebauer (1968) Galactic Center stellar photons	$P_e > 5$ GeV/c	0.04
Inverse Compton by nominal CR electrons on Shivandan *et al.* (1968) infra-red 8 K photons. Cowsik and Pal (1969), Shen (1969)	$P_e > 50$ GeV/c	~ 1
Extrapolated (3 decades) discrete X-ray sources Ogelman (1969)		~ 1

[a] By nominal cosmic ray protons and electrons we mean the measured intensity near the earth at solar minimum.

a clearer discrepancy with expectation because the complex galactic center region is removed from consideration.

The existence of γ-rays of galactic origin can hardly be questioned in view of the highly directional properties of the measured intensity. No such convincing evidence exists to prove the reality of the measured high galactic latitude and presumably

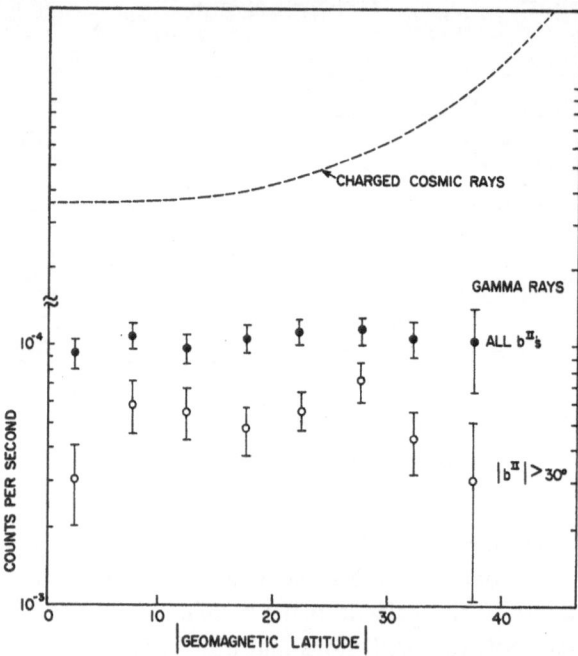

Fig. 8. Variation of γ-ray intensity with geomagnetic latitude of the satellite.

isotropic component. All conceivable forms of background are related to the charged cosmic ray flux incident on the orbiting instrument or on the atmosphere beneath it. Since the orbit of OSO-III traverses a range of geomagnetic latitudes between $+40°$ and $-40°$, and since the charged cosmic ray flux varies significantly over this range, any background should vary also with geomagnetic latitude. We have therefore examined our data for this type of dependence and the results are shown in Figure 8. Certainly neither the total γ-ray intensity nor the γ-ray intensity from high galactic latitudes have any obvious tendency to increase with geomagnetic latitude. In order to investigate the question quantitatively, we have computed, for the high galactic latitude component, the ratio of measured intensity for $|\lambda| > 20°$ to that for $|\lambda| < 20°$. We have

$$R = \frac{I(|\lambda| > 20°)}{I(|\lambda| < 20°)} = 1.14 \pm 0.18$$

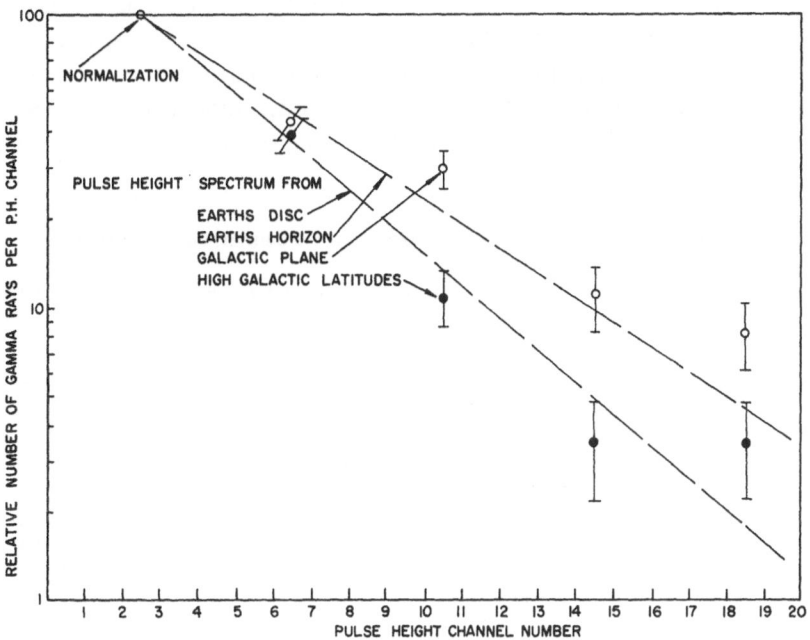

Fig. 9. Pulse height distribution of γ-rays from the earth's disc, from the earth's horizon, from the galactic plane and from high galactic latitudes.

The corresponding ratio for charged cosmic rays is 1.8, so the independence is established to about a 3.5σ level.

The instrument is equipped with a rather poor resolution γ-ray energy calorimeter. The results of the approximate energy measurements are still being studied but such preliminary results as are available are shown in Figure 9. The upper and lower dashed curves show pulse height distributions for γ-rays from the horizon of the earth and from the earth's disc, respectively. As is to be expected from simple kinematic arguments, γ-rays from the horizon, having followed the direction of the primary cosmic rays, have higher average energies. γ-rays from high galactic latitudes have a pulse height distribution similar to those from the earth's disc, while γ-rays from the galactic plane have a pulse height distribution similar to those from the horizon. We conclude that γ-rays from the galactic plane are on the average more energetic than those from high galactic latitudes. This qualitative statement is in agreement with the hypothesis that γ-rays from the galactic plane have a π°-decay (nuclear interaction) origin while those from high galactic latitudes have an electromagnetic origin. Our results cannot be taken to prove this, of course.

The values of the high energy γ-ray intensity are unchanged since our initial report. Fichtel *et al.* (1970) have recently flown their balloon-borne spark chamber instrument upside down so as to measure the upward moving γ-ray albedo intensity from the earth's disc. Their value for this intensity is about $\frac{1}{3}$ as large as ours. We feel it unlikely that our efficiency-solid angle calibration could be off by a factor as large as three, but the possibility has been recognized in preparing Figure 1. We and the GSFC group are currently planning a recalibration of both instruments in the same tagged γ-ray beam at the California Institute of Technology electron synchrotron.

In recent months a number of groups have provided supporting evidence, though at a marginal statistical level, for a narrow line of high energy γ-ray emission from the

TABLE II

Recent reports of high energy γ-ray detection via balloon-borne instruments

Cornell:	Delvaille, Albats, Greisen and Ogelman (1968) Spark Chamber; $E > 1$ GeV, $-1° < b^{II} < 1°$; $l^{II} \approx$ AC to Cygnus $I = (6 \pm 3) \times 10^{-4}$ (cm²-sec-sr)$^{-1}$
Minnesota:	Valdez and Waddington (1969) Emulsion-Spark Chamber, $E > 100$ MeV. $b^{II} \approx 0$, $l^{II} \approx 65°$ 2σ
GSFC:	Fichtel *et al.* (1970) Spark Chamber; $E > 50$ MeV, $-3° < b^{II} < 3°$, $l^{II} \sim -10°$ to $25°$ $J = (2.2 \pm 1.1) \times 10^{-4}$ (cm²-sec-rad)$^{-1}$
Case-Western Reserve:	Frye and Wang (preprint) Spark Chamber; $E > 100$ MeV, $-3° < b^{II} < 3°$, $l^{II} \approx 55°$ to $85°$ $J = (4 \pm 2) \times 10^{-5}$ (cm²-sec-rad)$^{-1}$
Imperial College:	Sood (preprint) Čerenkov Counters, $E > 50$ MeV, $b^{II} \approx 0$, $l^{II} \approx 30°$ $J = (1.5 \pm .5) \times 10^{-4}$ (cm²-sec-rad)$^{-1}$ (estimated)

galactic plane. These measurements are summarized in Table II. In addition, as reported in these Proceedings, Hutchinson *et al.* (1970) have detected a somewhat enhanced emission from the galactic plane with their spark chamber aboard OGO-5.

Acknowledgements

We acknowledge with gratitude the important contributions to the design and construction of our instrument by the Lincoln Laboratory of MIT and the able help with computational matters provided by Mrs. T. Thorsos.

References

Becklin, E. E. and Neugebauer, G.: 1968, *Astrophys. J.* **151**, 145.

Bleeker, J. A. M., Burger, J. J., Deerenberg, A. J. M., Scheepmaker, A., Swanenburg, B. N., Tanaka, Y., Hayakawa, S., Makino, F., and Ogawa, H.: 1968, *Canadian J. Phys.* **46**, S461.

Clark, G. W., Garmire, G. P., Kraushaar, W. L.: 1968, *Astrophys. J. Letters* **153**, L203.

Cowsik, R. and Pal, Y.: 1969, *Phys. Rev. Letters* **22**, 550.

De Jager, C. and De Loore, C.: 1970, this volume, p. 238.

Delvaille, J. P., Albats, P., Greisen, K. I., and Ogelman, H. B.: 1968, *Canadian J. Phys.* **46**, S425.

Fichtel, C. E., Kniffen, D. A., and Ogelman, H. B.: 1970, this volume p. 315.

Gorenstein, P., Kellogg, E. M., and Gursky, H.: 1969, *Astrophys. J.* **156**, 315.

Hoffman, W. F. and Frederick, C. L.: 1969, *Astrophys. J. Letters* **155**, L9.

Hutchinson, G. W., Ramsden, D., and Wills, R. D.: 1970, this volume p. 300.

Metzger, A. E., Anderson, E. C., Van Dilla, M. A., and Arnold, J. R.: 1964, *Nature* **204**, 766.

Ögelman, H.: 1969, *Nature* **221**, 753.

Peterson, L. E.: 1967, University of California at San Diego report SP68-1, July.

Peterson, L. E.: 1969, private communication.

Rocchia, R., Rothenflug, R., Boclet, D., Ducros, G., and Labeyrie, J., *Space Research* VII, North Holland Publishing Co., Amsterdam 1967, Vol. 1, pp. 1327–1333.

Shen, C. S.: 1969, *Phys. Rev. Letters* **22**, 568.

Shivanandan, K., Houck, J. R., and Harwit, M. O.: 1968, *Phys. Rev. Letters* **21**, 1460.

Stecker, F. W.: 1970, this volume p. 382.

Valdez, J. V. and Waddington, C. J.: 1969, *Astrophys. J. Letters* **156**, L85.

Vette, J., Matteson, J. L., Gruber, D., and Peterson, L. E.: 1970, this volume p. 335.

EVIDENCE FOR A GALACTIC COMPONENT OF THE DIFFUSE X-RAY BACKGROUND

B. A. COOKE, R. E. GRIFFITHS and K. A. POUNDS

X-ray Astronomy Group, Dept. of Physics, University of Leicester

1. Introduction

It is widely believed that the diffuse X-ray background, observed on several occasions over the energy range from 0.25 keV to above 1 MeV has an extragalactic origin. Evidence for this comes from the generally reported isotropy above several keV [1, 2, 3] and the observed galactic latitude dependence at 0.25 keV, believed to result from the interstellar attenuation of these low energy photons in passage through the Galaxy [4, 5].

The present paper reports some results from a recent high sensitivity sky survey in the 1.4–18 keV band, which show a clear excess of radiation in the galactic plane in the Vela-Carina-Centaurus region ($l^{II} = 220 - 320°$). No strong discrete X-ray sources are seen in this part of the Milky Way and the possibility of this flux representing a Galactic component of the diffuse background radiation is suggested. A study of the published results of previous rocket and balloon experiments has been made with the conclusion that the presently reported 'anistropy' could have remained unobserved in each case, for one or other of the following reasons. First, the great majority of observations have been concentrated on the strong source regions – in Scorpius, Sagittarius and Cygnus – where a narrow band of excess background radiation along the galactic equator would be obscured by the sources. Second, no other survey experiment of comparable sensitivity and resolution has searched the galactic longitudes covered in the present flight, where the absence of discrete sources is marked.

2. Flight details

The present observations were made from an unstabilised Skylark rocket, SL 723, launched from Woomera, South Australia, at 20^h00^m local time on June 12, 1968. Two proportional counter detectors, each of 1385 cm^2 effective area, were mounted viewing sideways from the nose-cone section of the rocket, detector C1 covering the energy range 1.4–3.6 keV in 7 energy channels and detector C2 the range 2–18 keV in 9 energy channels. The field of view of each detector was 28° and 4° (FWHM) with the greater collimator extension parallel to the longitudinal rocket axis for C2 and for six-tenths of C1, the remaining section of the C1 collimator being canted at 40° to the major rocket axis. During the flight, the rocket spun at a constant rate of 75° per second, while the spin axis precessed about a flat cone at 0.84° per second. Attitude information was provided by on-board moon sensors and magnetometers, with an estimated rms error of 1.8°.

L Gratton (ed.), Non-Solar X- and Gamma-Ray Astronomy, 280–288. All Rights Reserved

During the early part of the flight the detectors scanned roughly along the galactic equator, covering the zone from $l = 220°$ to $l = 320°$ and then across the Scorpius-Sagittarius region. These 'along scans' are shown as 1-16 in Figure 1. In the second half of the flight, part of the same low latitude region was scanned again in a perpendicular direction, these 'across scans' being $1'-20'$ in Figure 1.

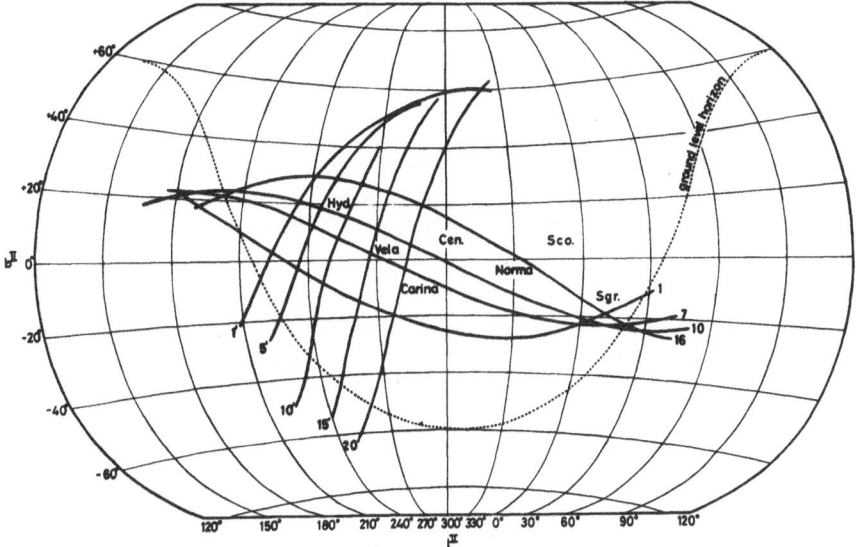

Fig. 1.   Sky map in l^{II}, b^{II} coordinates of the region of the galactic equator surveyed on June 12, 1968 by Skylark rocket SL 723.

3. Results

The accumulated data from groups of scans across the galactic equator are plotted in Figure 2, where separate scans have been added by using the galactic equator crossing as a common time origin. The occurrence of a persistent peak at the position of the galactic equator is clear in every group of scans, covering the galactic longitudes from 220 to 285°. The small peak to the left of the equator position, seen on scans $13'-20'$ is due to a discrete source in the Canopus region, as yet unreported, but confirmed in separate scans across this area of sky.

Evidence that the galactic equator peak does not arise from a *few* discrete sources is provided by the scans along the equator. Accumulated data from groups of these scans are shown in Figure 3. In the case of these almost parallel scans, the addition was carried out by summing each counting interval separated by a complete rocket spin period, the exactness of this procedure then being confirmed by plotting each scan on a globe. Scans 1–10 are plotted in Figure 3a, from which it may be seen that the X-ray count lies on the all-sky, isotropic background level in the Puppis region, while still North of the galactic equator, and then increases significantly as the equator is reached in the region of Vela. No strong source is seen in the position of Vel XR-1;

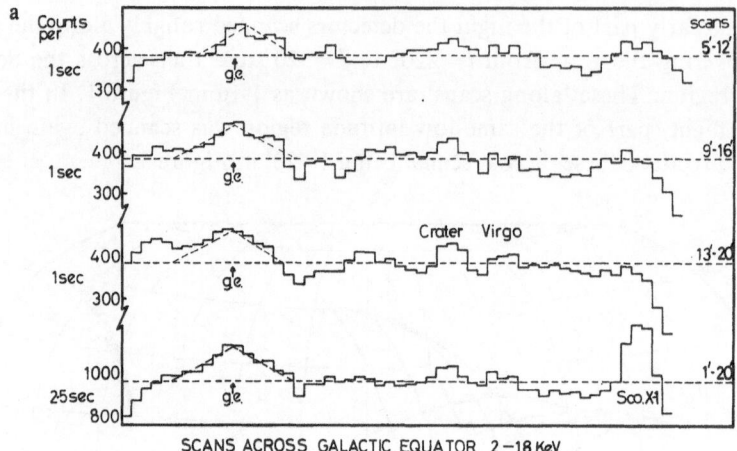

SCANS ACROSS GALACTIC EQUATOR 2–18 KeV

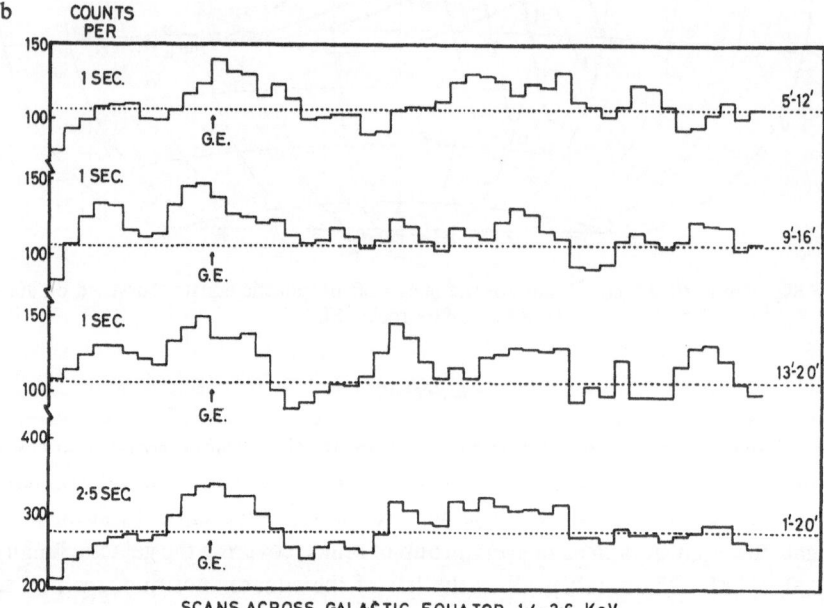

SCANS ACROSS GALACTIC EQUATOR 1·4–3·6 KeV

Figs. 2. Evidence for the X-ray enhancement at the position of the galactic equator produced by summation of groups of scans for detector C1 (1.4–3.6 keV) and C2 (2–18 keV).

the maximum intensity allowed by the present data being 0.15 photon cm^{-2} sec^{-1} above 2 keV. This is significantly less than the A.S. and E. Group's measurement [6], obtained in February 1968 and it seems probable, therefore, that Vel XR-1 is a variable or short-lived emitter. The estimated count rate profile that would have been seen by the present instrument from a source of the A.S. and E. strength is shown in Figure 3a. Further along these scans, there is some evidence for a source in the Centaurus region, which may be that previously reported as Cen XR-3 [7].

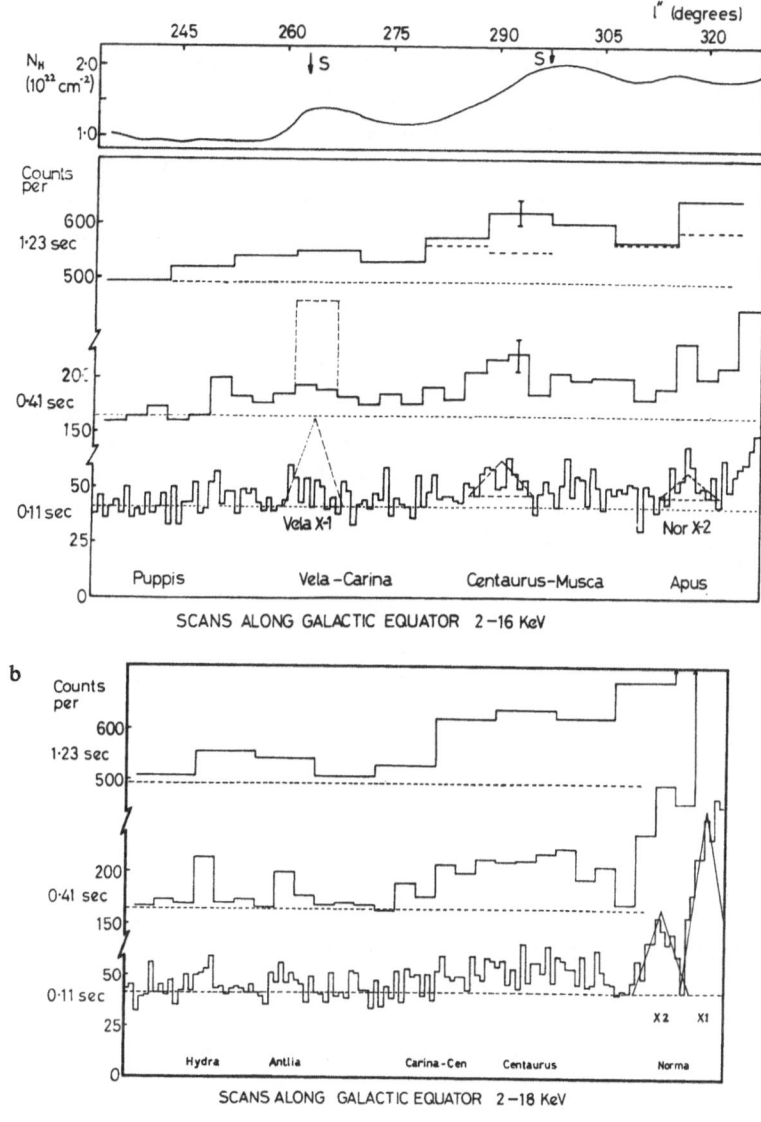

Figs. 3. Summed groups of scans along the galactic equator showing how the X-ray enhancement varies with galactic longitude in the Vela-Carina region (3a) and the Carina-Centaurus region (3b). The dashed triangle at Vela X-1 gives the strength measured by AS and E[6]. Those at Cen X-3 and Nor X1 and X2 give the strengths measured in this flight.

The measured intensity above 2keV is 0.25 photon cm^{-2} sec^{-1}, at a position 10^h15^m right ascension and $-74°$ declination, some $10°$ from the location of Cen XR-3 reported by the LRL Group. Of particular interest is the non-appearance of Cen XR-2 on our data. This source, which in April 1967 was stronger [3] than Sco X-1 faded below a level of 0.6 photon cm^{-2} sec^{-1} in September of that year [8]. The upper limit in June 1968, from the present data, is a factor of 4 lower still. Recently, however, the provisional results from two rocket launches from India in November, 1968 have been reported [9], indicating that at this time Cen XR-2 has re-appeared at a strength of about 1.0 photon cm^{-2} sec^{-1}, above 2 keV. Observations above 20 keV from balloons by the MIT group gave positive measurements in October 1967, at a level [10] roughly consistent with the LRL upper limit at lower energies. A preliminary report [11] of a new MIT balloon flight in April 1969, however, reports no evidence of Cen XR-2 to a level of one-third of the October 1967 flux! Beyond the Centaurus-Musca region, the scans plotted in Figure 3a pass South of Norma, a partial view of the NOR XR-2 source being obtained, and then the count rate rises steeply with the detection of the strong NOR XR-2 and ARA sources. It is perhaps important to record that the Norma, Ara, Scorpius and Sagittarius sources are all clearly seen in the continuation of these same scans, with intensities and positions close to those reported from previous flights by the NRL [12], A.S. and E. [13] and MIT [14] Groups. This gives us confidence in the reliability of the non-observation of several sources from the present flight, as discussed above. The spectra and intensities of the galactic sources from Norma to Sagittarius will be presented in a separate paper.

A re-examination of Figure 3a can now be made, allowing for the flux due to the observed discrete sources in Centaurus and Norma. This is done (short dashed lines) in the third panel up of Figure 3a. A gradual increase in the excess above the isotropic background level is now evident as the scans progress from galactic longitude 245–320°, in general conformity with the neutral hydrogen column density, shown in the top panel. Reference to Figure 3b confirms this trend. Here are plotted the counts accumulated in the 'along equator' scans 7–16 and reference to Figure 1 shows that these pass through Hydra and Antlia before converging on the galactic equator near longitude 280°, in the region of Carina-Centaurus. Once again, the count rate is seen to rise markedly from the isotropic background level when the galactic equator enters the field of view. These scans cross directly over the Norma sources which are shown at their full relative intensities in Figure 3b.

Thus, we observe a clear excess of X-rays from the galactic equator region whenever this is viewed. The width of the emission band can be derived from the across-scan data of Figure 2, to be 6° or less and its location indicates it is certainly of galactic origin, being either unresolved sources or a diffuse galactic flux, or both. The galactic excess near $l = 260°$ can be expressed as 0.3 photon cm^{-2} sec^{-1} rad^{-1}, rising to 0.5 photon cm^{-2} sec^{-1} rad^{-1}, in the Centaurus region, near $l = 300°$. Further towards the galactic centre it becomes very difficult to deduce the intersource background flux, since the Ara, Scorpius and Sagittarius sources are strong and numerous. A minimum count rate obtained near $l = 350°$ does not exclude a further factor of 2 increase in this galactic component, that is, to 1 photon cm^{-2} sec^{-1} rad^{-1}.

4. Discussion

It may be that the galactic equator excess is due to a number of unresolved sources which lie below the sensitivity threshold of the present survey and in this regard it is interesting to compare the rquired source distribution with the known galactic sources. It has been suggested [15] that the Scorpius-Sagittarius source may be 'typical' galactic sources, yielding a flux at earth's distance of between 1 and 2 photons cm^{-2} sec^{-1} above 2 keV and also that their most probable location is in the Sagittarius spiral arm, at a mean distance of 2 or 3 kiloparsecs. Again, the Cygnus group of sources [13], probably located in one extension of the local spiral arm, would lie at a similar, and reasonable, distance if they are of the same intrinsic strength. If it is accepted that the local arm extends in the opposite direction through Vela [16], then a very different situation must exist in this extension. The presently reported flux and the smooth distribution of excess counts as shown in Figure 3 require at least 10 sources to be spread out over the Vela-Centaurus region of the Milky Way. The average flux at the earth from each source is then only 0.04 photon cm^{-2} sec^{-1}, representing a 'typical' galactic source at an unreasonable large distance of 10 to 15 kpc. Of course, it may be that the Vela-Centaurus regions contain a much weaker class of galactic X-ray source. Higher resolution and more sensitive studies, using stabilised rockets, will be required to pursue this possibility.

An alternative explanation, considered briefly below, lies in the possibility that the low latitude enhancement is due mainly to a diffuse galactic X-ray flux, connected with the concentration of matter, radiation or cosmic rays in the galactic plane. The gradual increase of the excess X-radiation towards the galactic centre, in conformity with the increasing neutral hydrogen [17, 18] column density, is shown in Figure 3a. The spectral distribution of the enhancement is shown in Figure 4 and may be fitted with a power law, similar to the isotropic background spectrum [19] perhaps indicating a similar production mechanism. Several possible processes that could contribute towards a significant diffuse galactic X-radiation are briefly discussed below.

(A) *Suprathermal proton bremsstrahlung*, arising from the collisions of 2–20 MeV cosmic ray protons with ambient electrons in the spiral arms, has been considered [20] as a source of 1–10 keV X-rays. This flux will exhibit the (probable) power law spectrum of the protons and will be correlated in intensity with the neutral hydrogen. However, the heating of interstellar H I regions [21] places a limit on the proton flux which, in conjunction with the neutral hydrogen column densities shown in Figure 3a, yield an X-ray flux from the galactic plane more than an order of magnitude less than that observed here. Stecker has recently [22] presented evidence, however, that the interstellar gas density may be up to an order of magnitude higher than that estimated from 21 cm emission.

(B) *Electron capture* by heavy cosmic ray nuclei, colliding with the neutral hydrogen in the galactic disc, with the subsequent radiative cascade of the excited ions, yielding X-ray line emission. This process has recently been examined by Silk and Steigman [23], who show that it may yield fluxes comparable to the isotropic background X-radiation, in the 1–10 keV band. The strongest predicted lines, silicon and magnesium Ly-α, are reproduced from the above paper in Figure 4, Doppler broadening spreading the emission over the energy band 1–2.5 keV. The crude spectral data of the galactic X-ray enhancement does allow some steepening in the spectrum below 2.5 keV, but the statistical significance of the low energy points is clearly insufficient to prove the existence of this flux. It does seem, however, that the measured spectrum above 3 or 4 keV requires a different source mechanism, unless there is a much larger abundance of the heavier elements (P–Ni) in the few MeV range than is observed at higher energies [24].

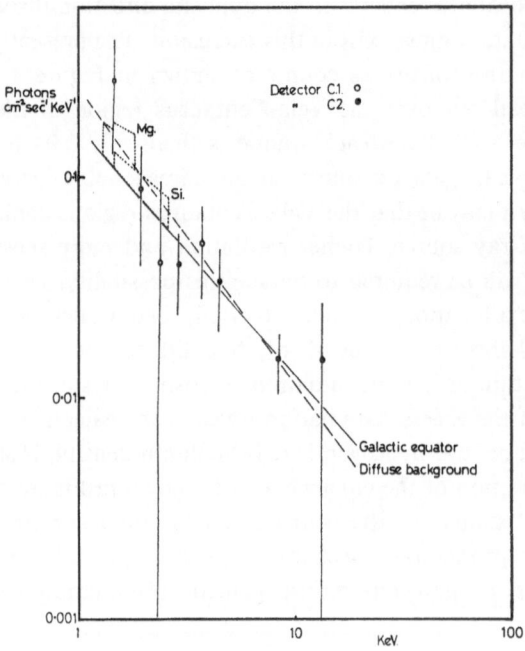

Fig. 4. The X-ray spectrum of the galactic equator enhancement compared with that of the diffuse X-ray background. The blocks labelled Mg and Si represent the line strength due to those elements from the paper of Silk and Steigman [23].

(C) *Inverse Compton radiation.* It is generally assumed that the most likely source of the isotropic diffuse X-radiation is by inverse Compton scattering of high energy electrons in the distant reaches of the universe, and the contributions from both dense galaxies and quasars [25] and the 'universal' black-body (3 K) radiation [26] have been estimated. The latter radiation falls short of explaining the present galactic excess by over an order of magnitude, however, and it is generally accepted that the flux of the appropriate GeV electrons in interstellar space is reasonably well known [27], since solar modulation is weak at these electron energies.

The Compton X-ray production could be raised to the observed intensity if the strong far infra-red radiation reported by Shivanandan *et al.* [28] is confirmed and pervades the whole galaxy. However, the existence of a widespread infra-red background of this intensity is hard to reconcile with the observed excitation of interstellar molecules [29].

Finally, there remains the possibility of inverse Compton scattering of starlight $(T \sim 6000 \, K)$ in the galactic disc. Rees and Silk [30] have recently put forward an explanation of the 100 MeV γ-ray flux, observed as an apparently diffuse, narrow band of radiation from the Milky Way [31, 32, 33, 34], in which it is proposed that the γ-rays arise as bremsstrahlung from the collisions of 100 MeV electrons with the interstellar gas atoms. The essential point of the Rees and Silk paper is that the required large flux of 100 MeV electrons does not contradict *any* available observational data. It is of particular interest here to note that an extension of their assumed electron spectrum from 100 MeV to 15 MeV would produce an excess of 2–10 keV X-rays in the galactic plane, by inverse Compton interactions with starlight, comparable with that indicated in the Vela-Centaurus region in the present paper. Assuming a starlight energy density of 0.5 eV cm^{-3}, and allowing for a factor of 2 increase in the scattering efficiency [35] due to the anisotropic starlight flux (travelling mainly in the galactic disc and directed outward from the centre), the energy density of the ~ 20 MeV electrons required is 0.25 eV cm^{-3}, to yield the 'observed' diffuse galactic X-radiation. It is noteworthy that this large electron flux would provide a heat input to the interstellar medium of the same order as that assumed by Field *et al.* [36], these authors attributing the heating to a low energy proton component, for which there is less direct evidence. Rees and Silk (private communication) also point out that the expected turn-over in the electron spectrum, at about 10 MeV, due to ionisation losses, could in addition explain the intensity and spectrum of 1–10 MeV γ-rays recently observed by Vette *et al.* [37].

Reference

[1] Gould, R. J.: 1967, Am. J. Phys **35**, 376.
[2] Gorenstein, P., Kellogg, E. M., and Gursky, H.: 1969, *Astrophys. J.* **156**, 315.
[3] Cooke, B. A., Pounds, K. A., Stewardson, E. A., and Adams D. J.: 1967, *Astrophys. J.* **150**, L189.
[4] Bowyer, C. A., Field, G. B., and Mack, J. E.: 1968, *Nature* **217**, 32.
[5] Kraushaar, W. L.: 1970, this volume.
[6] Gursky, H., Kellogg, E. M., and Gorenstein, P.: 1969, *Astrophys. J.* **154**, L71.
[7] Chodil, G., Mark, H., Rodrigues, R., Seward, F., Swift, C. D., Hiltner, W. A., Wallerstein, G., and Mannery, E. J.: 1967, *Phys. Rev. Letters* **19**, 681.
[8] Chodil, G., Mark, H., Rodrigues, R., and Swift, C. D.: 1968, *Astrophys. J.* **152**, L45.
[9] Rao, U. R., Chitnis, E. V., Prakasarao, A. S., and Jayanthi, U. B.: 1970, this volume p. 88.
[10] Lewin, W. H. G., Clark, G. W., and Smith, W. B.: 1968, *Astrophys. J.* **152**, L49.
[11] Lewin, W. H. G.: 1970, this volume p. 144.
[12] Friedman, H., Byram, E. T., and Chubb, T. A.: 1967, *Science* **150**, 374.
[13] Gursky, H., Gorenstein, P., and Giacconi, R.: 1967, *Astrophys. J.* **150**, L75.
[14] Bradt, H., Naranan, S., Rappaport, S., and Spada, G.: 1968, *Astrophys. J.* **152**, 1005.
[15] Giacconi, R., Gursky, H., and Van Speybroeck, L. P.: 1968, *Ann. Rev. Astron. Astrophys.* **6**, 373.
[16] 'Review of Galactic Structure', in *Landolt-Börnstein* – Group VI **1**, 623. (ed. by H. Voigt), Springer-Verlag, Berlin, 1965.
[17] Kerr, F. J.: 1967, *Radio Astronomy and the Galactic System* (ed. by H. van Woerden), Academic Press, p. 239.
[18] Schmidt-Kaler Th.: *Radio Astronomy and the Galactic System* (ed. by H. van Woerden), Academic Press, p. 171.

[19] Seward, F., Chodil, G., Mark, H., Swift, C., and Toor, A.: 1967, *Astrophys. J.* **150**, 845.
[20] Boldt, E. and Serlemitsos, P.: 1968, NASA/GSFC report X-611-68-385.
[21] Pikel'ner, S. B.: 1968, Soviet Astron. **11**, 737.
[22] Stecker, F. W.: 1969, Nature **222**, 865.
[23] Silk, J. and Steigman, G.: 1970, this volume p. 385.
[24] Comstock, C. M., Fan, C. Y., and Simpson, J. A.: 1969, *Astrophys. J.* **155**, 609.
[25] Rees, M. J. and Setti, G.: 1968, *Nature* **219**, 127.
[26] Brecker, K. and Morrison, P.: 1967, *Astrophys. J.* **150**, L61.
[27] Webber, W. R.: 1968, *Australian J. Phys.* **21**, 845.
[28] Shivanandan, K., Hauck, J. R., and Harwit, M. O.: 1968, *Phys. Rev. Letters* **21**, 1460.
[29] Bortolot, N. J., Clauser, J. F., and Thaddeus, P.: 1969, *Phys. Rev. Letters* **22**, 307.
[30] Rees, M. J. and Silk, J.: 1969, *Astron. Astrophys* **3**, 452.
[31] Clark, G. W., Garmire, G. P., and Kraushaar, W. L.: 1968, *Astrophys. J.* **153**, L203.
[32] Hutchinson, G. W., Ramsden, D., and Wills, R. D.: 1970, this volume p. 300.
[33] Sood, R. K.: 1969, *Nature* **222**, 650.
[34] Valdez, J. V. and Waddington, C. J.: 1969, *Astrophys. J.* **156**, L85.
[35] Baylis, W. E., Schmid, W. N., and Lüscher, E.: 1967, *Z. Astrophys.* **66** 271.
[36] Bowyer, C. S., Field G. B., and Mack J. E.: 1968, *Nature* **217**, 32.
[37] Vette, J., Matteson, J. L., Gruber, D., and Peterson, L. E.: 1970, this volume p. 335.

OBSERVATIONS ON DIFFUSE COSMIC X-RAYS IN THE
ENERGY RANGE 20–120 keV

P. C. AGRAWAL, S. BISWAS, G. S. GOKHALE, V. S. IYENGAR,
P. K. KUNTE, R. K. MANCHANDA and B. V. SREEKANTAN

Tata Institute of Fundamental Research, Bombay, India

In this paper we present observations of the diffuse background X-rays in the energy range 20–120 keV, based on two balloon experiments carried out from Hyderabad (latitude 17.6 °N, longitude 78.5 °E), India. The flights were made on April 28, 1968 and December 22, 1968. The detector used was a NaI(Tl) crystal of effective area 97.3 cm^2 and thickness 4 mm. The crystal was surrounded both by active and passive collimators. The passive collimator was a cylindrical graded shield of lead, tin, and copper, and the active collimator was a plastic scintillator surrounding the shield. The FWHM of the telescope was 18.6° and the geometrical factor for isotropic radiation 13.2 cm^2 sr. The pulses from the NaI crystal were sorted into ten contiguous channels extending from 17 to 124 keV. An Am^{241} source came into the field of view of the telescope periodically and provided in-flight calibration of the detector. All the information was recorded on photographic film.

The X-ray telescope was mounted on a oriented platform which was programmed to look in specified directions in the sky. The axis of the telescope was inclined at an angle of 25° with respect to the zenith in the first flight and at an angle of 32° in the second flight. In the 28th April flight, the plan was to look alternately in the North and South directions for about 10 min. However the orientor did not function quite the way it was planned. While it did point for considerable time in the prescribed directions, it suffered oscillations and also got locked in directions in which it was not programmed to look. Since a pair of orthogonal flux gate magnetometers provided information on the aspect continuously, it was possible to retrieve the data unambiguously and obtain results on both the background X-rays and discrete sources.

In the flight on 22nd December, the orientor worked perfectly. The telescope was programmed to point in four specific directions i.e. N ($\phi = 0$), S ($\phi = 180°$), SW ($\phi = 110°$), and NE ($\phi = 310°$). In this flight, the telescope picked up Sco X-1 in the South direction, and Cygnus X-1 in NE ($\phi = 310°$). and a new source in the direction SW ($\phi = 110°$). In the North ($\phi = 0$), there was no source during the period of observation and therefore information on background X-rays, was obtained from this direction.

Table I summarises the experimental results from the two flights. The atmospheric background counts were obtained by plotting the counting rate vs. altitude between 50 g/cm^2 and 10 g/cm^2 and fitting a least square line through the experimental points, and extrapolating to the floating altitude. The attenuation factors given in the table include the attenuation due to the intervening atmosphere in the direction of the axis of the telescope and also the attenuation in the aluminum wall of the

L. Gratton (ed.), Non-Solar X- and Gamma-Ray Astronomy, 289–296. All Rights Reserved

TABLE I

Data on the diffuse X-ray background from flights on 28.4.68 and 22.12.68

Date of flight	Energy interval keV	Mean energy keV	Total counts per sec	Atmospheric background counts per sec	Cosmic X-ray counts per sec	Attenuation factor (Air + Al)	Flux Photons/keV cm² sec sr.	Flux keV/keV cm² sec sr.
28th April, 1968 Ceiling alt. 5.3 g/cm²	23.4 – 31.50	27.5	1.97 ± 0.05	1.12 ± 0.10	0.85 ± 0.11	12.00	$(9.55 \pm 1.24) \times 10^{-2}$	2.70 ± 0.34
	31.50– 47.7	39.6	3.43 ± 0.07	2.17 ± 0.13	1.26 ± 0.15	4.70	$(2.76 \pm 0.33) \times 10^{-2}$	1.09 ± 0.13
	47.70– 72.0	59.8	6.96 ± 0.09	5.18 ± 0.08	1.76 ± 0.12	3.00	$(1.66 \pm 0.12) \times 10^{-2}$	$(9.92 \pm 0.71) \times 10^{-1}$
	72.0 – 97.4	84.6	6.74 ± 0.08	5.70 ± 0.07	1.04 ± 0.11	2.66	$(8.10 \pm 0.85) \times 10^{-3}$	$(6.85 \pm 0.68) \times 10^{-1}$
	97.4 –123.7	110.5	5.21 ± 0.05	4.93 ± 0.07	0.28 ± 0.08	2.42	$(1.92 \pm 0.54) \times 10^{-3}$	$(2.12 \pm 0.59) \times 10^{-1}$
22nd December, 1968 Ceiling alt. 7.5 g/cm²	29.9 – 52.30	40.1	5.22 ± 0.10	4.28 ± 0.11	0.94 ± 0.14	8.30	$(2.62 \pm 0.50) \times 10^{-2}$	1.05 ± 0.20
	52.30– 74.70	63.5	5.63 ± 0.10	5.23 ± 0.15	0.40 ± 0.17	4.92	$(6.59 \pm 2.83) \times 10^{-3}$	$(4.18 \pm 1.79) \times 10^{-1}$
	74.70–118.70	96.7	11.6 ± 0.15	11.30 ± 0.20	0.30 ± 0.25	3.97	$(2.07 \pm 1.68) \times 10^{-3}$	$(2.00 \pm 1.62) \times 10^{-1}$

Geometrical factor of the telescope for isotropic radiation: 13.2 cm² sr.

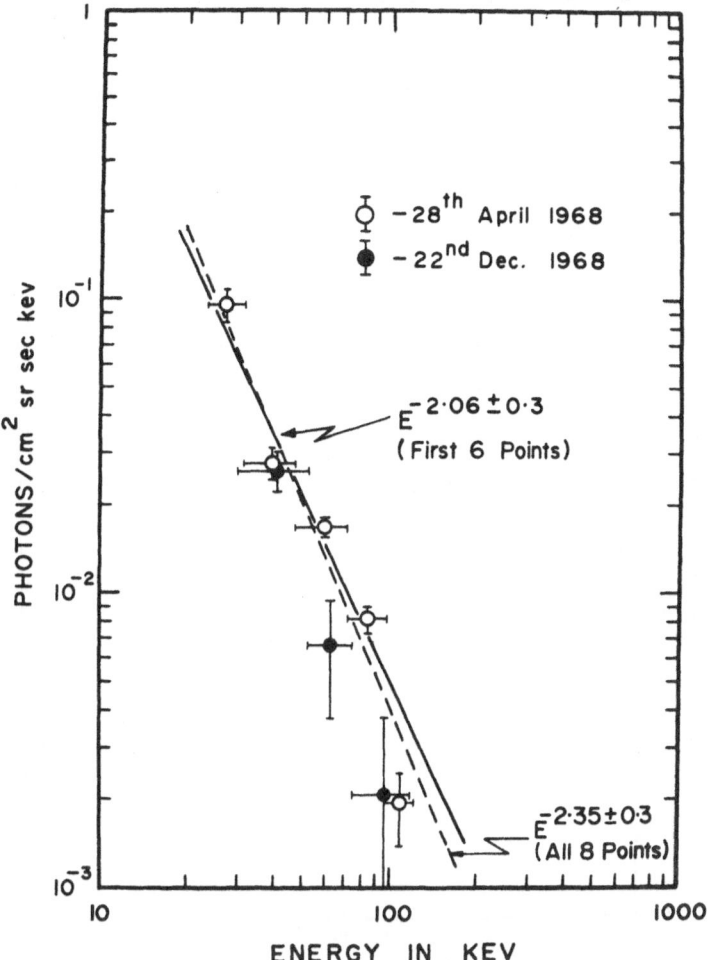

Fig. 1.   Spectrum of diffuse X-ray background measured on April 28, 1968 and December 22, 1968.

pressurised gondola in which the entire instrument including the oriented telescope
was housed.

In Figure 1, we have plotted the spectral data obtained in the two flights in the
energy range 23 to 124 keV, and in Figure 2, a comparison is made with other experimental results [1–3] obtained at balloon altitudes. In Figure 3, we have superimposed
our data on the composite spectral curve given by Gorenstein *et al* [4] which covers
the range from 0.1 keV to 10 MeV.

Our own data (Figure 1) may be fitted with a power law spectrum $dN/dE = 204$
$E^{-2.35 \pm 0.3}$ in the energy range 23.0 to 124 keV. However, it is evident from Figure 1,
the two points corresponding to mean energies greater than 90 keV, fall much below
the dotted curve, suggesting that a single slope may not fit the entire data very well.
The least square line corresponding to the first 6 points has a scope 2.06±0.30,

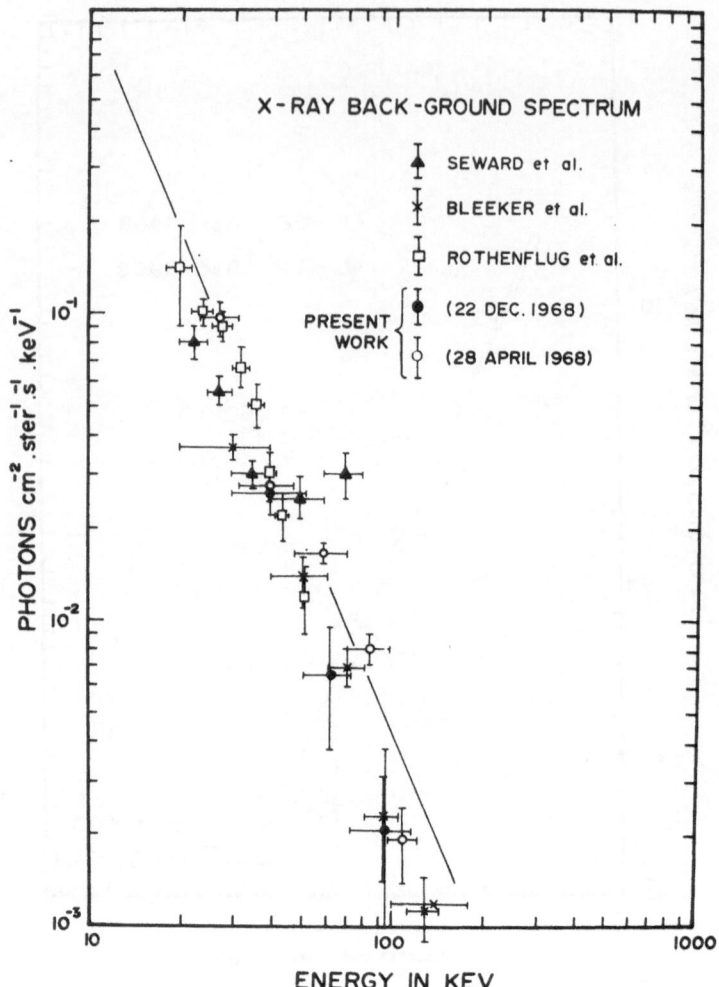

Fig. 2. The photon intensities of diffuse X-rays measured in the present experiments and by other investigators at balloon altitudes. The straight line of slope, −2.35, as in Figure 1, is shown for comparison only.

and the spectrum in the energy range 23 to 80 keV may be represented by $dN/dE = 69.4E^{-2.06 \pm 0.30}$ photons/keV cm² sec. sr.

Figure 2 brings out more clearly the existence of a break in the spectrum of the diffuse background X-rays, the slope changing somewhere around 60 to 80 keV.

As already pointed out, in the flight on 28th April, the telescope not only looked at the prescribed directions, but also scanned other directions. This enabled us to obtain information on the variation of the counting rate of the telescope as a function of azimuth. The data have been classified into two time intervals i.e. 2.10–3.00 IST (Indian Standard Time) and 3.00–3.33 IST, and presented in Table II and plotted in Figure 4. During the period 2.10–3.33 IST, the only known strong X-ray source

TABLE II

Date of flight: 28th April, 1968

Counting rates classified according to azimuth (counts/sec)

$\phi =$ 0 North
$\phi =$ 90 West
$\phi =$ 180 South

Time	Energy range (keV)	Azimuth 330–360	0–30	30–60	60–90	90–120	120–150	150–180	180–210	Grand average
2.10–3.00 (IST) 2040–2130 (UT)	17–48	7.67 ±0.14 (360)	7.39 ±0.16 (285)	7.87 ±0.20 (195)	7.77 ±0.25 (120)	8.04 ±0.18 (240)	7.15 ±0.28 (90)	7.19 ±0.18 (210)	8.35 ±0.15 (375)	7.70 ±0.06 (1875)
	48–124	19.04 ±0.23 (360)	18.40 ±0.25 (285)	19.37 ±0.31 (195)	18.10 ±0.39 (120)	18.74 ±0.28 (240)	16.78 ±0.43 (90)	18.63 ±0.29 (210)	18.72 ±0.22 (375)	18.48 ±0.09 (1875)
3.00–3.33 (IST) 2130–2203 (UT)	17–24 + 32–48	6.17 ±0.32 (60)	6.21 ±0.18 (195)	6.26 ±0.17 (225)	5.78 ±0.23 (105)	5.96 ±0.12 (435)	6.06 ±0.16 (225)	6.36 ±0.19 (165)	–	6.11 ±0.06 (1410)
	48–124	19.54 ±0.57 (60)	20.17 ±0.32 (195)	19.25 ±0.29 (225)	18.86 ±0.42 (105)	19.72 ±0.21 (435)	20.02 ±0.30 (225)	20.28 ±0.35 (165)	–	19.69 ±0.16 (1410)

Note: Figures in brackets indicate the observation time in sec.

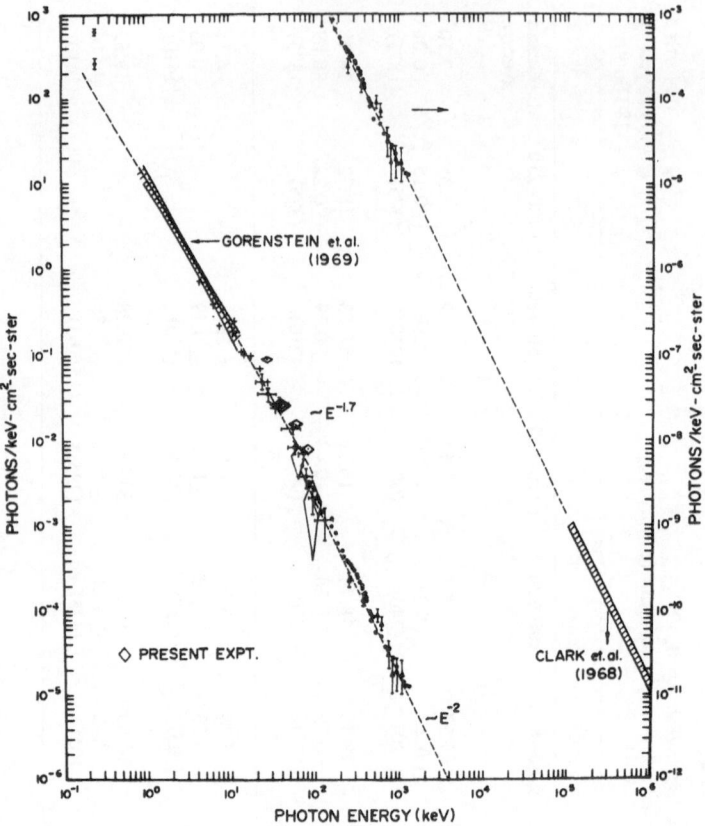

Fig. 3.   Our data on diffuse X-ray background super imposed on the composite spectral information
as given by Gorenstein *et al.* [4] (1968).

that was expected to be seen was Sco X-1 in the azimuthal range 150–210°. Some of
the sources in the Sagittarius region were also in the field of view in the same azimuthal
range, but the efficiency of the telescope was very poor for recording any contribution
from these sources, since the inclination of the telescope was unfavourable. In the
azimuthal range 330–0–150°, there was no known source in the field of view of
the telescope. However, the data indicated the possibility of a new source in the
azimuthal range 90–120° and the existence of this source has been confirmed in the
flight on December 22nd, as discussed in an accompanying paper (Agrawal *et al*,
[5]). The telescope did not scan the region 240° to 330°. It is seen from Figure 4, that
the total counting rate of the telescope (cosmic X-rays and atmosphere background)
as recorded at an altitude of 5.3 g/cm^2 and at an inclination of 25°, did not vary
by more than 8% in the entire azimuthal range 330–0–210°. However a puzzling
feature of Figure 4, is the large negative peak (4σ) observed in the azimuthal interval
120–150° during the period 2.10 to 3.00 LST in the energy interval 48–124 keV. This
observation cannot be simply interpreted as due to a statistical fluctuation since
there is a correlated negative peak (2σ) in the lower energy interval 17–48 keV also.

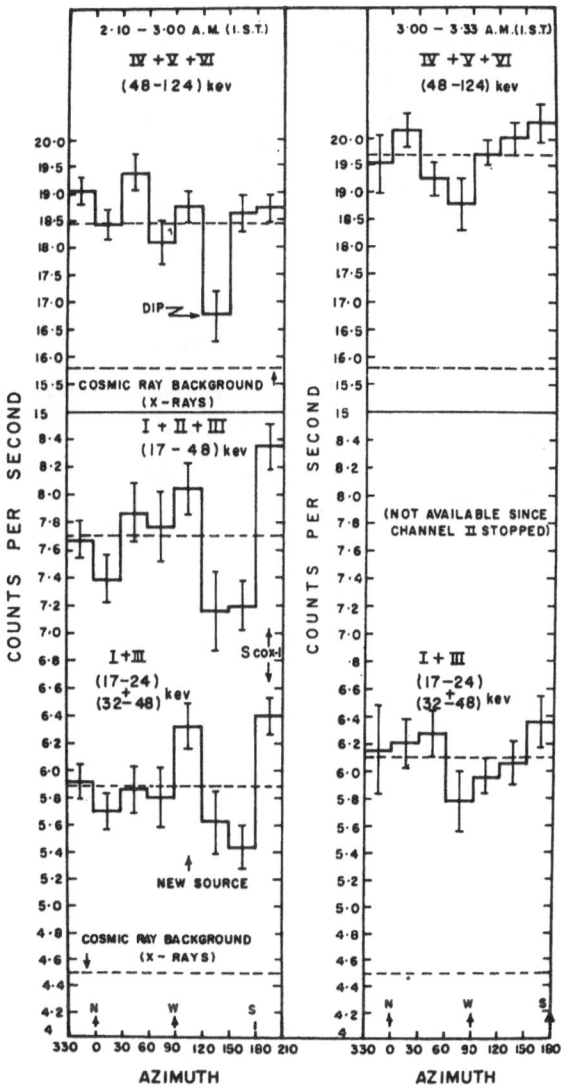

Fig. 4. Variation of the counting rates of the telescope as a function of azimuth on April 28, 1968, for two time intervals, 0210–0300 and 0300–0333 hrs. IST.

If the negative peak is due to physical causes, then we have to find them either in the atmospheric X-rays which are secondary to cosmic rays, or in the diffuse cosmic X-rays themselves. If it is due to cosmic rays, then there is no reason why the peak should be so conspicuously absent in the data corresponding to a later time i.e. 3.00 to 3.33 (Figure 4). Also, if the large variation in the counting rate arises as a result of geomagnetic effect of cosmic rays, then the effect should be more pronounced in the East–West direction, rather than in some odd South–East direction like 120–150°. If the negative peak is to be attributed to the diffuse background X-rays, then the consequences are rather drastic. There would have to be a depletion of the cosmic background X-rays by more than a factor of 2 in a localised celestial region around R.A. = 14 to 16 hrs. and $\delta = -14°$ to $+13°$. The depletion has to be more at higher energies (> 48 keV) than at lower energies. As far as we are aware there are no published experimental results that contradict the existence of localised regions of low X-ray intensity, at least at high energies. Clearly further experiments are necessary to establish unambiguously the existence of such regions of depleted X-ray intensity.

References

[1] Seward, F., Chodil, G., Mark, H., Swift, C., and Toor, A.: 1967, *Astrophys. J.* **150**, 845.
[2] Bleeker, J. A. M., Burger, J. J., Deerenberg, A. J. M., Scheepmaker, A., Swanenburg, B. N., Tanaka, Y., Hayakawa, S., Makino, F., and Ogawa, H.: 1968, *Suppl. Canadian J. Phys.* **46**, S461.
[3] Rothenflug, R., Rocchia, R., and Koch, L.: 1965, *Proc. Int. Conf. Cosmic Rays*, Vol. 1, p. 446.
[4] Gorenstein, P., Kellogg, E. M., and Gursky, H.: 1969, *Astrophys. J.* **156**, 315.
[5] Agrawal, P. C., Biswas, S., Gokhale, G. S., Iyengar, V. S., Kunte, P. K., Manchanda, R. K., and Sreekantan, B. V.: 1969, this volume, p. 289.

SEARCH FOR GALACTIC γ-RAYS WITH ENERGIES
GREATER THAN 500 MeV ON BOARD OGO-5

J. A. M. BLEEKER, J. J. BURGER, A. J. M. DEERENBERG,

H. C. VAN DE HULST, A. SCHEEPMAKER, and B. N. SWANENBURG,

Cosmic Ray Working Group, Kamerlingh Onnes Laboratorium, Leyden, The Netherlands

and

Y. TANAKA

Dept. of Physics, Nagoya University, Nagoya, Japan

A cosmic ray detector, sensitive to γ-rays with energies greater than 500 MeV is being flown on board the OGO-5 satellite. The spacecraft was launched into a highly eccentric orbit, apogee 145000 km, on March 4, 1968. γ-ray observations are restricted to altitudes higher than 80000 km, thereby excluding interference from the radiation belts and reducing the influence from the earth albedo flux. A description of the instrument is published in the literature (Rogowski *et al.*, 1969).

The peak sensitivity of the detector for a line source is 2–3 cm^2 rad with a FWHM of approximately 30°.

About 700 hrs of useful data have been analysed, covering the period from March 20 to July 1, 1968. During this period the orbital plane was almost perpendicular to the galactic equator. Part of the sky, effectively 30° wide, at $l^{II} = 60°$ and ranging from $b^{II} + 25°$ to $b^{II} = -35°$ has been scanned.

It is not possible to derive absolute values for the γ-ray intensities, because of the presence of cosmic ray induced background. Nevertheless, a search can be made for an increase in counting rate towards the galactic plane, provided the background rate does not vary with the position in orbit.

The observed counting rates of γ-ray events with energies greater than 500 MeV are listed in Table I for 5 intervals of b^{II}. The observed rates of charged particles, which could contribute to the background and the guard counter rates, are also listed. Changes in background, if any, are certainly smaller than the statistical uncertainties in the γ-ray rates. Contribution of earth albedo to the γ-ray events is estimated to be less than 10^{-5} counts/sec.

The present data only allow us to derive an upper limit for the galactic γ-ray intensity, namely:

$$N_\gamma (> 500 \text{ MeV}) \leqslant 7 \times 10^{-5} \text{ cm}^{-2} \text{ sec}^{-1} \text{ rad}^{-1}.$$

In Figure 1 the actual rates and the expected response to such a line intensity are plotted.

By analysis of all available data (> 7000 hrs) an intensity of $(2–4) \times 10^{-5}$ cm^{-2} sec^{-1} rad^{-1} expected from the measurement of Clark *et al.* (1968) above 100 MeV from the same region of the galactic plane should be observable.

TABLE I

Observed counting rates for γ-ray events and charged particles

	Directions of detector axis $l^{II} = 60° \pm 5°$				
	$25° > b^{II} > 20°$	$20° > b^{II} > 10°$	$10° > b^{II} > -10°$	$-10° > b^{II} > -20°$	$-20° > b^{II} > -35°$
Observation time (hrs)	29.0	83.6	305.8	135.9	104.2
'γ-rays' > 0.5 GeV (c/s) × 10³	0.24 ± .05	0.39 ± .04	0.30 ± .02	0.31 ± .02	0.29 ± .03
Charged particles (c/s) × 10³	15.2 ± .4	15.5 ± 0.2	15.1 ± .1	14.9 ± .2	15.3 ± .2
Guard counter rate (c/s)	687	679	664	660	671

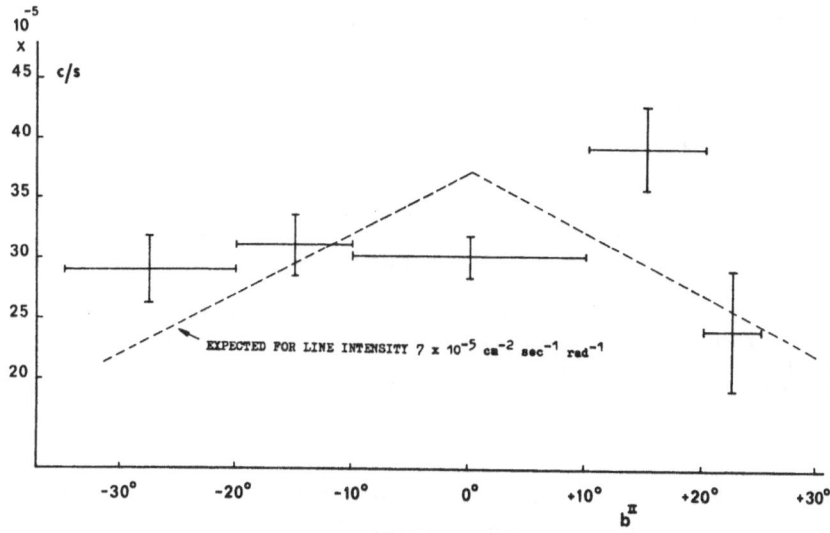

Fig. 1.   Counting rates for γ-ray events > 500 MeV at $l^{\text{II}} = 60°$
Beam width (FWMH): 30°.

References

Clark, G. W., Garmire, G. P., and Kraushaar, W. L.: 1968, *Astrophys. J. Letters* **153**, L203.
Rogowski, L. K., Hicks, D. B., Gilland, J. R., and Swanenburg, B. N.: 1969, *IEEE Trans. Nucl. Science* **NS–16**, 325.

SPARK-CHAMBER OBSERVATION OF GALACTIC γ-RADIATION

G. W. HUTCHINSON, A. J. PEARCE, D. RAMSDEN, and R. D. WILLS

Physics Department, University of Southampton, England

Abstract. A γ-ray telescope incorporating an acoustic spark chamber is included in the payload of the OGO-5 spacecraft. The performance of the instrument, which is sensitive to photons of energy 25 to 100 MeV, is discussed.

Observations are limited to a portion of the sky near Cygnus, but the first month's data indicate a variation of intensity showing a maximum in the direction of the galactic plane. If this plane contains a line source of radiation, its intensity is found to be $(9 \pm 5) \times 10^{-4}$ photons cm^{-2} sec^{-1} rad^{-1} above an energy of 40 MeV.

1. Introduction

The study of extraterrestrial γ-radiation is inhibited by the low intensity of this radiation compared with that of γ-rays produced in the earth's atmosphere by interactions of cosmic rays and with the much higher intensity of the charged cosmic rays themselves. A γ-ray astronomy experiment should therefore have a large geometrical factor (coupled with a good angular resolution for the observation of anisotropies) and be situated in an environment as free as possible of terrestrial radiation. The combination of wide acceptance solid angle and narrow angle of resolution in a device suitable for a satellite experiment suggests the spark chamber as an ideal instrument. The inclusion of such an experiment in an earth-oriented spacecraft in an eccentric orbit has advantages in the elimination of terrestrial interference, though imposing constraints on the field of view. The present experiment is the first application of the spark-chamber technique to satellite γ-ray astronomy.

2. Experiment Operation

The instrument is shown schematically in Figure 1 and has been described in detail by Dean *et al.* (1968). It includes a six-gap acoustic spark chamber of sensitive area 102 cm^2 surrounded by a scintillation-plastic anticoincidence counter. The chamber is triggered by a counter telescope, of acceptance solid angle 0.2 sr, consisting of a directional Čerenkov counter and a plastic scintillator. The anticoincidence veto may be relaxed on command to allow the recording of cosmic-ray protons for calibration purposes.

The experiment is included in the payload of the spacecraft OGO-5 which was launched on 4 March 1968 into an eccentric earth orbit with an apogee of 24 earth radii. The spacecraft is earth-oriented and the experiment is mounted so that the centre of its acceptance solid angle is directed away from the earth. Experiment operation is limited to the period when the spacecraft is outside the radiation belts because the high flux of charged particles renders the anticoincidence counter inoperative and could permanently change the characteristics of the photomultiplier

L. Gratton (ed.), Non-Solar X- and Gamma-Ray Astronomy, 300–305. All Rights Reserved

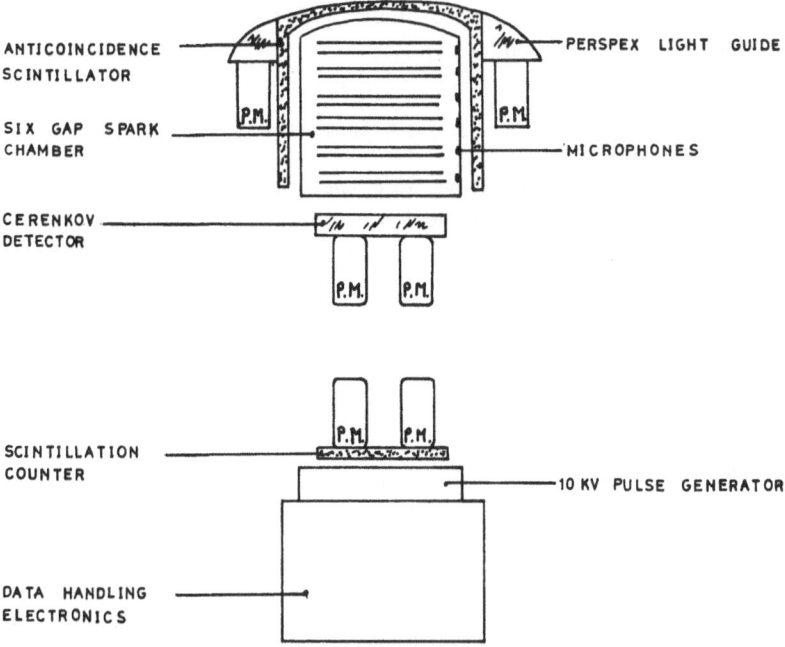

Fig. 1.   The experiment configuration.

tubes. Although this represents about 85% of the orbital period the region of sky scanned is restricted to a fairly small area near the direction of apogee. The axis of the telescope scans along the orbital plane from 18^h R.A., $+30°$ decl. to 22^h R.A., $+7°$ decl. but the aperture has a full width at half maximum approaching $30°$, in practice restricted to $18°$ by event-selection criteria (see Section 3).

In orbit it was found that the triggering rate of the experiment was much higher than expected. This was interpreted as an indication that the efficiency of the anti-coincidence system was much lower than had been indicated by preflight laboratory tests. This interpretation was confirmed by analysis of spark-chamber data which showed that the majority of events produced sparks along the whole length of their trajectory in the spark chamber and that the directions of incidence of the triggering particles were isotropic as would be expected if they were charged cosmic rays. The distribution of the points of entry of the particles over the surface of the anticoin-cidence counter was found to be uniform so it is believed that the fault is not in the detector assembly but in the operation of the veto electronics at high counting rates.

This malfunction imposed some restrictions on the operation of the experiment. The high rate of operation appears to have reduced the useful life of the spark chamber to about 5 months, during the last two of which a temporary data anomaly rendered subsequent analysis difficult. In addition the power supply limitations of the spark chamber high-voltage pulse system impose a dead time of 9.2 sec which is long compared with the actual triggering interval so that experiment live time can only be about 10% of real time. As a result of these two effects the total observation time is much less than had been hoped.

3. Data Analysis

In analysing the data genuine γ-rays were selected by using the upper gaps of the spark chamber as additional anticoincidence counters. The efficiency of the six spark gaps were measured from in-flight proton data to be better than 95%, with the exception of the 4th gap which failed 12 hours after the experiment was turned on.

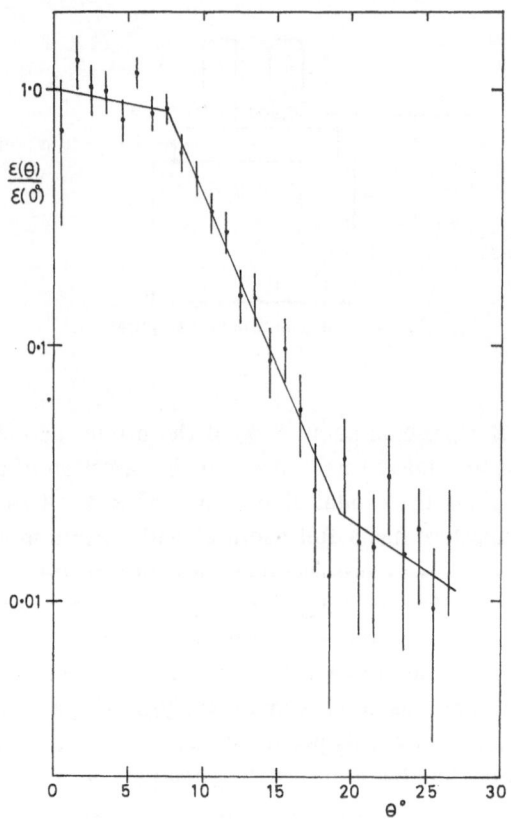

Fig. 2.   Variation of detection efficiency, E, with angle of incidence, θ, for particles incident through the top of the spark chamber.

Therefore selection of those events showing sparks in gaps 3, 5 and 6 on a track projecting through the sensitive area of gaps 1 and 2 would include about a 3% contamination of protons (taking the efficiency of the anticoincidence veto to be 75% as indicated by the relative counting rates in the γ- and proton modes). In addition to these requirements confidence in the direction measurement of the γ-rays required the sparks to be collinear within the errors of location, giving an accuracy of about 5°. In the first three months of operation 195 events satisfied these criteria and the present conclusions are derived from an analysis of the first 88 of these.

In order to know the effective viewing time as a function of direction it was necessary to have a measure of the off-axis response of the detector. This was obtained by relaxing the anticoincidence veto for a short period and recording cosmic-ray protons which were assumed to be isotropic. The variation of relative counting rate with

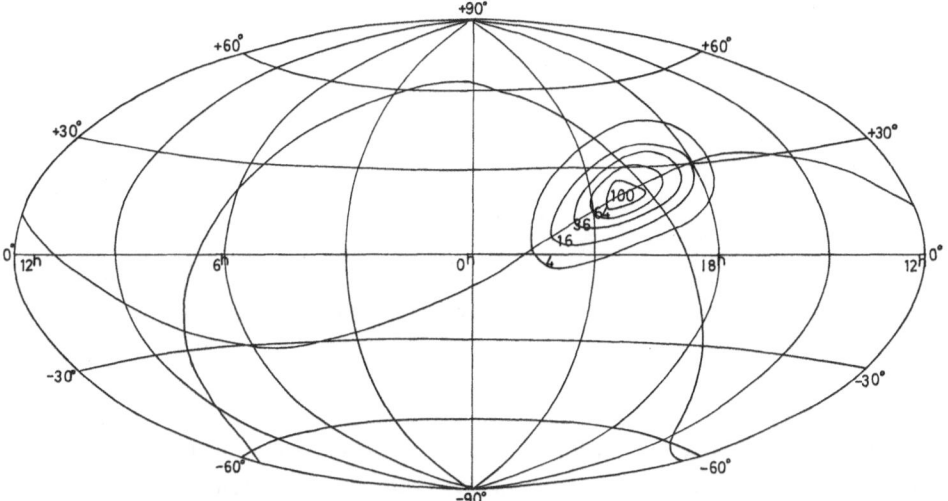

Fig. 3. Contours, in celestial coordinates, of equal effective viewing time, expressed as a percentage of the maximum.

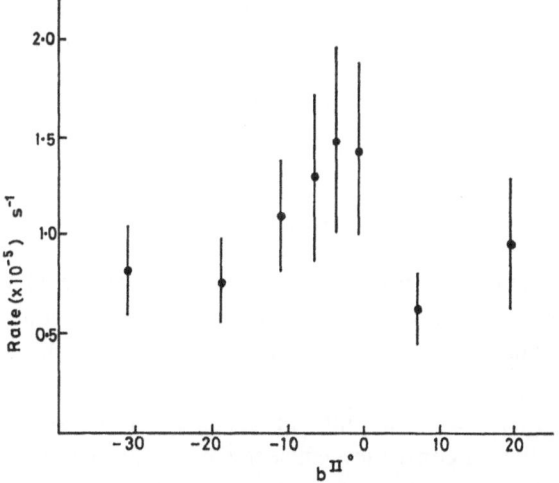

Fig. 4. Variation of counting rate with galactic latitude, averaged over the longitude range $45° < l^{\mathrm{II}} < 75°$.

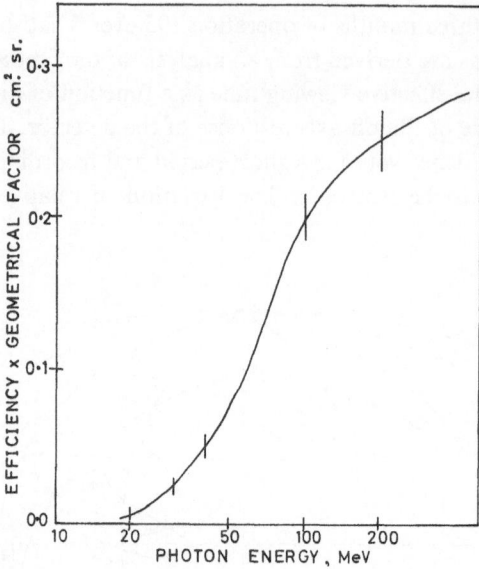

Fig. 5. Variation with energy of the detector response for γ-rays satisfying the selection criteria.

angle to the detector axis, for protons entering the chamber through the sensitive
area of the first gap, is shown in Figure 2. For simplicity of analysis three exponential
curves were fitted to the experimental points. This result was folded with the distribution of pointing-time for the direction of telescope axis (corrected for the large
dead-time effect) to obtain the effective viewing time, defined as the time a particular
direction would have to lie on the axis to obtain equivalent observation. Figure 3
shows contours of this parameter in celestial coordinates.

γ-ray events were divided amongst bins 3° square and the number of events
in each bin was divided by the effective viewing time for that direction. The results
are shown in Figure 4 as a function of galactic latitude, averaged over the range of
longitude viewed by the experiment ($45° < l^{II} < 75°$). There is evidence for an excess
flux from a direction close to the galactic plane.

4. Line-Source Interpretation

Since the width of the peak in Figure 4 is comparable with the angular resolution of
the experiment the suggestion of Clark *et al.* (1968) that an equivalent line-source
intensity may be defined can be followed. In order to quote an intensity at a specified
energy the energy response of the instrument must be known. This has been estimated
by means of a Monte Carlo computer program, the result of which is shown in
Figure 5 as the product of conversion efficiency, detection probability (subject to the
selection criteria set out in Section 3) and geometrical factor. If the detected γ-rays
have a differential spectrum proportional to E^{-2} the relative counting rate varies

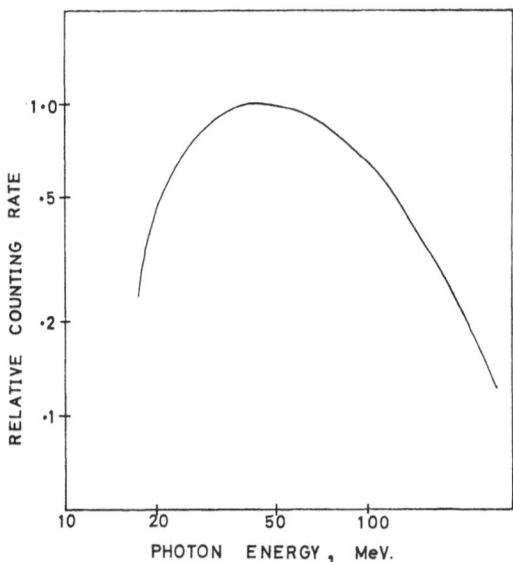

Fig. 6.   Relative counting rate of selected γ-rays as a function of energy, assume an energy spectrum of the form $N(E)\,dE \propto E^{-2}\,dE$.

with photon energy as shown in Figure 6. The most probable energy for detection is seen to be about 40 MeV.

The line source intensity for photons of energy greater than 40 MeV is then found to be $(9 \pm 5) \times 10^{-4}\ \mathrm{cm}^{-2}\ \mathrm{sec}^{-1}\ \mathrm{rad}^{-1}$. This result may be compared with the intensity of γ-rays above 100 MeV from the corresponding longitude measured by the OSO-3 experiment (Clark *et al.*, 1968; Kraushaar, 1970). The comparison supports the assumption, made in the derivation of both results, of an E^{-2} differential spectrum.

Acknowledgements

The authors wish to thank the National Aeronautics and Space Administration for the opportunity to participate in the OGO program and the University of Southampton Computing Service for the provision of facilities for reduction of the data. The project is supported by the Science Research Council and A. J. P. acknowledges the award of an S.R.C. research studentship.

References

Clark, G. W., Garmire, G. P., and Kraushaar, W. L.: 1968, *Astrophys. J.* **153**, L203–L207.
Dean, A. J., Hutchinson, G. W., Ramsden, D., Taylor, B. G., and Wills, R. D.: 1968, *Nucl. Instr. Meths.* **65**, 293–300.
Kraushaar, W. L.: 1970, this volume.

THE LOW ENERGY DIFFUSE COSMIC X-RADIATION

B. G. WILSON and A. J. BAXTER

Dept. of Physics, The University of Calgary, Calgary, Canada

Abstract. Recent measurements of the low energy diffuse cosmic X-radiation are reported. The results indicate that the large flux observed below 1 keV is inconsistent with an extrapolation of the power law spectrum observed above 2 keV. While the radiation between 2 and 10 keV appears largely isotropic, at least to the limits of resolution of the detector, the 0.27 keV radiation shows significant asymmetry with respect to galactic latitude.

We report here some of the results obtained from detectors flown in northern Canada during 1968. On one flight, using a Bristol Aerospace Black Brant IV vehicle, a 5-mil beryllium-windowed proportional counter reached an altitude of 880 km above Churchill, Manitoba, on April 5. The combination of large coning angle and 6-sec spin rate enabled a survey of almost all the northern sky to be made. The 38-cm² detector was covered by a hexagonal aluminum honeycomb limiting the field of view to 0.117 sterad. A ¼-mil mylar-windowed counter was flown over Resolute Bay, N.W.T., on October 7 by a Black Brant III to a height of 155 km. This detector was 30 cm² in area and contained 90% Argon-10% methane gas, with an identical field of view. Due to the peculiar motion of the freely spinning rocket, with the spin period almost exactly twice the coning period, the counter viewed the same quite narrow bands of the sky in successive rotations.

Detailed counter performance and the method of data reduction will be reported elsewhere. In a recent paper (Baxter *et al.*, 1969) we have provided a preliminary spectral analysis of the 0.20–4.0 keV radiation indicating a sharp increase in flux below 1 keV. In that paper the error bars shown are purely statistical. Further analysis, taking into account uncertainties in window thickness and calibration, provides the data in Figure 1. Although the efficiency of the mylar window is extremely low between 0.27 and 0.8 keV, photons were observed in this energy range, after full consideration of resolution effects. The experimental data appear quite consistent with a smooth increase in flux to the lowest energies observed.

In the 2–10 keV energy range the diffuse radiation is usually quoted as 'isotropic to within 10%'. The long flight on April 5 enabled a rather complete analysis of directional intensity to be made (Figure 2). Apart from the obvious contributions of the Crab and Cygnus sources, the sky appears to be rather even in X-ray brightness, to the resolution limit of the collimator (approximately 10° half-angle). Smaller scale irregularities would not be readily detectable. There is, however, some suggestion that the average intensity above 60 °N galactic latitude is greater than at lower latitudes.

When the carbon-K transmission band intensities are plotted, however, there is clear evidence of asymmetrical distribution with an approximately 7:1 intensity ratio between directions near the galactic pole and those towards the plane, taking the lower value of the two transit passes as more likely to be uncontaminated by sources of radiation (Figure 3).

L. Gratton (ed.), Non-Solar X- and Gamma-Ray Astronomy, 306–308. All Rights Reserved

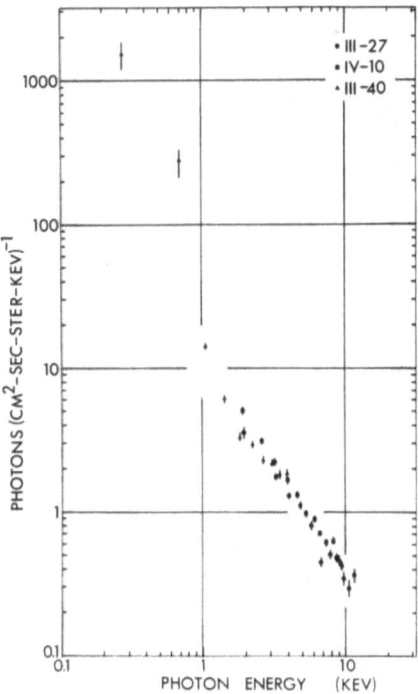

Fig. 1.   Spectral results for the intensity of the diffuse X-ray component measured by The University of Calgary series of experiments.

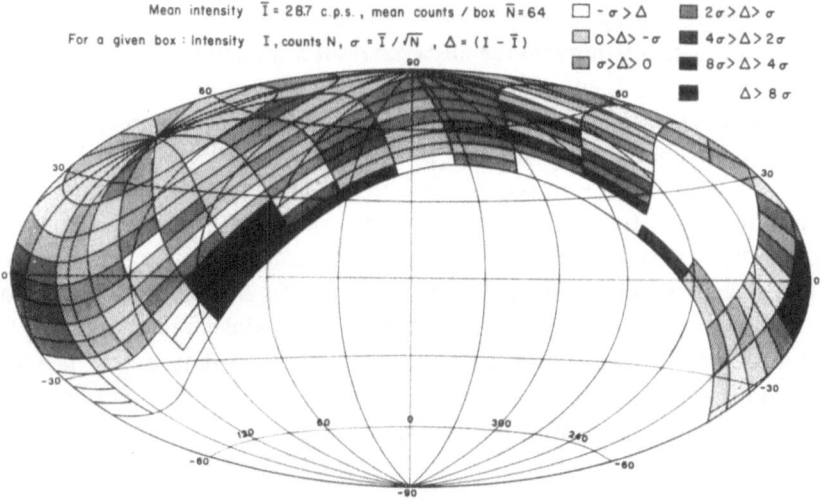

Fig. 2.   Distribution of counts in horizon coordinates in the energy range 1.6–10 keV (5 mil Be, IV-10), superimposed on galactic coordinates (l^{II}, b^{II}, 1950).

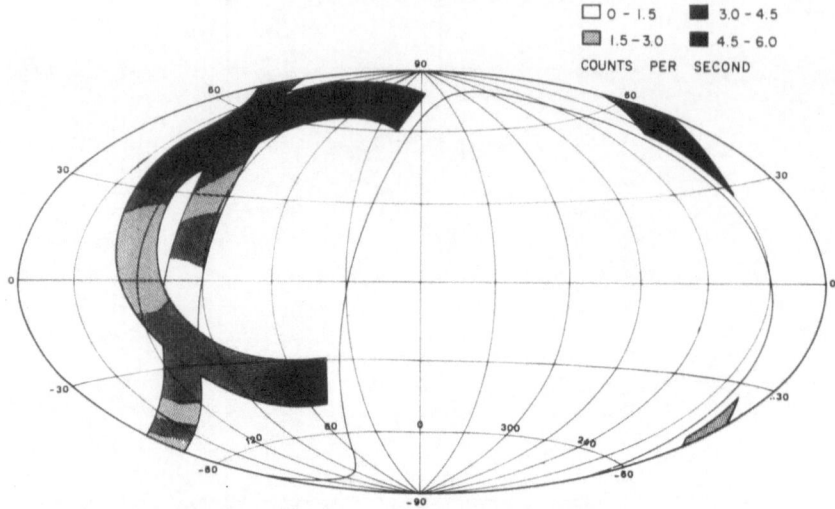

Fig. 3.   Distribution in galactic coordinates (l^{II}, b^{II}, 1950) of counts observed in the carbon K transmission band for the mylar counter (III-40), corrected for detector background.

During this flight the sun was about 20° below the rocket horizon. Some solar X-ray contributions were noted in the gross counting rate when the detector faced that direction during alternate rotations of the nose cone. No such contribution could be detected for the remaining 95% of the motion. This solar contribution was not detectable at energies below about 1 keV. The X-ray detector was enclosed by a scintillation-veto counter. The effectiveness of this combination in reducing local background effects was evident in the 15:1 reduction in intensity as the detector scanned the earth during each rotation of the nose cone. No horizon effects were noted in the data, suggesting that the observed results are unlikely to be contaminated by soft election precipitation in the polar sky. This conclusion is supported by the lack of anomalous results in a separate thin windowed Geiger Counter as well as by the consistent level of intensity observed by the proportional counter in 11 successive sky scans. The 2.0–4.0 keV results are also in excellent agreement with a wealth of experimental data in this part of the spectrum. Finally, the measured atmospheric absorption of the observed radiation is in good agreement with the predicted attenuation along the line of sight of the detector. (The 1350 Å aluminum coating of the mylar window makes the counter insensitive to ultra violet radiation.) We find it difficult therefore to resist concluding that the measured data are in fact due to cosmic X-rays which exhibit a surprising intensity below 1 keV and considerable anisotropy related to the gross galactic structure.

Reference

Baxter, A. J., Wilson, B. G., and Green, D. W.: 1969, *Astrophys. J. Letters* **155**, L145.

2–20 keV X-RAY SKY BACKGROUND

E. A. BOLDT, U. D. DESAI, S. S. HOLT, and P. J. SERLEMITSOS

NASA/Goddard Space Flight Center, Greenbelt, Md., U.S.A.

Abstract. The diffuse background of 2–20 keV X-rays over a band of the sky extending from Scorpius to the North galactic pole is found to be isotropic to within 5%, with a spectrum given by

$$10.3\, E^{-n} \text{ photons/(cm}^2\text{-sec-sr-keV)},$$

where $n = (1.35 \pm {}^{0.07}_{0.10})$.

A comparison with spectra at higher energies indicates that the lower energy spectrum is flatter, corresponding to an apparent unit change in spectral index within the band 20–80 keV. A spectral break in this energy region has been discussed in connection with the collisional energy loss lifetime for metagalactic protons that radiate X-rays via inverse bremsstrahlung collisions with the ambient electrons of the intergalactic medium (Boldt and Serlemitsos, 1969; Hayakawa, 1970).

Observations of the diffuse X-ray background above approximately 20 keV, made from balloon platforms (Rothenflug *et al.*, 1967; Agrawal *et al.*, 1970; Bleeker, 1970), indicate a spectral index for the photon flux of magnitude at least 2. Recent rocket observations (see review by Oda, 1970) of the background in a band extending down to about 1 keV suggest that the spectral index for this lower energy band is smaller than 2 and could be as low as 1.3. We report here on the results of a rocket flight launched from White Sands, New Mexico on March 3, 1969 in which we investigated the diffuse sky background within this band with a wide angle instrument especially suited to this purpose.

Figure 1 shows a schematic representation of the telescope, which consists of 6 argon-filled proportional counters of all-beryllium construction, enclosed in an anti-coincidence shield of plastic scintillator. All the counters were nominally identical except for the thin windows of the top counters which were of 2 mil beryllium. The pulses from these counters were in mutual anti-coincidence and the height of accepted pulses was analyzed by an on-board 128 channel analog to digital converter. We did not employ pulse shape discrimination.

A paddle mounted at the entrance of the telescope was rotated in discrete steps in order to vary the acceptance solid angle and thereby modulate a diffuse flux in a well-determined way. The telescope was contained in the rocket cylinder during the ascent and descent phases of the flight. During these phases the counters were calibrated with an Fe-55 source and the internal detector background spectrum was measured when the source was occulted by the paddle. The side door of the vehicle was opened at an altitude above 250000 feet and the telescope was deployed in a direction normal to the vehicle axis. The first portion of the exposure was devoted to a measurement of Sco X-1 (see Holt *et al.*, 1970). The second portion was a scan of about 1° per sec along a great circle from Sco X-1 to an extended stop at the North galactic pole and provided the data for the results reported here. The count rate at

L. Gratton (ed.), Non-Solar X- and Gamma-Ray Astronomy, 309–314. All Rights Reserved
Copyright © 1970 by the IAU

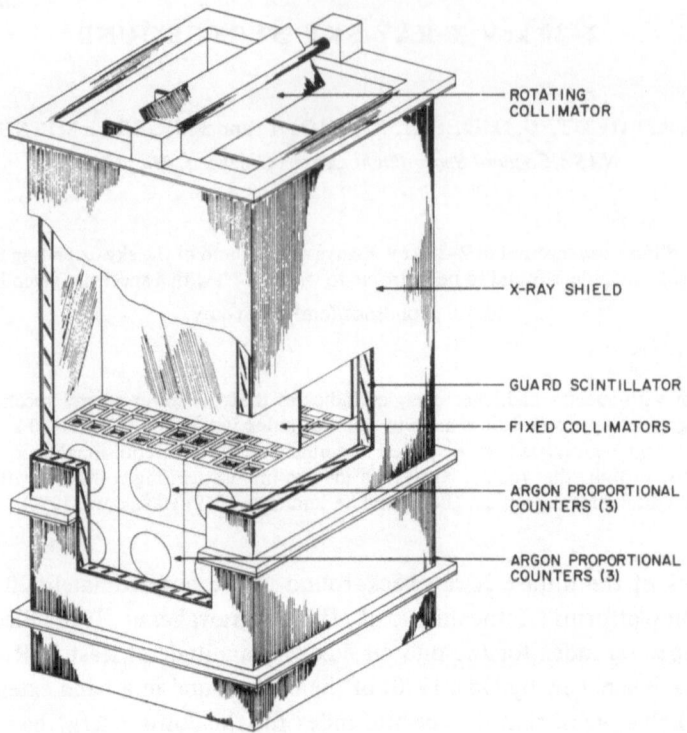

Fig. 1.   Schematic diagram of the X-ray telescope. The proportional counters are $2'' \times 2'' \times 6''$, filled with a P-10 Argon-Methane gas mixture at atmospheric pressure. A multilayer graded (Sn-Cu-Al) shield surrounds the plastic scintillator. A stepping motor rotates the paddle through 36° every 2.6 sec.

midgalactic latitudes was found to be within 5% of that measured at the North galactic pole.

The collimation angle in the scan direction is 14° FWHM, and 37° FWHM normal to it. The gross observed counting rate minus the measured detector background rate was verified to vary linearly with solid angle. Most genuine X-ray events occurred in the counters of the top layer, whereas extraneous background events induced by penetrating radiation in the environment occurred at the same rate in every counter, top or bottom. Figure 2 summarizes the modulating effect of the rotating paddle by exhibiting the observed counting rates per unit solid angle as a function of the solid angle of the telescope's field of view. We call attention to the lower graph of Figure 2 which shows the *net* counting rate per unit solid angle for the top counters to be independent of solid angle and indicates that the events from these counters used for the primary spectral analysis are due entirely to a diffuse flux of X-rays.

In order to evaluate the spectrum of the incident flux we assumed a power law for a range of spectral indices and folded these hypothetical spectra through the measured response functions of our detectors. After a normalization to the total count, the expected spectra were compared with the observations over intervals that are 4 chan-

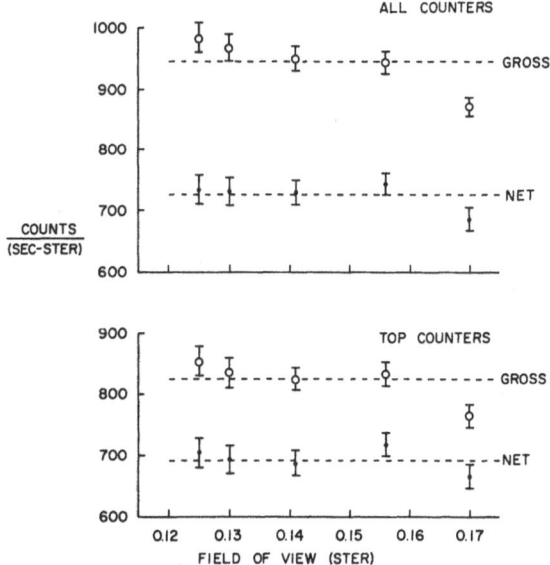

Fig. 2. The observed counting rate per unit solid angle as a function of the solid angle of the telescope. The lower graph is for the composite of the 3 top counters; the net rate is derived from the gross rate by subtracting the detector background measured when the telescope was enclosed within the rocket cylinder. The top graph is for the composite of all 6 counters, top and bottom. Each dotted line is at the level of the average counting rate for the 5 data points of the set considered. The 5 data points of each set correspond to the 5 distinguishable positions of the paddle. Note that the 4 sets of such points exhibited here are *not* statistically independent sets since there is overlap of data among these four categories.

nels wide, which amount to increments of about 1 keV. The single hypothesis that fits best for *all* three top counters is shown in Figure 3. The three histograms result from the same assumed incident spectrum as folded through the slightly different response functions measured for the 3 top counters A1, A2 and A3. The observed data are shown by diamonds. The χ^2 evaluation of these results indicated a good statistical fit for each of these counters taken separately as well as for all taken together.

We find limits 1.25 and 1.42 on the spectral index. These limits correspond to hypotheses that yield values of χ^2 within a 68% probability band centered about the most probable value. From the point of view of statistics, these results essentially rule out an index larger than 1.6.

In order to examine systematic errors in this spectral determination, we compared the normalizations obtained independently from the top and bottom counters. Since the bottom and top counters are internally identical, and since the flux traversing the top counters and entering the bottom counters is spectrally hardened, such a comparison constitutes a severe test for systematic errors. We assume incident spectra of the form CE^{-n}, fold them through the response functions of the counters, and normalize to the observed count in the range 2-20 keV. Figure 4 shows the

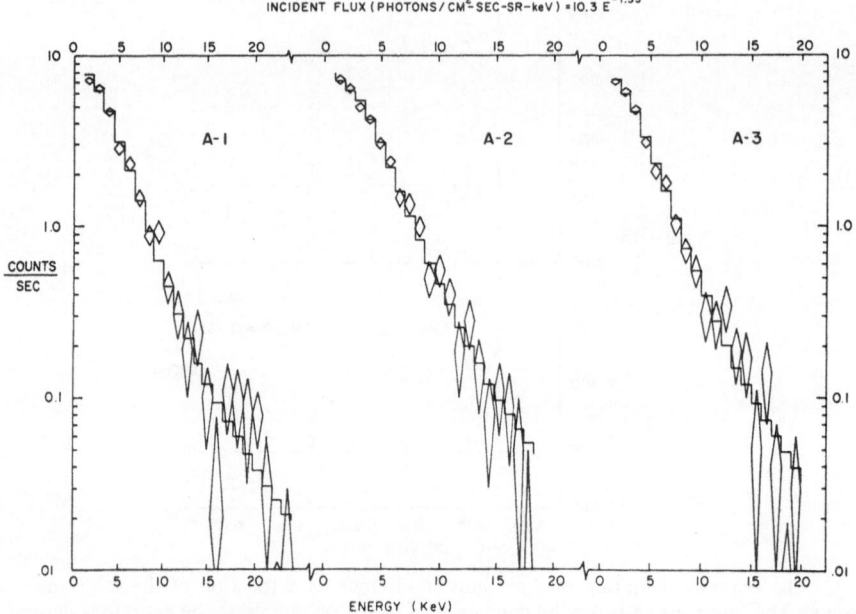

Fig. 3. The observed counts per sec (diamonds) for each of the top counters (A1, A2, A3) as a function of the detected energy (keV). The 3 histograms are the result of folding the indicated incident flux through the measured response function of each detector.

results of this analysis; we plot here the ratio of C obtained from the top counters to C obtained from the bottom counters for various assumed values of the spectral index (n) of the incident diffuse flux. If there were perfect precision in our measurement of the relative efficiencies of the top and bottom counters, then the value of the spectral index n that yields a ratio of unity would definitely be the appropriate one. However, the precision in our relative calibration is 3% and is shown by dotted lines. Considering all limits, this comparison constrains acceptable values of the spectral index to be within 1.4 and 1.5. Therefore, we conclude that the evidence presented here is sufficiently free of systematic as well as statistical errors to give a strong indication that the index is in fact close to 1.4. These results are in good agreement with those obtained above 1.5 keV by Henry *et al.* (1968) for the North galactic pole region and with those obtained for the anti-center of the galaxy by Ducros *et al.* (1969) as well as by us in a previous experiment (Boldt *et al.*, 1969).

In summary, our results give a spectrum for the diffuse background over the band 2–20 keV as

$$10.3\ E^{-n}\ \text{photons}/(\text{cm}^2\text{-sec-sr-keV}),$$

where E is the photon energy in keV, and $n = (1.35\ ^{+0.07}_{-0.10})$.

The value of the normalization constant (C) is determined from an absolute calibration of the detection efficiency of the telescope to a broad beam of X-rays of known intensity at 14.4 keV (Co-57) and 22.3 keV (Cd-109) from distant on-axis radio-active sources. The geometry factor required for a diffuse flux measurement involves knowing this efficiency at all angles. We have calculated this from the measured normal broad beam response on the basis of a detailed, but necessarily approximate, evaluation of our relatively complicated detector geometry. The OSO III satellite observations of hard diffuse X-rays with a NaI scintillation detector by Schwartz *et al.* (1969) extend down into our spectral region and overlap in energy with our observations for their bands 7.7–12.5 keV and 12.5–22 keV. The flux values observed by the scintillator are about 30% lower than would be expected for each of these two bands on the basis of our power law spectrum. This discrepancy is indicative of the magnitude of systematic errors made in determining the absolute detection efficiency for a diffuse X-ray flux at these energies.

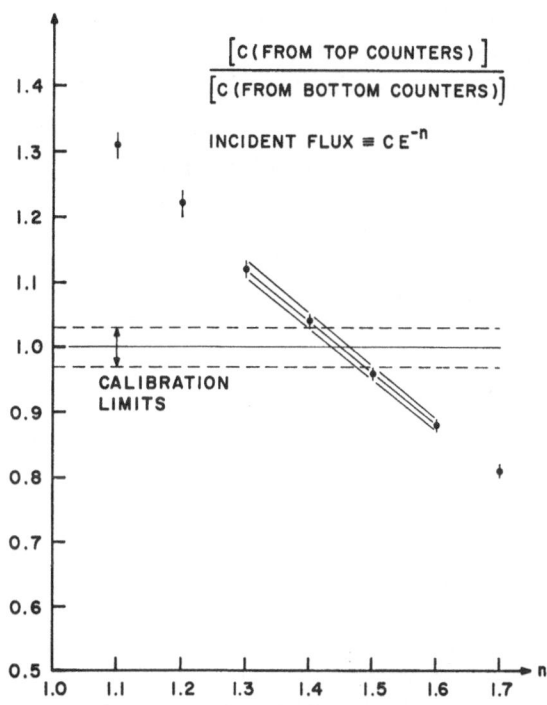

Fig. 4.   The ratio of the normalization constant (*C*) obtained from the top counters to (*C*) obtained from the bottom counters is shown for several assumed values for the spectral index (*n*) of the incident diffuse flux. The dispersions for these points correspond to rms deviations among the values obtained from the counters of each level.

A spectral picture that emerges is one where the band 2–20 keV ($\bar{E} \approx 6$ keV) may be described with $n \approx 1.4$, whereas $n \approx 2.0$–2.4 (Agrawal *et al.*, 1970; Rothenflug *et al.*, 1967) may be required for the band 20–80 keV ($\bar{E} \approx 36$ keV), and $n \approx 2.5$ (Bleeker, 1970) for the band 20–220 keV ($\bar{E} \approx 43$ keV). An intermediate energy band 4–40 keV ($\bar{E} \approx 10$ keV) has been studied from a rocket observation (Seward *et al.*, 1967) and may be described with $n \approx 1.6$, an intermediate value for the spectral index. Taken together, all these results suggest a change in spectral index of about unity within the interval 20–80 keV.

A possible unit change in the spectral index has been attributed by Henry *et al.* (1968) to a hypothetical break at > 1 GeV in the spectrum of metagalactic electrons generating X-rays via inverse Compton scattering with ambient low energy photons. If we are dealing with bremsstrahlung X-rays, then the electron spectrum would have to break at $\gtrsim 20$ keV (Silk and McCray, 1969). However, as pointed out by Boldt and Serlemitsos (1969) and emphasized at this meeting by Hayakawa (1970), the inverse bremsstrahlung of suprathermal protons with the ambient electrons of the intergalactic medium probably provides the only natural set of circumstances for a spectral break in the correct energy region. A break in the proton spectrum at 40 MeV would give a corresponding break in the inverse bremsstrahlung X-rays at 20 keV (Boldt and Serlemitsos, 1969); a 40 MeV proton has a lifetime of about 10^{10} years for collisional energy loss in a medium of 10^{-5} electrons/cm^3, and therefore exhibits a feature that is sufficient to provide the required spectral break for metagalactic protons.

Acknowledgements

It is a pleasure to thank messrs. F. Birsa, R. Bleach, M. Ziegler and the rocket instrumentation and operating crews at the Goddard Space Flight Center and at the White Sands Missile Range for their important contributions to the success of this experiment.

References

Agrawal, P. C., Appa Rao, M. V. K., Biswas, S., Gokhale, G. S., Iyenger, V. S., Kunte, P. K., Manchanda, R. K., and Sreekantan, B. V.: 1970, this volume, p. 289.
Bleeker, J. A. M.: 1970, this volume, p. 297.
Boldt, E. and Serlemitsos, P.: 1969, *Astrophys. J.* **157** (in press).
Boldt, E. A., Desai, U. D., and Holt, S. S.: 1969, *Astrophys. J.* **156**, 427.
Ducros, G., Ducros, R., Rocchia, R., and Tarrius, A.: 1969, (preprint) C.E.N., Saclay, France.
Hayakawa, S.: 1970, this volume, p. 121.
Henry, R. C., Fritz, G., Meekins, J. F., Friedman, H., and Byram, E. T.: 1968, *Astrophys. J.* (*Letters*) **153**, L11.
Holt, S. S., Boldt, E. A., and Serlemitsos, P. J.: 1970, this volume, p. 138.
Oda, M.: 1970, this volume, p. 260.
Rothenflug, R., Rocchia, R., Boclet, D., and Durouchoux, P.: 1967, *Space Research*, Vol. VII, North-Holland Publishing Co., Amsterdam.
Schwartz, D. A., Hudson, H. S., and Peterson, L. E.: 1969, private communication.
Seward, F., Chodil, G., Mark, H., Swift, C., and Toor, A.: 1967, *Astrophys. J.* **150**, 845.
Silk, J. and McCray, R.: 1969, preprint.

γ-RAY ASTRONOMY BALLOON RESULTS

C. E. FICHTEL, D. A. KNIFFEN and H. B. OGELMAN

NASA/Goddard Space Flight Center, Greenbelt, Md., U.S.A.

1. Introduction

The significance of high energy ($\gtrsim 30$ MeV) γ-ray astronomy and its relationship to cosmic rays and many of the high energy processes of the universe has been realized for over a decade. Through the last six to eight years, searches for point sources, mostly from balloon experiments, but also Explorers 11 and OSO-3, have been unsuccessful in clearly establishing the existence of any point source. Recently on OSO-3, Clark *et al.* (1968) have obtained positive evidence for a celestial γ-ray flux which is anisotropic with a higher intensity in the direction of the galactic center region. In this talk, I wish to summarize our balloon results relating to both of these questions and indicate what we hope to accomplish with our new large γ-ray detector over the coming months.

2. 6″ × 6″ Digitized Spark Chamber γ-Ray Telescope

Development of this smaller detector system began in January, 1964, and it has been flown on balloons over the last three years.

A. DETECTOR

The detector system itself is shown in Figure 1 and has been described in detail previously (Ehrmann *et al.*, 1967). The large plastic scintillator anticoincidence dome together with the directional Čerenkov counter is employed to restrict the analysis to downward moving particles and discriminate against charged particles. The spark chamber satisfies the need for a large volume high information content detector to permit selection of the γ-rays and measure the properties of the negatron-positron pair. The central plastic scintillator together with the Čerenkov counter in coincidence and the plastic dome in anticoincidence provides the information to determine whether or not the spark chamber would be triggered. Each of the grid wires in the spark chamber threads a magnetic core, which receives and contains its datum of information when the high voltage is pulsed to the chamber plates. The cores are then read out, and the 'picture' of the e^+, e^- event is telemetered to the ground station.

B. ANALYSIS

A picture of the type shown in Figure 2 can be reconstructed for each event. From an analysis of the coulomb scattering of the electrons in the plates, an estimate of the energy of the electrons and hence the γ-ray can be made. With this information and the direction data resulting from the balloon gondola aspect system, the direction of

L. Gratton (ed.), Non-Solar X- and Gamma-Ray Astronomy, 315–320. All Rights Reserved
Copyright © 1970 by the IAU

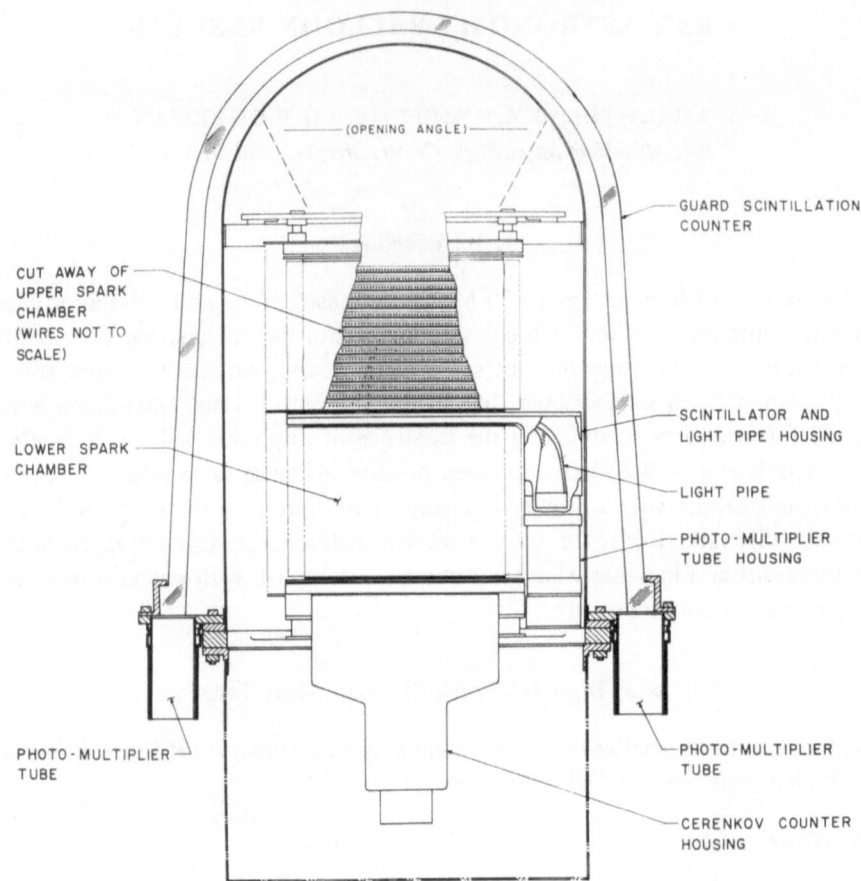

Fig. 1. Schematic diagram of the 6″ × 6″ spark chamber γ-ray telescope flown on balloons over the last three years.

the γ-ray can be determined. This analysis is described in detail in a recent paper (Fichtel *et al.*, 1969) and earlier papers by Pinkau (1966, 1967) and Kniffen (1967).

The uncertainty in the γ-ray energy is determined to be about 30% at 30 MeV and increases to about a factor of 2 at 150 MeV.

C. RESULTS

(a) *Point sources*: We have now obtained upper limits on a large number of possible point sources, and have no positive evidence for a flux. The limits are generally similar to or above straight line extrapolations of X-ray sources; so negative results are not surprising. The way in which these limits are calculated must be considered carefully. This procedure is described in detail in the paper by Fichtel *et al.* (1969) together with the limits obtained, and will not be reported here.

(b) *Galactic Center Region*: In the introduction it was mentioned that, with an experiment on OSO-3, Clark *et al.* (1968) detected γ-rays from the galactic plane with a flux which was strongest in the region near the galactic center. In an earlier paper, results of our group were published which set an upper limit for the flux from a point source assumed to be at the galactic center of about $3 \times 10^{-5}/(\text{cm}^2 \text{ sr sec})$ above 100 MeV. The balloon flight data including the galactic center region have now been reanalyzed in terms of a possible line source of finite width centered about the galactic plane.

In the analysis of the balloon flight data obtained in the flight on Dec. 10, 1966, four different spatial intervals were examined; these were contained within the contour for 50% of maximum detection efficiency and $+15°$ to $-15°$ galactic latitude. For each region, the number of observed γ-rays above 100 MeV and above 150 MeV was determined, the flux calculated, and the atmospheric background determined over a region away from the galactic plane subtracted. The resulting flux was then converted to a line intensity and the results are shown in Table I.

The results in Table I by themselves do not justify the claim of a detected flux, but

TABLE I

γ-ray line intensity with one standard deviation errors along the galactic plane from galactic longitude $-10°$ to $+25°$ in γ's/(cm² sec rad)

Energy interval	Galactic latitude interval			
	$-15°$ to $+15°$	$-10°$ to $+10°$	$-5°$ to $+5°$	$-3°$ to $+3°$
30–100 MeV	$0.9^{+2.2}_{-0.9} \times 10^{-4}$	$1.1^{+1.8}_{-1.1} \times 10^{-4}$	$0^{+1.0}_{-0} \times 10^{-4}$	
> 100 MeV	$2.4^{+2.4}_{-2.4} \times 10^{-4}$	$(2.9 \pm 2.0) \times 10^{-4}$	$(2.2 \pm 1.4) \times 10^{-4}$	$(2.2 \pm 1.1) \times 10^{-}$
> 150 MeV	$0^{+2.1}_{-0} \times 10^{-4}$	$0^{+1.7}_{-0} \times 10^{-4}$	$(1.2 \pm 1.2) \times 10^{-4}$	$(1.7 \pm 1.0) \times 10^{-4}$

the line intensity of $(2.3 \pm 1.2) \times 10^{-4}$ γ's/(cm² sec rad) above 100 MeV for the $+3°$ to $-3°$ interval is certainly consistent with a positive flux, and not in disagreement with a line intensity of $(4.1 \pm 0.7) \times 10^{-4}$ γ's/(cm² sec rad) observed in the OSO-3 γ-ray experiment.

An interesting comparison can be made between the balloon results discussed here and the OSO-3 results for the upcoming γ-rays from the earth's atmosphere. The flux of γ-rays coming directly upwards is essentially the same at the balloon altitude of this experiment (~ 3 g/cm²) as it is outside the atmosphere, basically because the interaction mean free path of both the charged cosmic rays and γ-rays is large compared to 3 g/cm². The flux of upward atmospheric γ-rays above 100 MeV measured in this experiment over a solid angle similar to the OSO-3 experiment was $(3.7 \pm 0.8) \times 10^{-3}$ γ-rays/(cm² sr sec).

The fluxes of upcoming γ-rays ($E > 100$ MeV) measured on Explorer 11 (Kraushaar et al., 1965) and on OSO-3 (Clark et al., 1969) at corresponding geometric latitudes

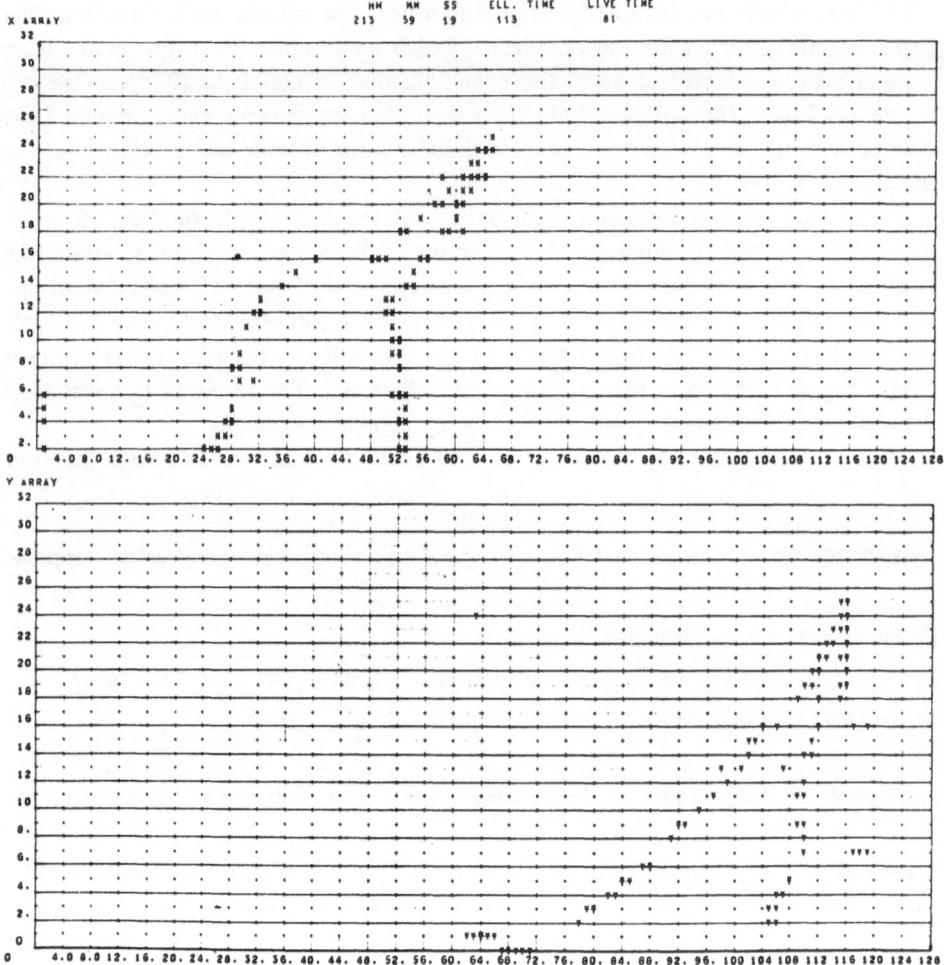

Fig. 2. Computer printout of the magnetic cores showing a γ-ray event in the 6″ × 6″ spark chamber. The top view shows the projection from one side view and the bottom from an orthogonal one. The vertical scale is compressed relative to the horizontal one by a factor of about 3. The wires of the 17th and 1st decks are at an angle of 45° with respect to the rest to remove the ambiguity of which track in the x-view is associated with which in the y-view. These decks are not shown in the x-array, but are added for information as the zero and first deck in the y-array.

are $(1.9 \pm 0.5) \times 10^{-3}$ and $(10.5 \pm 1.0) \times 10^{-3}$ γ-rays/(cm^2 sr sec) respectively. Notice that the Explorer 11 results fall below the measurement reported here, but the OSO-3 result is appreciably higher. The differences seem larger than would be expected on the basis of estimated errors, but are small compared to the differences of more than an order of magnitude between the galactic γ-ray flux observed by OSO-III and the predicted one. The differences may in part be due to the steep energy spectrum of the upward γ-rays which makes calibration more difficult.

3. Future Experiments

About two years ago work began on a γ-ray telescope with 10 times the area of the 6″ chamber and having a total of 25 200 cores. In order not to lose the advantage of the increased area to dead time resulting from chamber readout, the telemetry data rate was increased by a factor of 8 from 1.5 to 12 kilobits/sec. This instrument is

Fig. 3.   A picture of the ½ m by ½ m γ-ray spark chamber balloon telescope with the cover and anticoincidence dome removed.

now built, and Figure 3 shows a picture with the cover and anticoincidence dome removed. Unfortunately a balloon burst prevented us from presenting results here, but we hope to have a second balloon flight around the first of July, 1969.

With this detector system, the flux from the galactic plane will be easily detected and will be particularly striking if it is concentrated in a band with a width or the order of 6° or less about the galactic plane. Not only a measure of the flux, but also an estimate of the energy spectrum will be possible. This detector system is equipped to look for supernovae γ-ray bursts resulting from hydromagnetic shock waves reaching the supernovae surface. The experimental method used is described by Fichtel and Ogelman (1968). It will also be possible to search for pulsed radiation even from pulsars with periods as short as that of the Crab, since accurate timing is recorded to the nearest millisecond.

The SAS-B γ-ray satellite experiment has now been approved and work has commenced. Its area will be about three times as large as that of the 6" × 6" spark chamber telescope, and both the supernovae and the pulsar modes will be included. This experiment will be able to obtain the angular distribution of the γ-radiation measured by OSO-3 to an accuracy of about 1° and measure the flux and energy spectrum uncontaminated by atmospheric background. A search will also be made for point sources of γ-rays at the level detectable by the instrument which is significantly lower than balloon-borne instruments which are hindered by atmospheric background and collection time.

References

Clark, G. W., Garmire, G. P.. and Kraushaar, W. L.: 1968, *Astrophys. J.* **153**, L203.
Clark, G. W., Garmire, G. P., and Kraushaar, W. L.: 1969, private communication.
Ehrmann, C. H., Fichtel, C. E., Kniffen, D. A., and Ross, R. W.: 1967, *Nucl. Instr. Meth.* **56**, 109.
Fichtel, C. E. and Ogelman, H. B.: 1968, NASA TN D-4732.
Fichtel, C. E., Kniffen, D. A., and Ogelman, H. B.: 1969, *Astrophys. J.* **158**, 193.
Kniffen, D. A.: 1967, Ph. D. Thesis, Catholic University.
Kraushaar, W. L., Clark, G. W., Garmire, G., Helmken, H., Higbie, P., and Agogino, M.: 1965, *Astrophys. J.* **141**, 845.
Pinkau, K.: 1966, *Z. Phys.* **XCCVI**, 163.
Pinkau, K.: 1967, *Nucl. Instr. Meth.* **57**, 173.

MEASUREMENT OF THE COSMIC X-RAY BACKGROUND
IN THE 25–200 keV RANGE

D. BRINI, F. FULIGNI, and E. HORSTMAN-MORETTI

Space Research Group of the Lab. TESRE of the C.N.R.,
Institute of Physics of Bologna University, Italy

An apparatus to study the cosmic X-ray background was launched aboard an ESRO Skylark rocket on May 22, 1967 from the Perdas de Fogu range in Sardinia.

The basic detector consisted of a NaI(Tl) phoswich, 5 cm in diameter and 1.27 cm thick, surrounded by both passive (lead) and active (plastic scintillator) cylindrical collimators.

Four of these were placed at 45° from the rocket axis and arranged symmetrically around it, one of them being completely screened to give a direct evaluation of the instrumental background, while the large aperture (about 30° FWHM) of the other three allowed a wide scan of the sky.

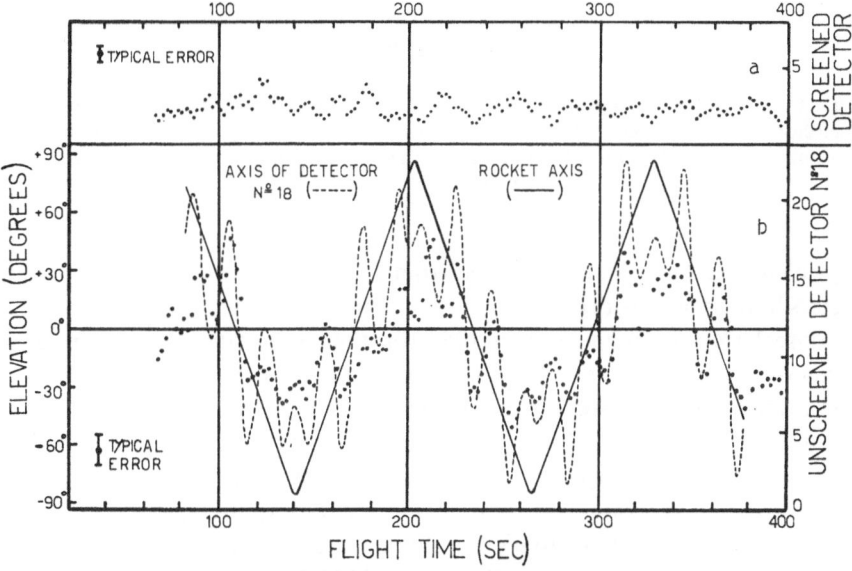

Fig. 1. (a) Integral (25–100 keV) counting rate of the screened detector as a function of time after launch. – (b) Integral (25–200 keV) counting rate of one of the unscreened detectors (No. 18) as a function of time after launch. Counting rates are obtained by averaging over periods of 8 sec. In contrast of what happens to the screened detector, which keeps constant, we note the correlation existing between the counting rate of the detector (No. 18), and the elevation of the detector axis (dotted line) and the rocket axis (full line).

The events from each detector were analyzed in four energy channels (nominally 25–40, 40–70, 70–110, and 110–210 keV) for which the product of the efficiency and the geometric factor gave, respectively, 6.4, 7.2, 7.3 and 6.5 cm^2 sr.

L. Gratton (ed.), Non-Solar X- and Gamma-Ray Astronomy, 321–324. All Rights Reserved
Copyright © 1970 by the IAU

The flight, started at 0107 UT and reaching 192 km, was a successful one but for malfunctions of the longitudinal magnetometer and lunar cells, which caused doubts and delay in the attitude determination. We believe, however, that our data behaviour itself, as shown in Figure 1 for one of the detectors together with the corresponding elevation vs. time, provides a satisfactory check of the final aspect solution.

While the screened detector counting rate stays constant after cone ejection, the three unscreened detectors show a consistent variation with elevation in the lower

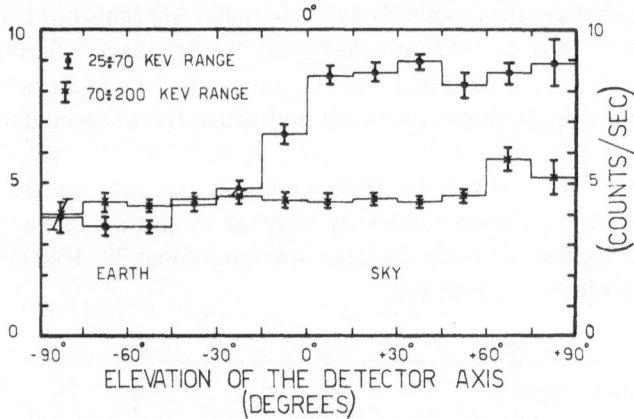

Fig. 2. Counting rates averaged over the three unscreened detectors as a function of the elevation of the detector axis. Please note the sky-earth asymmetry for the counting rate in the 20–70 keV energy range. This shows that the spectrum of the cosmic X-Ray background is steeper than that of the albedo X-rays.

energy channels. This can also be seen from Figure 2 where the data from all three detectors have been considered.

Because the X-ray horizon is about 105° from the zenith at our altitudes and energies, the data with positive elevation have been considered to be of extra-atmospheric origin and were analyzed with respect to celestial coordinates.

Because of the motion of the despun rocket, most of the sky above the horizon was seen, although the coverage was not uniform. Many X-ray sources were scanned, namely those in the Scorpio-Sagittarius region and the Cygnus region.

Our counting rate for both zones showed increases well above the general average. We also saw an increase from the direction of M87, but it was not statistically significant. Further analysis is needed for the sources, although the wide aperture makes it a difficult and ambiguous task. We could not, however, include the counting rates from these regions in our collection of background data, even if this reduces the explored area and statistics (especially at low galactic latitudes).

So, we did not consider data for which the detector's axis direction was less than 18° away from one of the ten major sources, taking as a parameter the low energy fluxes quoted by NRL (1966). We also excluded the M87 region even though it would have had a small effect on the background average.

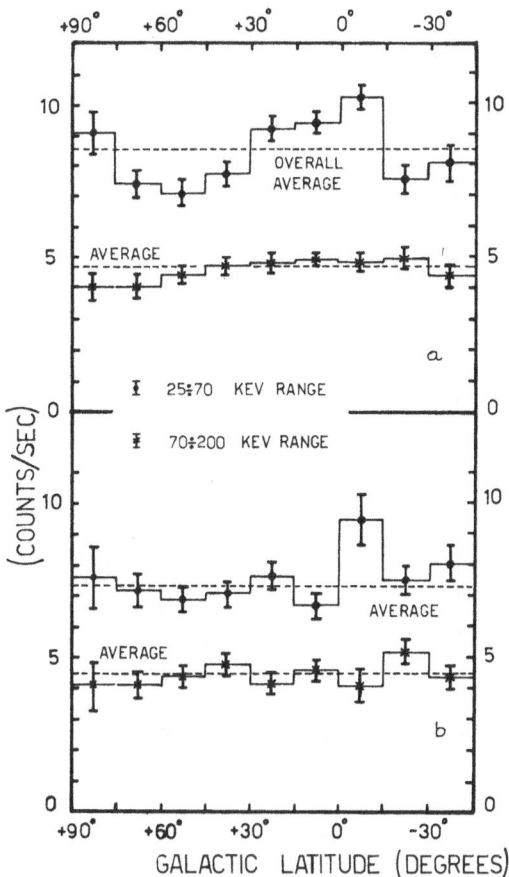

Fig. 3. Counting rates of the three unscreened detectors as a function of the galactic latitude. (a) All data referring to elevation $> 0°$. – (b) The same but excluding the regions of X-ray sources.

Figure 3 shows the results of the analysis vs. galactic latitude, both including and excluding data from the 'hot' regions. For the latter, no systematic trend appears in the background behaviour. We compared the total counting rates (corrected for instrumental background) of the galactic equator (from $-15°$ to $+15°$) and the pole (above $+60°$) to evaluate the anisotropy as given by $A = (E-P)/(E+P)$.

We obtained $A = 0.03 \pm 0.04$ thus setting an upper limit of 11% at the 2σ level for the anisotropy in our energy range.

The counting rates outside the source regions, corrected for efficiency, channel width, instrumental background and geometric factor, give intensities of (1.29 ± 0.04) cm^{-2} sr^{-1} and (0.76 ± 0.04) cm^{-2} sr^{-1} respectively for sky and earth fluxes, while the corresponding energy spectra are plotted in Figure 4.

Only statistical errors are shown, to which an uncertainty of about 10% over the energy thresholds should be added.

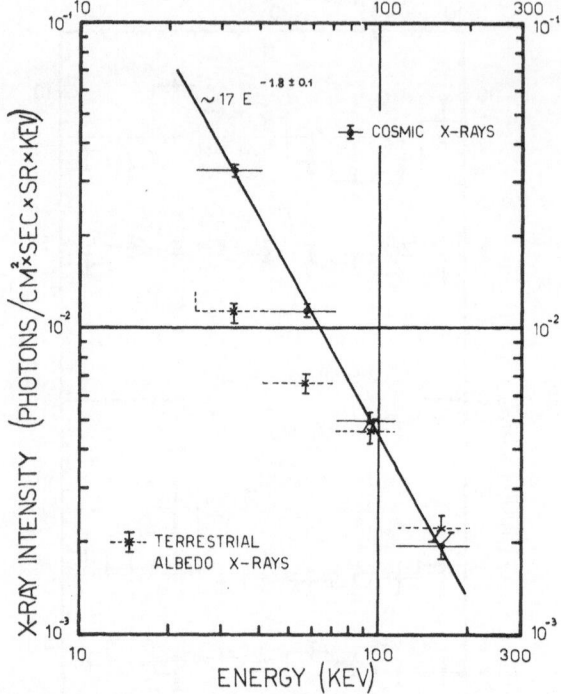

Fig. 4.   Energy spectrum of the diffuse cosmic X-ray background as measured in this experiment. Also shown is the spectrum of the terrestrial X-rays albedo (dotted line).

As can be seen, the good agreement of the data with a power law $E^{-1.8\pm0.1}$ does not call for any break or substantial bend in the spectrum.

Since the subtraction of the instrumental background was negligible for all but the higher energy channel, we give more importance, however, to the results from the first three channels. If there is indeed an abrupt change in the slope, we believe it must be outside the range 25–110 keV.

The albedo spectrum appears different from what has been measured inside the atmosphere.

We think that it shows a contribution of scattered primary X-rays, which should be taken into account in the extrapolations necessary in balloon measurements.

This could be a reason, together without direct measurement of the instrumental background, for the disagreement between our results and other measurements made in the same energy region.

Looking at the data for all energies studied, one feels, however, that an overall agreement could be found with a power law spectrum similar to our result.

RESULTS OF TWO BALLOON FLIGHTS FOR THE DETECTION OF HIGH ENERGY γ-RAYS

M. NIEL, G. VEDRENNE,

Centre d'Etude Spatiale des Rayonnements

and

R. BOUIGUE

Observatoire de Toulouse

After many years of fruitless research on primary γ-rays, the results obtained by Clark *et al.* [1] with the OSO-3 satellite, proved for the first time the existence of a primary γ-ray flux. The study of the distribution of this radiation showed a strong anisotropy in the direction of the galactic disk and, more precisely, in the direction of the galactic center. Now the production of γ-rays in the spatial medium is related to high energy processes and to the presence of relativistic electrons. The high energy processes bring about an emission of γ-rays essentially by decay of the π° mesons created for instance, by the interaction of cosmic radiation with interstellar matter or by matter-antimatter annihilation. The relativistic electrons can lead to an emission of γ-rays by various processes: in particular, by bremsstrahlung of the electrons in the interstellar matter or by the Inverse Compton effect with the photons of the stellar light or of the infrared radiation background.

These different production processes which have been analyzed in detail [2-3-4-5] enable the estimation of the γ-rays fluxes at the level of our galaxy and at that of the metagalactic medium. However, because the values obtained by OSO-3 are of an order of magnitude greater than the theoretical estimations, it becomes necessary to explain this disagreement:

either by greatly modifying certain parameters that were used, such as the density of matter. An increase of an order of magnitude in relation to the estimations given by radio observations of the 21 cm wavelength ray is thus necessary [6].

or, for example, by assuming the contribution of discrete sources [7] without modification of the basic parameters (density, cosmic-ray flux ...)

In the light of these results, it therefore seems very desirable to continue the measure of the energetic γ-rays because the astrophysic implications linked to their proof are considerable, as much from the cosmological point of view as from that of the study and origin of the primary radiation. This paper gives the results of two flights carried out with a balloon using a detection system for the γ-rays of energies greater than 100 MeV.

1. Essential Characteristics of the Detection System

The detector which has already been described [8] is essentially made up of a spark chamber which is set off by a fast coincidence between a Čerenkov detector and a

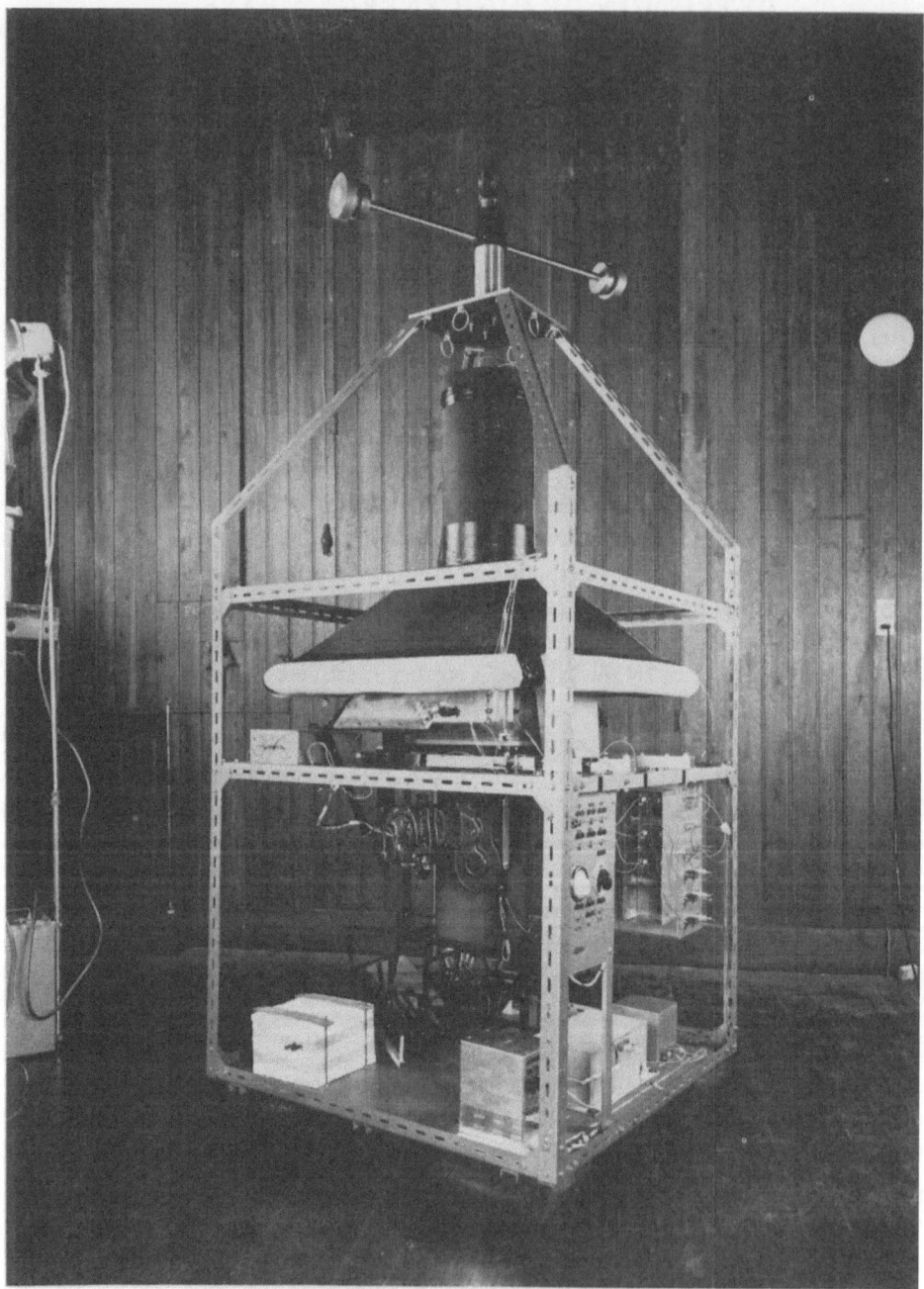

Fig. 1.   Detection assembly for the search of γ-rays with energy above 100 MeV.

plastic scintillation counter. This system is completed by a plastic scintillator which surrounds the spark chamber as much as possible and which enables the elimination of most of the spurious events, and, in particular, those produced by the interaction of particles with the lateral walls of the detector (Figure 1).

In fact, for the two flights whose results are discussed in this paper, the experimental apparatus, although based on the same principle, differs essentially from that in Figure 1 because of the position of the different scintillators. The schematic block diagram of the system used is given in Figure 2. The plastic scintillator, which is in

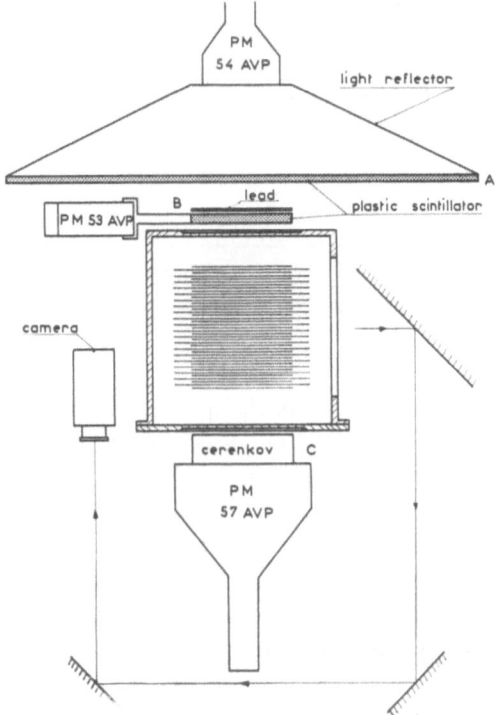

Fig. 2. Synoptic scheme of the detector used for the flights of May and July 1968.

coincidence with the Čerenkov detector, is placed above the chamber (17 cm × 17 cm) and the conversion of the γ-rays into (e⁺, e⁻) pairs takes place in a lead plate. The efficiency of the detector varies from 20% to 25% between 100 MeV and 1000 MeV. The half angle of aperture of the detection system is 23° and the geometrical factor is 33 cm² sr.

2. Analysis of the Results

The two flights in May and July 1968, were carried out from AIRE SUR L'ADOUR (43°70 N, 0°30 E) and GAP (44°5 N, 6°E). The launching times were chosen to explore the same zone of the sky crossing the galactic plane. Presented as an

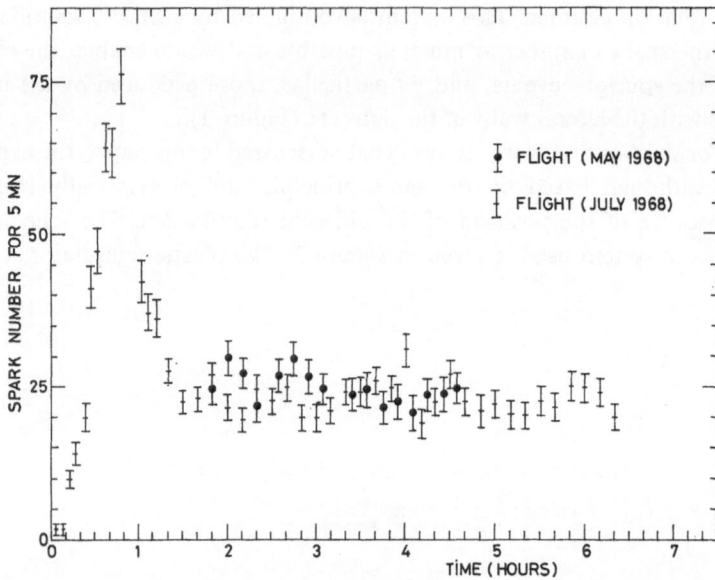

Fig. 3.   Variation of the sparks counting rate with time for the 2 flights.

example, Figure 3 shows the comparison of the sparks counting rate during these two flights for a residual pressure of 8 mb. The good agreement of the measurements makes it possible to put together the count rates obtained and thus to improve the statistical accuracy of the measured fluxes. In order to interpret these results, however, it is necessary to define the different types of tracks which are to be analyzed and then to reconstitute the arrival direction of the γ-rays materialized by one or two tracks in the spark chamber. After rejecting the unexploitable negatives, two classes of events were retained, single and double trajectories.

In the first category, the tracks are either straight or curved. The straight tracks can correspond to high energy γ-rays (e^+ and e^- tracks confused), to protons that the anticoincidence system has not eliminated, or to particles associated with the interaction of neutral particles in the chamber or in the walls of the chamber.

The curved trajectories can correspond, for example, to a single prong of an (e^+, e^-) fork if one of the two electrons does not cross the spark chamber or to a Compton interaction in the lead or in the upper part of the chamber.

In the second category, the trajectories are double; that is to say that the e^+ and e^- electron tracks are brought to light, and the presence of an associated γ-ray is deduced without ambiguity.

This classification leads to the analysis of two parameters:

(a) the arrival direction of the γ-ray which is either that of the single trajectory or that of the bisector of the tracks corresponding to the (e^+, e^-) pair;

(b) the apparent aperture angle of the fork which allows the measure of the energy of the γ-ray.

The analysis of the photographs leads to the classification of the different types of events. A program then gives their characteristics: azimuthal angle, zenithal angle, aperture angle of the forks, restitution of the γ-rays arrival directions in the (α, δ) coordinates and then in the (l_{II}, b_{II}) coordinates.

A. ATMOSPHERIC BACKGROUND

From this analysis and from the variation of the sparks counting rate as a function of time, some verifications that the apparatus was working correctly were undertaken.

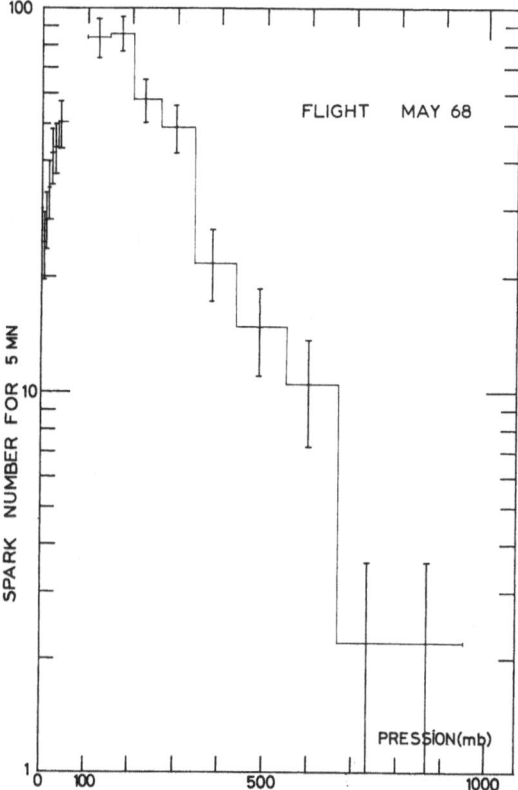

Fig. 4.   Variation of γ-rays flux with residual pressure.

For example, the variation of the sparks counting rate as a function of the pressure between 8 and 1000 mb is in good agreement with the previous experimental results [9], [10], and the curve's maximum is around 180 mb (Figure 4). The detailed study of the part of the curve ranging from 8 to 25 mb cannot, by extrapolation to 0 mb, lead to a precise determination of an isotropic primary flux. Moreover, a flux value included between 5.10^{-4} and 10^{-3} γ/cm^2 s sr is consistent with the experimental data.

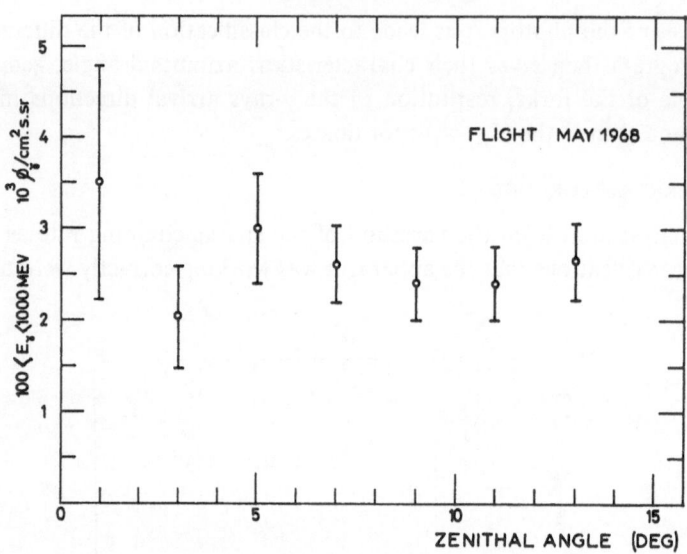

Fig. 5.   Variation of γ-rays flux (100 MeV ≤ E_γ ≤ 1000 MeV) with zenith angle.

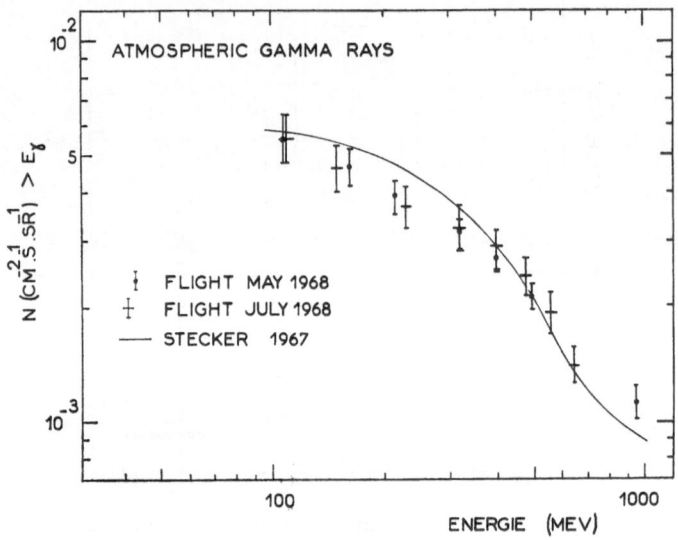

Fig. 6.   Integral γ-rays spectrum observed during the two flights at 8 g/cm².

In other respects, for the different types of analyzed events and being given a constant value for the solid angle, the variation of the γ-ray flux as a function of the zenithal angle ($0 < \theta < 15°$) (Figure 5) is isotropic. The anisotropy of this flux of essentially secondary origin comes to light only for greater values of θ [11], and our results are in good agreement in the common zone ($\theta < 15°$).

Finally, the energy spectrum of the observed γ-rays can be deduced from the measurement of the apparent aperture angle of the forks. For this we used the calculations developed by Olsen [12], which take into account the modifications of the aperture angle of the forks by the multiple diffusion process. Figure 6 shows the spectrum obtained; it is in good agreement with the previously published [13], [14], [15] theoretical and experimental results.

B. RESEARCH OF A PRIMARY FLUX

The sky being divided into elements of constant solid angles, and considering only the forks of angles of less than 6° the distribution of the γ-rays detected in 6° × 6° squares, can be deduced either in the (α, δ) or in the (l_{II}, b_{II}) coordinates. For each of the sky elements, the exposure time and the secondary origin background have been calculated. The comparison of this background and of the number of γ-rays actually observed is presented in Figure 7; it does not show any significant flux associated with a definite region. Moreover, the presence of an anisotropy when crossing the galactic plane does not appear in this display. Nevertheless, in order to bring to light an eventual anisotropy, and to make a comparison with the results

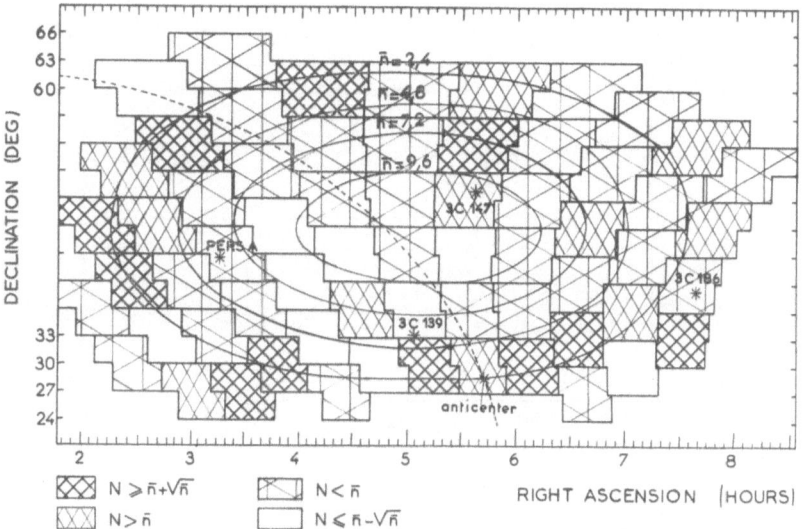

Fig. 7.   Angular distribution over the celestial sphere of γ-rays trajectories. Contour lines represent $\bar{n}$ the average number of events in 6° × 6° square bins.

already obtained in this area [1], we plotted the distribution of γ-rays in the (l_{II}, b_{II}) galactic coordinates system. For that the number of sparks observed was counted as a function of b_{II} for a wide range of galactic longitude ($130° < l_{II} < 183°$) and for the same area-time factor (Figure 8).

No conclusion can be drawn in respect to a flux increase when crossing the galactic

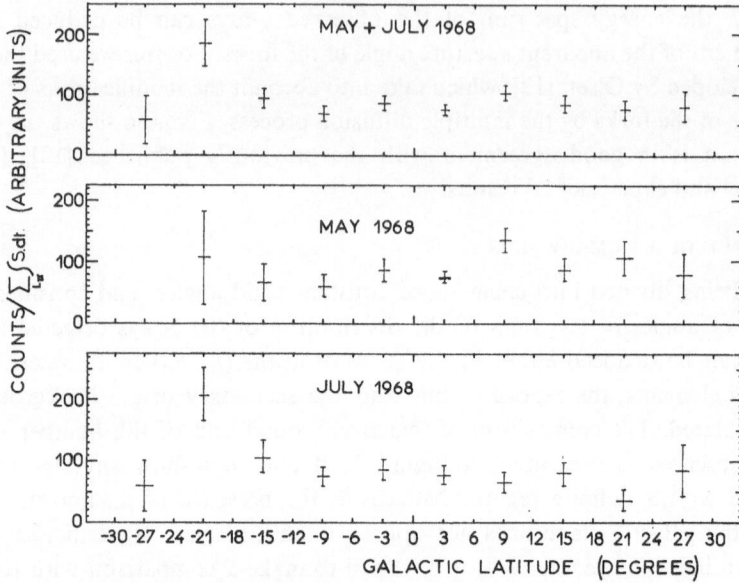

Fig. 8. γ-Ray intensity plotted against galactic latitude b_{II}.

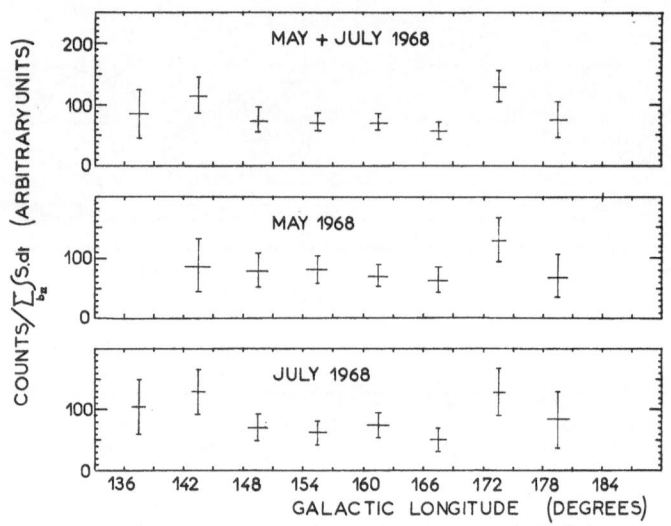

Fig. 9.   Variation with galactic longitude l_{II} of the rate of celestial events for $-9 \leqslant b_{II} \leqslant +9°$.

plane. Nevertheless an anomalously high number of events can be noticed in the vicinity of $b_{II} = -20°$ for the 5 July flight but it is not really statistically significant.

The analogous curve as a function of the galactic longitude does not lead to a sure conclusion (Figure 9). We can notice that the anticenter remained during the flight on the edge of the aperture cone of the detector, and the corresponding exposure time was short. Beyond the analysis of the crossing of the galactic plane, the two flights carried out make it possible to give several limit fluxes associated with the punctual sources in transit in the apparatus field during the flight. Unfortunately, the analysis zone which was chosen for the study of the crossing of the galactic plane is not very rich in eventual γ sources; thus, only the anticenter and several strong radiosources could be analyzed. The cutting up of the sky dome, as shown in Figure 7 is used to that effect. The average background $\bar{n}$ in one of these elements is thus:

$$\bar{n} = \Phi_{is} \cdot \varepsilon \cdot dw \int Sdt$$

Φ_{is} = isotropic flux averaged for the whole flight; ε = detection efficiency; dw = solid angle element; and $\int Sdt$ = integral of the surface presented to the source during the flight. This integral is calculated for a direction corresponding to the center of a square bin.

For the May, 1968, flight, a pressure correction had to be added as the balloon descended steadily during the flight. The curves plotted in Figure 7 correspond to a constant background flux.

The limit fluxes associated with the different elements analyzed are calculated from:

$$\left(m - \frac{\bar{n}}{f}\right) f = \Phi_{Lim} \cdot \varepsilon \cdot \int Sdt$$

m is the average value such that there exists a 5% measurement probability of obtaining a count rate less than the count rate N actually observed (N = total number of forks observed in the bin).

f = total number of forks/number of forks whose aperture angle is less than 6°.

The limit fluxes obtained are presented in Table I. This method as noticed by Hearn [16] is not entirely satisfactory; thus, he used the integrated likelihood method as suggested by Greisen. The limit fluxes deduced by this method are presented in Table II and are not very different from the preceding ones, for the actual number of counts observed per square is relatively important.

	TABLE I		TABLE II
Object	$\int Sdt$ (cm²s)	Limit fluxes (cm²s)⁻¹	Limit fluxes [16] (cm²s)⁻¹
Pers. A	3.84×10^5	1.4×10^{-5}	2.3×10^{-5}
3C 147	6.16×10^5	3.10^{-5}	3.1×10^{-5}
$\alpha = 83.3°$ $\delta = 39.1°$	5.15×10^5	1.2×10^{-5}	1.9×10^{-5}
3C 139	2.99×10^5	2.9×10^{-5}	3.4×10^{-5}
3C 186	0.92×10^5	5.4×10^{-5}	7.10^{-5}
Anticenter	1.79×10^5	5.6×10^{-5}	5.9×10^{-5}

Finally, the inaccuracy on the direction of the incident γ-ray in the case of the forks, taken into account, we have worked out a classification program of the events so that the γ-rays can be associated with all squares situated inside the aperture angle of the fork it creates. In this way the same γ-ray is counted in several elements thus assuming an equal emission probability of the γ-rays in the totality of the aperture angle of the pair. Under these conditions, the distributions obtained in the zones where the background is constant are in all cases comparable to Poisson distributions and thus no anomaly in the fluxes can be considered for the whole of these zones.

In conclusion the analysis of these flights, because of too important a residual atmosphere, makes it impossible to find an anisotropy in crossing the galactic plane. For the same reason, no source could be brought to light and only upper limits of fluxes were associated with several objects. The use of a new detection system (Figure 1) with a greater sensitivity and with a better identification of the γ-rays ought to lead to positive results for the presence of a primary origin γ-rays flux.

References

[1] Clark, G. W., Garmire, G. P., and Kraushaar, W. L.: 1968, *Astrophys. J. Letters* **153**, L 203.
[2] Ginzburg, V. L. and Syrovatsky, S. I.: 1964, *Origin of Cosmic Rays.*
[3] Garmire, G. and Kraushaar, W. L.: 1965, *Space Sci. Rev.* **4**, 123.
[4] Gould R. J. and Burbidge, G. R.: 1966, *Handbuch der Physik* **46**, No. 2.
[5] Fazio, G. G.: 1967, *Ann. Rev. Astron. Astrophys.* **5**, 481.
[6] Stecker, F. W.: 1969, *Nature* **222**, 865.
[7] Ögelman, H.: 1969, *Nature* **221**, 753.
[8] Niel, M., Cassignol, M., Vedrenne, G., and Bouigue, R.: 1969, *Nucl. Instr. Methods.* **69**, 309.
[9] Cline, T. L.: 1961, *Phys. Rev. Letters.* **7**, 109.
[10] Duthie, J. G., Cobb, R., and Stewart, J.: 1966, *Phys. Rev. Letters* **17**, 263.
[11] Fichtel, C. E., Kniffen, D. A., Ögelman, H. B.: COSPAR 69 (Rome), to be published.
[12] Olsen, H.: 1963, *Phys. Rev.* **131**, 406.
[13] Stecker, F. W.: 1968, Nasa TMX 63 181, April.
[14] Fazio, G. G., Helmken, H. F., Cavrak, S. J., and Hearn, D. R.: 1968, *Can. J. Phys.* **46**, S427.
[15] Fichtel, C. E., Kniffen, D. A., and Ögelman, H. B.: 1968, NASA Report X 611, 69 58, February.
[16] Hearn, D.: 1969, *Nucl. Instr. Methods* **70**, 200.

THE COSMIC γ-RAY SPECTRUM NEAR 1 MeV
OBSERVED BY THE ERS-18 SATELLITE

JAMES I. VETTE

Goddard Space Flight Center, Greenbelt, Md., U.S.A.

and

DUANE GRUBER, JAMES L. MATTESON, and LAURENCE E. PETERSON

University of California, San Diego, La Jolla, Calif., U.S.A.

In this paper, we wish to report new data obtained on the total, or omnidirectional, cosmic γ-ray flux in the 250 keV to 6 MeV range. Flux in this energy range up to 2 MeV was first detected on the Ranger 3 (Metzger *et al.*, 1964) and has been confirmed up to 0.5 MeV and extended to lower energies by a series of rocket (Gorenstein *et al.*, 1969; Seward *et al.*, 1967) and balloon measurements (Bleeker *et al.*, 1966; Rocchia *et al.*, 1967). Observation of the spectrum at higher energies has become of considerable importance since the recent successful detection of galactic 100 MeV γ-rays by the OSO-3 (Clark *et al.*, 1968).

The Environmental Research Satellite-18 was launched April 28, 1967, along with the Vela 4 satellites into a highly elliptical orbit of 117 500 km apogee and 15 000 km perigee. This 7.8 kg satellite, in addition to carrying a number of small radiation detectors designed to measure magnetospheric particles and cosmic rays, had a large detector to measure the cosmic γ-ray spectrum in the energy range 0.25–6 MeV. This detection system consisted of a 7.65 cm long × 6.35 cm diameter NaI (Tl) scintillation counter with a 1 cm thick plastic anti-coincidence shield for charged particle rejection. The shield counter covered all but one circular face of the NaI crystal. A drawing of the detector arrangement is given in Figure 1. Six channels of pulse height analysis permitted spectral measurement into the 5 energy bands given in Table I, as well as a lower energy integral channel, and an integral channel > 6 MeV. The anti-coincidence was switched between on and off by the spacecraft commutator about every 500 sec in order to access the counting rates due to cosmic ray energy losses. The difference in counting rate for these two modes can serve as a calibration using the cosmic ray beam.

A more complete description of the satellite and its experiments has been given elsewhere (Peterson *et al.*, 1968). Although the information bandwidth of the ERS-18 was only 6 Hz, and only a small fraction of this was devoted to the γ-ray experiment, the measurement is soon limited by systematic effects rather than counting statistics. The satellite spends a large fraction of its 48-hour orbit in regions where magnetospheric particles are not detectable by the charged particle detectors with threshold energies of 40 keV for electrons and 380 keV for protons.

Data reported here were obtained primarily during May and June, 1967, when the local time of satellite apogee was generally directed toward the dusk meridian.

L. Gratton (ed.), Non-Solar X- and Gamma-Ray Astronomy, 335–341. All Rights Reserved
Copyright © 1970 by the IAU

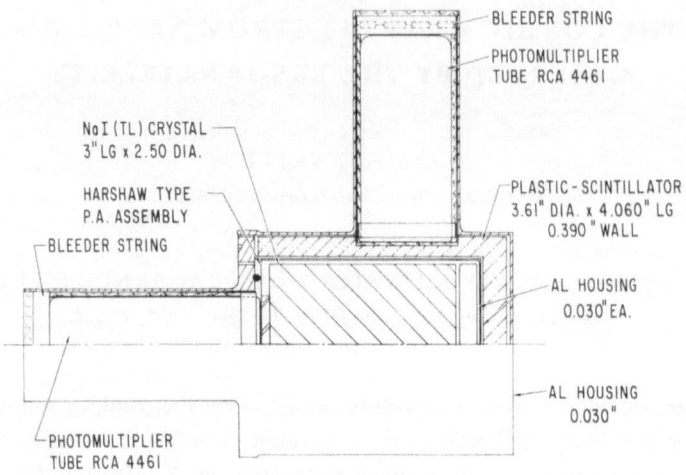

Fig. 1. The isotropic γ-ray detector which was included in the ERS-18 payload and which detects
energy losses in the 0.25 to 6.0 MeV range.

TABLE I

ERS-18 counting rates

Channel	Energy loss range range (MeV)	Rate, anti-coincidence on (counts/cm²-sec-MeV)	Rate, anti-coincidence off (counts/cm²-sec)
1	> 0.025	–	31.7
2	0.25–0.6	0.940	0.970/MeV
3	0.6–1	0.286	0.328/MeV
4	1–2	0.144	0.173/MeV
5	2–3.7	0.0526	0.079/MeV
6	3.7–6	0.0355	0.0637/MeV
7	> 6	–	2.62

Near apogee at 18 earth radii the earth subtends a solid angle of about 10^{-2} steradian, and earth albedo γ-rays produce a completely negligible flux. Furthermore, because of the high perigee (2.4 earth radii) the satellite does not appreciably penetrate the high energy proton zones which can induce radioactivity in the satellite and the NaI crystal. This effect has plagued previous experiments (Peterson, 1965; Peterson *et al.*, 1968) but does not contribute an observable background rate to this experiment. Furthermore, the data presented here were selected from intervals during which there was no evidence of detectable particle fluxes in interplanetary space above those attributable to galactic cosmic rays. Data obtained during the solar proton events of late May and early June have been excluded from the analysis.

The rates obtained with the anti-coincidence on and off are shown in Table I and Figure 2 after converting to energy loss spectra by dividing by the channel width and by the isotropic geometry factor, 54 cm². Clearly, with the anti-coincidence off cosmic ray energy losses dominate the spectrum at the higher energies. The difference between the two rates must be due to charged particles. The expected spectrum due to edge effects of minimum ionizing cosmic rays isotropic on the detector has been computed using a Monte Carlo method and interplanetary cosmic-ray intensity of 3.07/cm²-sec, as measured on the Vela satellites at the time (J. R. Asbridge, private

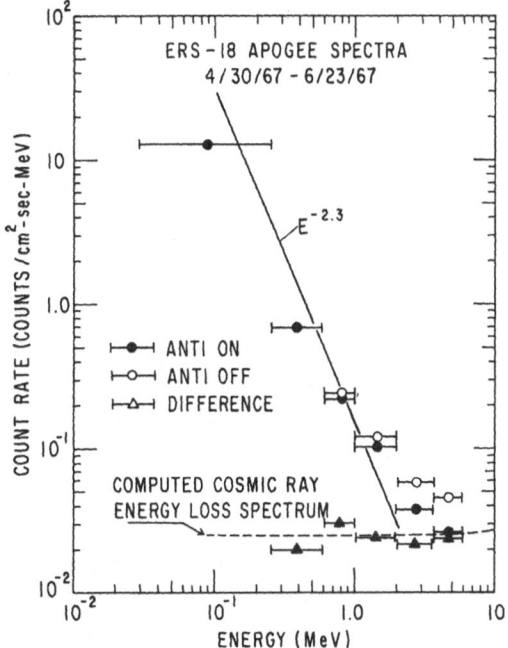

Fig. 2.   The spectrum measured in interplanetary space with the anti-coincidence shield turned on and off. The difference is compared with that predicted due to cosmic-ray energy losses in the crystal edges. The anti-on rates are interpreted as due to cosmic γ-rays.

communication). This computed spectrum, shown as a dotted line in Figure 2, is flat at energy losses below that corresponding to ionization losses by particles traversing the length or diameter of the crystal. That the theoretical spectrum due to particle losses agrees so closely with the observed one is taken as substantial evidence that the anti-coincidence performed properly and that the calibration had not appreciably changed in orbit. Furthermore, the integral rate (>6 MeV) measured in the central detector agrees with the independently measured cosmic ray flux by Asbridge to within 17%, giving one considerable confidence in the entire system operation.

The energy loss spectrum with the anti-coincidence on must therefore be due to γ-rays incident on the detector. No known process has sufficient cross-section for cosmic rays to produce an appreciable γ-ray flux in passing through the 8 gm/cm² of average satellite material adjacent to the γ-ray counter. Approximately $\frac{2}{3}$ of the 4π steradian solid angle is shielded only by the guard counter and its container plus the solar cell panels. Only about 5% of the cosmic rays will interact in traversing the satellite material to begin the electromagnetic and nucleonic cascades which develop to their maximum in about 100 gm/cm^2. The γ-rays in our region of observation are produced by secondary interactions in the cascade process. A detailed Monte Carlo calculation of the cascade development in a small amount of matter has not been carried out. However, we can obtain some experimental evidence for the lack of coupling due to cosmic rays by searching for correlations in the γ-ray channels associated with the modulation of galactic cosmic rays by solar activity associated with the events of May–June 1967. During this period the sea level cosmic-ray indices varied some 10%, and the intensity at the satellite by 30%. The correlation coefficients between the integral (>6 MeV) channel and the 1–2, 2–3.7, and 3.7–6.0 MeV channels were typically 0.3 with the anti-coincidence on and 0.5 with the anti-coincidence off. We have also summed the total number of counts in each γ-ray channel over two specific time periods; one when the >6 MeV channel was counting less than 145 c/sec and one when it was counting this value or greater. These sums were compared with the integrated counts of the >6 MeV channel and the neutron monitor counts over the same time period. From these results we can infer that the contribution from γ-rays produced in the satellite is less than $10\pm10\%$ for any channel. The comparison of our cosmic ray rates with the Vela results and the fluctuations of the γ-rates of 5% over the time period reported here lead us to assign a systematic error to these results of not more than 20%.

The ERS-18 data are shown in Figure 3, compared with other positive measurements of the diffuse component, over the entire X- and γ-ray range. In addition to the balloon, satellite and OSO-3 γ-ray data referenced previously, the 250 eV flux recently reported by Bowyer *et al.* (1968) and by Henry *et al.* (1968) is indicated. Also shown are measurements obtained on the OSO-3 over the 7.7–200 keV range (Hudson *et al.*, 1969). The fluxes near 1 MeV are not corrected for efficiency or photo fraction, since this correction is dependent on the spectral shape to the highest energies. At energies below about 100 keV, the diffuse component is known to be isotropic to within 10%; at 100 MeV the galactic disk is the strongest emitter, and only an upper limit to the diffuse (non-galactic) component at this energy has been presented (Clark *et al.*, 1968).

The data in the 30 keV to 1 MeV range obey a power law spectrum with a number index of -2.2 (Gould, 1967). This component has been interpreted as due to intergalactic electrons scattered on the 3K radiation (Felten and Morrison, 1966). If this is indeed the case, the power law should continue at least as steep to the highest energies. Two other possible production mechanisms, bremsstrahlung (Silk and McCray 1969; Stecker and Silk, 1969), and proton-antiproton annihilation (Stecker, 1969a) would predict an even steeper power law above 1 MeV. This is clearly inconsistent with the ERS-18 data above about 1 MeV. We therefore conclude that a second component of cosmic γ-rays contributes an appreciable flux above about 1 MeV.

The ERS-18 data contribute little to the detailed spectra of this component and nothing about its directionality. A possible origin is nuclear γ-rays, of either galactic or extragalactic origin. Although quantitatively, with the present understanding of

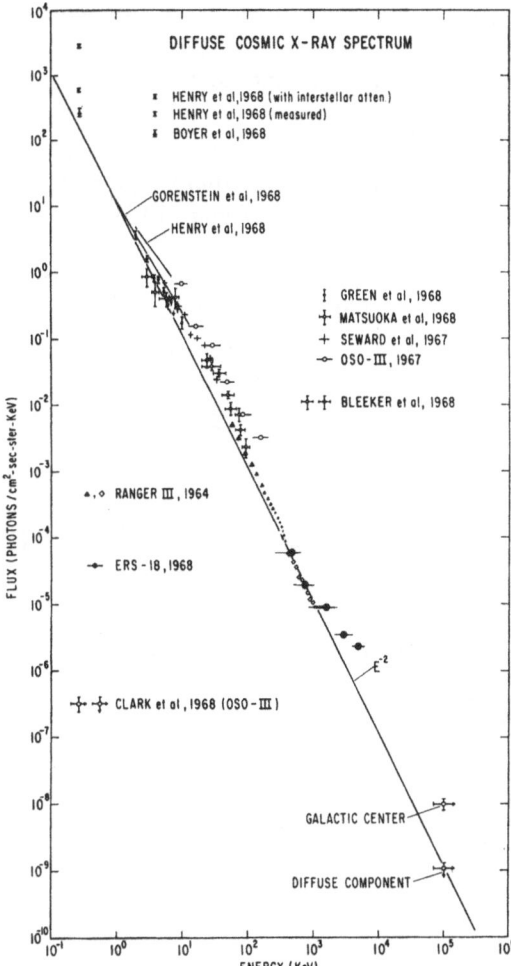

Fig. 3.   Measurements of the total cosmic background spectra over the 250 eV to 100 MeV range. The fluxes are known to be generally isotropic at the lower energies and originate in the galactic plane at 100 MeV. The ERS-18 data on the total, on 4π flux, lie considerably above an extension of the lower energy power law spectrum.

galactic cosmic ray processes, this source seems unlikely (Ginzburg and Syrovatskii, 1964). Such a spectrum would drop steeply above about 10 MeV. Another possibility would be due to extension of the 100 MeV galactic component to lower energies with a harder spectrum than observed below 1 MeV. The 1–10 MeV range then becomes the crossover range where the spectrum hardens, and anisotropy becomes evident.

One hypothesis which fits the data has been advanced by Stecker (Stecker, 1969b; Stecker and Silk, 1969). He has suggested an additional isotropic component due to photons produced by π° decays when cosmic rays were first accelerated during the expansion following the 'big-bang'. Such photons would now be red-shifted and depending on the epoch, would appear in the 1–10 MeV range. The cosmological interpretation of our data following this hypothesis is presented in the accompanying paper by Stecker.

The actual conversion of the ERS-18 data to a photon spectrum depends upon the shape of this spectrum at higher energies. We have considered the total flux to be composed of two components (a) a power law spectrum, which fits the data at low energy and (b) a peaked spectrum, similar to that proposed by Stecker to fit the data above 1 MeV. These spectra, which both extend to high energies, were then used as

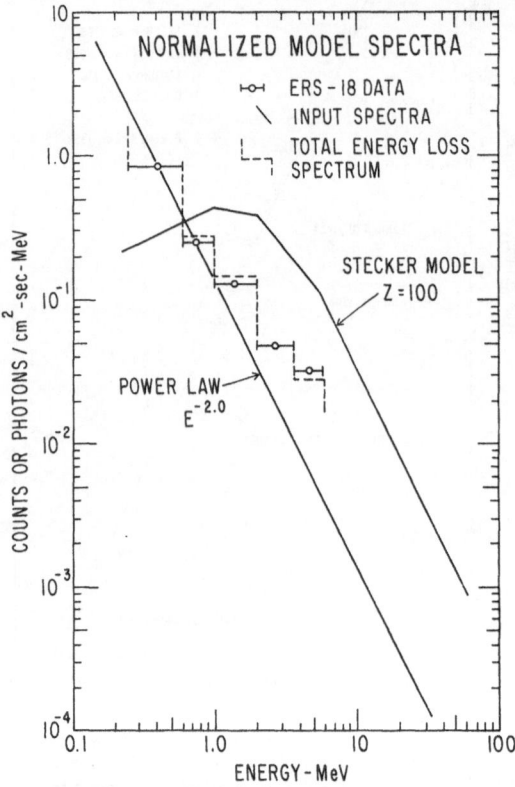

Fig. 4. The results of a computation in which a theoretical power law spectrum, and a red-shifted π^0 spectrum corresponding to a Stecker model of $Z = 100$ were inputed to a Monte Carlo program which takes into account the detector response at all energies. The resultant energy loss spectrum is compared with the ERS-18 data.

input for a Monte Carlo program which accounted for the relevant path-length distribution, cross-sections, efficiencies, etc. in a 7.5×6.35 cm diameter NaI(Tl) detector, and produced an energy loss spectrum. Parameters were adjusted to fit the measured energy loss spectrum. The power law flux requires an index of -2.0 to reproduce the measured energy loss spectrum up to a few hundred keV. A total flux of 1.3 photons/cm^2-sec at 1 MeV in the $Z = 100$ Stecker spectrum added to the power law contribution produces the required flatness near 1 MeV. The final results of this process are shown in Figure 4.

We believe the ERS-18 experiment has clearly measured cosmic γ-rays above 1 MeV. The spectrum is flatter than that observed at lower energies and suggests that a new component is present. These results are consistent with a component having the cosmological $\pi°$-decay origin suggested by Stecker. Although our results cannot definitely establish this process as the source mechanism, it is an exciting possibility which, if true, means we are viewing photons which have been red shifted by a factor of about 100. This is considerably greater than the red shift observed in quasars.

It is clear that additional measurements in this energy range are needed to obtain directional information as well as more detailed spectral data. This coupled with additional theoretical developments should lead to a more detailed understanding of the source mechanisms which clearly have cosmological implications.

References

Bleeker, J. A. M., Burger, J. J., Scheepmaker, A., Swanenburg, B. N., and Tanaka, Y.: 1966, The Royal Dutch Academy of Sciences Preprint.

Bowyer, C. S., Field, G. B., and Mack, J. E.: 1968, *Nature* **217**, 32.

Clark, G. W., Garmire, G. P., and Kraushaar, W. L.: 1968, *Astrophys. J.* **153**, L203.

Felten, J. E. and Morrison, P.: 1966, *Astrophys. J.* **146**, 686.

Ginzburg, V. L. and Syrovatskii, S. I.: 1964, *Usp. Fiz. Nauk* **84**, 201. Translation in 1965: *Soviet Phys. Uspekhi* **7**, 696.

Gorenstein, R., Kellogg, E. M., and Gursky, H.: 1969, *Astrophys. J.* **156**, 322.

Gould, R. J.: 1967, *Am. J. Phys.* **35**, 376.

Henry, R. C., Fritz, G., Meekins, J. F., Friedman, H., and Byram, P. T.: 1968, *Astrophys. J.* **153**, L11.

Hudson, H. S., Peterson, L. E., and Schwartz, D. A.: 1969, *Solar Phys.* **6**, 205.

Metzger, A. E., Anderson, E. C., Van Dilla, M. A., and Arnold, J. R.: 1964, *Nature* **204**. 766.

Peterson, L. E.: 1965, *J. Geophys. Res.* **70**, 1762.

Peterson, L. E., Matteson, J. L., Huszar, L., and Vette, J. I.: 1968, UCSD-SP-68-7.

Rocchia, R., Rothenflug, R., Boclet, D., Ducros, G., and Labeyrie, J.: 1967, *Space Research VIII*, North-Holland Publishing Co., Amsterdam, Vol. I, pp. 1327–1333.

Seward, F., Chodil, G., Mark, H., Swift, C., and Toor, A.: 1967, *Astrophys. J.* **150**, 845.

Silk, J. and McCray, R.: 1969, *Astrophys. Letters* (in press).

Stecker, F. W. and Silk, J.: 1969, *Nature* **221**, 1229.

Stecker, F. W.: 1969a, *Nature* **221**, 425; corrections: *Nature* (in press).

Stecker, F. W.: 1969b, *Astrophys. J.* (in press).

INTENSITY AND GALACTIC ABSORPTION OF
SOFT BACKGROUND X-RAYS

A. N. BUNNER, P. L. COLEMAN, W. L. KRAUSHAAR, D. McCAMMON,

T. M. PALMIERI, A. SHILEPSKY, and M. ULMER

Dept. of Physics, University of Wisconsin, Madison, Wis., U.S.A.

Abstract. Measurements of the soft X-ray background intensity show that the intensity is largest in that part of the sky where the columnar atomic hydrogen density is smallest. In the direction $b^{II} = -60°$, $l^{II} = 10°$ the intensity is found to be 195 ± 20 photons $cm^{-2} sec^{-1} ster^{-1} keV^{-1}$ near 0.26 keV and 20 ± 3 photons $cm^{-1} sec^{-1} ster^{-1} keV^{-1}$ near 0.9 keV. The intensity in the galactic plane is unexpectedly large.

We wish to report here recent new results on the intensity, angular distribution and possible origin of the soft X-ray background radiation. Our data are in partial agreement with the already-reported results of the Berkeley group (Bowyer *et al.* [1]) and the NRL group (Henry, *et al.* [2]), but are in rather serious disagreement with the results of the Calgary group (Baxter and Wilson [3]).

Two separate counter systems were flown, each of 250 cm² effective area and each with 2.6° × 23° (FWHM) collimation. The principle axes of the two collimators were 180° apart and the counters were mounted so as to view perpendicular to the long axis of the slowly spinning and unguided Aerobee 150 rocket. The X-ray counters were separated from their veto counters by arrays of 0.005″ ground wires. This arrangement is very effective in reducing charged particle and γ-ray induced background. The window material of one counter was 4 micron 'Mylar', for the other 2 micron 'Kimfol'. Kimfol is a trade name for polycarbonate. The windows were coated on the inside with a very thin layer of colloidal carbon. Laboratory tests showed that the window material itself amply eliminated possible ultraviolet sensitivity. But at somewhat longer wavelengths where the windows became transmitting, we found that photoelectrons could be ejected from thin aluminum coatings. Carbon, with its high photoelectronic work function eliminated this potential source of malfunction. That ultraviolet and light sensitivity did not confuse our measurement was verified in flight through comparison of our data with the data from photometers prepared by the University's Space Astronomy Laboratory and flown on the same rocket.

The efficiencies of the two counters, plotted as functions of incident X-ray energy are shown in Figure 1. Notice that at 284 eV, the carbon K-edge, the Kimfol transmission is nearly three times that of the Mylar. In addition, the Kimfol transmission is large enough in the 500 eV to 1 keV region to permit measurements there, too. The data telemetered from each counter were accumulated counts for pulse heights corresponding to X-ray energies of 0.15 to 0.48, 0.48 to 0.96, 0.96 to 1.6, 1.6 to 2.65, 2.65 to 6.5 and 6.5 to 8.4 keV. An on-board retractable Fe^{55} calibration source demonstrated that the entire system including electronic gas regulator, high voltage, amplifier gains and level discriminators were stable to within 5% in measured X-ray

L. Gratton (ed.), Non-Solar X- and Gamma-Ray Astronomy, 342–351. All Rights Reserved
Copyright © 1970 by the IAU

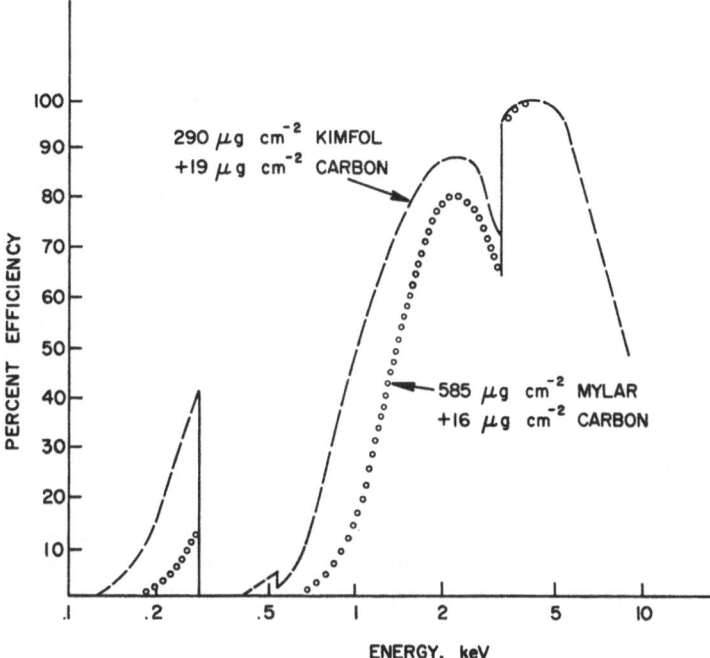

Fig. 1.   Efficiency of the two counters vs. X-ray energy. Kimfol is $C_{16}H_{14}O_3$ and Mylar is $C_{10}H_8O_4$.

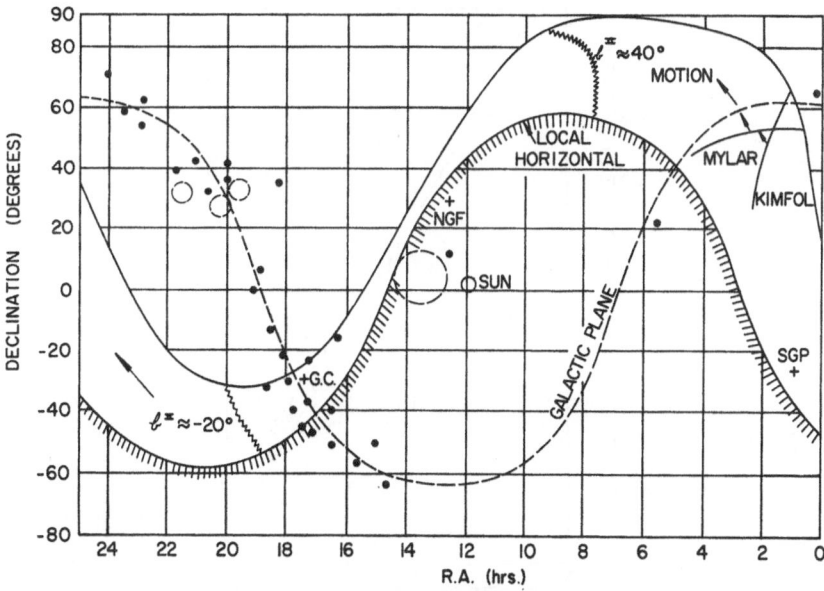

Fig. 2.   Scan path in celestial coordinates. The black dots are approximate positions of known X-ray sources. The broken circles indicate regions of suspected enhanced soft X-ray emission. The large region near 14^{hr} is discussed by Bowyer et al. [1]; the three small regions near 20^{hr} are discussed by Henry et al. [2].

energy throughout the flight. The counters were flushed and refilled just prior to launch, and for several hours prior to launch counting gas was continuously forced through the payload section under a slight positive pressure.

The portion of the sky scanned is shown in Figure 2. Some discrete sources were crossed and their energy spectra are being studied. Only the region between $b^{II} \approx -20°$, $l^{II} \approx 340°$ and $b^{II} \approx 40°$, $l^{II} \approx 160°$ was used for the diffuse X-ray analysis to be discussed here.

We have used only the data taken near the south galactic pole to study the diffuse X-ray intensity in the region 1–8 keV. We find that in this region our data fit best an $E^{-1.5}$ power law and the absolute intensity as found by us is in good agreement with that found by others. See, for example, the summary plot of Gorenstein *et al.* [4]. As will be discussed further below, our data in the two lowest energy pulse height channels indicate the need for some sort of soft X-ray source in addition to the extrapolated power law. To emphasize this point, we have forced a fit of an $E^{-1.7}$ power law to the data for $E > 1$ keV and show the result together with all the high galactic latitude data points in Figure 3. The plotted points represent X-ray flux incident on the detector. *No corrections for interstellar absorption have been made.* Also shown in Figure 3 is the power law which *best* fits our data above 1 keV, $E^{-1.5}$. In any case, particularly because the data themselves show clearly the presence of interstellar absorption, it seems clear that our soft X-ray data cannot be explained by a power law extrapolation alone.

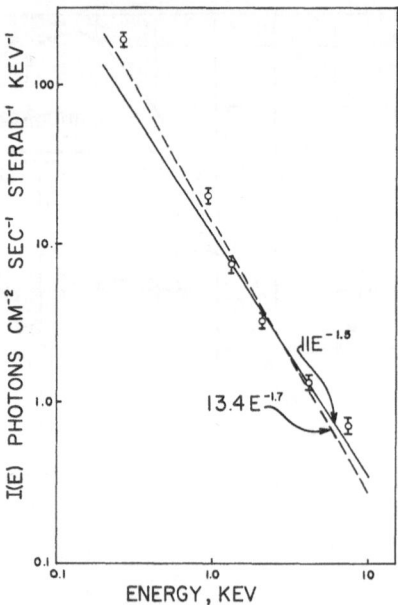

Fig. 3. X-ray intensity incident on the detectors vs. X-ray energy. The data used here were taken at high galactic latitudes, and no absorption effects have been included. The $E^{-1.5}$ power law fits our data best for $E > 1$ keV. The $E^{-1.7}$ power law is shown for comparison.

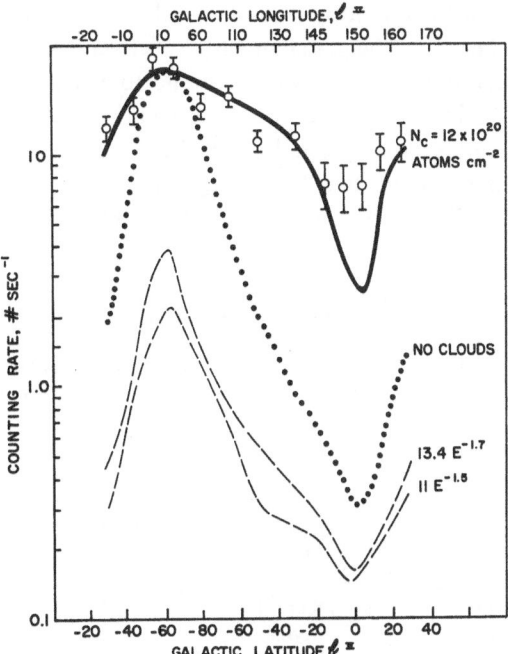

Fig. 4. Kimfol counter data in the 0.15 to 0.48 keV channel plotted vs. scan path in galactic coordinates. A background of 2.9 ± 0.3 sec^{-1} has been subtracted. Predicted curves are described in the text.

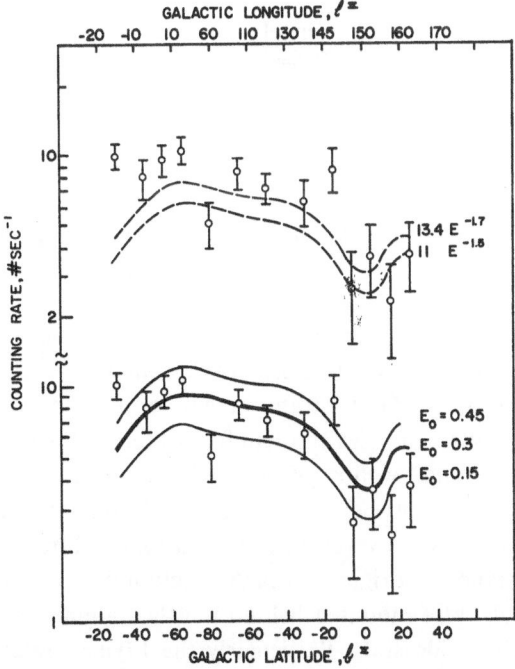

Fig. 5. Kimfol counter data in the 0.48 to 0.96 keV channel plotted vs. scan path in galactic coordinates. A background of 1.8 ± 0.2 sec^{-1} has been subtracted.

Figures 4 and 5 show the counting rates (Kimfol window counters) in the 0.15 to 0.48 and 0.48 to 0.96 keV pulse height channels respectively, plotted vs. position along the scan path. Galactic latitude is shown on the bottom scale, galactic longitude on the top. Data from the Mylar window counter are entirely similar, but the rates and statistical accuracy are less. Background rates of 2.9 ± 0.3 sec^{-1} and 1.8 ± 0.2 sec^{-1} have been subtracted from the two low energy channels. These background rates ware taken when the collimators viewed the earth. Table I shows a summary of the

TABLE I

Measured background rates

Altitude	Direction	Detector window	Measured rates sec^{-1}	
			0.15–0.48 keV	0.48–0.96 keV
80 km[a]	(Doors on)	Kimfol	3.8 ± 0.6	2.3 ± 0.4
80 km[a]	(Doors on)	Mylar	3.0 ± 0.5	2.4 ± 1.4
80 km[a]	up	Kimfol	2.9 ± 0.7	1.6 ± 0.5
80 km[a]	up	Mylar	2.5 ± 0.5	2.3 ± 0.4
80 km[a]	down	Kimfol	2.9 ± 0.4	1.7 ± 0.3
80 km[a]	down	Mylar	1.5 ± 0.4	1.5 ± 0.4
80 km[b]	up	Kimfol	3.9 ± 0.6	1.4 ± 0.5
80 km[b]	up	Mylar	4.1 ± 0.6	2.1 ± 0.5
80 km[b]	down	Kimfol	2.3 ± 0.4	1.9 ± 0.4
80 km[b]	down	Mylar	1.9 ± 0.4	1.3 ± 0.3
> 140 km	down	Kimfol	2.9 ± 0.3	1.8 ± 0.2
> 140 km	down	Mylar	2.8 ± 0.3	1.6 ± 0.2

[a] Rocket ascending.
[b] Rocket descending.

background rates, evaluated in different ways. The scatter is perhaps a little greater than what one would expect if all were independent measures of the same quantity, but we regard the consistency and small relative magnitude of the background as good evidence that our soft X-ray measurements are not seriously contaminated with, for example, an anamolous charged particle background.

The measured ratio of the Kimfol transparency to that of the Mylar is 2.7 ± 0.3 at the carbon edge. At lower energies the Kimfol is more transparent than the Mylar by an even larger factor and for most of the spectra we have considered the predicted counting rates in the first channels of the Kimfol and Mylar counters are in the ratio 5 to 1. The measured rates, however, after background subtraction, are in the ratio 2.7 ± 0.2. We understand the origin of a large fraction of this apparent discrepancy. The window materials were mounted behind polished aluminum frames into which had been milled 3/16″ wide slots. Particularly the Mylar stretched and bulged up under pressure into these slots and so produced, in effect, a non-uniform window thickness. The centers of the bulges were more transparent than the original flat material on which the transparencies were measured. We have evaluated the effect in the laboratory and when all uncertainties are included the predicted ratio becomes 3.6 ± 0.6 (rather than 5) to be compared with the measured value in flight of 2.7 ± 0.2. This state of affairs is admittedly not entirely satisfactory and will be investigated further.

The data plotted in Figure 5 and particularly in Figure 4 show clearly the effects of interstellar absorption. The highest counting rate encountered along the scan path occurs at the position of least atomic hydrogen, as measured in the 21 cm emission surveys [5]. In both figures the broken lines represent predicted counting rates given (i) the extrapolated power law spectra incident from outside the Galaxy, (ii) galactic absorption calculated from the effective cross sections compiled by Bell and Kingston [6] and (iii) columnar hydrogen density as summarized by Kerr and Westerhout [5]. (An outline of the procedures used in making these and other predictions is shown in Table II.) The dotted curve in Figure 4 shows the effect of simply in-

TABLE II

Analysis procedure

1. X-ray spectrum incident on galaxy
   $$A E^{-\gamma} + B/E \exp(-E/E_0).$$
2. Attenuation by interstellar gas
   (a) Kerr and Westerhout [5] N_H contours;
   (b) Bell and Kingston [6] σ;
   (c) Evaluate $e^{-\tau}$ at 9 points along collimator.
3. Evaluate atmospheric attenuation (exclude if $> 10\%$ effect).
4. Results of (1, 2 and 3 above) is predicted X-ray spectrum incident on counter windows.
5. Multiply 4 above by counter window and gas efficiency.
6. Evaluate transfer effects in counter (Argon escape peak).
7. Result is predicted pulse height spectrum in counter with perfect resolution.
8. Fold 7 above with counter resolution.
9. Integrate 8 above over pulse height bins.

I	II	III	IV	V	VI
0.15–0.48	0.48–0.96	0.96–1.6	1.6–2.65	2.65–6.5	6.5–8.4

10. Result is predicted counting rate in each pulse height bin at one time in flight.
11. Procedure repeated at 0.5 sec intervals ($\simeq 9°$) through useful portion of flight.
12. Average predictions over repeated scans of same part of sky to compare with observations.

creasing the assumed incident soft X-ray intensity so as to fit the observations at the high counting rate point.* The dependence of the observations on galactic latitude is much weaker than predicted. That is, there is too little apparent absorption, as already pointed out by Bowyer et al. [1].

---

* Note that this dotted curve is not just the lower curve scaled by a constant factor. A portion of the counting rate in the 0.15 to 0.48 keV channel, particularly near the galactic plane, has its origin in the argon K_α escape phenomena excited by X-rays near 3 keV. This effect with the Kimfol window counter is relatively small. With the Mylar window counter it is large and tends to obscure deep absorption features.

1. Anomalous Apparent Absorption

We have considered in our analysis a number of possible explanations for the unexpected weak dependence of the soft X-ray intensity on galactic latitude.

(a) If there were less actual gas than indicated by the 21 cm surveys or if the helium content of the interstellar medium were less than was assumed by Bell and Kingston [6], there should be too little apparent absorption in the 0.48 to 0.96 keV measurements also. To within the rather poor statistical accuracy of our measurements in this region (see Figure 5) this seems not to be the case.

(b) If the interstellar gas were concentrated in small dense clouds rather than distributed more or less smoothly as was assumed in making the previous predictions, the X-ray absorption would be less. Bowyer and Field [7] have also considered this possibility. The apparent absorption of X-rays by gas of columnar density N_H distributed not uniformly but in randomly-positioned clouds of columnar density N_c each is described by an optical depth

$$\tau_e = (N_H/N_c)\left[1 - \exp\left(- N_c\sigma\right)\right], \tag{1}$$

where σ is the energy-dependent absorption cross section. We find that hypothetical clouds of $N_c \approx 12 \times 10^{20}$ atoms cm^{-2} are required to produce absorption in reasonable agreement with that observed. The bold solid curve of Figure 4 has been computed under this assumption. (The incident spectrum assumed for this case was $11\, E^{-1.5} + (100/E) \exp(- E/0.3)$, and will be discussed later.) This value of N_c is very large and while there is perhaps some support for dense clouds structures [8, 9, 10], we feel it appropriate to examine less severe assumptions.

(c) If there were some unsuspected source of isotropic soft X-rays of terrestial or solar origin, an otherwise highly anisotropic flux would appear less so. In this case we should subtract the observed flux level near $b^{II} = 0$ from all observations, and the data treated in this way are shown in Figure 6. As in Figure 4, the two dashed lines show predicted rates for extrapolated power law spectra, normal interstellar absorption having been assumed. The dotted curve, again as in Figure 4, shows the effect of increasing the assumed incident soft X-ray intensity so as to fit the observations at the high counting rate point. Even here, the observed and predicted dependence on galactic latitude are not in agreement. There is too little absorption. Now, however, clouds of only $N_c \approx 4 \times 10^{20}$ atoms cm^{-2} are required to match the absorption, and the bold solid curve has been calculated for clouds of this thickness. The incident spectrum is the same as that assumed before, $11\, E^{-1.5} + (100/E) \exp(- E/0.3)$. That the second term is the same as before is accidental.

(d) The very steep dependence of the predicted soft X-ray intensity on galactic latitude is a direct consequence of our assumption that the soft X-ray source is extragalactic. If the soft X-ray emission were of galactic origin the intensity variation with galactic latitude would depend on how the supposed emission was distributed relative to the gas. In order to examine this possibility we have assumed gaussian distributions in height, z, above the galactic plane for the gas and emission and have for the predicted intensity

$$I = S_0 \int_0^\infty \exp - \left(\frac{r}{z_s \sin b} \right)^2 e^{-\tau} \, dr \,, \qquad (2)$$

where

$$\tau = \sigma n_0 \int_0^r \exp - \left(\frac{r'}{z_g \sin b} \right)^2 dr' \,.$$

S_0 is the source strength, z_g and z_s characterize the gas and emission distributions in z, respectively, σ is the soft X-ray absorption cross section and n_0 is the gas density

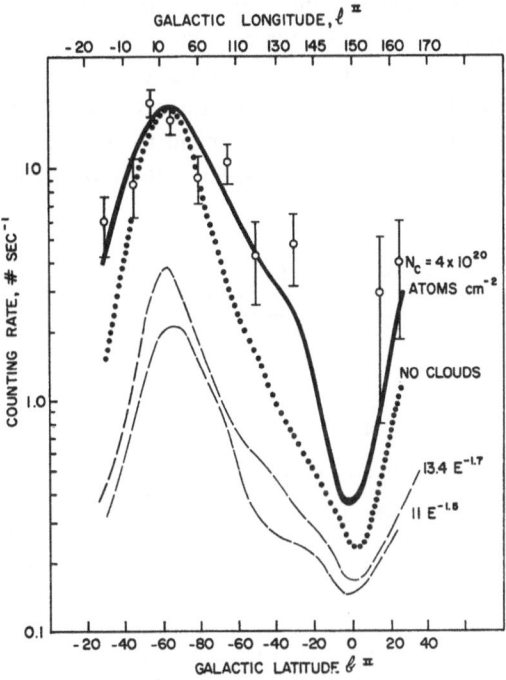

Fig. 6.   Kimfol counter data in the 0.15 to 0.48 keV channel plotted as in Figure 4. Here in addition to the 2.9 ± 0.3 sec^{-1} background, the observed rate near $b^{II} = 0$ has been subtracted as background. Predicted curves are discussed in the text.

in the galactic plane. Approximate fits to the latitude dependence of the data in our lowest energy channel (0.15 to 0.48 keV) can be obtained with $n_0 = 1$ cm^{-3}, $z_s = 400$ pc, $z_g = 94$ pc and $S_0 = 4 \times 10^{-19}$ photons cm^{-3} sec^{-1} sterad^{-1} keV at 0.26 keV. The value of S_0 is very large and corresponds to about 13 objects of the soft X-ray luminosity of the sun (quiet sun at solar maximum) per cubic parsec. The value of z_g corresponds to about 160 pc between half-density points either side of the galactic plane and while appreciably smaller than Schmidt's [1] value of 220 pc, is comparable

A. N. BUNNER ET AL.

with more recent estimates [5]. The value of z_s, 400 pc, is entirely reasonable for
Population II objects.

2. Soft X-Ray Spectrum

With only two broad band energy windows we cannot, of course, determine a precise
spectral shape. There is, however, evidence for an intensity above that attributable to
an extrapolated power law in the 0.48–0.96 keV pulse height channel as well as the
lowest energy channel and this provides some spectral information. We have chosen
rather arbitrarily to fit our data to incident spectra of the form $AE^{-\gamma} + BE^{-1}$ exp
$(-E/E_0)$ – a power law, the normalization and index of which are established by
measurements at $E > 1$ keV, and a free-free spectrum with constant Gaunt factor.
Under a given set of assumptions, data from the first or low energy channel determine
possible combinations of B and E_0 and data from the second channel determine E_0,
or a temperature. Quite accidentally, the required values of B and E_0 are the same
whether we treat the data as in (b) above (subtract just the background as found
from earth scans) or as in (c) above (subtract the observed intensity near $b^{II} = 0$ as a
'background'.) The sensitivity of the measurement to assumed values of E_0 is shown
by the bottom set of curves of Figure 5. Because the input spectra are the same for

TABLE III
Values of B and E_0

Case (b), $N_c = 12 \times 10^{20}$ atoms cm^{-2}

γ	B	E_0
1.5	100	0.30
1.7	68	0.30

Case (c), $N_c = 4 \times 10^{20}$ atoms cm^{-2}

γ	B	E_0
1.5	100	0.30
1.7	70	0.30

cases (b) and (c), and because even clouds of $N_c = 12 \times 10^{20}$ atoms cm^{-2} have very
little effect on the absorption of X-rays that contribute to the observations in the
second channel, the curves in the bottom part of Figure 5 are appropriate to both
cases. We conclude that under these assumptions, E_0 is near 0.30 keV and values as
small as 0.15 keV or as large as 0.45 keV are probably excluded.

We have repeated the entire procedure for a power law of index 1.7. The temperature or E_0 sensitivity is about the same as with a 1.5 power law index, and values
of B and E_0 are shown in Table III.

3. Summary

The maximum soft X-ray intensity we observe is in the direction $b^{II} \approx -60°$, $l^{II} \approx 10°$

and coincides with the direction in which the columnar atomic hydrogen density is smallest. Near 0.26 keV the measured intensity is

$$I = 195 \pm 20 \text{ photons cm}^{-2} \text{ sec}^{-1} \text{ ster}^{-1} \text{ keV}^{-1}$$

and near 0.9 keV is

$$I = 20 \pm 3 \text{ photons cm}^{-2} \text{ sec}^{-1} \text{ ster}^{-1} \text{ keV}^{-1}.$$

The corresponding values near the galactic plane ($b^{II} \approx 0$, $l^{II} \approx 155°$) are, near 0.26 keV

$$I = 57 \pm 8 \text{ photons cm}^{-2} \text{ sec}^{-1} \text{ ster}^{-1} \text{ keV}^{-1}$$

and near 0.9 keV

$$I = 6 \pm 1.5 \text{ photons cm}^{-2} \text{ sec}^{-1} \text{ ster}^{-1} \text{ keV}^{-1}.$$

The quoted uncertainties include counting statistics and estimated systematic errors such as uncertainties in window thickness, background subtraction, etc.

If we ignore for the moment the possibility that galactic objects contribute to the soft X-rays observed, our data imply an extragalactic component with an intensity given by

$$I = 11E^{-1.5} + 100E^{-1} \exp(-E/0.30) \text{ [photons cm}^{-2} \text{ sec}^{-1} \text{ ster}^{-1} \text{ keV}^{-1}]$$
(E in keV)

It should be emphasized that we have by no means established the analytic form of this expression. The first term is just the power law spectrum extrapolated to low energies, the second term a convenient computational device.

Acknowledgements

It is a pleasure to acknowledge the essential contributions of the Sounding Rocket Branch of the Goddard Space Flight Center and the financial support provided by NASA under grant NGR 50-002-044.

References

[1] Bowyer, C. S., Field, G. B., and Mack, J. E.: 1968, *Nature* **217**, 32.
[2] Henry, R. C., Fritz, G., Meekins, J. F., Friedman, H., and Byram, E. T.: 1968, *Astrophys. J. Lett.* **153**, L199.
[3] Baxter, A. J., Wilson, B. G., and Green, D. W.: 1969, *Astrophys. J. Lett.* **155**, L145.
[4] Gorenstein, P., Kellogg, E. M., and Gursky, H.: 1969, *Astrophys. J.* **156**, 315.
[5] Kerr, F. J. and Westerhout, G.: in *Galactic Structure*, Vol V. of *Stars and Stellar Systems* (ed. by A. Blaauw and Schmidt), University of Chicago Press, Chicago and London.
[6] Bell, K. L. and Kingston, A. E.: 1967, *Monthly Notices Roy. Astron. Soc.* **136**, 241.
[7] Bowyer, C. S. and Field, G. B.: 1969, *Nature* **223**, 573.
[8] Heiles, C.: 1967, *Astron. J.* **72**, 1040.
[9] Makova, S. P.: 1965, *Soviet Astron-A.J.* **8**, 485.
[10] Grahl, B. H.: 1960, *Mitt. Univ.-Stern. Bonn*, No. 28.
[11] Schmidt, M.: 1957, *Bull. Astron. Inst. Neth.* **13**, 247.

ORIGIN OF THE COSMIC X-RAY BACKGROUND

(Invited Discourse)

G. SETTI

Istituto di Fisica 'A. Righi', Osservatorio Astronomico Universitario, University of Bologna, Italy

and

M. J. REES

Institute of Theoretical Astronomy, University of Cambridge, England

Abstract. In this paper we review the theories which have been proposed to account for the extra-galactic X-ray background. Although there is still no detailed theory, one may devise reasonable models which account in a natural way both for the intensity and the spectral shape over the whole energy band, provided that cosmological evolutionary effects are included. A model based on Compton scattering of cosmic black body photons by relativistic electrons in radio sources at large redshifts ($z \gtrsim 4$) seems to give the most satisfactory explanation. However, the data are not yet good enough to discriminate against alternative models.

A discussion of the recent observations in the soft X-ray region (< 1 keV), and their relevance to the physics of interstellar and intergalactic gas, is given. The available data are somewhat confusing, but it seems that this part of the spectrum may still be consistent with a simple extrapolation of the non-thermal spectrum at higher energies, though various workers have claimed the detection of a new component probably due to hot intergalactic gas. If this interpretation is correct one may deduce interesting conclusions about the state of ionization and composition of the intergalactic gas, because of the importance of the absorption effects in this energy band.

Also it appears that the Galaxy is more transparent than one would deduce from 21-cm observations. However, due to the lack of observational data, no firm conclusions can be reached.

1. Introduction

In this review of theoretical interpretations of the X-ray background we shall concentrate on the astrophysical aspects of the problem. The relevant emission mechanisms and some aspects of the interpretation have already been extensively discussed in a number of reviews (Hayakawa and Matsuoka, 1964; Ginzburg and Syrovatsky, 1964a; Gould and Burbidge, 1965; Gould, 1967; Weymann, 1968; Sciama, 1968). The observations are fully reviewed by Oda (paper in this volume). The observed isotropy of the X-ray background indicates that it is predominantly extragalactic in origin (Seward *et al.*, 1967). Some enhancement is, however, observed towards the galactic plane (Cooke *et al.*, 1969), and several proposals have been made which could account for this contribution. We shall here consider primarily the isotropic component, which has been tentatively fitted by two power law spectra of different slopes (Gorenstein *et al.*, 1969):

$$\frac{dN}{dE} = 12.4\, E_{\text{keV}}^{-1.7 \pm 0.2} \text{ photons } (\text{cm}^2 \text{ ster sec keV})^{-1} \qquad (1')$$

L. Gratton (ed.), Non-Solar X- and Gamma-Ray Astronomy, 352–371. All Rights Reserved
Copyright © 1970 by the IAU

from 1 to 40 keV, and approximately

$$\frac{\mathrm{d}N}{\mathrm{d}E} \simeq 20\, E_{\mathrm{keV}}^{-2} \text{ photons } (\mathrm{cm}^2 \text{ ster sec keV})^{-1} \qquad (1'')$$

above 60 keV.

There is a substantial spread in the fluxes reported by various experimental groups, so that these precise spectra should not be taken too literally, but it definitely seems that the slope changes around 20–40 keV. Bleeker and Deerenberg (paper in this volume), who have a homogeneous set of observations between 20 and 220 keV, obtain a straight spectrum with the much steeper slope 2.45 ± 0.1 over this range, and conclude that any spectral break must occur at $\lesssim 20$ keV.* In view of all the uncertainties, it is difficult to specify either the position of the break, or the slope of the spectrum in various energy ranges, with any precision. As we shall see, detailed knowledge of the spectrum would be important in deciding between various models.

A background X-ray flux has also been reported below 1 keV, but at the moment it is impossible to give any reliable figures either for the intensity or the behaviour of the spectrum (Oda, paper in this volume). In view of the astrophysical and cosmological interest which attaches to these soft X-ray observations, we devote a special section of our review to this topic.

In the following sections we attempt to present a critical discussion of current ideas on these topics.

2. Energy Requirements for Universal X-Ray Background

The energy density of the isotropic background from 2 to 1000 keV, with spectrum (1), is

$$\sim 6 \times 10^{-5} \text{ eV cm}^{-3}. \qquad (2)$$

Neglecting cosmological effects, this figure represents the mean energy input per unit volume required to generate the flux. If the bulk of this radiation were emitted at redshifts $z \simeq \bar{z}$, the energy input per unit *coordinate* volume would be increased by $(1 + \bar{z})$ (see Section 4).

It is interesting to compare this with the energy densities of background radiation in other energy bands. The estimated metagalactic energy density at radio wavelengths (1–1000 MHz) is $\sim 10^{-7}$ eV cm^{-3}; the energy density of 2.7 K black-body radiation is ~ 0.3 eV cm^{-3}; and the metagalactic background at optical wavelengths is $\sim 10^{-2}$ eV cm^{-3} (the infrared background is highly uncertain, but is unlikely to exceed that at optical wavelengths (Low and Tucker, 1968)). It is thus clear that the energy going into X-rays is small compared with the output of typical galaxies at longer wavelengths. The problem raised by the X-ray background is therefore not primarily an energetic one – the problem is rather to understand how $\sim 0.5\%$ of the

---

* Brini *et al.* (paper in this volume), who obtain a much less steep spectrum more in agreement with (1″), also do not find any appreciable break down to 25 keV.

radiation from sources* can be channelled into hard photons with the observed spectrum.

3. Models not Involving Cosmological Evolution

It is convenient to divide the proposed models into two categories:

(a) those which interpret the X-ray background as the integrated effect of discrete sources;

(b) those which invoke emission mechanisms that operate throughout intergalactic space, or at least in very extended diffuse regions.

In our discussion of (a), we shall merely consider the likely contribution from various classes of objects. (The emission mechanisms for discrete sources are reviewed by Felten, paper in this volume.) We shall consider (b) at greater length, because it is in this case possible to attempt to fit the observed spectrum. We adopt this same procedure in Section 4, where evolutionary models are discussed.

A. DISCRETE SOURCE MODELS

The X-ray emission from sources in the Galaxy is 10^{39}–10^{40} ergs sec^{-1} (Oda, 1965; Silk, 1969). In order to estimate the likely contribution from all normal galaxies, it is convenient to express this output as 5×10^{-6}–5×10^{-5} ergs gm^{-1} sec^{-1}. Assuming all galaxies to have the same mean X-ray emission per unit mass, and that the smoothed-out density of galaxies has the value $\sim 3 \times 10^{-31}$ estimated by Oort (1958), we obtain an X-ray energy density of $3 \times 10^7 - 3 \times 10^{-6}$ eV cm^{-3}. Even taking the optimistic estimate of $\sim 10^{40}$ ergs sec^{-1} for the output of our Galaxy, we still fall short of the observed X-ray background intensity by a factor ~ 20. Unless other galaxies typically emit X-rays more strongly than our own, or unless the density of faint galaxies greatly exceeds Oort's estimate, it seems unlikely that the observed background could be produced. However, in view of our ignorance of the X-ray properties of other galaxies, we cannot exclude this possibility completely. One would have to invoke galactic sources with power-law spectra (e.g. the Crab) rather than sources like Sco X-1 which are better fitted by exponential spectra. (The isotropy of the X-ray background is also evidence against the idea that it arises from normal galaxies – at visible wavelengths the background anisotropy is very large because starlight from our Galaxy swamps the extragalactic contribution.)

The X-rays may of course come from a rarer class of sources more powerful than ordinary galaxies – e.g. quasars or radio galaxies. No quasars have yet been definitely detected as X-ray sources, but the flux from 3C 273 could be $\sim 7 \times 10^{45}$ ergs sec^{-1} in the 1–10 Å band (Friedman and Byram, 1967). If all quasars in fact emitted a flux of this strength, the required space density would be $\sim 10^{-6}$ Mpc^{-3}. On the other hand, Schmidt (1968) estimates the actual local space density of objects like 3C 273 to be only $\sim 10^{-9}$ Mpc^{-3}, so that in this case also there appears to be a large discrepancy. Alternatively one may consider radio galaxies as candidates. The local

---

* Since any radiation present in the early dense phases of an evolving universe would have been thermalized, the X-ray background clearly cannot be primeval.

space density of radio galaxies emitting $\gtrsim 10^{42}$ ergs sec^{-1} at radio wavelengths is a few times 10^{-6} Mpc^{-3} (Schmidt, 1966). The mean X-ray flux from these would need to be ~ 1000 times greater than their radio output. Present observations show that this cannot be true for the nearby radio galaxies Centaurus A (Haymes *et al.*, 1969) and Virgo A (Friedman and Byram, 1967). We would therefore have to appeal exclusively to the strongest radio sources ($\gtrsim 10^{44}$ ergs sec^{-1} in the radio band), which, according to Schmidt (1966) have a local space density of 10^{-8}–10^{-9} Mpc^{-3}.

It thus appears difficult to account for the X-ray background in terms of known types of discrete sources, without appealing to cosmological evolutionary effects.

B. DIFFUSE PROCESSES

1. *Compton Radiation*

The bulk of the radiation energy in intergalactic space resides in the microwave region, and has been interpreted as having the spectrum of a 2.7 K black body. We confine attention to Compton scattering of these microwave photons. (Scattering of optical photons is relatively unimportant (Felten and Morrison, 1966).)

When isotropic radiation is scattered by electrons of Lorenz factor γ, the energy of the photons is raised by a factor $\sim \gamma^2$. The black-body photons have typical energies $3kT \simeq 10^{-3}$ eV. In order to obtain X-rays in the range 1 keV – 1 MeV, electrons are needed with γ in the range 10^3–$10^{4.5}$. These same electrons would emit radio frequency synchrotron radiation in magnetic fields with the strength 10^{-4}–10^{-6} G, which are believed typical of radio galaxies. The X-ray and radio spectra would have similar shapes, a power-law electron spectrum $N(E)$ d$E \propto E^{-\beta}$ dE giving rise to radiation spectra with slope $-(\beta-1)/2$ (or *photon* number spectra with slope $-((\beta+1)/2)$. This immediately suggests the possibility of relating X-ray and radio observations.

The radio background emission in the direction of minimum sky brightness is ~ 80 K at 178 MHz (Turtle and Baldwin, 1962). This gives a definite upper limit to the extragalactic radio background. It is usually assumed, in fact, that the true extragalactic contribution is no more than 20 or 30 K, the remainder being due to the galactic halo and disc (Bridle, 1967a). About half of this extragalactic component is contributed by sources already seen in the source counts (Pooley and Ryle, 1968). Therefore, if the electrons producing the X-ray background via Compton scattering are not simultaneously to produce a radio background exceeding ~ 10 K at 178 MHz – and this is a conservative supposition – we may evaluate an upper limit to the magnetic field by comparing the synchrotron and Compton fluxes at two frequencies ν_R and ν_X:

$$\frac{F(\nu_X)}{F(\nu_R)} = Q(\beta, w_{\text{ph}}) \langle H \rangle^{-\frac{\beta+1}{2}} \left(\frac{\nu_X}{\nu_R}\right)^{-\frac{\beta-1}{2}}, \tag{3}$$

where β is the exponent in the differential electron spectrum, $\langle H \rangle$ the average magnetic field, w_{ph} the energy density in the black-body radiation, and Q is a function of β and w_{ph} (Bergamini *et al.*, 1967). For $\beta \approx 3$ it turns out that

$$\langle H \rangle \lesssim 1.7 \times 10^{-7} \, \text{G} . \tag{4}$$

More generally, this sets a limit to the mean-square field seen by an electron over its lifetime. This is far below the estimated magnetic field in our Galaxy, or in any discrete sources, so we conclude that the number of electron actually within these sources is insufficient to produce the X-ray background by Compton scattering of 2.7K microwave photons. E.g., if the field were $\sim 10^{-6}$ G, the number of electrons required to produce the whole radio background would give an associated X-ray background only $\sim 1\%$ of what is required. (For this reason, X-ray emission from the galactic halo would be negligible.)

Felten and Morrison (1966) were the first to propose that there may be an inter-galactic flux of relativistic electrons sufficient to generate the X-ray background. The time t_{comp} in which an electron loses half its energy by scattering black-body photons is

$$t_{comp} \approx 10^{12}/\gamma \, \text{years} . \tag{5}$$

In intergalactic space, Compton losses would dominate all others, so the relevant electrons have lifetimes $10^9 - 3 \times 10^7$ years. The fact that this is short compared with the Hubble time means that almost all the energy produced in the form of relativistic electrons with $\gamma \gtrsim 10^3$ could have been converted into X-rays.

Equation (4) requires that, if radio sources with magnetic fields of, say, 3×10^{-6} G, dimensions ~ 50 kpc, and radio power $\sim 10^{42}$ erg sec^{-1}, generate enough electrons to give the X-ray background, the electrons must escape from the radio volume in less than 10^6 years. Taking Schmidt's (1966) estimate of $\sim 10^9$ years for the lifetime of such radio sources, the energy requirements amount to $\sim 10^{61}$ ergs in relativistic electrons alone. Even if the energy of the associated protons were ~ 100 times greater, this could not be definitely excluded on energetic grounds, provided that the accelera-tion mechanism is highly efficient (e.g., the collapse of $\sim 10^{10} \, M_\odot$ would yield 10^{63} ergs for $\sim 5\%$ efficiency).

Such models can be further tested by comparing the spectra of radio sources and the X-ray background. The average radio spectral index of radio galaxies is $\alpha = 0.8$. (The corresponding differential electron spectrum has $\beta \approx 2.6$.) If electrons escaped into intergalactic space with this spectrum and were thereafter subject only to Comp-ton losses, the X-ray background would have a *steeper* slope $\alpha = 1.3$ right down to ~ 10 eV (Felten and Morrison, 1966), since the lifetimes of the electrons which radiate at higher energies are shorter than the Hubble time (from (5)). Therefore, if the X-ray background is indeed flatter below ~ 20 keV, further consideration of the escape mechanism, or of the low energy spectrum within the source, is required in order to reconcile this model with the observations.

Brecher and Morrison (1969) have considered whether normal galaxies, rather than radio galaxies, could generate the required relativistic electrons. The energy which must be injected amounts to

$$\sim 3 \times 10^{-22} \, \text{eV cm}^{-3} \, \text{sec}^{-1} \tag{6}$$

In estimating the contribution from an individual galaxy one needs to know the space density of galaxies involved. These authors assumed the radio luminosity function proposed by Sholomitskii (1968):

$$n(P)\,dP = 1.3 \times 10^{23}\,P^{-2.18}\,dP^*, \tag{7}$$

where $P(W(Hz\,ster)^{-1})$ is the radio power at 178 MHz, and $n(P)$ the source number density in Mpc^{-3}. This luminosity function is not well established in the range $P \simeq 10^{20}\text{--}10^{22}$ W $(Hz\,ster)^{-1}$ spanned by normal galaxies. However, even granting its validity, Brecher and Morrison's work seems to entail several difficulties.

If one cuts off the luminosity function(7) below a value P_1 such that

$$\int_{P_1}^{\infty} n(P)\,dP \simeq 0.03, \tag{8}$$

the density of normal galaxies given by Allen (1963), one finds that $P_1 \simeq 8 \times 10^{20}$ W $(Hz\,ster)^{-1}$. (Sholomitskii chose P_1 in this manner, but appears to have made an arithmetical error in evaluating it.) Brecher and Morrison, however, extrapolate (7) right down to $P = 10^{20}$ W $(Hz\,ster)^{-1}$, which gives a source density $\sim 0.27\,Mpc^{-3}$. Even taking this high density of galaxies, we find that each has to produce

$$\sim 5 \times 10^{40}\,erg\,sec^{-1}. \tag{9}$$

This is a measure of the rate at which relativistic electrons must be produced. The observed energy density of relativistic electrons above 1 BeV in the disc of our Galaxy is $\sim 10^{-2}$ eV cm^{-3}. Thus, if they are generated at the rate (9), a volume of $\sim 1 \times 10^{68}\,cm^3$ could be filled or replenished in $\sim 10^6$ years. This volume is comparable with that of the whole galactic halo. (If the electrons are confined only in the disc, the escape time would be no more than $\sim 10^5$ years.) As Brecher and Morrison point out, these energy requirements are not impossible, even allowing 100 times as much energy for the replenishment of the associated protons, which presumably escape in the same timescale. We note, however, that in this model the cosmic-ray primaries could not have passed through the 3–4 gm cm^{-2} (Shapiro and Silberberg, 1968) of matter implied by the composition data above ~ 1 BeV per nucleon**; for the usually assumed density of ~ 1 atom cm^{-3} in the disc and $\lesssim 10^{-2}\,cm^{-3}$ in the halo, this requires confinement for $\sim 10^6$ years or $\gtrsim 10^8$ years respectively.

The difficulties with this model are aggravated when we remember that most of the galaxies included in the luminosity function (7) are smaller than our own. If we follow Brecher and Morrison in deducing the electron density in these galaxies from an equipartition argument, assuming their storage volume to be similar to that of our Galaxy, we find that the average galaxy contributes only $\sim 10\%$ of the electron flux

* The source counts in fact indicate that the luminosity function depends on redshift, the proportion of powerful sources being higher in the past (Longair, 1966). When this is allowed for, it is no longer true that 95% of the radio background would come from normal galaxies, as is stated by Brecher and Morrison.
** Unless one postulates that all this matter has been traversed within the sources.

from our own. (If, on the other hand, we assume that the storage volumes are proportional to P and the escape times similar, we lose *two* orders of magnitudes.) Consequently, even if (7) can be validly extrapolated down to $P \simeq 10^{20}$ W (Hz ster)$^{-1}$, we have severe doubts about this model.

The slope of the electron spectrum in the Galaxy changes at ~ 3 BeV. This is associated with a break in the galactic synchrotron background spectrum at a few hundred MHz. A similar bend is observed in the radio spectra of certain other normal galaxies (Lang and Terzian, 1969). It is interesting that the corresponding bend in the Compton X-ray spectrum would occur at ~ 40 keV, where such a break indeed appears (but see Section 1). Brecher and Morrison point out that the observed X-ray background spectrum could be explained if there were a break in the electron spectrum at 3 BeV in all galaxies. They have to suppose that this break is characteristic of the injected spectrum, and therefore that all the sources (e.g. supernova remnants) in all galaxies inject identical spectra. The break has a natural interpretation in terms of losses if one takes the traditional estimates for the confinement time; but of course we can no longer invoke this explanation in the theory of Brecher and Morrison, where the electrons must escape more rapidly.

2. Thermal Bremsstrahlung

Plasma at temperatures $\gtrsim 10^6$ K emits thermal bremsstrahlung in the X-ray band. A significant background flux could arise from a general intergalactic medium with density $\sim 10^{-5}$ particles cm^{-3} or from isolated regions of higher density (e.g. gas within clusters of galaxies). As the X-ray background above 1 keV has a typically non-thermal spectrum, we cannot attribute it to thermal bremsstrahlung without postulating either an *ad hoc* distribution of temperatures up to $\sim 10^9$ K, or a very special thermal history for the intergalactic gas.

This process may, however, contribute to the X-ray background at energies $\lesssim 1$ keV, and we shall discuss it further in Section 5.

3. Non-Thermal Bremsstrahlung

a. *Electrons*. – A suprathermal electron of energy ε moving through a cool plasma emits bremsstrahlung photons with a spectrum which falls off sharply above a frequency such that $h\nu \simeq \varepsilon$ and is approximately flat at lower energies. Non-thermal electrons with a power-law spectrum would give rise to radiation which also obeys a power-law. The radiation spectral index is the same as that of the electrons when the latter are non-relativistic, and is one power flatter when they are relativistic. So, in order to explain the X-ray background, one may postulate a non-relativistic electron flux with spectral index 0.7 below 20 keV and ~ 1 above 20–40 keV (the electron number spectrum *per unit volume* would be half a power steeper).

The lifetime of an electron of energy ε_{keV} in a gas of density n is

$$t_L \approx \frac{6 \times 10^7}{n} \varepsilon_{keV}^{3/2} \tag{10}$$

For $n = n_0 \simeq 10^{-5}$ cm^{-3} this is less than the Hubble time for energies $\lesssim 1000$ keV. However, it is important to note that most of the electrons energy is transferred directly to the thermal gas, and only a small proportion appears as bremsstrahlung emission. The actual fraction is directly proportional to ε, and is $\sim 10^{-2}$ for $\varepsilon \simeq 10$ keV. This means that the energy requirements are $\sim 10^2$ times higher than in the Compton case. Also, the energy input into the intergalactic gas would heat it to $\gtrsim 10^6$ K, in which case thermal X-rays would also be emitted. Also, the only way of interpreting the break in the spectrum (apart from considering evolution, as we shall in the next section) is to take a gas density $n \ll 10^{-5}$ cm^{-3}, so that the Hubble time is $\sim t_L$ for $\varepsilon \simeq 20$–40 keV. The energy requirements then amount to > 1 eV cm^{-3} in sub-relativistic electrons, which seems an exorbitant value for intergalactic space (Silk, 1969).

b. *Protons.* – It is also possible to obtain photons of energy $h\nu$ from collisions of suprathermal protons of energy $\varepsilon \gtrsim (m_p/m_e) h\nu$ with thermal electrons. This is the so-called 'innerbremsstrahlung' process described by Hayakawa and Matsuoka (1963). Taking $n = 10^{-5}$ cm^{-3} the energy requirements are ~ 20 times greater than for electrons[*], quite apart from the difficulty of fitting the break in the spectrum unless an intergalactic gas density substantially greater is assumed. Therefore this process seems unlikely to play much of a role as far as the isotropic background is concerned, though it may make a small contribution ($\lesssim 10\%$) to the diffuse X-ray emission from the galactic disc (Boldt and Serlemitsos, 1969).

4. Models Incorporating Cosmological Evolution

The preceding discussion demonstrates the difficulty of accounting for the observed intensity and spectrum of the X-ray background in terms of processes occurring at the present epoch, without making rather extreme assumptions regarding the energy sources. Also there is no obvious interpretation for the bend in the spectrum. Various workers have consequently been led to propose models which relegate the production of the X-rays to a remote epoch. In evolutionary cosmologies, the intergalactic gas would have been denser in the past, and the primeval radiation field more intense. Moreover, there is evidence from the distribution of quasars (Véron, 1966a, b; Schmidt, 1968; Rowan-Robinson, 1968) and from the radio source counts (Davidson and Davies, 1964a, b; Longair, 1966; Rowan-Robinson, 1967; and the excellent review paper by Ryle, 1968) that violent events were more frequent at earlier epochs, so that the energy problem may be less embarrassing.

All of the models mentioned in Section 3 have been extended to incorporate evolutionary effects, and we shall discuss each briefly. But first we remember, as mentioned in Section 2, that the energy requirements per unit volume are *increased* in proportion to the mean redshift factor $(1 + \bar{z})$. The observed photons would of course have been generated with energies higher by $(1 + \bar{z})$. The expansion timescale t_{exp} for the universe would have been shorter in the past (except in Lemaître-type models).

[*] It is true that the energy of a proton is ~ 1800 that of an electron, but the proton range is correspondingly greater.

For the Einstein-de Sitter model (with deceleration parameter $q_0 = \frac{1}{2}$) $t_{exp} \propto (1+z)^{-3/2}$, and for a low-density model ($q_0 \simeq 0$), $t_{exp} \propto (1+z)^{-1}$. These two effects require that the power output in X-rays would need to have been *much* greater in the past in order for the dominant contribution to come from large redshifts (in fact the power input must depend on z more steeply than $(1+z)^{2.5}$ for an Einstein-de Sitter model).

A. DISCRETE SOURCE MODELS

Silk (1968, 1969) has considered evolution of the X-ray power of normal galaxies with redshift. It is plainly possible to account for any observed flux by postulating the requisite amount of evolution. The only basis for the assumed evolution is an analogy with the inferred evolution of strong radio galaxies. However, the radio power of normal galaxies could not evolve in the manner hypothesised for the X-ray power without violating the observed limit on the extragalactic radio background.

There is strong evidence for a sharp increase in the coordinate density of powerful quasars, at least out to $z \simeq 2$. According to Schmidt (1968) this density is ~ 1000 times greater at $z \simeq 2$ than it is locally. Even if this strong evolution continued out to $z \simeq 3$, and all QSS emit X-rays at the maximum level possible for 3C 273, one fails by a factor ~ 10. However, if one includes all the radio-quiet objects and assumes that they evolve equally strongly, one may perhaps fit the observed flux. It is not worthwhile to pursue this possibility further until the X-ray intensities and spectra of at least a few QSS are known. At present there is still no confirmation of X-ray emission from 3C 273, which is, intrinsically as well as apparently, one of the strongest known QSS.

In the case of ordinary radio galaxies, we cannot assume such strong evolution as for QSS and other very strong sources, because of the limit set by the radio background (Longair, 1966). Their contribution to the X-ray background would be unimportant unless their X-ray power evolved more strongly than their radio power.

B. DIFFUSE PROCESSES

1. *Compton Radiation*

In a 'big bang' cosmology, the black-body radiation temperature $T \propto (1+z)$ and the energy density

$$w_{ph} \propto (1+z)^4. \tag{11}$$

This means that at large redshifts the energy of relativistic electrons can be converted more rapidly into X-rays. In fact we can generalise (5) to

$$t_{comp} \approx \frac{10^{12}}{\gamma(1+z)^4} \text{ years}. \tag{5'}$$

In consequence, the permitted magnetic energy density given by (4) can be increased in proportion to w_{ph}, so that, if the X-rays come from redshifts $\bar{z}$, we have

$$\langle H \rangle \lesssim 1.7 \times 10^{-7}(1+\bar{z})^2 \tag{4'}$$

(we assume a spectral index $\alpha \simeq 1$).

Therefore, for $\bar{z} \gtrsim 2$, values of $\langle H \rangle \gtrsim 10^{-6}$ G are allowed. These magnetic fields are strong enough to make the radio galaxies themselves (at large z) good candidates as sources of the X-ray background. We may then connect the X-rays to observations of the evolution and spectra of radio sources, as proposed by Bergamini *et al.* (1967). Similar conclusions have been reached independently by Brecher and Morrison (1967) and Rees (1967).*

The steep slope of the $\log N/\log S$ curve can be explained if a class of powerful sources were more numerous in the past. A consequence of this interpretation is that a large proportion of the observed sources which contribute to the radio background, should have large redshifts. The bulk of the emission could even come from redshifts $z \simeq 5$. At the moment it is not clear whether quasars alone, or whether radio galaxies also partake in this evolution (Ryle, 1968). However, it is certainly consistent with the present data to assume that a large proportion of the extragalactic radio background comes from extended sources (though the extent to which such sources contribute to the counts depends on their individual power). If these sources are to contribute the X-ray background the magnetic field must, from (4'), be $\lesssim 10^{-6}$ G at $z \simeq 2$ and $\lesssim 6 \times 10^{-6}$ G at $z \simeq 5$ (though higher fields would be permitted if the electrons escape into a weaker field before losing all their energy). There are no reliable estimates of the magnetic fields within any extended radio volumes, but these required fields are well below the equipartition field in the strong sources. However, there is no basis other than the assumption of minimum energy for assuming an equipartition field. Indeed, observations of sources where synchrotron self-absorption occurs show that – at least in these compact objects – the field *is* in fact far below the equipartition strength (Hornby and Williams, 1966; Williams, 1966; Bridle, 1967b; but see Kellermann and Pauliny-Toth, 1969). There is no problem as far as the basic energetics are concerned in assuming weaker magnetic fields in the sources. We turn now to consider the interpretation of the observed X-ray background spectrum in the Compton theories.

Observations of individual sources with $z \lesssim 1$ indicate that the injected electron spectrum has index $\beta \simeq 2.6$, corresponding to a radio spectral index $\alpha \simeq 0.8$ (Kellermann, 1966). If we suppose tentatively that this is also the injection spectrum in the sources at larger z, then the integrated radio and X-ray spectra would both be steepened by Compton losses, and we would expect the α to be closer to 1.3, at least at the higher energies where the Compton losses would dominate.

To account for the position of the break, one needs a more specific source model allowing for the competing adiabatic losses (which are relatively more important at low energies). Such a model was proposed by Rees and Setti (1968), and we summarise it briefly below.

The model is based on the hypothesis that the evolution of extended radio sources is controlled by two main factors: (i) dynamical interaction with the intergalactic medium, and (ii) Compton losses. The development of a source from the initial explosion is divided into two phases: Phase I, a free relativistic expansion to a radius

---

* It was not possible to make this connection in the previous discussion of radio galaxies, since we did not then specify the emission mechanism, and could not predict its dependence on z.

$R_1 \simeq 10^4$ pc, at which stage the internal field is assumed to be $H_1 \simeq 10^{-5}$ G; and Phase II, a slower (but still rapid in cosmic time) continuing expansion against the 'ram pressure' of the intergalactic gas. For a model with internal energy $\sim 10^{60}$ ergs at the beginning of Phase II which is expanding into an external gas of density $2 \times 10^{-29}(1+z)^3$ gm cm^{-3}, the timescale for the expansion from R_1 to, say, $2R_1$ is

$$t(2R_1) \simeq 1 \times 10^5 (1 + z)^{3/2} \text{ years}. \tag{12}$$

During this time the fast particles must give up much of their energy in the adiabatic expansion (this energy may eventually heat the intergalactic gas). A break in the spectrum at 20 keV could be explained if the sources making the main contribution to the background are at a redshift such that t_{comp} (given by (5')) for the relevant electrons (which have $\gamma \simeq 4 \times 10^3$),* is equal to $t(2R_1)$ (Felten and Rees, 1969). The redshift derived in this way is ~ 4.5.** The parameters of the model – the intergalactic density, the internal energy, and the initial magnetic field – are very uncertain, so it would be unjustifiable to attach too much significance to this particular redshift. Nevertheless it is gratifying that a plausible choice for these parameters can explain not only the intensity, but also the spectrum, of the X-ray background.

It is worth pointing out that at present it is very hard to predict any more precise relationship between the radio and X-ray background spectra, because (a) the various classes of source may not make the same relative contributions to the X-ray and radio background, and (b) even a single source may make its main contribution to the two bands at different stages in its evolution. The spectrum of the extragalactic radio background is not in fact known with any precision: one may be mistaken in supposing that it should have the same slope as the mean spectra of resolved sources.

An attractive feature of this model is the way in which it relates the X-ray background to cosmology, to the physics of radio sources, and to radio studies of sources and the background. Painstaking studies of the extragalactic radio background, spectral information on a sample of very faint radio sources, and further understanding of the sources contributing to the counts, should eventually permit us to test the Compton model.

2. Non-Thermal Bremsstrahlung

Silk and McCray (1969) have attempted to fit the observed X-ray spectrum by a nonthermal electron bremsstrahlung model in an evolving universe. The process is more efficient at large redshifts because of the higher density of the intergalactic gas. These authors postulate that sub-relativistic electrons with a suitable spectrum are injected at a redshift ~ 10. The break in the spectrum results from Coulomb losses, which flatten the electron spectrum at low energies. In order to fit the observed position of

---

* Note that the energy of the electrons which give rise to X-ray photons of a given energy is independent of the redshift at which the emission occurs, because the cosmological redshift is cancelled by the higher energy of the black body photons which are being scattered.

** In the paper of Felten and Rees (1969), a redshift ~ 3 was derived, because the break was presumed to occur at ~ 60 keV rather than ~ 20 keV.

the break, they require a cosmology in which the intergalactic gas now has a density of only $\sim 10^{-7}$ cm^{-3}. They cannot fit the spectrum if the density is higher than this, because losses would then flatten it in the 100 keV range.

The energy requirements of the postulated burst of electron injection around $z \simeq 10$ are equivalent, in terms of energy input per unit volume, to the production of 6×10^{-2} eV cm^{-3} at the present epoch. The injected spectrum is $N(E) \propto E^{-2.5}$ particles (cm^2 sec ster keV)$^{-1}$. The sources of this enormous flux – for which there is no other evidence – are not specified, and the spectrum is chosen in an *ad hoc* fashion. Also, it is not easy to imagine how the slow electrons could escape from their sources into the low density intergalactic gas before being degraded in a denser medium (in which case the resulting bremsstrahlung spectrum would be altered). If the same sources (radio galaxies or quasars?) also produce relativistic electrons with even 1% of this energy density, *more* X-rays would be generated by the Compton process. This is because bremsstrahlung emission is inefficient, whereas Compton scattering may be almost 100% efficient. A corollary of this low efficiency is that a large amount of heat goes into the thermal gas. It is true that the X-ray emission from a low density intergalactic gas would not be detectable. However the model obviously requires that the thermal particles should be less energetic than the relevant non-thermal ones (i.e. $T \lesssim 10^7$ K). If there were 100 times as much energy in protons as in electrons, the gas temperature T would rise to $\sim 10^9$ K – in this case the distinction between 'thermal' and 'non-thermal' particles would be lost, and there would be no possibility of reproducing the observed X-ray spectrum. Finally, the calculated spectrum, even after all these assumptions have been made, does not seem an adequate fit over the whole range from 1 keV to 1 MeV.

Hayakawa (paper in this volume) has developed an analogous model based on bremsstrahlung from protons. As we did not receive any detailed report of this work, we cannot discuss it further. However, because of the similarity to the electron bremsstrahlung process, some of the above remarks apply to this work also. In particular, the problem of the energy requirements seems even more severe.

5. Soft X-Rays

There have recently been several observations of diffuse X-rays at energies $\lesssim 1$ keV. The inconsistencies between the various measurements, which reflect the experimental difficulties, are discussed by Oda (paper in this volume). Soft X-ray observations are especially interesting because of the important role played by galactic (and perhaps also intergalactic) absorption, and also because there may be an additional thermal contribution to the background at these energies.

A straight extrapolation of the power-law spectrum observed at higher energies (1–10 Å) yields a flux incident on the galaxy of ~ 40 counts at 0.5 keV and ~ 115 counts at 0.27 keV (using the spectrum (1')). The flux at the earth attributable to this component would be reduced as a consequence of absorption. The galactic absorption can be estimated either from 21-cm measurements, or directly from the dependence of the observed soft X-ray flux on galactic latitude. We return to this problem later.

We may contrast this extrapolation with the flux quoted by Bunner *et al.* (paper in this volume) which, after their correction for galactic absorption has been applied, is ~50 counts at 0.5 keV and ~120 counts at 0.27 keV. The corrected flux at 0.28 keV quoted by Bowyer *et al.* (1968) is ~200 counts.* Although other observers (Henry *et al.*, 1968; Baxter *et al.*, 1969) have claimed that the flux incident on the Galaxy is somewhat higher, we do not consider that the observations have yet established the definite existence of any component in addition to the non-thermal background observed at higher energies.

If there *were* such a component, however, it could be interpreted as bremsstrahlung emission from an ionised gas at a temperature ~10^6 K. Indeed Henry *et al.* (1968), who quoted a corrected flux of 600 counts at 0.28 keV – too high to be a plausible extrapolation of the non-thermal spectrum – attributed the excess to an intergalactic medium with a density ~10^{-5} cm^{-3}. If correct, this would be the first positive evidence for the existence of a quantity of material sufficient to close the universe, and would plainly be exceedingly important for cosmology. Weymann (1967) has considered in detail the 'thermal history' of an intergalactic gas, assuming it to consist of hydrogen and helium (in the ration 10 : 1 by number). Present observations set two important constraints on the state of such a gas: it must be very highly ionized so as to account for the lack of absorption shortward of the redshifted Lyman α wavelength in the spectra of distant quasars (Gunn and Peterson, 1965), and so (at least if it is collisionally ionised) it must be hot; on the other hand, it must not be so hot that it emits thermal X-rays with an intensity exceeding the observed background at any wavelength. Weymann showed that if the present density of this gas were ~10^{-5} cm^{-3}, the 'thermal history' was constrained within very narrow limits by the data available at that time (1967). Although one of Weymann's models did indeed predict a soft X-ray flux comparable with that reported by Henry *et al.*, we wish to emphasise that this is only one of several possible origins for thermal bremsstrahlung of this intensity. (A fuller discussion of this and related topics is given by Rees *et al.*, 1968). All that is required is a gas with emission measure ~0.3 cm^{-6} pc in a typical direction, at a temperature ~5×10^5 K (or ~$5 \times 10^5 (1+z)$ K if the main emission comes from a substantial redshift z). This emission measure could be furnished by gas of density ~10^{-4} cm^{-3} within clusters of galaxies. Another possibility would be emission from the halo of our Galaxy (with density ~10^{-2} cm^{-3} and dimensions ~10^4 pc), or possibly smaller and denser regions of gas with $T \sim$ a few times 10^6 K, such as might be expected behind shock fronts associated with the high velocity clouds (Savedoff *et al.*, 1967). In the latter cases the required emission measure could be even lower than 0.3 cm^{-6} pc, because we would expect heavy elements also to be present, and these emit more efficiently than hydrogen and helium when $T \simeq 10^6$ K (Pottasch, 1966).

Shklovskii (1969) has proposed that it could be due to emission lines of O VII and O VIII (with $\lambda \simeq 20$ Å) from supernovae in galaxies with $z \simeq 1$. Finally, the excess soft

prAn additional uncertainty in these quoted fluxes arises because the figure itself depends on the
* esumed spectrum over the broad energy band covered by the detector.

X-ray flux could be the integrated effect of a population of extragalactic sources which mimic a thermal spectrum.

An indirect method of estimating how the soft X-ray spectrum extends into the ultraviolet has been proposed by Sunyaev (1969). He argues that the low density H I which has been detected in the outlying parts of M 31 and other nearby galaxies would have been dispersed by an inward-propagating ionisation front if the extragalactic ultraviolet radiation were too intense. Sunyaev claims that the bremsstrahlung curve derived by Weymann (1967), which was consistent with the 0.28 keV flux measured by Henry et al., can be ruled out by this argument. However, Bunner et al., whose observations in two energy bands gave them some spectral information, found that any thermal component of the background would not have such a steep slope as the spectrum given by Weymann's models, and, when extrapolated down to the Lyman limit, is actually compatible with the upper limit given by Sunyaev.

Also, it may be possible to evade Sunyaev's conclusions by supposing that the observed neutral hydrogen is surrounded by an ionised region (perhaps filling the volume of the Local Group) which shields the H I from the external ultraviolet flux. We have considered this point further in another paper in this volume.

A. GALACTIC ABSORPTION

A straightforward estimate of the X-ray opacity of the Galaxy can be made by using 21-cm observations of the column density of neutral hydrogen as a measure of the amount of matter along the line of sight. The most accurate calculations of the absorption cross-sections have been carried out by Bell and Kingston (1967). At energies below the K-edge of carbon (0.28 eV) the absorption is mainly due to hydrogen and helium. Following Bowyer et al. (1968), we may write the absorption cross-section per hydrogen atom in the form

$$\sigma(E) = \sigma_0 \left(\frac{E}{E_0}\right)^{-3.2} \text{cm}^2,$$ (13)

where

$$\sigma_0 = (0.84 + 21.4\xi) \times 10^{-21} \text{cm}^2$$ (14)

σ_0 is the effective cross-section at 0.28 keV, and $\xi = n(\text{He})/n(\text{H})$. If we assume that ξ has the 'cosmical' value of 0.1, $\sim 70\%$ of the absorption would be due to helium. (There may be some extra opacity from molecular hydrogen.) According to Bowyer et al. and Bunner et al., the background at 0.28 keV decreases more slowly with angular distance from the galactic pole than would be expected on the basis of the H I column density derived from 21-cm data. Unless interstellar space is almost devoid of helium, one must assume that either (a) the straightforwardly calculated galactic opacity is misleadingly high, or (b) local sources are responsible for some of the background at low galactic latitudes.

(a) (i) 21-cm maser action.

In an attempt to account for the weak Lyman α absorption in some stellar spectra (Morton, 1967; Jenkins and Morton, 1967; Carruthers, 1968; Morton and Jenkins,

1968) Fischel and Stecher (1967) suggested that 'pumping' by blue-shifted Lyman α photons could lead to population inversion in the hyperfine states of H I. Because 21-cm radiation would be enhanced by the ensuing maser action, a lower column density of H I is then sufficient to produce a given intensity in the 21-cm line. (A similar suggestion had earlier been made by Shklovskii (1967), following Varshalovich (1967), for the specific case of the high-velocity clouds.) A more detailed study (Van Bueren and Oort, 1968; Storer and Sciama, 1968), however, shows that, because of the short mean free path of Lyman α photons in H I regions, this effect only occurs in thin layers, and so could not significantly affect the 21-cm intensity along a whole line of sight through the galactic disc. This proposed explanation for the anomalous latitude-dependence of the X-ray background must consequently be discarded.

(ii) 'Clumpiness' of interstellar hydrogen.

High resolution 21-cm studies at low galactic latitudes reveal irregularities which suggest that interstellar hydrogen may be in clouds, the precise nature of which is, however, still unclear (Van Woerden, 1967). A given amount of hydrogen would obviously tend to absorb X-rays less if it were concentrated into clouds. This is because part of the hydrogen which contributes to the 21-cm emission would be embedded in clouds which were already so opaque that its additional X-ray absorbing power is not used effectively. For this to be an important effect, a substantial fraction ($\gtrsim 60\%$) of the H I must be in large clouds. Bowyer and Field (1969) considered clouds with column densities up to $(1-5) \times 10^{21}$ cm^{-2}, and showed that the absorption, averaged over a solid angle large enough to contain a statistical sample of clouds, would be only 0.33 ± 0.18 of the value if the same amount of hydrogen were smoothly distributed. This reduction would be sufficient to account for the latitude-dependence which Bowyer et al. observed. The beam used by these workers was $8° \times 8°$. However, the large clouds discussed by Bowyer and Field would be ~ 25 pc across (Van Woerden, 1967), so that at distances 100–300 pc they would subtend angles larger than the beam. This fact seems to us to raise questions regarding the validity of Bowyer and Field's discussion. Bunner et al. included the effects of 'clumpiness' in their analysis, and considered clouds with column density 1.2×10^{21} cm^{-2}, which would be opaque at 0.28 keV but not at 0.5 keV. They used a beam $2.6° \times 23°$ at half power, but it is not clear to us, from the paper of Bunner et al., whether their work is subject to the above objection.

As is remarked by Bunner et al., the large hydrogen clouds may be opaque in the 21-cm line, particularly if the spin temperature is low. Should this be so, their column density may have been underestimated. If this effect were properly allowed for, it is by no means obvious whether the assumption that the observed H I is in clouds actually reduces the calculated X-ray opacity at all.

(b) Any galactic contribution to the soft X-ray flux must come effectively from distances $\lesssim 200$ pc, because most 0.28 keV photons emitted at greater distances would be absorbed before reaching us. The existence of a galactic component could account for the observed latitude-dependence of the diffuse soft X-rays without the necessity to invoke anomalous absorption. If such a background were the integrated effect of unresolved sources, each source must be much fainter than the known discrete galactic sources. For none to have been resolved by Bunner *et al.*, the density of sources would need to be $\gtrsim 10^{-2}$ pc^{-3}.

Whether such a component is due to sources or is genuinely diffuse, its spectrum must be steeper than that of the extragalactic background. Otherwise the anisotropy above 2 keV (at which energies the whole disc is transparent) would be much greater than that reported by Cooke *et al.* (paper in this volume). The latitude-dependence of the expected galactic flux depends on whether the scale height of the emission is larger or smaller than the scale height of the interstellar gas responsible for the absorption (Rees, paper in this volume).

B. INTERGALACTIC ABSORPTION

All the theories for the X-ray background which we have discussed imply that the bulk of the observed radiation originates at cosmological distances ($z \simeq 1$): in models involving evolution, the observed photons may even have originated at redshifts $z \gtrsim 5$. If a substantial part of the X-ray background at 0.28 keV is an extension of the non-thermal background at high energies, we may infer upper limits to the amount of intergalactic absorption. The condition that the optical depth out to $z \simeq 1$ should be $\lesssim 1$ requires that the mean density of neutral hydrogen must be $\lesssim 10^{-7}$ cm^{-3}, so that, if the intergalactic gas has the 'magic' density of $\sim 10^{-5}$ cm^{-3} it must be $\gtrsim 99\%$ ionised.* Although the Lyman α observations of quasars set a much more stringent limit, this is a non-trivial additional constraint, because (a) it is independent of the cosmological interpretation of quasar redshifts, and (b) some theoretical 'thermal histories' would allow the gas to partially recombine by the present time, even if the ionisation at $z \simeq 2$ were essentially complete.

If the X-rays originate at redshifts much greater than 2, we can infer something about the ionisation level of the gas beyond the remotest observed quasars. (In calculating the absorption at very large redshifts, we must allow for the facts that the gas density $\propto (1+z)^3$, that the cross-section for absorption of a photon of observed wavelength λ by hydrogen or helium is roughly proportional to λ^3 (i.e. to $(1+z)^{-3}$), and that the effective path length is $(1+z)^{-3/2}$ in an Einstein-de Sitter model and $(1+z)^{-1}$ in a low-density model.) The combined result of these effects is that the limits we can place on the degree of ionisation are somewhat less sensitive than for smaller redshifts. These conclusions would not alter significantly if the gas were concentrated in clusters, because there would be several clusters along every line of sight out to a redshift $z \simeq 2$.** If an accurate spectrum of the X-ray background from, say, 1–50 Å were available, one might be able to infer whether the long wavelength flux was being attenuated and, if so, where the absorption was taking place.

Limits can also be placed on the intergalactic abundance of heavy elements. At

---

* Some possible effects of scattering by ionised intergalactic gas, which may be important if the X-rays originate at $z \gtrsim 5$, are discussed by us elsewhere (paper by Rees and Setti, in this volume).
** Specially large clusters may be detectable in absorption at 1–3 keV if they contain a large amount of gas.

temperatures $\lesssim 10^6$ K, the K-electrons of elements such as CNO would not be stripped off by collisions with electrons. If the optical depth out to $z \simeq 1$ is $\lesssim 1$ at 0.28 keV, these elements could not have more than $\sim 2\%$ of their 'cosmical' abundance if the total density is $\sim 10^{-5}$ cm^{-3} (with correspondingly less sensitive limits for lower densities). While it would be surprising if these elements were present with their full local abundance relative to hydrogen, this is nevertheless a significant result, since it limits the cumulative contamination of intergalactic space by galaxies and exploding objects during the expansion of the universe.*

6. γ-Rays

Various physical processes which might give rise to cosmic γ-ray emission have been described in various papers (Ginzburg and Syrovatskii, 1964a, b; Gould and Burbidge, 1965; Hayakawa *et al.*, 1964; Oda, 1965; Fazio, 1967; Garmire and Kraushaar, 1965). The present data indicate an isotropic flux at ~ 100 MeV (Clark *et al.*, 1968) and also a flux at 1–10 MeV observed with a non-directional detector (Vette *et al.*, paper in this volume). There are also upper limits to the background at much higher energies (Oda, 1965). The 100 MeV flux lies on an extrapolation of the X-ray spectrum (1″), and is possibly due to the same process. Although the Compton spectrum must cut off at some energy, there seems no reason why it should not extend up to 100 MeV. Electrons with $\gamma = 3 \times 10^5$ would be required to produce these photons via Compton scattering of black body photons.

From time to time, various other processes have been proposed for producing γ-rays. Among these we may mention decay of π_0 produced in collisions of high-energy cosmic rays with intergalactic matter. Even if the gas density is 10^{-5} cm^{-3}, a rather high intergalactic flux of cosmic rays would be needed to produce the observed γ-rays (Gould, 1967). Another possibility is matter-antimatter annihilation (Stecker, 1969). It would be premature to discuss these processes in detail until more data are available.

7. Conclusions

Our primary conclusion is that there is not yet any reliably established theory for the X-ray background. It should be superfluous to emphasise the necessity for further

---

* In a recent preprint, Longair and Sunyaev (1969) have proposed a unified model to account for the spectrum of the whole X-ray background. They suppose that relativistic electrons are generated in a class of sources, which they tentatively identify with Seifert galaxies. It is envisaged that the electrons are confined in the nucleus for a period during which losses cause a break to develop in their spectrum. During this stage they emit X-rays by Compton scattering of infrared radiation, and it is suggested that the X-ray background above ~ 1 keV is the integrated effect of such emission in all sources of this class. The electrons then escape, and emit soft X-rays by Compton scattering of the 3 K background. In order to obtain the observed energy in the soft X-ray background, it must be assumed that the electrons avoid serious adiabatic losses during their escape. Also, the model requires remarkably 'standardised' conditions in all the sources, otherwise the bends in the X-ray background spectrum would have been smeared out. Finally if these sources were indeed Seyferts, and since 1–2% of all the galaxies are of the Seyfert type, NGC 1068, e.g., would be likely to appear at least as strong as M 87 at X-ray energies.

observations before a firmer understanding can be achieved. At present it is impossible to discriminate decisively between alternative models, any of which can be made compatible with the present data. We have concentrated our discussion on the astrophysical implications of the various possibilities, and on the plausibility of the assumptions made in each of them. At the moment it seems that the model which entails the minimum number of *ad hoc* assumption is Compton scattering of black body photons by relativistic electrons in radio galaxies at large redshifts. We are doubtless guilty of oversimplification in seeking an explanation in terms of any single mechanism: indeed, we cannot discount the possibility that *all* the processes discussed here may contribute significantly to the X-ray background.

There is need for an improved spectrum in the 1 keV – 1 MeV range, in order to pin down the position of the apparent 'break'. More observations are required in order to clarify the confusing situation below 1 keV, and, at the other extreme, information in the γ-ray region is still lacking. High-resolution studies below 1 keV, in conjunction with improved 21-cm maps at high galactic latitudes, should indicate whether the galactic absorption is due to small and dense clouds. It should then be possible to estimate what contribution to the flux at low galactic latitudes comes from local sources, or from diffuse processes occurring within the disc.

There is also a need for high-resolution observations, and studies of the level of isotropy on various angular scales.

Apart from M 87, and perhaps the Large Magellanic Cloud (Mark *et al.*, 1969) and 3C 273 (Friedman and Byram, 1967) there are no data on the X-ray emission from any discrete objects beyond our own Galaxy. Until this state of affairs has changed, it is difficult to estimate the likely contribution of such sources to the background.

Attempts to detect X-rays from rich clusters (e.g., Virgo or Coma) would provide a better estimate of the mean output of normal galaxies. (Low energy observations could also yield information about gas within the clusters.) If sources do indeed contribute appreciably to the X-ray background, observations with good angular resolution may tell us the number and power of the objects involved. (Some relevant figures are quoted by Weymann, 1968). Some 'patchiness' would also be expected in the so-called 'diffuse' processes, because neither the fast particles (nor, in the case of bremsstrahlung, the thermal gas) is likely to be distributed smoothly enough for the emission to appear completely uniform. In particular, Compton scattering of the black-body background, normally thought of as a diffuse process, would in fact give rise to apparently discrete sources, simply because at large redshifts the Compton lifetimes may be only $\sim 10^6$ years, and the electrons do not have time to escape far from their point of origin. It may be possible to resolve such sources with $z \simeq 2\text{--}5$ (Felten and Rees, 1969).

Acknowledgements

We have benefited from discussions with many colleagues, and would especially like to thank Drs. J. E. Felten and D. W. Sciama. One of us (G.S.) acknowledges support from CNR.

References

Allen, C. W.: 1963, *Astrophysical Quantities*, The Athlone Press, University of London.
Baxter, A. J., Wilson, B. G., and Green, D. W.: 1969, *Astrophys. J.* **155**, L145 (see also paper in this Symposium).
Bell, K. L. and Kingston, A. E.: 1967, *Monthly Notices Roy. Astron. Soc.* **130**, 373.
Bergamini, R., Londrillo, P., and Setti, G.: 1967, *Nuovo Cimento* **52B**, 495.
Boldt, E. and Serlemitsos, P.: 1969, *Astrophys. J.* **157**, 557.
Bowyer, C. S., Field, G. B., and Mack, J. E.: 1968, *Nature* **217**, 32.
Bowyer, C. S. and Field, G. B.: 1969, *Nature* **223**, 573.
Brecher, K. and Morrison, P., 1967: *Astrophys. J.* **150**, L61.
Brecher, K. and Morrison, P.: 1969, *Phys. Rev. Letters* **23**, 802.
Bridle, A. H.: 1967a, *Monthly Notices Roy. Astron. Soc.* **136**, 219.
Bridle, A. H.: 1967b, *Observatory* **87**, 263.
Carruthers, G. R.: 1968, *Astrophys. J.* **151**, 269.
Clark, G. W., Garmire, G. P., and Kraushaar, W. L.: 1968, *Astrophys. J.* **153**, L203.
Cooke, B. A., Griffiths, R. E., and Pounds, K. A.: 1969, *Nature* **224**, 134. (See also paper in this volume.)
Davidson, W. and Davies, M.: 1964a, *Monthly Notices Roy. Astron. Soc.* **127**, 241.
Davidson, W. and Davies, M.: 1964b, *Monthly Notices Roy. Astron. Soc.* **128**, 363.
Fazio, G. G.: 1967, *Ann. Rev. Astron. Astrophys.* **5**, 481.
Felten, J. E. and Morrison, P.: 1966, *Astrophys. J.* **146**, 686.
Felten, J. E. and Rees, M. J.: 1969, *Nature* **221**, 924.
Fischel, D. and Stecher, T. P.: 1967, *Astrophys. J.* **150**, L51.
Friedman, H. and Byram, E. T.: 1967, *Science* **158**, 257.
Garmire, G. P. and Kraushaar, W. L.: 1965, *Space Sci. Rev.* **4**, 123.
Ginzburg, V. L. and Syrovatskii, S. I.: 1964a, *Usp. Fiz. Nauk. SSSR* **84**, 201 [transl. *Soviet Phys. – Usp.* **7**, 1965, 696].
Ginzburg, V. L. and Syrovatskii, S. I.: 1964b, *Origin of Cosmic Rays*, Pergamon Press, Oxford.
Gorenstein, P., Kellog, E. M., and Gursky, H.: 1969, *Astrophys. J.* **156**, 315.
Gould, R. J. and Burbidge, G. R.: 1965, *Ann. Astrophys.* **28**, 171.
Gould, R. J.: 1967, *Amer. J. Phys.* **35**, 376.
Gunn. J. E. and Peterson, B. A.: 1965, *Astrophys. J.* **142**, 1633.
Hayakawa, S. and Matsuoka, M.: 1964, *Progr. Theoret. Phys.* (Kyoto), Suppl. **30**, 204.
Hayakawa, S., Okuda, H., Tanaka, Y., and Yamamoto, Y.: 1964, *Progr. Theoret. Physics*, Suppl. **30**, 153.
Haymes, R. C., Ellis, D. V., Fishman, G. J., Glenn, S. W., and Kurfess, J. D.: 1969, *Astrophys. J.* **155**, L31.
Henry, R. C., Fritz, G., Meekins, J. F., Friedman, H., and Byram, E. T., 1968, *Astrophys. J.* **153**, L11.
Hornby, J. M. and Williams, P. J. S.: 1966, *Monthly Notices Roy. Astron. Soc.* **131**, 237.
Jenkins, E. B. and Morton, D. C.: 1967, *Nature* **215**, 1257.
Kellermann, K. I.: 1966, *Astrophys. J.* **146**, 621.
Kellermann, K. I. and Pauliny-Toth, I. I. K.: 1969, *Astrophys. J.* **155**, L71.
Lang, K. R. and Terzian, Y.: 1969, *Astrophys. Letters* **3**, 29.
Longair, M. S.: 1966, *Monthly Notices Roy. Astron. Soc.* **133**, 421.
Longair, M. S. and Sunyaev, R. A.: 1969, preprint No. 84, Lebedev Physical Institute, Moscow.
Low, F. J. and Tucker, W. H.: 1968, *Phys. Rev. Lett.* **21**, 1538.
Mark, H., Price, R., Rodrigues, R., Seward, F. D., and Swift, C. D.: 1969, *Astrophys. J.* **155**, L143.
Morton, D. C.: 1967, *Astrophys. J.* **147**, 1017.
Morton, D. C. and Jenkins, E. B.: 1968, *Astron. J.* **73**, S110.
Oda, M.: 1965, *Proc. Int. Conf. Cosmic Rays*, London, Vol. I, p. 68.
Oort, J. H.: 1958, 'La structure et l'évolution de l'Univers', Brussels, Solvay Conference.
Pooley, G. G. and Ryle, M.: 1968, *Monthly Notices Roy. Astron. Soc.* **139**, 515.
Pottasch, S. R.: 1966, *Bull. Astron. Inst. Netherl.* **18**, 156.
Rees, M. J.: 1967, *Monthly Notices Roy. Astron. Soc.* **137**, 429.
Rees, M. J., Sciama, D. W., and Setti, G.: 1968, *Nature* **217**, 326.
Rees, M. J. and Setti, G.: 1968, *Nature* **219**, 127.

Rowan-Robinson, M.: 1967, *Nature* **216**, 1289.
Rowan-Robinson, M.: 1968, *Monthly Notices Roy. Astron. Soc.* **138**, 445.
Ryle, M.: 1968, *Ann. Rev. Astron. Astrophys.* **6**, 249.
Savedoff, M. P., Hovenier, J. W., and Van Leer, B.: 1967, *Bull. Astron. Inst. Netherl.* **19**, 107.
Sciama, D. W.: 1968, Review paper, presented at the X-Ray Meeting of the Royal Society, London, November 1968. To appear in *Phil. Trans. Roy. Soc.*
Schmidt, M.: 1966, *Astrophys. J.* **146**, 7.
Schmidt, M.: 1968, *Astrophys. J.* **151**, 393.
Seward, F. D., Chodil, G., Mark, H., Swift, C. D., and Toor, A.: 1967, *Astrophys. J.* **150**, 845.
Shapiro, M. H. and Silberberg, R.: 1968, Proc. 10th Internat. Conf. Cosmic Rays, Calgary, *Canadian Jour. Phys.* **46**, S561.
Shklovskii, I. S.: 1967, *Soviet Astron. – A.J.* **11**, 240.
Shklovskii, I. S.: 1969, *Astrophys. Letters* **3**, 1.
Sholomitskii, G. B.: 1968, *Soviet Astron. – AJ* **11**, 756.
Silk, J.: 1968, *Astrophys. J.* **151**, L19.
Silk, J.: 1969, *Nature* **221**, 347.
Silk, J. and Mc Cray, R.: 1969, *Astrophys. Letters* **3**, 59.
Stecker, F. W.: 1969, *Nature* **221**, 425.
Storer, S. H. and Sciama, D. W.: 1968, *Nature* **217**, 1237.
Sunyaev, R. A.: 1969, *Astrophys. Letters* **3**, 33.
Turtle, A. J. and Baldwin, J. E.: 1962, *Monthly Notices Roy. Astron. Soc.* **124**, 459.
Varshalovich, D. A.: 1967, *Soviet Phys. – JETP* **25**, 157.
Van Bueren, H. G. and Oort, J. H.: 1968, *Bull. Astron. Inst. Netherl.* **19**, 414.
Van Woerden, H.: 1967, in *Radio Astronomy and the Galactic System*, I.A.U. Symp. No. 31 (ed. by H. Van Woerden), p. 3.
Véron, P.: 1966a, *Nature* **211**, 724.
Véron, P.: 1966b, *Ann. Astrophys.* **29**, 231.
Weymann, R. J.: 1967, *Astrophys. J.* **147**, 887.
Weymann, R. J.: 1968, in *Highlights of Astronomy*, I.A.U. Publication (ed. by L. Perek), Reidel Publ. Comp., Dordrecht, Holland, p. 220.
Williams, P. J. S.: 1966, *Nature* **210**, 285.

COSMIC BACKGROUND X-RAYS PRODUCED BY
INTERGALACTIC INNERBREMSSTRAHLUNG

SATIO HAYAKAWA

Dept. of Physics, Nagoya University, Nagoya, Japan

It has been found difficult to explain both the absolute intensity and energy spectrum of the background component of cosmic X-rays in terms of a superposition of X-ray sources in distant galaxies [1] and of the inverse Compton collisions of metagalactic electrons with microwave photons [2]. Although the absolute intensity could be obtained by choosing suitable values of parameters which are not well known as yet, the gradual bending of the X-ray spectrum around 30 keV provides a critical test of theories. In this respect innerbremsstrahlung of intergalactic protons colliding with electrons is a candidate of background X-rays, since the X-ray spectrum is closely related to the proton spectrum and the latter usually bends in the non-relativistic region.

Innerbremsstrahlung is associated with the production of knock-on electrons by protons. The latter is a part of the ionization process which is responsible for heating the intergalactic medium [3]. Thus the background component of cosmic X-rays may provide a means of studying the thermal history of the universe.

The proton spectrum may bend due either to ionization energy loss or to an energy dependence of the acceleration rate or to both. If the former is the case, the spectrum bends at about [4]*

$$E_c = \left(\frac{3an_0t_0}{2Mc^2}\right)^{2/3} mc^2 \approx 6.5 \text{ keV},$$

where $a \approx 3 \times 10^{-7}$ eV cm^3 sec^{-1} is an ionization energy loss rate, n_0 the present electron density, t_0 the present age of the universe, and M and m the proton and electron masses, respectively. The numerical value is obtained for $n_0 = 10^{-5}$ cm^{-3} and $t_0 = 3 \times 10^{17}$ sec; this is too small compared with the observed bending energy of about 30 keV. The acceleration at a rate proportional to total energy gives a power law in the relativistic region, but the kinetic energy spectrum in the non-relativistic region becomes less steep as energy decreases. Although we do not know a precise shape of the spectrum, unless detail of the acceleration process is known, it is quite likely that the proton spectrum in kinetic energy changes its slope at several tens of MeV. This may be transferred to the X-ray spectrum which also changes its slope as observed.

If the cosmic ray production rate varies with cosmic age as t^{-2}, as expected from radio star count, and the production started at the redshift parameter $z = 10$, the absolute intensity of background cosmic rays can be explained for the present value

---

* The X-ray energy relative to the proton energy given in [4] should be reduced by a factor of 4.

L. Gratton (ed.), Non-Solar X- and Gamma-Ray Astronomy, 372–373. All Rights Reserved
Copyright © 1970 by the IAU

of the cosmic ray energy density times matter density of about 10^{-20} erg cm^{-6}. For $n_0 = 10^5$ cm^{-3}, the cosmic ray energy density of 10^{-15} erg cm^{-3} is consistent with that given in [3].

References

[1] Fujimoto, M., Hayakawa, S., and Kato, T.: 1969, *Astrophys. Space Sci.* **4**, 64.
[2] Fukui, M. and Hayakawa, S.: 1969, *Progr. Theor. Phys.* **42**, 119.
[3] Ginzburg, V. L. and Ozernoi, L. M.: 1966, *Soviet Astron.-AJ* **9**, 726.
[4] Hayakawa, S.: 1969, *Prog. Theor. Phys.* **41**, 1592.

A MODEL OF γ-RAY SOURCES IN THE GALAXY

S. HAYAKAWA and Y. TANAKA

Dept. of Physics, Nagoya University, Nagoya, Japan

In a recent observation Clark *et al.* [1] have found that the intensity of γ-rays of energies around 100 MeV is strong in the direction of the galactic plane; the observed intensity of the order of 10^{-4} cm^{-2} sec^{-1} rad^{-1} is greater by an order of magnitude than that predicted on the neutral pion decay hypothesis [2]. It should, however, be remarked that the theory is based on the assumption that cosmic rays are uniformly distributed in our galaxy. The assumption seems doubtful in view of the experimental result that the γ-ray intensity depends rather strongly on the galactic longitude.

If cosmic rays were uniform in the galactic disk, the intensity of γ-rays produced by neutral pion decays would have to be proportional to the column density of interstellar matter. Although the observed longitude dependence of the γ-ray intensity qualitatively shows such a tendency, it is much stronger than the longitude dependence of the column density of hydrogen atoms obtained from 21-cm radio observations [3]. The

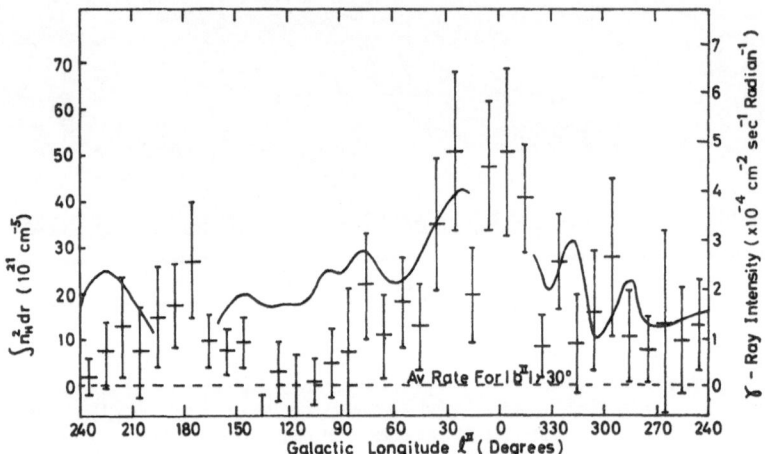

Fig. 1.   Longitude dependences of the γ-ray intensity and the column square density of hydrogen atoms.

γ-ray intensity is better correlated with the column square density of hydrogen atoms,

$$\langle n_H^2 R \rangle = \int n_H^2 dr , \tag{1}$$

as shown in Figure 1.

The correlation shown in Figure 1 seems to indicate that the density of γ-ray sources is proportional to the square density of the interstellar gas. The same relation has been suggested to hold also for X-ray sources, and the reason for this has been attributed to the fact that the star formation rate is proportional to the square density of gas and that most X-ray sources are associated with stars of ages younger than 10^9 years [4]. The same argument may hold for γ-ray sources. It is thus assumed that most γ-ray sources are associated with supernova remnants, a majority of which belong to type II supernovae and are identified with galactic non-thermal radio sources. The radio intensity of these sources [5] is compared with the γ-ray intensity in each longitude range in Figure 2. The agreement between these two is fairly good, if a few strong radio sources such as Cas A and Tau A are discarded.

The correlations shown above seem to suggest that the galactic γ-rays are produced mainly by neutral pion decays in supernova remnants. If this is the case, electrons are

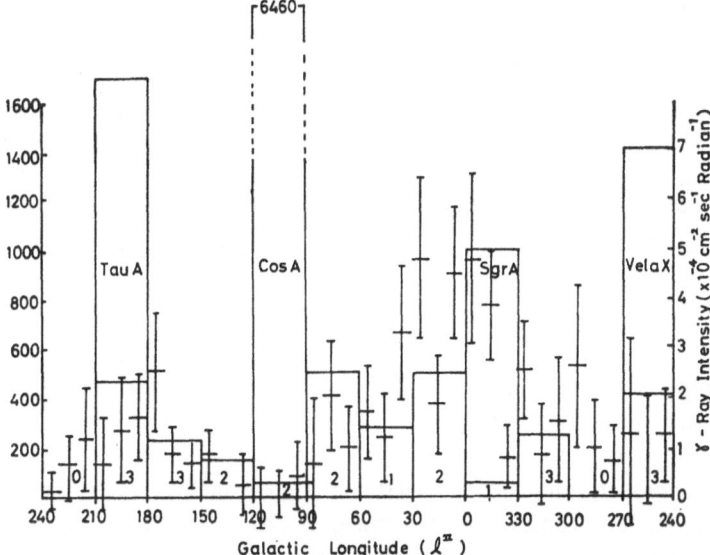

Fig. 2. Correlation between the γ-ray intensity and the radio intensity of supernova remnants. The numbers of supernova remnants adopted are indicated. Four strongest sources are separately indicated.

produced as well by charged meson decays and the radio waves emitted by these electrons would be too strong. This difficulty can be avoided, if strong magnetic fields are concentrated in an expanding shell rather than they are distributed uniformly in the whole volume of a remnant. This feature is in fact seen in Cygnus loop and other supernova remnants, and also is favored in the radio intensity distribution in the Crab Nebula. In what follows we shall demonstrate that this may be the case.

From Figure 2 we see that the intensity of γ-rays is about $J_\gamma = 1 \times 10^{-4}$ cm^{-2} sec^{-1} rad^{-1}. This is related to the production rate at a source, q_γ, and the surface density, D, of the sources in the galactic disk of thickness, d, as

$$J_\gamma = \frac{q_\gamma D}{4\pi}\left[1 + \ln\left(\frac{R_m}{d/2b_m}\right)\right], \qquad (2)$$

where the γ-rays coming from galactic latitudes within $\pm b_m$ and distances within R_m are included. Since $b_m = 15°$ [1] and $D = 0.85$ kpc^{-2} [5], we obtain

$$q_\gamma \simeq 3 \times 10^{39} \text{ sec}^{-1}. \tag{3}$$

On the other hand, the radio yield at 1 GHz is evaluated by reference to the compilation in [5] as

$$Y \simeq 2 \times 10^{23} \text{ erg sec}^{-1} \text{ Hz}^{-1}. \tag{4}$$

The radio yield can be accounted for in terms of synchrotron radiation of relativistic electrons in magnetic fields. The intensity of the electrons is assumed as $1/K$ times the proton intensity. If the radio emitting region is restricted to the shell of thickness Δr, which is smaller than the radius of a supernova remnant, r, we obtain

$$q_\gamma / Y \simeq 0.7 \times 10^{14} K n_H H^{-(1+\alpha)} (r/3 \Delta r), \tag{5}$$

where n_H is the hydrogen density in cm^{-3}, H is the magnetic field strength in μ gauss and α is the radio spectral index which is in most cases 0.7. This is compared with the observed value of 1.5×10^{16} erg^{-1} Hz. For a typical set of values, $K = 10^2$, $n_H = 10$, and $H = 10$, we find that $r/\Delta r = 30$ can account for the observed ratio of γ-ray and radio intensities.

A question arises whether or not the value of q_γ estimated in (3) is consistent with the intensity of electrons. Since there are about 600 radio emitting supernova remnants, the γ-ray production rate in our galaxy is estimated as

$$Q_\gamma \approx 600 \, q_\gamma \approx 2 \times 10^{42} \text{ sec}^{-1}. \tag{6}$$

This is compared with the intensity of electrons observed. A recent measurement of positrons with energies greater than 100 MeV gives an intensity of about 1×10^{-3} cm^{-2} sec^{-1} sr^{-1} [6]. Since most of such low energy electrons from meson decays are positive and the electron spectrum tends to be flat below a few hundred MeV, the total intensity of meson-decay electrons may not exceed 2×10^{-3} cm^{-2} sec^{-1} sr^{-1}. This would require that the electron production rate in our galaxy should be about 3×10^{41} sec^{-1}, which is by an order of magnitude smaller than that given in (6), if the electrons were uniformly distributed in our galaxy and they were not subject to solar modulation. If, however, the solar modulation is appropriately taken into account [7], the interstellar intensity of electrons with energies around and below 100 MeV expected from the production rate (6) is not inconsistent with the observed intensity of positions. It should also be remarked that the former is neither inconsistent with the general galactic radio emission.

References

[1] Clark, G. W., Garmire, G. P., and Kraushaar, W. L.: 1968, *Astrophys. J.* **153**, L203.
[2] Hayakawa, S., Okuda, H., Tanaka, Y., and Yamamoto, Y.: 1964, *Prog. Theor. Phys. Suppl.* No. 30, 153.
[3] Kerr, F. J.: 1962, *Monthly Notices Roy. Astron. Soc.* **123**, 327.
[4] Fujimoto, M., Hayakawa, S., and Kato, T.: 1969, *Astrophys. Space Sci.* **4** 64.
[5] Poveda, A. and Woltjer, L.: 1968, *Astron. J.* **73**, 65.
[6] Fanselow, J. L., Hartman, R. C., Hildebrand, R. H., and Meyer, P.: 1968.
[7] Webber, W. R.: 1968, *Aust. J. Phys.* **21**, 845.

THE X-RAY EMISSION OF A HOT DENSE
INTERGALACTIC PLASMA

J. BERGERON*

Institute of Theoretical Astronomy, Cambridge, England

The implications of the existence of a dense, hot, intergalactic plasma are discussed for friedman universes with a zero cosmological constant. The points of interest are the temperature of the intergalactic gas and the degree of ionization of its constituents. The soft X-ray emission of this dense intergalactic gas must be in agreement with the actual observations of isotropic X-ray background around 1–3 keV and 0.27 keV. Moreover the model cannot be in contradiction with the lack of Lyman α absorption in the emission spectra of the quasi-stellar objects.

We take the intergalactic gas to be 90% hydrogen and 10% helium, we assume that this gas is heated and that the heating period occupies a time short compared to the lifetime of the universe. We have solved the coupled evolution equations for the temperature and degree of ionization of hydrogen and helium vs. redshift (Bergeron, 1969). In the equations radiative recombinations, electronic collisional ionizations and photoionizations by free-free radiation are considered. In the energy balance equation, the energy losses of the gas are free-free, free-bound, spontaneous Lyman α and two photons emissions; the energy gains are the photoionization by free-free radiation and by the radiation of recombination when the optical depth in the Balmer continuum is large. We need to know the optical depth in the Lyman continuum, at each redshift z, to determine the rate of radiative recombination and the free-free intensity. The coupled equations are solved by iterating the density from which the optical depth is obtained, at each redshift. The energy losses and gains, by radiation emission and by photoionization of the intergalactic gas, are negligible compared to the energy loss due to the adiabatic expansion of the universe for temperatures $T \gtrsim 2 \times 10^5 \, \mathrm{K}$.

The study of the emission spectra of the QSO's sets a lower limit on the temperature of the intergalactic gas, for the lack of Lyman α absorption in the emission spectra of QSO's gives an upper limit on the density of neutral hydrogen in their neighbourhood $(z \gtrsim 2)$. The optical depth $z(v_0)$, for the present time, at a frequency v_0, is

$$z(v_0) = \int_0^{z_{em}} n_H(x)\, \sigma(v_0(1+x)) \frac{dl}{dx}\, dx, \qquad (1)$$

where $\sigma(v_0(1+x))$ is the cross-section of radiative excitation for the Lyman α transition

$$\sigma(v) = \frac{\pi e^2}{mc}\, f g(v - v_{\mathrm{Ly}\alpha}), \qquad (2)$$

* On leave from the Institut d'Astrophysique de Paris, France.

L. Gratton (ed.), Non-Solar X- and Gamma-Ray Astronomy, 377–381. All Rights Reserved

where g is the normalized profile of the line. Using a δ function for the profile, (1) becomes

$$z\left(v_0 = \frac{v_{Ly\alpha}}{1 + z_{em}}\right) \simeq \frac{\pi e^2 f_{Ly\alpha}}{m H_0 v_{Ly\alpha}} \frac{n_H(z_{em})}{(1 + z_{em})(1 + 2q_0 z_{em})^{1/2}},$$ (3)

where H_0 is the present Hubble constant and q_0 the acceleration parameter.

We consider cases where the sudden heating occurs at a redshift running from 2 to 9. The study is done both for a closed universe $(q_0 = 1)$ and an open universe $(q_0 = 0.2)$. For the closed model the minimum temperature at the heating period vs. the redshift of this period is given in Figure 2. For the open model a detailed study has been made only for the heating period taking place at a redshift of 3, taking an optical depth of 0.5, the lower limit of the temperature at this redshift is

$$T(z = 3) > 5.4 \times 10^6 \, \text{K}.$$

A comparison of the theoretical emission field of the hot dense intergalactic plasma with the isotropic X-ray background at 1–3 keV and 0.27 keV gives an upper limit, or a possible value for the temperature of the intergalactic gas. A study of the X-ray emission of the intergalactic plasma for temperature $T > 2 \times 10^5$ K shows that the only important contribution is the free-free emission. The theoretical results are compared to the recent observations of Henry *et al.* (1968), Gorenstein *et al.* (1969), Baxter *et al.* (1969), Bowyer *et al.* (1968), Kraushaar (1969), Oda (1969). The results are shown in Figure 1. For energies greater than 1 keV, the observed spectrum obeys a power law, and since the theoretical spectrum obeys an exponential law the comparison between theory and observations gives an upper limit to the temperature. The recent measurements around 0.27–1 keV might show a break in the X-ray background spectrum and from this we could obtain a value for the temperature of the intergalactic gas. Yet the results of the observations differ by more than a factor 10 and so we obtain a number of possible temperatures, the largest of these gives us another upper limit for the temperature of the intergalactic gas. The theoretical X-ray intensity, at the present epoch is

$$I_{v_0} = 4\pi \int_0^{z_{max}} j_v e^{-\tau(v, x)} (1 + x)^{-3} \frac{dl}{dx} \, dx,$$ (4)

where j_v is the free-free emissinity per unit volume, per steradian, $\tau(v, x)$ is the optical depth at the frequency v and at a redshift x. Equation (4) can be written

$$I_{v_0} = I_0 \int_0^{z_{max}} e^{-\tau(v, x)} e^{-a/(1+x)} (1 + 2q_0 x)^{-1/2} \, dx$$ (5)

where

$$I_0 = 4\pi K (n_e \sum_{i, z} Z^2 n_{i, z})_0$$

$$a = h v_0 / k T_0$$

with the index 0 indicating the present epoch.

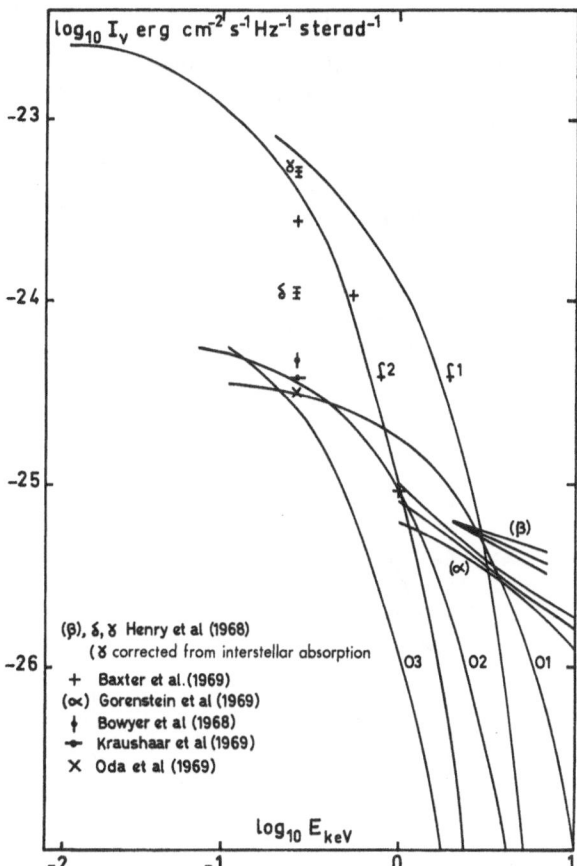

Fig. 1.   The X-ray intensity vs. the energy of the photons. The theoretical model: $f1$: $T(z=3)$ $= 3.7 \times 10^7\,\mathrm{K}$, $q_0 = 1$, $f2$: $T(z=3) = 1.6 \times 10^7\,\mathrm{K}$, $q_0 = 1$, the X-ray flux $f2$ is obtained by taking into account the variations of the gaunt factor with temperature and energy, 01: $T(z=3) = 1 \times 10^8\,\mathrm{K}$, $q_0 = 0.2$, 02: $T(z=3) = 4 \times 10^7\,\mathrm{K}$, $q_0 = 0.2$, 03: $T(z=3) = 1.7 \times 10^7\,\mathrm{K}$, $q_0 = 0.2$.

The temperature is then obtained by taking into account the variation of the gaunt factor vs. frequency and temperature (Karzas and Latter, 1961) in the computation of the free-free intensity, and by taking the observations at 0.27 keV corrected for interstellar absorption. The importance of this correction is not fully known since it depends on the interstellar density of hydrogen. Therefore we will give the possible values of the intergalactic gas for different observational intensities of the X-ray spectrum at 0.27 keV.

For a closed universe model ($q_0 = 1$) the theoretical intensity is given in Figure 1 (curves $f1, f2$) for a heating period taking place at a redshift of 3. But then the upper limit for the temperature is not determined by the measurements at 0.27 keV but by the ones at $E \gtrsim 1$ keV and this will be the case for every model with a heating period at a redshift $z \lesssim 4$. The results are plotted in Figure 2. Possible temperatures at the heating period are given vs. this epoch; the possible X-ray intensities at 0.27 keV are 5, 3 and 1×10^{-24} erg cm^{-2} sec^{-1} Hz^{-1} sterad^{-1}, and the lower limit for the temperature is given for an optical depth in the Lyman α line equal to 0.4 and 0.1.

For a closed model ($q_0 = 1$) the hypothesis of a dense intergalactic plasma is not in contradiction with observations provided that the gas temperature is of the order of

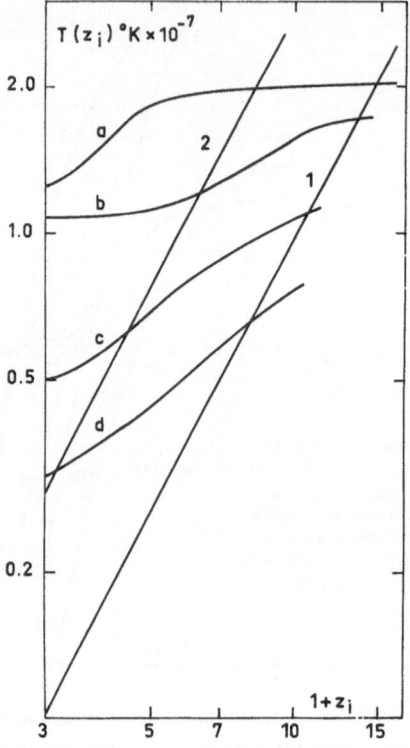

Fig. 2. Temperature of the intergalactic gas at the heating time of this gas (defined by z_t) vs. the redshift z_t. $1 = \tau$ (Lyman $\alpha(z=2)) = 0.4$; $2 = \tau$ Lyman $\alpha(z=2)) = 0.1$; $a = z_t > 4$ I_ν (0.27 keV) $= 5 \times 10^-$erg cm^{-2} s^{-1} Hz^{-1} sterad^{-1}; $z_t < 4$ I_ν (1 keV) $= 10^{-25}$ erg cm^{-2} s^{-1} Hz$^-$ sterad^{-1}; $b = I_\nu$ (0.27 keV) $= 3 \times 10^{-24}$ erg cm^{-2} s^{-1} Hz^{-1} sterad^{-1}; $c = I_\nu$ (0.27 keV) $= 1 \times 10^{-24}$ erg cm^{-2} s^{-1} Hz^{-1} sterad^{-1}; $d = I_\nu$ (0.27 keV) $= 3 \times 10^{-25}$ erg cm^{-2} s^{-1} Hz^{-1} sterad^{-1}.

10^7 K at the heating period, and that the heating time of the intergalactic gas occurs at redshifts less than 7.

For an open model ($q_0 = 0.2$) there is no possibility of explaining the high intensity values, $I_\nu \gtrsim 1 \times 10^{-24}$ erg cm^{-2} sec^{-1} Hz^{-1} sterad^{-1}, of the observed X-ray spectrum at 0.27 keV. But the model can account for the measurements at 0.27 keV of Bowyer et al. (1968), Kraushaar (1969), Oda (1969). A detailed study has been done for a heating period at a redshift of 3, and the range of temperatures allowed by the observations, at that redshift is

$$5.4 \times 10^6 \, \text{K} < T(z = 3) \lesssim 4 \times 10^7 \, \text{K}$$

for other heating periods, occurring at a redshift running from 2 to 9, only an approximate study has been done. Nevertheless we can conclude that for an open model ($q_0 = 0.2$) there is no contradiction between the observations and the model regardless of the time the heating occurs. If we consider a continuous heating beginning at a redshift z, $2 \lesssim z \lesssim 9$, the number of theoretical models in agreement with the observations is less than previously. The theoretical X-ray spectrum is more strongly dependent on the present temperature of the intergalactic gas than on its past history, on the beginning of the heating period. In order to fit the observed soft X-ray background spectrum, all models must give the same order of present-day temperature; accordingly the temperatures at the beginning of the heating period are smaller in the case of continuous heating than in that of sudden heating. If the heating is mainly collisional and not radiative the temperature at a redshift of 2 has to be the same as in the previous case and the range of the possible temperatures is then reduced.

For a closed model ($q_0 = 1$), if the heating is such that the variation of the gas temperature goes as $(1 + z)$, the hypothesis of a dense intergalactic gas is not in contradiction with the observations provided that the heating of the intergalactic gas starts at a redshift less than 3. If the heating is so efficient that the temperature remains constant, independent of the redshift, there is no closed model consistent with both observations of the soft X-ray background and the lack of Lyman α absorption in the emission spectrum of the QSO's.

References

Baxter, A. J.: this volume, p. 306.
Bergeron, J.: 1969, *Astron. Astrophys.* **3**, 42.
Bowyer, C. S. Field, G. B., and Mack, J. E.: 1969, *Nature* **217**, 32.
Gorenstein, P., Kellog, E. M., and Gursky, H.: 1968, *Astrophys. J.* **156**, 315.
Henry, R. C., Fritz, G., Meekins, J. F., Friedman, H., and Byram, E. T., 1968, *Astrophys. J.* **153**, L11.
Karzas, W. J. and Latter, R.: 1961, *Astrophys. J. Suppl.* **6**, 167.
Kraushaar, W.: this volume.
Oda, H.: this volume, p. 260.

POSSIBLE INITIAL EVIDENCE OF EXTRAGALACTIC COSMIC-RAY PROTONS AND THE AGE OF EXTRAGALACTIC COSMIC-RAY SOURCES

F. W. STECKER

Theoretical Studies Branch, NASA Goddard Space Flight Center, Greenbelt, Md., U.S.A.

Abstract. We have compared the recent cosmic background γ-ray observations with spectra predicted by various possible cosmic interactions. We find that the observed isotropic γ-rays with energies > 1 MeV can best be explained as being due to the decay of π°-mesons produced in extragalactic cosmic-ray collisions. This interpretation indicates that extragalactic cosmic-ray sources were more active (or prevalent) in the past and started to form at a redshift of ~ 100 corresponding to 10^7–10^8 years after the 'big-bang'.

For a present extragalactic gas density of 10^{-7}–10^{-5} cm^{-3}, the present extragalactic cosmic-ray flux is inferred to be 10^{-5}–10^{-3} the galactic value.

Recent theoretical studies by the author [1–4] have indicated the importance of observing isotropic cosmic-γ-radiation in the 1–100 MeV energy region. These predictions of isotropic γ-ray spectra from metagalactic inelastic strong interactions [1, 3, 4], matter-antimatter annihilation [2], and bremsstrahlung [4], along with studies of metagalactic Compton γ-rays [5] and bremsstrahlung γ-rays below 1 MeV energy [6] have indicated the following qualitative points:

(1) Bremsstrahlung and Compton processes may be possible alternative explanations of the observed isotropic X-ray spectrum below 1 MeV. The Compton process, however, requires constant regeneration of cosmic-ray electrons [7].

(2) Inelastic proton-proton interactions may account for the observed isotropic γ-ray flux of Clark *et al.*, [8], if the observed flux is considered to be real, rather than an upper limit. Extrapolations of predicted bremsstrahlung $(\sim E_\gamma^{-3.6})$ and Compton $(\sim E_\gamma^{-2.3})$ photon spectra, normalized to fit the X-ray observations, would only be compatible with the measurement of Clark *et al.* if that measurement is taken as an upper limit due to a spurious signal.

(3) When the predicted γ-ray spectra were normalized to fit the observations below 1 MeV and above 100 MeV (Clark *et al.*), it became apparent that a determination of the dominant process, or combination of processes which produce the observed X- and γ-rays, would only be made possible by a determination of the γ-ray spectrum between 1 and 100 MeV.

The recent observations of Vette *et al.* [9], have now provided us with measurements of background γ-rays up to 6 MeV. These data, along with some of those of Metzger *et al.* [10], are shown in the accompanying figure.* The differential intensity at 100 MeV is found from the integral measurement of Clark *et al.* by assuming that above 100 MeV the spectrum can be approximated by a power law with an index of

---

* We have also included an upper limit set by a balloon flight of the Rochester group and updated by a recent recalibration (G. Share, private communication).

L. Gratton (ed.), Non-Solar X- and Gamma-Ray Astronomy, 382–384. All Rights Reserved

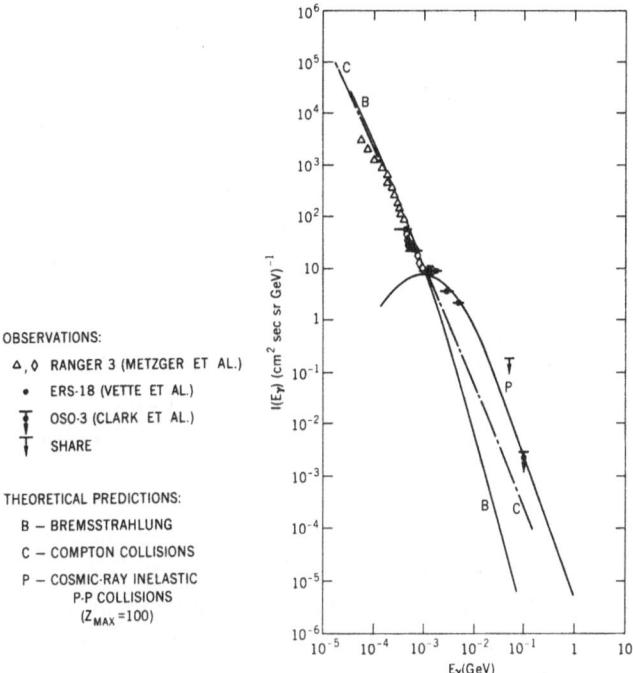

Fig. 1.   Extragalactic high energy photon spectra.

~3 as shown for the theoretical p–p spectrum. Also shown in the accompanying figure, are predicted γ-ray spectra due to the various possible metagalactic interactions. These spectra have been discussed in detail in References [1–4] and such detailed discussion will not be repeated here.

The new data of Vette *et al.*, are consistent with the power law trend below 1 MeV as indicated by the Ranger 3 measurements and other observations [11]. However, they indicate a marked departure from the power law above 1 MeV. For example, the 6 MeV point is an order of magnitude higher than what would be expected on the basis of a power law extrapolation of the X-ray data. These data, taken with the data of Clark *et al.*, being interpreted as a real flux, fit the shape of the theoretical γ-ray spectrum from p–p interactions integrated to a maximum redshift of ~100 for a burst or evolving sources model where cosmic-ray production was higher in the past. [1, 4]. They do not seem consistent with the other theoretical spectra for energies above 1 MeV.

These suggestive results make it even more imperative to obtain other γ-ray observations in the 1–100 MeV region in order to confirm the data of Vette *et al.*, and to extend the measurements to higher energies. However, on the basis of these first results we present the following interpretation.

Comparison of the predicted spectra with the γ-ray observations indicates that extragalactic γ-radiation may be due to the decay of neutral pi-mesons produced in inelastic collisions of metagalactic cosmic-ray protons and gas. The peak in the spectrum, which normally occurs at ~ 70 MeV, is redshifted down to ~ 1 MeV energy. This effect is due to the increased collision rate at larger redshifts when our expanding universe was in a more compact state as well as increased cosmic-ray production at large redshifts. A cosmic-ray production rate which is constant over all reshifts will not account for the new observations [3].

Either a burst model or evolving sources model for the time-dependence of cosmic-ray production in the past will fit the predicted spectrum; the position of the peak depends primarily on the maximum redshift at which γ-rays are produced [3]. However, the assumption of various time-dependence models for cosmic-ray production leads to different requirements for the present metagalactic flux needed to produce the observed γ-rays [1, 4]. The maximum redshift needed to produce the observations is ~ 100, which corresponds to an epoch when the age of the universe was 10^7–10^8 years and the temperature of the universal radiation field was ~ 270 K. This may correspond to the epoch when objects of galactic mass were beginning to form from the metagalactic medium [12]. There is mounting evidence that radio sources were more active (or prevalent) at earlier epochs [13], and it is plausible to speculate that in these sources, where electrons are accelerated to cosmic-ray energies, protons may also be accelerated to these energies. Whereas the electrons have short lifetimes at these redshifts due to Compton interactions with the universal radiation field [7, 14] possibly restricting their radio emission stage to redshifts of ~ 10 or less, the protons do not undergo significant depletion from Compton interactions. If we consider present extragalactic gas densities of 10^{-5} to 10^{-7} cm^{-3}, and assume increased cosmic-ray production in the past, we find that the present intergalactic cosmic-ray flux need only be $\sim 10^{-3}$ -10^{-5} of the galactic value in order to account for the observed γ-ray intensity. Such a flux has been strongly advocated by Ginzburg and Syrovatskii [15].

References

[1] Stecker, F. W.: 1968, *Nature* **220**, 675; Corrections: *Nature* **222**, 1157 (1969).
[2] Stecker, F. W.: 1969, *Nature* **221**, 425; Corrections: *Nature* **222**, 1157 (1969).
[3] Stecker, F. W.: 1969, *Astrophys. J.* **157**, 507.
[4] Stecker, F. W. and Silk, J.: 1969, *Nature* **221**, 1229.
[5] Felten, J. E. and Morrison, P.: 1966, *Astrophys. J.* **146**, 686.
    Fazio, G. G., Stecker, F. W., and Wright, J. P.: 1966, *Astrophys. J.* **144**, 611.
    Gould, R. J.: 1965, *Phys. Rev. Letters* **15**, 511.
    Hoyle, F.: 1965, *Monthly Notices Roy. Astron. Soc.* **120**, 338
[6] Silk, J. and McCray, R.: 1969, *Astrophys. Letters* **3**, 59.
[7] Brecher, K. and Morrison, P.: 1969, *Astrophys. J. Letters* **150**, L61.
[8] Clark, G. W., Garmire, G. P., and Kraushaar, W. L.: 1968, *Astrophys. J. Letters* **153**, L203.
[9] Vette, J. I., Gruber, D., Matteson, J. L., and Peterson, L. E.: 1970, this volume, p. 335.
[10] Metzger, A. E., Anderson, E. C., van Dilla, M. A., and Arnold, J. R.: 1964, *Nature* **204**, 766.
[11] See, for example, References in Gould, R. J.: 1967, *Am. J. Phys.* **35**, 376.
[12] Weymann, R.: 1967, *Astrophys. J.* **147**, 887.
[13] Longair, M. S.: 1966, *Monthly Notices Roy. Astron. Soc.* **133**, 421.
    Rowan-Robinson, M.: 1968, *Monthly Notices Roy. Astron. Soc.* **138**, 445.
    Schmidt, M.: 1968, *Astrophys. J.* **151**, 393.
[14] Bergamini, R., Londrillo, P., and Setti, G.: 1967, *Nuovo Cimento* **52B**, 495.
[15] See discussion and references in Ginzburg, V. L.: 1968, *Astrophys. Space Sci.* **1**, 125.

GALACTIC LINE EMISSION FROM 1–10 keV

GARY STEIGMAN and JOSEPH SILK

Institute of Theoretical Astronomy, Cambridge, England

Abstract. Calculations are given of K-series X-rays produced by interaction of both low energy cosmic rays and diffuse X-rays above 1 keV with heavier ions present in HI regions. We further consider electron capture to excited states by cosmic ray nuclei of heavy elements, followed by cascades down to the ground state. It is found that the electron capture process may yield appreciable line intensities in the 1–10 keV range in the galactic plane.

Hayakawa [1] originally suggested that low-energy cosmic rays (in the range 1–100 MeV) may be an important heat source in the interstellar medium. Cosmic ray heating has been extensively studied by Hayakawa *et al.* [2] and more recently by Pikel'ner [3], Balasubrahmanyan *et al.* [4], and Spitzer and Tomasko [5], who find substantial agreement with the observed properties of the interstellar medium. However, this is at best an indirect argument for the presence of sub-cosmic rays. Indeed the observed diffuse soft X-ray flux may be of comparable significance as a heating mechanism [6]. In order to distinguish between heating by sub-cosmic rays and other possible mechanisms, it is clearly of great importance to attempt to observe low-energy cosmic rays. Direct observations at low energies are unreliable because of the substantial degree of solar modulation [7].

A more promising approach is to investigate the interactions of low-energy cosmic rays with HI regions. Observations yield an upper limit on the heating due to low energy cosmic rays [8], in the form of an integral over the cosmic ray spectrum. Greenberg [9] has discussed the radio emission lines produced by electron cascades following recombination to highly excited states. Proton inner-bremsstrahlung radiation by low energy cosmic rays interacting with the interstellar medium has been discussed by Hayakawa and Matsuoka [10], and more recently by Boldt and Serlemitsos [11]. A power-law spectrum is produced, of energy spectral index equal to that of the non-relativistic proton spectrum at photon energies above about 20 keV. However, the flux falls some two orders of magnitude or more below the observed diffuse background at 1 keV. Note that this result depends only on the heat input assumed for the low-energy cosmic rays.

The fluxes of characteristic X-ray lines from K-series transitions following K-shell ionizations by fast protons in the direction of the galactic centre have previously been discussed by Gould and Burbidge [12] and Hayakawa and Matsuoka [10]. In this paper, we extend their work to include calculations of the K-series X-rays produced by interaction of both low-energy cosmic ray protons and diffuse X-rays above 1 keV with heavier ions present in HI regions, and of cosmic ray nuclei with atomic hydrogen. The processes that we consider are collisional ionization of K-electrons of interstellar atoms by cosmic nuclei and subsequent K-transitions; photo-ionizations of K-shell electrons by diffuse X-rays; and charge exchanges involving electron capture to

L. Gratton (ed.), Non-Solar X- and Gamma-Ray Astronomy, 385–391. All Rights Reserved
Copyright © 1970 by the IAU

excited states by cosmic ray nuclei of heavy elements followed by cascades down to the ground state.

The observed X-ray background is isotropic over 4–40 keV to within 10% [13], and is thought to be of extragalactic origin. The power-law spectrum is suggestive of a non-thermal mechanism, and several mechanisms have been proposed which produce continuous spectra [10]. We shall assume that the isotropic background flux is continuous for purposes of comparison with our results.

The line fluxes that we calculate are correlated with the distribution in the galaxy of neutral hydrogen, and so are most readily detectable along lines of sight lying within the galactic plane. Our results are presented in Tables I and II.

TABLE I

K-transitions following photo- and collisional ionizations

Element	E(keV) [26]	J^Z_{photo} (cm^{-2} sec^{-1} st^{-1})	J^Z_{Coll} (cm^{-2} sec^{-1} st^{-1})
Na	1.1	6.0×10^{-5}	1.1×10^{-5}
Mg	1.3	2.4×10^{-3}	4.0×10^{-4}
Al	1.5, 1.6	2.1×10^{-4}	3.2×10^{-5}
Si	1.7, 1.8	4.4×10^{-3}	6.1×10^{-4}
S	2.3, 2.5	5.2×10^{-3}	5.0×10^{-4}
Ar	3.0, 3.2	1.2×10^{-4}	1.1×10^{-5}
Ca	3.7, 4.0	4.5×10^{-4}	2.8×10^{-5}
Fe	6.4, 7.1	1.8×10^{-3}	3.6×10^{-5}

E = Energy of K X-ray

J^Z_{photo} = Flux of K X-rays emitted after photo-ionizations

J^Z_{Coll} = Flux of X-rays emitted after collisional ionizations

TABLE II

$2p$–$1s$ and $3p$–$1s$ transitions following electron capture by cosmic ray nuclei

Element	$E\alpha$(keV)	J^Z_α (cm^{-2} sec^{-1} st^{-1})	$E\beta$(keV)	J^Z_β (cm^{-2} sec^{-1} st^{-1})
Ne	1.02	0.20	1.21	0.06
Na	1.23	0.32	1.46	0.01
Mg	1.47	0.65	1.74	0.20
Al	1.72	0.09	2.04	0.02
Si	2.00	1.12	2.37	0.31
P-K	2.29–3.68	0.41	2.72–4.37	0.07
Ca-Cr	4.08–5.88	0.43	4.84–6.97	0.08
Mn-Ni	6.38–8.00	0.69	7.56–9.49	0.13

E_α = Energy of $2p$–$1s$ transition

E_β = Energy of $3p$–$1s$ transition

J^Z_α = Flux of X-rays with energy E_α

J^Z_β = Flux of X-rays with energy E_β

Note: For $1 \leqslant E \leqslant 2$ (keV) $\dfrac{J\alpha + J\beta}{J\gamma} = 0.31$

For $2 \leqslant E \leqslant 10$ (keV) $\dfrac{J\alpha + J\beta}{J\gamma} = 0.37$,

where $J\gamma$ is the diffuse X-ray flux [19] in the appropriate energy interval.

The low-energy cosmic ray flux is assumed to be mono-energetic, at 2 MeV per nucleon. A similar assumption is made by Spitzer and Tomasko [5] and Field *et al.* [8] in their discussions of the heating of the interstellar medium. In the following, we describe the details of our calculations and we also mention the effects of taking different forms for the low-energy cosmic ray spectrum.

Photo-ionization. An X-ray from the diffuse background $(E \gtrsim 1\text{keV})$ may knock a K-shell electron out of an atom whose nuclear charge is Z. The cross-section for such K-shell photo-ionizations, valid near threshold, is [14, 15]

$$\sigma_z(E) = 4.23 \times 10^{-23} \frac{Z_K^4}{E^3} \left\{ 1 - 4.40 \times 10^{-3} \frac{Z_K^2}{E} \right\} \text{cm}^2 . \tag{1}$$

The X-ray energy E is in keV. The effective charge $Z_K = Z - 0.3$ corrects for inner screening. The probability W_K^Z, that the K-shell vacancy will be filled via X-ray emission (rather than by an Auger transition) is given by [16]

$$W_K^Z = \left(1 + \frac{A}{Z^4} \right)^{-1} ; \quad A = \begin{cases} 1.19 \times 10^6, & 10 \leqslant Z \leqslant 18 \\ 1.27 \times 10^6, & Z > 18 \end{cases} . \tag{2}$$

The flux of line X-rays from atom Z may be expressed as

$$J_{\text{photo}}^Z = \langle N_H R \rangle \left(\frac{N_Z}{N_H} \right) W_K^Z \int_{I_K Z}^{\infty} \sigma_z(E) j_\gamma(E) \, dE \text{ cm}^{-2} \text{ sec}^{-1} \text{ st}^{-1} . \tag{3}$$

The number of H atoms along a line of sight $\langle N_H R \rangle$ is taken to be $3 \times 10^{22} \text{ cm}^{-2}$ in the direction of the galactic centre, for X-rays above 2 keV. We have neglected absorption of X-rays at these energies. In the range 1–2 keV, we have corrected for interstellar absorption using the results of Bell and Kingston [18] to provide an effective value of $\langle N_H R \rangle$ corresponding to optical depth unity. For directions away from the galactic centre the number of H atoms along a line of sight $\langle N_H R \rangle$ is less than $3 \times 10^{22} \text{ cm}^{-2}$ and the fluxes are correspondingly reduced. I_K^Z is the experimental K-shell ionization potential [27]. Below 10 keV, the diffuse X-ray photon flux may be fitted by a power-law spectrum [19].

$$j_\gamma(E) = 12.4 \, E^{-\alpha} \, (\text{cm}^{-2} \text{ sec}^{-1} \text{ keV}^{-1} \text{ st}^{-1}); \quad \alpha = 1.7 \pm 0.2 . \tag{4}$$

The line X-ray fluxes have been evaluated for the most abundant atoms whose K X-rays lie in the range from 1–10 keV. The results appear in Table I.

Collisional ionization. The flux of X-ray photons from atoms of charge Z whose K-electrons have been knocked out in collisions with low energy cosmic ray protons is given by

$$J^Z = \langle N_H R \rangle \left(\frac{N_Z}{N_H} \right) W_K^Z \int j(E) \sigma_Z(E) \, dE . \tag{5}$$

For the low-energy cosmic ray flux we shall adopt the 2 MeV flux employed by Field *et al* [8], i.e. $4\pi j(E) = 24\delta(E-2)$. The cross-section for K-shell ionization may be written as [20, 21]

$$\sigma_Z(E) = \frac{S(\eta)}{Z_K^4}; \quad \eta = \frac{40}{Z_K^2} E(\text{MeV}). \tag{6}$$

For $\eta \lesssim 10^{-1}$, $S(\eta) \sim \eta^4$. For $\eta \gg 1$, $S(\eta) \sim \eta^{-1} \ln \eta$. For intermediate values of η, $S(\eta)$ is given graphically by Merzbacher and Lewis [20]. Note that the maximum value of $S(\eta)$ occurs at $\eta \sim 1$, and is $\sim 4.5 \times 10^{16}$ cm^2.

If the projectile is a bare nucleus of charge Z', (instead of a proton) the flux of K X-rays is given by

$$J_{Z'}^Z = \left(\frac{N_{Z'}}{N_H}\right)_{CR} (Z')^2 J^z \tag{7}$$

The quantity $(N_Z/N_H)_{CR}$ is the abundance of nuclei with charge Z relative to protons in the 2 MeV cosmic ray flux.

The total X-ray flux caused by all cosmic ray nuclei colliding with atoms Z is

$$J_{Coll}^Z = \sum_{Z'} \left(\frac{N_{Z'}}{N_H}\right)_{CR} (Z')^2 J^z. \tag{8}$$

We estimate the relative abundances of cosmic ray nuclei at 2 MeV from the results of Comstock *et al.* [22], and obtain $\sum (N_{Z'}/N_H)_{CR} (Z')^2 = 2.65$. $J_{Coll}^Z \equiv 2.65 J^Z$ has been evaluated for the same elements as before. The resulting line intensities are given in Table I.

Electron capture. In an encounter with a hydrogen atom, a low-energy cosmic ray nucleus may capture the electron by the charge exchange process

$$H(1s) + A^{Z(+)} \rightarrow A^{(Z-1)(+)}(nl) + H^+.$$

The electron thus captured in an excited state, will cascade down to the ground state. If, in this process, the electron finds itself in the $2p$ state, it will emit the analog of the Ly-α line and jump down to the ground state. The energy of the line thus emitted is

$$E_\alpha^Z = 1.02 \left(\frac{Z}{10}\right)^2 \text{keV}, \tag{9}$$

where Z is the charge of the cosmic ray nucleus.

Similarly, if the electron finds itself in the $3p$ state, it will emit the analog of the Ly-β line and jump down to the ground state, 88% of the time. The other 12% of the time it will end up in the metastable $2s$ state. The energy of this line is

$$E_\beta^Z = 1.21 \left(\frac{Z}{10}\right)^2 \text{keV}. \tag{10}$$

The cross-section for Ly-α emission may be expressed as

$$\sigma_\alpha^Z = \sum_{l=0}^{\infty} \sum_{n=l+1}^{\infty} f_{nl}^\alpha \sigma_{1s-nl}^Z. \tag{11}$$

f_{nl} is the probability that an electron, initially in the state nl, will ultimately make the $2p$–$1s$ transition. σ_{1s-nl}^Z is the cross-section for a nucleus of charge Z to capture an electron from the ground state of hydrogen into the nl state. The corresponding cross section for Ly-β emission is

$$\sigma_\beta^Z = \sum_{l=0}^{\infty} \sum_{n=l+1}^{\infty} f_{nl}^\beta \sigma_{1s-nl}^Z. \tag{12}$$

f_{nl}^β is the probability an electron captured into the state nl will ultimately make the $3p-1s$ transition.

The probabilities f_{nl}^α, f_{nl}^β have been evaluated by Bethe and Salpeter [23]. Bates and Dalgarno [24] have evaluated the electron cros-ssections for all states through $n=4$, $l=3$. Hiskes [25] has given expressions for σ_{1s-nl} for all states through $n=15$, $l=14$.

The flux of line emission from electron capture is

$$J_\alpha^Z = \langle N_H R \rangle \left(\frac{N_Z}{N_H} \right)_{CR} \int j(E) \, \sigma_\alpha^Z(E) \, dE,$$

$$J_\beta^Z = \langle N_H R \rangle \left(\frac{N_Z}{N_H} \right)_{CR} \int j(E) \, \sigma_\beta^Z(E) \, dE. \tag{13}$$

As previously, we employ the flux $4\pi j(E) = 24\delta(E-2)$ and take $(N_Z/N_H)_{CR}$ from Comstock et al. [22]. The heavy nuclei are assumed to be present at an energy of 2 MeV/nucleon. As before, we take $\langle N_H R \rangle = 3 \times 10^{22}$ cm^{-2}.

For $E=2$ MeV and $10 \leqslant Z \leqslant 26$, the cross-sections σ_α^Z and σ_β^Z were estimated from (11) and (12) by including all contributions through the $4f$ state. The resulting cross-sections are therefore lower limits to the exact cross-sections. They are most accurate for low Z. For $Z \gtrsim 20$ they may be too small by an order of magnitude. The results appear in Table II. The lines below 2 keV have been corrected for interstellar absorption, as previously described.

Although none of the K-lines due to collisional or photo-ionizations should be observable, it appears from Table II that it would be feasible to look for the $2p-1s$ transitions following electron capture by low energy cosmic ray nuclei. In our calculation of the line fluxes for the $2p-1s$ and $3p-1s$ transitions we have used cross-sections which in fact are lower limits to the exact cross-sections. Measurements of line intensities (or null results) would therefore yield upper limits to the flux of cosmic ray nuclei with $10 \leqslant Z \leqslant 28$ at 2 MeV per nucleon. It should be noted that these lines will be Doppler broadened. Since $\Delta E/E \simeq 0.13$, the lines with $E \gtrsim 2$ keV will begin to overlap, making them more difficult to resolve.

Throughout the discussion of X-ray lines produced by low-energy cosmic rays we have used the 2 MeV flux employed by Field et al. [8]. It is of interest to know how these results are modified if a more realistic spectrum (say, an extrapolation of the demodulated spectrum of high energy cosmic rays) is used. A detailed investigation of this is in progress but some general remarks may be made. In the case of collisional ionization, the reduced cross-section $S(\eta)$ (see Equation (7)) is rather slowly varying in the neighbourhood of its maximum $(0.1 \lesssim \eta \lesssim 10)$. A rough estimate of J_{Coll}^Z may be obtained as

$$J_{\text{Coll}}^Z \approx 2.65 \langle N_H R \rangle \left(\frac{N_Z}{N_H} \right) W_K^Z \langle \sigma_Z \rangle \int\limits_{\eta=0.1}^{\eta=10} j(E) \, dE \, ; \quad \eta = \frac{40E}{Z^2} \, . \quad (14)$$

Such estimates of J_{Coll}^Z indicate order of magnitude agreement with the previous results.

For energies $\geqslant 2$ MeV, the cross-sections σ_α^Z, σ_β^Z are very small $(\sigma \sim E^{-6})$. Thus, most of the contributions to σ_α^Z and σ_β^Z come in an energy interval around or less than 2 MeV. However, even if the cosmic ray spectrum peaks at a somewhat higher energy than 2 MeV, we would still predict an appreciable flux at 2 MeV. Ionization losses tend to produce a spectrum of positive slope at low energies, and the spectrum would be unlikely to peak much above 15 MeV, otherwise the cosmic ray heating would be drastically reduced. Although this gives an effective reduction in the flux at 2 MeV, the integration over a cosmic-ray spectrum tends to enhance the line intensities. Moreover, even a small flux below 2 MeV makes an important contribution to the lines because of the steep dependence on energy of the cross-sections for $2p-1s$ and $3p-1s$ transitions. The net result appears to be that the line fluxes are not significantly reduced by inclusion of a more realistic spectrum.

We would like to thank Drs. G. B. Field, M. J. Rees and M. W. Werner for helpful comments on this work.

References

[1] Hayakawa, S.: 1960, *Publ. Astron. Soc. Japan*, **12**, 110.
[2] Hayakawa, S., Nishimura, S., and Takayanagi, K.: 1961, *Publ. Astron. Soc. Japan* **13**.
[3] Pikel'ner, S. B.: 1968, *Soviet Astron.-AJ* **11**, 737.
[4] Balasubrahmanyan, V. K., Boldt, E., Palmeira, R. A. R., and Sandri, G.: 1968, *Can. J. Phys.* **46**, 5633.
[5] Spitzer, L., and Tomasko, M. G.: 1968, *Astrophys. J.* **152**, 971.
[6] Silk, J. and Werner, M. W.: 1969, *Astrophys. J.* **158**, 185.
[7] Gloeckler, G. and Jokipii, J. R.: 1967, *Astrophys. J.* **148**, 241.
[8] Field, G. B., Goldsmith, D. W., and Habing, H. J.: 1969, *Astrophys. J.* **155**, L149. The flux of 2 MeV cosmic rays used by these authors is about half that assumed in Reference [5], as heavy nuclei were included. This increases the heat input of a given flux of low-energy cosmic rays at 2 MeV/nucleon by a factor $\Sigma_Z (N_{Z'}/N_H)_{\text{CR}} (Z')^2$.
[9] Greenberg, D. W.: 1969, *Astrophys. J.* **155**, 451.
[10] Hayakawa S. and Matsuoka M.: 1964, *Supp. Prog. Theor. Phys.* **30**, 204.
[11] Boldt, E. and Serlemitsos, P.: 1969, paper presented at 128th meeting of A.A.S., Austin, Texas (December).
[12] Gould, R. J. and Burbidge, G. R.: 1963, *Astrophys J.* **138**, 969.
[13] Seward, F., Chodil, G., Mark, H., Swift, C., and Toor, A.: 1967, *Astrophys. J.* **150**, 845.
[14] Heitler, W.: 1954, *The Quantum Theory of Radiation*, Clarendon Press, Oxford, Sec. 21.

[15] Bethe, H. A. and Salpeter, E. E.: 1957, *Quantum Mechanics of One-And-Two Electron Atoms*, Springer-Verlag, Berlin, Sec. IVb.
[16] Fink, R. W., Jopson, R. C., Mark, Hans, and Swift, C. D. 1966, *Rev. Mod. Phys.* **38**, 526.
[17] Allen, C. W.: 1963, *Astrophysical Quantities*, The Athlone Press, London.
[18] Bell, K. L. and Kingston, A. E.: 1967, *Monthly Notices Roy. Astron. Soc.* **136**, 241.
[19] Gorenstein, P., Kellogg, E. M., and Gursky, H.: 1969, *Astrophys. J.*, **156**, 315
[20] Merzbacher, E., and Lewis, H. W.: 1958, *Handbuch der Physik*, Vol. 34, 166.
[21] Bates, D. R. and Griffing, G.: 1953, *Proc. Phys. Soc.* **A66**, 961.
[22] Comstock, G. M., Fan, C. Y., and Simpson, J. A.: 1969, *Astrophys. J.* **155**, 609.
[23] Bethe, H. A. and Salpeter, E. E.: 1957, *Quantum Mechanics of One-and-Two Electron Atoms* Springer-Verlag, Berlin, Sec. IVa.
[24] Bates, D. R. and Dalgarno A.: 1953 *Proc. Phys. Soc.* **A66** 972.
[25] Hiskes, J. R.: 1965, *Phys. Rev.* **137A**, 361.
[26] Bearden, J. A.: 1967, *Rev. Mod. Phys.* **39**, 78.
[27] Bearden, J. A.: 1967, *Rev. Mod. Phys.* **39**, 125.

DIFFUSE COSMIC X-RAYS FROM NON-THERMAL
INTERGALACTIC BREMSSTRAHLUNG

JOSEPH SILK*

Institute of Theoretical Astronomy, Cambridge, England

Abstract. The diffuse X-ray background between 1 keV and 1 MeV is interpreted as non-thermal bremsstrahlung in the intergalactic medium. The observed break in the X-ray spectrum at ~ 40 keV yields the heat input to the intergalactic medium, the break being produced by ionization losses of sub-cosmic rays. Proton bremsstrahlung is found not to yield as satisfactory an agreement with observations as electron bremsstrahlung: excessive heating tends to occur. Two alternative models of cosmic ray injection are discussed, one involving continuous injection by evolving sources out to a redshift of about 3, and the other model involving injection by a burst of cosmic rays at a redshift of order 10. The energy density of intergalactic electrons required to produce the observed X-rays is $\sim 10^{-4}$ eV/cm³. Assuming a high density ($\sim 10^{-5}$ cm⁻³) intergalactic medium, the energy requirement for cosmic ray injection by normal galaxies is $\sim 10^{58-59}$ ergs/galaxy in sub-cosmic rays. The temperature evolution of the intergalactic medium is discussed, and we find that a similar energy input is also required to explain the observed high degree of ionization (if 3C9 is at a cosmological distance).

1. Introduction

The diffuse X-ray background between 1 keV and 1 MeV has been measured by many different experimenters over the past 5 years. Paying particular attention to the most recent results presented at this Symposium, we note two significant conclusions that may be drawn from the data. Out of the galactic plane, the diffuse background appears to be remarkably isotropic. Also, there appears to be a change in slope, between 20 and 60 keV.

The isotropy of the X-ray background has been an important feature of earlier attempts at interpretation. These have generally assumed a metagalactic origin, and may be divided into either discrete source models or models involving X-ray production in the intergalactic medium. It was soon apparent that normal galaxies, if assumed to have X-ray luminosities comparable to that of our own galaxy, would be inadequate to account for the diffuse background (Oda, 1965), and evolving discrete source models were proposed (Bergamini *et al.*, 1967, subsequently denoted by BLS; Silk, 1968). One of the less appealing aspects of these models is that, in order to enhance the X-ray background at the cost of the radio background one has to appeal in essence to a new class of X-ray sources, emitting at redshifts exceeding those attainable by radio sources. A somewhat different approach involves inverse Compton scattering of 3 K black-body radiation in the intergalactic medium (Felten and Morrison, 1966), and is essentially independent of evolutionary assumptions (Brecher and Morrison, 1967).

However, realization of the break in the X-ray spectrum at $\varepsilon_b \sim 40$ keV has prompted reconsideration of these models. In particular, the discrete source models are found to impose stringent requirements on source evolution (Silk, 1969; Felten and Rees, 1969). For example, in the BLS interpretation, ε_b has a strong z-dependence: with the

---

* Present address: Department of Astronomy, University of California, Berkeley, California 94720.

L. Gratton (ed.), Non-Solar X- and Gamma-Ray Astronomy, 392–401. All Rights Reserved
Copyright © 1970 by the IAU

radio source model suggested by Rees and Setti (1968) one requires $\varepsilon_b \alpha (1+z_*)^{-11}$, where z_* is the limiting redshift at which evolutionary effects are important. Now z_* is found to be ~ 4 from studies of radio source counts, and in a Friedmann cosmology there is no reason to expect a sharp cut-off in source distribution with z. Only in somewhat esoteric circumstances, such as occur, for example in the Lemaître cosmology, could one hope to reconcile the BLS model with observation.

Intergalactic inverse Compton radiation produces a break at ε_b provided that the metagalactic relativistic electron spectrum has a corresponding break at $\gamma_b \simeq (\varepsilon_b / \langle \varepsilon \rangle)^{1/2}$, where $\langle \varepsilon \rangle \simeq 3.6 kT$ and $T = 2.7$ K. Brecher and Morrison (1969) have recently suggested that this requirement may in fact be met by electrons injected from normal galaxies. However, an intrinsic break in the electron source spectrum at $\sim \gamma_b$ is hypothesized by these authors without proposing any physical mechanism which would somehow constrain γ_b to be similar for all normal galaxies.

Clearly, none of these interpretations can be regarded as accounting in a satisfactory manner for the diffuse X-ray spectrum. We wish to discuss in the present contribution a mechanism which provides a natural (i.e. model-independent) explanation of the 40 keV break. Non-thermal bremsstrahlung in the intergalactic medium by a power-law spectrum of fast electrons (or protons) will produce a power-law photon spectrum with the same differential energy spectral index (for non-relativistic particles). The stopping of the cosmic rays in the intergalactic medium by ionization losses produces a break in their spectrum, and this is reflected by a corresponding curvature in the X-ray bremsstrahlung spectrum. We show below that the spectral region at which this effects occurs depends on the density of the intergalactic medium n_0 and only very weakly, if at all, on z_*. There is essentially no dependence on the properties of individual sources, although the mode of cosmic ray injection enters as a parameter which is determined by the X-ray spectrum at $\varepsilon < \varepsilon_b$. A consequence of this model is that one may also account for the thermal properties of the intergalactic medium. Indeed, the heat input to the intergalactic medium is essentially determined by our interpretation of ε_b.

2. Relaxation of Cosmic Ray Spectra

We first discuss how one may study the relaxation of a cosmic-ray spectrum, subject to losses by adiabatic expansion of the metagalaxy and coulomb losses in the intergalactic medium. The relaxation of a cosmic-ray flux $J(E)$ $(\text{cm}^2 - \sec - \text{st} - \text{keV})^{-1}$ is described by a Fokker-Planck equation for a spatially uniform system of the form (cf. Kardashev, 1962)

$$\frac{\partial}{\partial t} \left\{ \frac{J(E, z)}{(1+z)^3 V(E)} \right\} + \frac{\partial}{\partial E} \left\{ \frac{b(E, z) J(E, z)}{(1+z)^3 V(E)} \right\} = \frac{S(E, z)}{V(E)}. \tag{1}$$

Here $V(E)$ is the velocity of a cosmic ray, $b(E, z)$ is an energy loss term, and $S(E, z)$ is the source function, for which we shall consider two models. For case A, we take

$$S_A(E, z) = J(E, z) \delta(z - z_*).$$

This describes an initial burst of cosmic rays, at redshift z_*. For case B we write

$$S_{\mathrm{B}}(E, z) = q(z) E^{-\gamma},$$

where we now consider the effects of continuous injection of cosmic rays. Possible evolutionary effects are contained in the z-dependence of $q(z)$. Cases A and B, in an appropriate linear combination, may be taken to represent a realistic model for cosmic ray injection.

In order to solve (1), it is necessary to specify the energy loss term $b(E, z)$. Consider first the relaxation of a non-relativistic electron spectrum, for which $b(E, z)$ takes the form

$$b(E, z) \equiv \frac{dE}{dt} = -\frac{2E}{t(z)} - \frac{E_{\mathrm{cr}}^{3/2}}{E^{1/2}} \frac{(1 + z)^3}{t_0}. \tag{2}$$

The first term on the right-hand side represents the effect of adiabatic expansion and the second term describes Coulomb losses. In deriving the second term we have made use of the energy loss formula for a fast non-relativistic electron in ionized hydrogen with electron density $n_0 (1+z)^3$ cm^{-3}, namely (Montgomery and Tidman, 1964)

$$\frac{dE}{dt} = -\xi \frac{n_0(1 + z)^3}{\sqrt{E_{\mathrm{keV}}}} \text{ keV sec}^{-1}, \tag{3}$$

where

$$\xi = 7.35 \times 10^{-10} \ln(24\pi\eta_0 L_{\mathrm{D}}^3),$$

and we shall neglect the slight variation of the logarithmic factor. After a time t_0 electrons below a critical energy $E_{\mathrm{cr}}(z) = (1+z)^2 \times (\xi n_0 t_0)^{2/3}$ are substantially depleted. If t_0 is identified with the characteristic time-scale for energy loss by adiabatic expansion, we have $t_0 = H_0^{-1} \alpha (1+z)^{-\alpha}$, where $H_0^{-1} = 4 \times 10^{17}$ sec (Sandage, 1968). We consider two cosmological models, defined by $\alpha = 1.5$ (an Einstein-de Sitter cosmological model in which $q_0 = \frac{1}{2}$ and $n_0 \simeq 10^{-5}$ cm^{-3}), and $\alpha = 1$ (a low density-universe with $q_0 = 0.02$ and $n_0 \simeq 10^{-7}$ cm^{-3}). Hence we may write $E_{\mathrm{cr}}(z) = (1+z)^{2-2\alpha/3} E_{\mathrm{cr}}$ where $E_{\mathrm{cr}} = (\xi n_0 H_0^{-1})^{2/3}$. Numerically we obtain $E_{\mathrm{cr}} \simeq 1$ MeV with $n_0 = 10^{-5}$ and $E_{\mathrm{cr}} \simeq 50$ keV with $n_0 = 10^{-7}$. The corresponding values of E_{cr} for a cosmic ray proton spectrum are larger by a factor $\sim (m_p/m_e)^{1/3}$.

We have derived analytic solutions for $J(E, z)$ in a few cases of interest, both for the burst and continuous injection source functions. For the burst source function $S_{\mathrm{A}}(E, z) = K(z) E^{-\gamma} \delta(z - z_*)$, we find that the power law shape is preserved above $E_{\mathrm{cr}}(z)$. In this region, ionization losses are negligible compared with the adiabatic expansion losses. For $E \ll E_{\mathrm{cr}}(z)$, the spectrum relaxes to $J(E) \alpha E$, and the complete solution is

$$J(E, z) = KE^{-\gamma}\left(\frac{1 + z}{1 + z_*}\right)^{2(\gamma + 1)}$$

$$\times \left[1 + \frac{3}{2}\left(\frac{E_{\mathrm{cr}}}{E}\right)^{3/2} \frac{(1 + z)^{3-\alpha}}{\alpha}\left\{1 - \left(\frac{1 + z}{1 + z_*}\right)^{\alpha}\right\}\right]^{-2(\gamma + 1)/3}$$

With the continuous injection source function S_B, the power-law spectrum is also maintained for $E \gg E_{cr}(z)$. However, for $E \ll E_{cr}(z)$, we find that $J(E, z) \propto E^{-(\gamma-3/2)}$. Exact solutions have been found in analytic form only for an evolutionary function $q(z) = q_0(1+z)^m$, for specific values of γ or m. Here it will suffice to give the asymptotic solutions, valid for arbitrary m. For $E \ll E_{cr}(z)$

$$J(E, z) \simeq \frac{q_0 t_0}{(\gamma-1)\,\alpha\,(1+z_*)^{2(\gamma-1)-m}} \frac{(1+z)^{2\gamma}}{} \left(\frac{E}{E_{cr}}\right)^{3/2} E^{-\gamma}$$

$$\times \left[1 - \left\{1 + \frac{3(1+z)^{3-\alpha}}{2} \frac{}{\alpha} \left(\frac{E_{cr}}{E}\right)^{3/2} \left(1 - \left(\frac{1+z}{1+z_*}\right)^{\alpha}\right)\right\}^{-2(\gamma-1)/3}\right] \quad (5)$$

and for $E \gg E_{cr}(z)$,

$$J(E, z) \simeq \frac{t_0 q_0 (5-2\alpha)/3}{\alpha + 2\gamma - m - 2} (1+z)^{m+3-\alpha} E^{-\gamma}. \quad (6)$$

The flattening by $\frac{3}{2}$ power of energy is due to the fact that the stopping time $t_0 E^{3/2}$ determines the equilibrium spectrum, to a first approximation. In order to take proper account of relativistic effects, especially for the $n_0 = 10^{-5}$ cm^{-3} model, it is necessary to seek exact numerical solutions of (1). Further details of these solutions will be given elsewhere (Arons et al., 1970). One may then include additional energy losses, such as inverse Compton losses for relativistic electrons. However, we shall here restrict ourselves to a qualitative discussion based on the asymptotic relaxation spectra.

3. Cosmic Ray Electron Bremsstrahlung

A noteworthy feature of cosmic-ray bremsstrahlung as a mechanism for producing diffuse X-rays is that the cosmological expansion predicts a significant enhancement at earlier epochs. At the present epoch, non-thermal bremsstrahlung is an extremely inefficient process, and we find that unrealistically high fluxes of metagalactic cosmic rays would be required to explain the X-ray background. The effects of cosmology are apparent from the expression for the X-ray flux (Silk and McCray, 1969)

$$I(\varepsilon) = c H_0^{-1} n_0 \int_0^{z_*} j\{\varepsilon(1+z), z\} \frac{dz}{(1+z)^{\alpha+1}} \text{ keV}$$

$$\times (\text{cm}^2\text{-sec-st-keV})^{-1} \quad (7)$$

where the bremsstrahlung emissivity is given by

$$j(\varepsilon, z) = \int_\varepsilon^\infty \chi(\varepsilon, E) J(E, z)\, dE \ \text{ keV(sec-st-keV)}^{-1}. \quad (8)$$

Here, $\chi(E, \varepsilon) = 2\pi\alpha\sigma_T\beta^{-2}\,G(E, \varepsilon)/\sqrt{3}$ and $G(E, \varepsilon)$ is the Gaunt factor (Bekefi, 1966). A power-law spectrum of non-relativistic cosmic-ray electrons $J(E) = KE^{-\gamma}$ gives rise to an X-ray flux $I(\varepsilon)$, which is a power law with the same exponent γ. The slope of the X-ray spectrum resulting from bremsstrahlung of relativistic particles is $\gamma - 1$. It then follows that photons of energy $\varepsilon > \varepsilon_{cr} \equiv E_{cr}(z)/(1 + z)$ originate from that part of the electron spectrum not affected by stopping. Note that ε_{cr} has only a weak dependence on redshift, varying at most as $(1 + z)^{1/3}$ if $n_0 = 10^{-7}$ cm^{-3}.

The effects of cosmology on the cosmic-ray requirements of this model are best seen by considering the energy range where stopping is unimportant. We may then obtain an analytic expression for the X-ray flux (neglecting the energy dependence of the Gaunt factor)

$$I(\varepsilon) = I_0(\varepsilon) f(z_*),$$

where $I_0(\varepsilon) = cH_0^{-1} j(\varepsilon)$. Approximate expressions for $f(z_*)$ are, in model A,

$$f(z_*) = [(1 + z_*)^{\gamma + 2 - \alpha} - 1]/(\gamma + 2 - \alpha) \tag{9}$$

and in model B,

$$f(z_*) = [(1 + z_*)^{m + 3 - \alpha - \gamma} - 1]/(m + 3 - \alpha - \gamma). \tag{10}$$

The required electron flux is reduced at the present epoch from that required if the effects of cosmology were negligible (i.e. $0 < z_* \ll 1$) by $\sim [f(z_*)]^{-1}$.

For $\varepsilon > \varepsilon_{cr}$, stopping is unimportant, and the X-ray energy spectral index is γ. At photon energies $\varepsilon < \varepsilon_{cr}$, however, stopping is significant, and the burst model produces an essentially flat X-ray spectrum ($\gamma \simeq 0$). The continous injection model gives an X-ray spectrum of slope $\sim (\gamma - \frac{3}{2})$, for $\gamma > 2.5$, below ε_{cr}. These remarks are not exact, because relativistic bremsstrahlung is one power flatter than non-relativistic bremsstrahlung, for the same electron spectrum. Hence relativistic effects, especially in the high-density model, will modify the spectra.

In the burst model, the observations are best fitted with a low-density intergalactic medium ($n_0 = 10^{-7}$). With $z_* = 10$, we require a metagalactic electron density of $\sim 10^{-4}$ eV/cm^3. The injected electron spectral index is 2.5, and above 100 keV, the X-ray slope is approximately 1.5 (Silk and McCray, 1969). Alternatively, we might consider the continuous injection model. This would enable us to utilize a high density intergalactic medium ($n_0 = 10^{-5}$). With burst injection, the X-ray spectrum would flatten below ε_{cr}, and with $\varepsilon_{cr} \sim 1$ MeV, too flat a spectrum would result. However, by continuous injection of an electron spectrum with $\gamma \simeq 2$, the X-ray spectrum is steepened sufficiently at low energies to fit the observations, which indicate a slope of ~ 0.5 below 40 keV.

4. Cosmic Ray Proton Bremsstrahlung

We have hitherto been concerned mainly with bremsstrahlung by cosmic-ray electrons. However, much of our discussion is valid for the analogous case of inner-bremsstrahlung by cosmic-ray protons, originally suggested as a possible mechanism by Hayakawa and Matsuoka (1964). Identification of ε_b as due to the stopping of fast protons in the intergalactic medium would then be an alternative possibility. In this section, we explore the feasibility of this suggestion.

The cross-section for bremsstrahlung by fast protons is reduced by m_e/m_p relative to that for electrons, for non-relativistic particles of the same energy (the centre of momentum of the proton-electron collision being approximately that of the fast proton). Thus an electron of energy E and a proton of energy $E(m_p/m_e)$ produce a similar bremsstrahlung spectral intensity. Hence with burst injection, we require the proton spectrum to break at ~ 30 MeV in order to produce ε_b at ~ 15 keV. Since $E_{cr}^{proton} \sim (m_p/m_e)^{1/3} E_{cr}$, we see that a high density intergalactic medium may just allow this, although if n_0 were less than 2.10^{-5} the X-ray spectrum would remain steep at too low an energy. Continuous injection does not modify this requirement on n_0. Moreover, the proton spectral index must be ~ 1.2 at injection to produce the required X-ray slope above the break.

5. Heating of the Intergalactic Medium

The lack of Ly α absorption in 3C9 offers compelling evidence, provided we accept that this quasar is at a cosmological distance, that the intergalactic medium is highly ionized. Sciama (1964) realized that ionization losses by sub-cosmic rays would be an important heat source for the intergalactic medium, and a detailed discussion was subsequently given by Ginzburg and Ozernoi (1966). In an analogous manner we now estimate the heating due to the cosmic-ray flux $J(E, z)$ derived earlier.

In our interpretation of the X-ray background the heat input to the intergalactic medium is essentially an observable quantity, determined by ε_b. The energy transferred by sub-cosmic rays to the intergalactic medium is comparable to that remaining in the relaxed spectrum, which is itself of order

$$W(E_{cr}, z) \simeq 4\pi \int_{E_{cr}(z)} EJ(E, z) \frac{dE}{V(E)}. \tag{11}$$

The rate of heat input is $L = W(E_{cr}, z)/t(z)$. Denoting by L_0 the value of L at $z = 0$, we find that $L = L_0(1 + z)^{5+p}$, where $p = 2\alpha\gamma/3$ in the burst model, and $p = m + 1 + 2\alpha\gamma/3 - \alpha - 2\gamma$ in the continuous injection model. One would have guessed $p = 0$ without carrying out the calculation, since the energy density of subcosmic rays varies as $(1 + z)^5$. The difference is due to the fact that we are taking proper account of the effect of ionization losses on the injected spectrum of cosmic rays. Values of $W_0 (\equiv L_0 t_0$, the energy density of cosmic rays at $z = 0$) are given in Table I for cosmic ray electrons, with $n_0 = 10^{-5}$ and 10^{-7}, and $z_* = 3$ and 10. Similar results for protons are given in Table II.

Following Ginzburg and Ozernoi (1966), we have solved the energy equation for the kinetic temperature $T(z)$ as a function of epoch. We find (if $p \neq \alpha$)

$$\frac{T(z)}{(1+z)^2} = T_* + T_1 \frac{\alpha}{\alpha - p} [(1 + z)^{-(\alpha - p)} - (1 + z_*)^{-(\alpha - p)}], \tag{12}$$

TABLE I

Parameters for electron bremsstrahlung
Injected electron spectral index $\gamma = 2.5$

	continuous injection ($m = 3$)		burst	
n_0 (cm^{-3})	10^{-5}		10^{-7}	
E_{cr} (keV)	1300		60	
z_*	3	10	3	10
W_0 (eV/cm^3)	10^{-4}	10^{-5}	10^{-2}	10^{-4}
T_1 (K)	10^5	10^4	10^9	10^7
P_{inj} (ergs/galaxy)	10^{59}	$10^{58.5}$	$10^{62.5}$	10^{62}

TABLE II

Parameters for proton bremsstrahlung.
Injected proton spectral index $\gamma = 1.2$
$n_0 = 2 . 10^{-5}$ cm^{-3}, $E_{cr} = 30$ MeV

	continuous injection ($m = 3$)			burst	
z_*	10	30	100	10	30
W_0 (eV/cm^3)	10^{-2}	10^{-3}	10^{-4}	10^{-3}	$10^{-4.5}$
T_1 (K)	$5 . 10^6$	$5 . 10^5$	$5 . 10^4$	$5 . 10^5$	10^4
P_{inj} (ergs/galaxy)	$10^{61.5}$	10^{61}	$10^{60.5}$	10^{63}	$10^{62.5}$

where $T_1 = L_0 t_0 / n_0 k$. Numerical values for T_1 are given in Table I. This expression indicates that cosmic-ray heating allows T to decrease less rapidly with z than in adiabatic expansion. A steeper cosmic-ray spectrum would produce more heating at $z \sim z_*$, but the temperature decrease with z would rapidly tend towards its adiabatic dependence on z, if $p > \alpha$. The effect of enhanced evolution (increasing m) is similar in the case of continuous injection of cosmic rays. However, for $m \simeq 3$, as suggested by the radio source counts (Longair, 1966) and $\gamma = 2.5$, we find $p < \alpha$, and the temperature satisfies

$$T(z) \simeq T_1 (1 + z)^{2 + p - \alpha} \tag{8}$$

for $z \ll z_*$. Quasar observations (Schmidt, 1968) indicate a somewhat larger value for m, of about 5. Consequently if $p > \alpha$, we have

$$T(z) \simeq T_1 (1 + z)^2 (1 + z_*)^{p - \alpha}. \tag{9}$$

Now observations of diffuse soft X-rays (< 1 keV) enable limits to be set on the thermal emission from the intergalactic medium. In fact recent observational data set the limit (Bunner *et al.*, 1969) $T(0) \lesssim 10^6$ K. (If the measured flux is interpreted as thermal bremsstrahlung of the intergalactic medium, then $n_0 \simeq 2 \times 10^{-6}$ cm^{-3}; however, possible contributions from local sources have not yet been eliminated). Hence a constraint on possible injection models is $T_1 \lesssim 10^6$ K, in the high density model. This imposes severe limitations on the proton bremsstrahlung mechanism. From Table II it appears that one has to take $z_* \gtrsim 30$ to avoid excessive thermal bremsstrahlung by over-heating the intergalactic medium. Electron bremsstrahlung, however, enables one to choose $z_* \sim 3$ in the case of continuous injection (see Table I). On the other hand, with a low density intergalactic medium and burst injection, we require $z_* \gtrsim 10$, otherwise a power-law spectrum would not be produced below 100 keV.

6. Conclusion

Various properties of the non-thermal bremsstrahlung interpretation of diffuse X-rays are summarised in Tables I and II. Choice of n_0 determines E_{cr}, and the required present energy density of intergalactic cosmic rays W_0 is determined once z_* is chosen. The temperature $T_1 = W_0/n_0 k$ is related by (8) or (9) to the present kinetic temperature of the intergalactic medium. We have also estimated the energy requirements at injection, defining the energy requirement per galaxy to be $P_{inj} = W(z_*)/n_g(z_*)$. $n_g(z) = n_g(0)(1+z)^3$, and $n_g(0) \simeq 2 \times 10^{-75}$ cm^{-3} is the observed number density of normal galaxies.

It is apparent that P_{inj} is only slightly reduced by appealing to large values of z_*. If we regard it as desirable to minimize the energy requirements, then the most favourable model involves continuous cosmic-ray electron injection in a high density universe. An ionized intergalactic medium is maintained by heating by supra-thermal electrons for $3 \lesssim z_* \lesssim 10$, and between 10^{59} and $10^{58.5}$ ergs/galaxy are required at z_* in 10 or 100 MeV electrons.

The situation does not change significantly even if the injected energy is in the form of subcosmic ray protons. This is because the subcosmic rays lose energy by production of plasma waves which are extremely efficient at accelerating thermal electrons in the Maxwellian tail (cf. Pikel'ner and Tsytovich, 1969). These electrons are then stopped by interacting with the thermal plasma. Hence with injection into the IGM at z_* of 10^{59} ergs/galaxy in 100 MeV protons, one still has almost as much energy present in fast electrons. Pikel'ner and Tsytovich show that the electron spectrum has a power-law behaviour, cutting off above $m_e c^2$. Hence proton inner-bremsstrahlung in the IGM would necessarily imply far more X-radiation via electron bremsstrahlung from electrons produced in this manner for protons of energy $\sim E_{cr}^{proton}$. For the two mechanisms to give equal contributions, the proton energy must exceed $\sim 10\ E_{cr}^{proton}$. One can therefore only hope to account for the break at ε_b by the stopping of supra-thermal protons if the X-radiation is predominantly electron bremsstrahlung.

It is of interest to note that a recent interpretation of diffuse galactic γ-rays by electron bremsstrahlung requires that our galaxy contains ~ 1 eV/cm^3 in 10 MeV electrons, or about $10^{58.5}$ ergs averaged over 10^{10} years (Rees and Silk, 1969). Uncertainty in solar modulation allows one to assume a low-energy electron flux of this magnitude. If this suggestion is correct, then ionization losses by these electrons are an important heat source for interstellar HI regions. The possible excess γ-ray flux over the extrapolated X-ray background at ~1 MeV reported at this Symposium by Vette *et al.* may be attributed in a bremsstrahlung interpretation to the transition from nonrelativistic to relativistic electron energies, where the photon spectrum flattens by one power. This explanation requires an injection spectral index of about 1.5. At electron energies above a few MeV, inverse Compton losses become important. Provided that in this energy range, the injected electron spectrum has a spectral index no flatter than 2.5, the inverse Compton radiation at keV energies produced by ultrarelativistic electrons interacting with microwave photons will be an order of magnitude lower in intensity than the bremsstrahlung X-rays. Proton inner-bremsstrahlung seems to be less plausible than electron bremsstrahlung as a mechanism for the X-ray background because z_* must exceed ~30. Improved spectral measurements of the diffuse component should enable us to choose between different injection models, since the degree of flattening below ε_b is a direct measure of continuous injection of cosmic rays.

Measurement of small-scale fluctuations in the diffuse X-ray background would also provide a means of distinguishing between different theories. Electrons with energy $E \ll E_{\rm cr}$ do not travel far from their sources, and the stopping distance increased with increasing electron energy. Consequently, with a diffuse bremsstrahlung origin of the X-ray background, fluctuations in the background should be present at $\varepsilon \ll \varepsilon_b$, and their amplitude should *decrease* with increasing photon energy. However if the diffuse X-rays are produced by inverse Compton scattering of microwave photons by ultrarelativistic electrons in the IGM, background fluctuations should *increase* with increasing photon energy (since the more energetic an electron, the shorter its mean free path against inverse Compton losses). If the diffuse X-rays are produced by discrete sources, there seems to be no reason to expect any correlation between fluctuation scales and photon energy observed.

References

Arons, J., McCray, R., and Silk, J.: 1970, (to be published).

Bekefi, G.: 1966, *Radiation Processes in Plasmas*, John Wiley, New York.

Bergamini, R., Londrillo, P., and Setti, G.: 1967, *Nuovo Cimento* **52B**, 495.

Brecher, K. and Morrison, P.: 1967, *Astrophys. J.* **150**, L61.

Brecher, K. and Morrison, P.: 1969, *Phys. Rev. Letters* **23**, 802.

Bunner, A. N., Coleman, P. C., Kraushaar, W. L., McCammon, D., Palmieri, T. M., Shilepsky, A., and Ulmer, M.: 1969, *Nature* **223**, 1222.

Felten, J. E. and Morrison, P.: 1966, *Astrophys. J.* **146**, 686.

Felten, J. E. and Rees, M. J.: 1969, *Nature* **221**, 924.

Ginzburg, V. L. and Ozernoi, L. M.: 1966, *Soviet Astron. - AJ* **9**, 726.

Hayakawa, S. and Matsuoka, M.: 1964, *Supp. Prog. Theor. Phys.* **30**, 204.

Kardashev, N. S.: 1962, *Soviet Astron. - AJ* **6**, 317.

Longair, M. S.: 1966, *Monthly Notices Roy. Astron. Soc.* **133**, 421.

Montgomery, D. C. and Tidman, D. A.: 1964, *Plasma Kinetic Theory*, McGraw-Hill, New York.

Oda, M.: 1965, *Proc. Internat. Conf. Cosmic Rays* (London).

Pikel'ner, S. B. and Tsytovich, V. N.: 1969, *Sov. Astron. - A.J.* **13**, 5.

Rees, M. J. and Setti, G.: 1968, *Nature* **219**, 127.

Rees, M. J. and Silk, J.: 1969, *Astron. Astrophys.* **3**, 452.
Sandage, A.: 1968, *Astrophys. J.* **152**, L149.
Schmidt, M.: 1968, *Astrophys. J.* **151**, 393.
Sciama, D. W.: 1964, *Quart. J. Roy. Astron. Soc.* **5**, 196.
Silk, J.: 1968, *Astrophys. J.* **151**, L19.
Silk, J.: 1969, *Nature* **221**, 347.
Silk, J. and McCray, R.: 1969, *Astrophys. Lett.* **3**, 59.

INTERACTIONS OF NON-THERMAL X-RAYS AND
ULTRAVIOLET RADIATION WITH THE INTERGALACTIC GAS

M. J. REES

Institute of Theoretical Astronomy, University of Cambridge, England

and

G. SETTI

Institute of Physics, University of Bologna, Italy

1. Introduction

We wish to discuss two ways in which non-thermal background X-rays and ultraviolet radiation may interact with intergalactic matter: (i) low energy photons may be intense enough to photoionize the intergalactic medium and maintain it as an H II region; (ii) the recoil associated with Compton scattering of hard X-rays provides an additional heat input into the medium, as well as causing a characteristic modification of the background radiation spectrum. We shall merely summarize our results here, as fuller details are appearing elsewhere (Rees and Setti, 1969).

2. Ionization of Intergalactic Gas by Soft Photons

Observations of quasars with redshifts $z \simeq 2$ indicate that the density of intergalactic neutral hydrogen at the corresponding epoch must have been exceedingly low. For the optical depth shortward of the redshifted Lyman α wavelength to be $\lesssim \frac{1}{4}$, as is observed,

$$n_H \lesssim 3 \times 10^{-11} \, \text{cm}^{-3} \tag{1}$$

(Gunn and Peterson, 1965). This means that if there is a substantial amount of intergalactic gas it must be almost completely ionized. For it to be *collisionally* ionized to the required extent, gas temperatures $\gtrsim 10^6$ K are needed, for present total gas densities $\bar{n} \gtrsim 10^{-6} \, \text{cm}^{-3}$. There is then the problem of avoiding thermal bremsstrahlung emission in excess of the observed soft X-ray background.

If there were an adequate flux of intergalactic ultraviolet radiation, this ionization could be maintained without the necessity for a high gas temperature. No thermal X-rays would then arise from the gas, and an interesting new range of thermal histories for intergalactic matter would be compatible with all the data. The extension of the non-thermal X-ray background towards longer wavelengths may constitute such a flux.

Galactic absorption precludes reliable observations of the X-ray background at energies much below 1 keV. However there seems no reason why the background intensity incident on the Galaxy should not continue to rise towards longer wavelengths,

L. Gratton (ed.), Non-Solar X- and Gamma-Ray Astronomy, 402–405. All Rights Reserved

unless *intergalactic* absorption is important. If the non-thermal background above ~ 1 keV arises from inverse Compton scattering of black body photons by relativistic electrons with $\gamma > 10^3$, the intensity would definitely be expected to rise at least as steeply as $\propto \nu^{-1/2}$, because this is the spectrum which results when the high energy electrons are degraded. If additional electrons are injected with $\gamma < 10^3$, the radiation spectrum would of course be steeper than this. In order to abbreviate our discussion and to emphasize the main points, we take a simple 'standard' spectrum

$$F(\nu) \propto \nu^{-1} \tag{2}$$

right down to 10 eV, the intensity being normalized to give 20 photons $(\text{cm}^2 \text{ sec ster keV})^{-1}$ at 1 keV. This spectrum lies within 50% of all the observations above 1 keV. The flux at the Lyman limit would then be

$$F_{\text{Ly}} = 10^{-23} \text{ erg } (\text{cm}^2 \text{ sec } (\text{c/sec}) \text{ ster})^{-1}. \tag{3}$$

Although we have extrapolated by a factor ~ 100 in wavelength beyond the band where reliable data exist, this corresponds to a factor only ~ 10 in electron energy (i.e. down to $\gamma \simeq 100$). The actual intensity at the Lyman limit is obviously very uncertain. However (3) is unlikely to exceed the emitted inverse Compton flux at this energy by as much as an order of magnitude. It may, however, be a more serious *underesti-mate*, especially if additional components make large contributions to the background radiation below 1 keV. (3) greatly exceeds the predicted ultraviolet flux from galaxies, and is higher than even optimistic estimates of the contribution from quasars, based on extrapolation of their optical spectra (Noerdlinger, 1969).

If this radiation all originated at a redshift z^* (which we shall assume to exceed 2), then the value of F_{Ly} at a redshift $z < z^*$ is $\propto (1+z)^4$. We find that, at $z \simeq 2$, this intensity would be capable of maintaining a level of ionization satisfying (1) if the electron density were as high as $\sim 3 \times 10^{-5} \text{ cm}^{-3}$ (for an electron temperature $\sim 20000 \text{ K}$). This corresponds to a present gas density $\bar{n} \simeq 10^{-6} \text{ cm}^{-3}$.

According to the usual 'big bang' cosmologies, the gas would have been cool and neutral (cooler even than the primeval radiation) until objects condensed and released the energy to reheat it. We now consider whether non-thermal radiation could have accomplished this hearing. For a gas density $\bar{n} \simeq 10^{-6} \text{ cm}^{-3}$, our 'standard' spectrum yields $20(1+z) \text{ eV}$ per particle in the form of photons with energy 10–100 eV. The heating must have occurred at $z > 2$, and so sufficient energy would have been available to ionize all the H and He atoms, and to raise the temperature to any value up to $\sim 10^5 \text{ K}$ (we return to the question of the temperature later). The details of the heating would be complex. However, the results are unlikely to differ much from the idealised situation in which the photons are generated suddenly and uniformly at some redshift z^*. The heating would occur without much distortion of the spectrum (2), and the temperature immediately after the ionization (which is simply related to the mean energy of the absorbed photons) would be $\sim 25000 \text{ K}$. (An important feature of this type of heating is that it naturally 'switches off' once the medium has been ionized. One of the difficulties with other processes – e.g. cosmic ray heating – is that the heat input has to be very nicely adjusted in order to avoid the temperature shooting up to $\sim 10^8$ K, in which case there would be too much thermal X-ray emission even if $\bar{n} \simeq 10^{-6}$ cm^{-3}, once the gas breaks through the 'thermal barrier' at 10^4–10^5 K.)

In the subsequent expansion, the heat input via photoionizations partially compensates for the radiative and adiabatic losses. The temperature at $z \simeq 2$ would depend on when the heating occurred, but, for $\bar{n} \simeq 10^{-6}$ cm^{-3}, would be 10000–20000 K. Assuming free expansion down to the present epoch, the temperature could now be as low as ~ 4000 K, even if the ionization were still maintained. It is interesting to see whether the medium could have recombined by the present time (this is not possible when the ionization at $z \simeq 2$ is collisional). The optical depth at the Lyman limit would be ~ 3 times greater than at $z \simeq 2$. However it must then have been $\lesssim \frac{1}{4}$ (corresponding to an optical depth $\lesssim \frac{1}{4}$ in Lyman α), so that recombination seems rather unlikely. The number of recombinations during the expansion time-scale decreases only slightly more slowly during the expansion than the flux of ionizing photons (even when one allows for the falling temperature).

This situation would change if the gas did not expand freely, so that its local density was not proportional to $(1+z)^3$. For example, it is possible that the intergalactic gas is now all concentrated in clusters of galaxies. Since clusters occupy a few per cent of the volume of space, they could not have separated out before $z \simeq 2$. We may approximate the situation by supposing that up to $z \simeq 2$ they expand freely, but afterwards the density of the gas stays constant. The photoionization rate would still be $\propto (1+z)^4$, whereas the recombination rate would in fact *increase*, because radiative cooling decreases the temperature. The gas within clusters could therefore have become predominantly neutral. (In the case of collisional ionization, the concentration of the gas into clusters greatly aggravates the situation regarding the bremsstrahlung X-rays; the temperature remains high because of the absence of adiabatic cooling, and the higher density promotes more efficient X-ray emission.)

If the gas had now recombined, no ultraviolet background radiation beyond the Lyman limit would survive – indeed, there may be significant absorption right up to 1 keV. Sunyaev (1969) has argued that the existence of low density H I bridges between galaxies, and in an extended region around M31, is inconsistent with the presence of an intense background in the far ultraviolet. The flux (3) is actually consistent with Sunyaev's limit, although a gas with $\bar{n}$ much greater than $\sim 10^{-6}$ cm^{-3} could not have been maintained ionized at $z \simeq 2$ without F_{Ly} violating this limit (unless the gas has now recombined). However it appears to us that Sunyaev's argument can be evaded. The incident radiation is only capable of maintaining a layer of gas of density n_e and thickness L cm ionized if

$$L \lesssim (10^{23} F_{\text{Ly}}) \cdot 2 \times 10^{14} \, T^{1/2} n_e^{-2} \,. \tag{4}$$

The observed H I discussed by Sunyaev may be surrounded by ionized regions of enhanced density whose parameters correspond to approximate equality between the two sides of (4). A stationary situation would be possible if, for example, the Local Group contained ionized hydrogen of density $\sim 10^{-3}$ cm^{-3}.

3. Scattering of Hard X-Rays

Hard X-rays have an insignificant chance of being absorbed photoelectrically, but they also may affect the intergalactic gas, at least if they come from very large redshifts.

When a photon of frequency v is scattered, it transfers a fraction $\sim hv/mc^2$ of its energy to the electron. Consequently electron scattering has no effect on the spectrum or intensity of background radiation at optical or radio wavelengths, though it obviously obscures discrete sources. However a photon with $hv \simeq mc^2$ which is scattered just once gives up $\sim \frac{1}{2}$ its energy to the electron. Thus, if radiation with spectrum (2) were emitted at a redshift $z_{\tau=1}$ such that the universe had optical depth ~ 1 to Thomson scattering, we would observe an attenuation of $\sim 50\%$ below the power law at energies $\sim mc^2 (1 + z_{\tau=1})^{-1}$. The energy lost by the photons would have gone into the gas, but would not be capable of ionizing it completely. Note that this process is not affected by whether the gas is ionized or neutral, since the energy transferred by the relevant photons in the recoil is enormously higher than the binding energy of hydrogen or helium.

4. Conclusions

A plausible extrapolation of the non-thermal X-ray background into the ultraviolet yields a flux of photons around the Lyman limit which exceeds the likely flux arising from other processes. The background radiation probably originated at $z \gtrsim 2$, and could ionize an intergalactic gas of present density $\sim 10^{-6}$ cm^{-3}. If the gas were smoothly distributed through the universe, it would probably still be ionized, but if it were now concentrated in clusters it could have recombined. The electron temperature can be calculated in terms of the spectrum of the ionizing photons, and would be $\lesssim 20000$ K. A stronger ultraviolet flux than our assumed extrapolation could ionize a denser gas, and still the temperature would be too low for thermal X-rays to be emitted.

Hard X-rays are degraded when they are scattered by thermal electrons. This modifies their spectrum, and transfers energy to the gas.

References

Gunn, J. E. and Peterson B. A.: 1965, *Astrophys. J.* **142**, 1633.
Noerdlinger, P. D.: 1969, *Astrophys. J.* **156**, 841.
Rees, M. J. and Setti, G.: 1969, *Astron. Astrophys.* (submitted for publication).
Sunyaev, R. A.: 1969, *Astrophys. Letters* **3**, 33.

SOFT X-RAYS FROM THE GALAXY

M. J. REES

Institute of Theoretical Astronomy, University of Cambridge, England

Below 1 keV, analyses of X-ray background data are complicated by galactic absorption effects, which cause the received intensity to vary with galactic latitude. Bowyer *et al.* (1968) observed that the diffuse background did not fall off as rapidly as was expected towards the galactic plane. One plausible interpretation of their data would be to suppose that a significant flux of soft X-rays emanates from the disc itself. I wish to discuss what could be inferred about the latter component from improved observations of its latitude-dependence, and by indirect methods.

Consider first an idealised model of the disc in which the absorbing interstellar matter is smooth, and the galactic X-ray emission – whatever its origin may be – is likewise smoothly distributed. The optical depth at energy ε is then $\tau_\perp(\varepsilon) \operatorname{cosec}\theta$, where $\tau_\perp$ is the optical depth towards the galactic pole and θ is the galactic latitude. (At low energies, where absorption is mainly due to H and He, $\tau_\perp \propto \varepsilon^{-3}$; and $\tau_\perp \simeq 1$ for $\varepsilon \simeq 0.3$ keV.) At energies for which $\tau_\perp < 1$, we may define an angle $\theta_{abs}(\varepsilon)$ at which the optical depth would be unity. The extragalactic X-ray flux would be severely attenuated at latitudes $\theta < \theta_{abs}$, and the galactic X-rays would then be relatively more important.

Absorption also affects the *galactic* soft X-rays, but the expected latitude-dependence of this component depends on the relative scale heights of the emission and the absorption. Suppose that the density of the interstellar gas is proportional to $f(z)$, where z is the coordinate perpendicular to the plane. The precise form of $f(z)$ is uncertain, but the scale height is ~ 100 pc. The X-ray emission per unit volume will depend differently on z: let us suppose it is proportional to $f(k^{-1}z)$ (i.e. the same distribution, but a scale height k times greater). At energies for which $\tau_\perp < 1$, the intensity of this component will be roughly proportional to $\operatorname{cosec}\theta$ for $\theta > \theta_{abs}$, but for $\theta \simeq 0$ the intensity will be k^{-1} times as great as for $\theta \simeq \theta_{abs}$. Thus, if $k > 1$, this component has a *maximum* intensity at the intermediate latitude $\theta \approx \theta_{abs}$; if $k < 1$ the maximum intensity occurs in the plane. Obviously the actual emission will depend on z in a manner which cannot be accurately fitted by a function $f(k^{-1}z)$. However, the important point is that, if the galactic emission is important, and has a scale height exceeding that of the gas, the diffuse X-ray background should display a prominent maximum intensity at a galactic latitude $\theta \approx \theta_{abs}$.

The appropriate value of k depends on what process is responsible for the galactic X-rays. They may be the integrated contribution from numerous intrinsically weak sources, some of which may be observed individually by detectors with better resolution. All types of stars would have values of $k > 1$, and so the maximum at $\theta \simeq \theta_{abs}$ should exist if the emission is associated with a stellar population of sources. We also expect $k > 1$ if the X-rays are due to inverse Compton scattering. One type of process

L. Gratton (ed.), Non-Solar X- and Gamma-Ray Astronomy, 406–407. All Rights Reserved
Copyright © 1970 by the IAU

that gives rise to emission which is *more* concentrated towards the plane than the absorbing material (i.e. $k < 1$) is one which involves some interaction of nonthermal particles (whose density would presumably not *increase* with z) with the gas itself – e.g. innerbremsstrahlung, or the line emission mechanism considered by Steigman and Silk (1969). Further investigation of the latitude-dependence may therefore indicate which type of galactic emission process is dominant. Accretion of interstellar gas by neutron stars or collapsed objects may yield a significant flux of soft X-rays, for which we would expect $k < 1$.

If a substantial fraction of the diffuse X-rays in the range 0.25–0.5 keV are of galactic origin, some limits can already be set to the emission spectrum $F(\varepsilon)$ of this component. The photons received at ~ 0.25 keV come effectively from distances $\lesssim 200$ pc, because of the absorption. If, say, 50% of the X-rays at this energy were galactic, and if this component had the same spectrum as the extragalactic background, there would be a tenfold enhancement towards the plane at energies $\gtrsim 2$ keV, for which the effective pathlength may be 5–15 kpc. Since this vastly exceeds the observed galactic contribution (Cooke *et al.*, 1969) we conclude that $F(\varepsilon)$ must fall off towards higher energies at least one power more steeply than the extragalactic spectrum. A thermal spectrum at $\sim 10^6$ K would be consistent with this. At even lower energies (below 0.25 keV) the observed radiation would be entirely galactic, and its spectrum would be $\propto \varepsilon^3 F(\varepsilon)$, because of the more efficient absorption at lower energies. The shape of $F(\varepsilon)$ in this range is important because these very soft galactic X-rays may constitute a major heat input into the gas. Indirect evidence on the contribution of X-rays to the heating of interstellar matter could come from detailed studies of the ionization equilibrium. For example, because of their larger photoionization cross-sections, helium and heavy elements in H I regions would be *more* highly ionized, relative to hydrogen, than would be the case if cosmic rays provided the main heat input. Such effects may be studied by means of the radio combination lines of H and He, or by searching for (e.g.) N II $\lambda 1085$ in absorption. X-rays may also modify the ionization equilibrium of interstellar Ca and Na.

The cloudy and irregular nature of interstellar H I may prohibit any clear-cut conclusion emerging from low-resolution observations. When detailed X-ray maps are available, it will be possible to correlate them with 21-cm data. One should then be able to decide conclusively whether the X-rays originate beyond the galactic hydrogen (in which case the X-rays would be attenuated exponentially by hydrogen clouds) or whether they originate within the disc, and to inter something about the nature and properties of discrete galactic sources.

References

Bowyer, C. S., Field, G. B., and Mack, J. E.: 1968, *Nature* **217**, 32.
Cooke, B., Griffiths, R., and Pounds, K.: 1969, in this volume, p. 280.
Steigman, G. and Silk, J. I.: 1969, in this volume, p. 385.

X-RAY EMISSION FROM THE SOLAR WIND

E. BOLDT

NASA, Goddard Space Flight Center, Greenbelt, Md., U.S.A.

and

A. KLIMAS* and G. SANDRI*

Aeronautical Research Associates of Princeton, Inc., Princeton, N.J., U.S.A.

1. Introduction

Recent developments have made it possible to detect cosmic X-rays with energies as low as several hundred eV. Several measurements of the diffuse X-radiation in this range have been reported (Baxter *et al.*, 1969; Bowyer *et al.*, 1968; Henry *et al.*, 1968). In this note we investigate the possibility that these observers have detected X-radiation emitted by the solar wind. We conclude that they probably have not. However, we also find that bremsstrahlung may be detectable from a region of the sky near the sun. If this measurement is possible, it would represent an important method for determining some characteristics of the solar wind away from the ecliptic plane of the solar system. Even though the mean free path for electron collisions is large compared with the astronomical unit, the collision frequency, electron density, and energy released per encounter are sufficient to yield detectable soft X-radiation for lines of sight close to the sun. We have estimated the expected X-ray intensity in the vicinity of the earth on the basis of two models of the solar wind flow pattern; in the first, the flow is radial in all directions away from the sun, and in the second, the flow is confined to a disc of uniform thickness near the ecliptic. In both cases, we neglect temperature gradients for the electrons and compute the total flux received from interplanetary plasma along the line of sight. Most of the received intensity comes from the segment of the line of sight which is nearest the sun. The results are insensitive with respect to the position of the boundary of the solar cavity. Accordingly, we neglect the boundary and consider an infinitely large solar cavity.

The X-radiation is assumed to arise from bremsstrahlung in an optically thin, fully ionized hydrogen gas. In the solar wind, bremsstrahlung contributions of higher mass elements (helium, in particular) should represent only a small correction. We use a standard bremsstrahlung emission formula, as presented in Tucker and Gould (1966) or Greene (1959); thus we assume that the distribution of electrons is approximately Maxwellian. The electron distribution function in interplanetary space is not well known. From the measured ion distribution, we might expect anisotropies which are characterized by different temperatures parallel and perpendicular to the interplanetary magnetic field. The anisotropy in the ion temperature is not very large, however, and

* This work was supported by National Aeronautics and Space Administration, Goddard Space Flight Center, Greenbelt, Md. under Contract No. NAS 5-11104.

for our purpose of estimating the magnitude of the expected radiation, a more detailed analysis is not justified.

We have estimated the recombination radiation from the solar wind on the basis of the ionization equilibrium calculation done by Tucker and Gould (1966) and a universal abundance model. Recombination radiation can be, at best, comparable in intensity to the bremsstrahlung. In the following, we consider the bremsstrahlung contribution only.

2. The Bremsstrahlung Intensity

The differential bremsstrahlung emission rate for a volume element, dV, of fully ionized hydrogen with density, n, as a function of frequency, ω, and temperature, T, is

$$\frac{dE}{dt\, dV\, d\omega} = n^2 \frac{2^4}{3^{3/2}} \frac{e^6}{m^2 c^3} \left(\frac{2\pi m}{kT}\right)^{1/2} \bar{g}(\omega, T)\, e^{-\hbar\omega/kT} \tag{1}$$

where $\bar{g}(\omega, T)$ is the average Gaunt factor (Greene, 1959). For the degree of accuracy desired here, we may set $\bar{g}(\omega, T) = 1$. The power received at a detector of surface area, dA, with beam width solid angle, $d\Omega$, from a distributed source in which the detector is embedded is

$$\frac{dE_R}{dt\, d\omega\, dA\, d\Omega} = \frac{1}{4\pi} \int_0^\infty dl \left(\frac{dE}{dt\, dV\, d\omega}\right) \tag{2}$$

where the integration is carried out along the line of sight. Assuming a uniform temperature for the electrons, we obtain

$$\frac{dE_R}{dt\, d\omega\, dA\, d\Omega} = \frac{2^4}{3^{3/2}} \frac{e^6}{m^2 c^3} \left(\frac{2\pi m}{kT}\right)^{1/2} e^{-\hbar\omega/kT} n_E^2 R_E \cdot \frac{1}{4\pi} \int_0^\infty d\left(\frac{l}{R_E}\right) \left(\frac{n}{n_E}\right)^2 \tag{3}$$

where n_E is the electron density at the earth and R_E is 1 AU. Equation (3) can be written numerically as

$$\frac{dE_R}{dt\, d(\hbar\omega)\, dA\, d\Omega} = 1.24 \times 10^{-4} n_E^2 \left(\frac{mc^2}{kT}\right)^{1/2} e^{-\hbar\omega/kT}$$
$$\cdot F(\hat{l}) \quad \text{(keV/keV sec cm}^2 \text{ ster)} \tag{4}$$

where

$$F(\hat{l}) = \frac{4}{\pi} \int_0^\infty d\left(\frac{l}{R_E}\right) \left(\frac{n}{n_E}\right)^2 \tag{5}$$

can only be a function of the direction of the line of sight, symbolized by $\hat{l}$. In Table I we list the value given by Equation (4), if $F(\hat{l}) = 1$, for some pertinent values of ω, $T_6 = 10^{-6} T$ (K) and n_E. In the remainder of this paper we calculate $F(\hat{l})$ for two models of the solar wind flow. The power received from a given direction at a particular temperature, density, and frequency can be obtained by multiplying the appropriate value of $F(\hat{l})$ from Figure 1 by the appropriate number from Table I.

TABLE I

$\hbar\omega$ (eV)	T_6	n_E	$dE_R/dt d(\hbar\omega) dA d\Omega$ keV/(keV sec cm² ster)
100	0.5	1	1.34×10^{-3}
		5	3.36×10^{-2}
		80	8.60
100	1	1	3.01×10^{-3}
		5	7.53×10^{-2}
		80	19.3
100	2	1	3.80×10^{-3}
		5	9.50×10^{-2}
		80	24.3
250	0.5	1	4.09×10^{-5}
		5	1.02×10^{-3}
		80	2.62×10^{-1}
250	1	1	5.28×10^{-4}
		5	1.32×10^{-2}
		80	3.38
250	2	1	1.60×10^{-3}
		5	3.99×10^{-2}
		80	10.2

3. The Models

In both models of the solar wind flow we neglect the small variation of solar wind speed near the sonic point (Parker, 1963). Thus, the radial dependence of the solar wind density can be obtained from kinematic considerations. If the flow is spherical, $n \propto R^{-2}$, and if the flow is cylindrical, $n \propto R^{-1}$. The effect of neglecting wind speed variation near the sun is to underestimate the expected X-ray intensity from a small region of the sky very near the sun.

In the cylindrical flow model we calculate $F(\hat{l})$ for directions in the ecliptic. Thus, the line of sight remains in the gas disc. For directions out of the ecliptic, we would expect the X-ray intensity to decrease more rapidly with increasing angular distance from the sun, but the details of this rapid decrease would depend on the disc structure. In any case, we would expect the actual data to clearly differentiate the cylindrical from the spherical flow.

We set

$$n = n_E (R_E/R)^\alpha \tag{6}$$

where $\alpha = 1$ is the cylindrical flow and $\alpha = 2$ is the spherical flow. Then

$$F_\alpha(\hat{l}) = \frac{4}{\pi} \int_0^\infty d\left(\frac{l}{R_E}\right)\left(\frac{R_E}{R}\right)^\alpha . \tag{7}$$

If θ is the angle between the line of sight in the ecliptic and the earth–sun line, then

$$\frac{R}{R_E} = 1 + \left(\frac{l}{R_E}\right)^2 - 2\frac{l}{R_E}\cos\theta . \tag{8}$$

We find

$$F_1(\theta) = \frac{4}{\pi}\left(\frac{\pi - \theta}{\sin \theta}\right)$$

and

$$F_2(\theta) = \frac{2}{\pi \sin^2 \theta}\left[\cos \theta + \frac{\pi - \theta}{\sin \theta}\right]. \tag{9}$$

For $\theta \ll 1$,

$$F_1(\theta) \propto 1/\theta \quad \text{and} \quad F_2(\theta) \propto 1/\theta^3 \tag{10}$$

and, for $\theta = \pi/2$,

$$F_1(\theta) = 2 \quad \text{and} \quad F_2(\theta) = 1. \tag{11}$$

$F_1(\theta)$ and $F_2(\theta)$ are plotted in Figure 1.

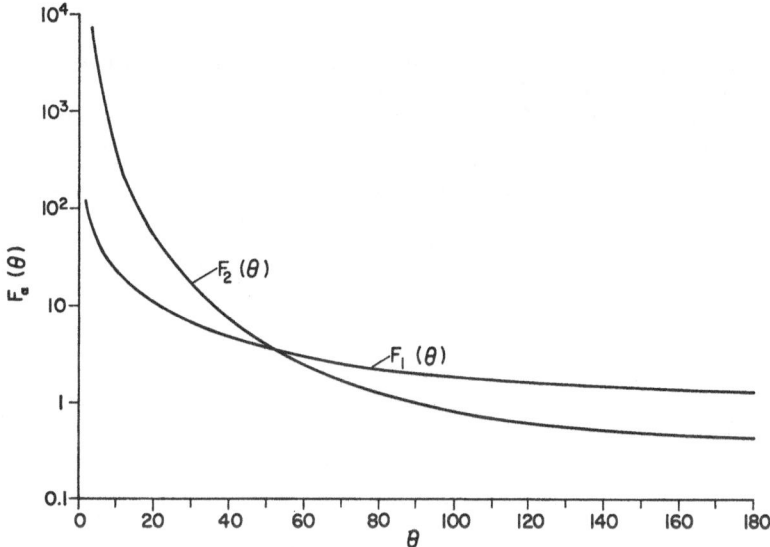

Fig. 1. The functions $F_1(\theta)$ and $F_2(\theta)$.

4. Discussion

We see that, except for small θ, $F_1(\theta)$ and $F_2(\theta)$ are slowly varying functions which remain close to the value, one. A scan of Table I for $\hbar\omega = 250$ eV clearly reveals that the X-ray intensity expected from the solar wind over most of the sky is small compared to the 50 keV/keV cm^2 sec ster which have been observed from the diffuse X-radiation (Kraushaar, this volume). On the other hand, the steep rise of $F_1(\theta)$ and $F_2(\theta)$ for small θ makes it likely that bremsstrahlung from a small region of the sky near the sun could be detected. To reduce the extragalactic background, observations should be made when the sun is at low galactic latitudes in the sky, where absorption in the interstellar gas will strongly attenuate the diffuse radiation. The necessary sensitivity to observe a predetermined region near the sun can be obtained from Table I and Figure 1. For example, consider the following quiet sun conditions:

$$n_E = 5 \text{ cm}^{-3}$$
$$T_6 = 1.$$

For $\hbar\omega = 250$ eV, at approximately $\theta = 5°$, we obtain the intensity, 50 keV/keV cm^2 sec ster. This intensity is comparable to the diffuse background intensity quoted above; however, the strong interstellar absorption at low galactic latitudes should prevent confusion of the two sources. During periods of high solar activity, when larger n_E are observed, the X-radiation from the solar wind would be more intense, and observable over a larger portion of the sky.

References

Baxter, A. J., Wilson, B. G., and Green, D. W.: 1969, *Astrophys. J. (Letters)* **155**, L145.

Bowyer, C. S., Field, G. B., and Mark, J. E.: 1968, *Nature* **217**, 32.

Greene, J.: 1959, *Astrophys. J.* **130**, 693.

Henry, R. C., Fritz, G., Meekins, J. F., Friedman, H., and Byram, E. T.: 1968, *Astrophys. J. (Letters)* **153**, L11.

Parker, E. N.: 1963, *Interplanetary Dynamical Processes*, Interscience, New York and London, pp. 51–72.

Tucker, W. H. and Gould, R. J.: 1966, *Astrophys. J.* **144**, 244.

MECHANISM FOR X-RAY PRODUCTION IN EXTARS

O. P. MANLEY

*American Science and Engineering**

and

S. OLBERT

Massachusetts Institute of Technology

This presentation attempts to describe in very qualitative terms a theory of production of high energy radiation (soft and hard X-rays) in magnetoactive plasmas of astrophysical interest. Special emphasis has been placed on the application of our model to extars and in particular to Sco X-1. More rigorous arguments may be found elsewhere [1] and the interested reader is urged to consult that reference for more details.

Because the construction of our model for X-ray production is independent of evolutionary considerations, in the body of the paper we have steered clear of such admittedly important problems as the origin and nature of the forces holding an extar together. Nor have we dealt adequately with the ultimate source of energy in extars. Our guiding philosophy has been somewhat akin to that taken by an earlier generation of astrophysicists in the context of stellar structures. It will be recalled that there too was an 'insurmountable problem', that of the energetics of stellar luminosities. Nevertheless the major results for a large class of stellar structures were obtained by applying rather elementary physical concepts. To this day, they stand relatively unchanged by the discovery of nuclear synthesis as the source of stellar radiation. It is quite possible that the principles governing the origin and dynamics of extars as well as other new and strange objects in galactic and extragalactic space, such as quasars and pulsars, lie outside of contemporary physics. If that is so, then these recent astronomical discoveries are bound to serve as a stimulus for a new synthesis in physics, much as other, long past astronomical discoveries led to past syntheses. Nonetheless it is tempting to apply some simple ideas stemming from contemporary physics, and see how well one can account for the observed properties of extars to date. It is this course that we propose to explore below.

For reasons dealt with at great length elsewhere [1] our point of departure is that the bulk of extar radiation, viz. X-rays, is due to synchrotron emission by relativistic electrons in a preexisting magnetic field. For the present purposes we assume that the energy required to account for the luminosity of extars is that lodged in the magnetic field. In this context, it is irrelevant whether or not the energy content of this field is replenished by a more fundamental, underlying celestial body – here this consideration would merely change the estimate of the effective lifetime of typical X-ray sources. What is important, however, is that a magnetic field is the only feasible means of storing significant amounts of energy in space without encountering unduly large

---

* Present address: Visidyne, Inc., 169 Merrimac Street, Woburn, Mass. 01801, U.S.A.

opacity over a large (~ 10 decades) photon energy range. And almost all observations of the known galactic X-ray sources militate against unduly large optical thicknesses.

In the study of ultrarelativistic (UR) electrons with nearly isotropic distribution, as in the case of cosmic rays [2], we are concerned with the equation for the differential number density, $N(E, \mathbf{x}, t)$ of the UR electrons. Here Nd^3xdE represents the number of UR electrons with energies between E and $E+dE$ within the volume element d^3x at position $\mathbf{x}$ and at time t. To the extent that we may disregard convection and diffusion in ordinary space we disregard the x-space dependence of N. One can then show that N obeys the following Fokker-Planck Equation [1, 2]:

$$\frac{\partial N}{\partial t} = \frac{\partial^2}{\partial E^2}(DN) + \frac{\partial}{\partial E}\left[\left(W - \left\langle\frac{\Delta E}{\Delta t}\right\rangle\right)N\right], \tag{1}$$

where W represents the average rate of the energy loss due to dissipative forces, here assumed to be restricted to synchrotron radiation losses; and the average rate of energy gain $\langle \Delta E/\Delta t \rangle$ is related to the 'diffusion coefficient' D by

$$\left\langle\frac{\Delta E}{\Delta t}\right\rangle = D\frac{d}{dE}\left[\ln(E^2D)\right]. \tag{2}$$

On substituting Equation (2) in Equation (1) and rearranging terms we find the corresponding diffusion equation in energy space:

$$\frac{\partial N}{\partial t} = \frac{\partial}{\partial E}\left\{E^2\left[D\frac{\partial}{\partial E}\left(\frac{N}{E^2}\right) + W\left(\frac{N}{E^2}\right)\right]\right\}. \tag{3}$$

If N is a sufficiently slowly varying function of time we may set Equation (3) equal to zero. Thence, a double integration with respect to the energy yields the quasi-steady state differential number density

$$N = \text{const } E^2 \exp\left(-\int\frac{W}{D}\,dE\right). \tag{4}$$

The first integration constant is easily shown to vanish for the condition of conservation of particles which we take to be valid here.

If N, as given above, is to be valid for a wide variety of non-thermal radiation sources, including extars, the ratio W/D must have certain easily determined attributes insofar as its energy dependence is concerned.

Thus in order for N to be effectively cut-off beyond some critical energy, W/D could be in part a polynomial in energy. On the other hand, to achieve some flexibility in the low energy behavior of N (i.e. energy dependence other than quadratic), and especially to obtain negative spectral power indices characteristic of non-thermal sources, a necessary and sufficient condition on W/D is that it contains as an additive term a member inversely proportional to energy. Inasmuch as the mean energy loss is limited here to synchrotron radiation, $W \sim E^2$. Hence a possible form for the diffusion coefficient D is

$$D \sim \frac{E^3}{P_3(E)},$$

where $P_3(E)$ is a cubic. The polynomial $P_n(E)$ could well be of higher order; however more detailed analysis shows that the higher order terms serve merely to cut N off even faster at high energies – their neglect does not seem to affect the generality of our presentation.

We shall now show how the simple considerations presented above lead to a fairly well defined picture of the environment required for generating non-thermal radiation such as that emanating from extars.

To this end it is instructive to recall here Fermi's ideas of the stochastic acceleration of cosmic-ray particles. By putting

$$\left\langle \frac{\varDelta E}{\varDelta t} \right\rangle = E \frac{u^2}{c^2} v_f \qquad (2a)$$

we see that $E(u^2/c^2)$ represents the net energy gain per 'collision' with a magnetic irregularity moving with speed u, and v_f is the average rate at which these random 'collisions' take place. (To be more specific, v_f represents the rate of effective momentum transfer) [2]. It is readily appreciated that the 'collision' rate, v_f, is intrinsically related to the statistics of the moving magnetic irregularities – i.e. to the fluctuations in the medium. In practice, a convenient way of describing fluctuations is the so-called power spectrum which gives a measure of the energy content of the fluctuations at a given frequency ω.

Note here that in order for a function of frequency to be a power spectrum of a realizable stochastic process [3], the function must be (a) symmetric in frequency; (b) integrable on the entire real frequency axis; and (c) non-negative for all real frequencies.

Fermi and others have taken v_f to be independent of the particle energy, E. An appropriate analysis [1] shows that the corresponding fluctuations must be such as to have a power spectrum, $S(\omega) \sim \omega^{-2}$. This is easily understood in the light of Fermi's acceleration mechanism: the particles were assumed by him to encounter the magnetic clouds at relatively infrequent intervals, the actual interaction time being short compared with the time between collisions. This, of course, is very much like the picture usually employed in the kinetic theory of dilute gases, where the gas particles are in effect assumed to undergo impulsive acceleration at random intervals, the intervals being Poisson-distributed. As is well known, the fluctuations characterized by Poisson statistics have in fact associated with them power spectra which at large frequencies behave as ω^{-2}.

A priori there is no reason why the intervals between interactions should always be very much longer than the interaction times themselves. One can easily imagine a situation in which the spacing between magnetic irregularities is comparable with the radius of gyration, R_g, of the particles within the irregularities – here then the interaction time is comparable with the time between interactions, and of the order of R_g/c. We can use the inverse of this quantity as a measure of the 'collision' rate v_f, which then turns out to be inversely proportional to E because $R_g \sim E$. Again an appropriate analysis [1] shows that if v_f has such a behavior with energy the corresponding fluctuations as seen by the accelerated particles must have a power spectrum $S(\omega) \sim \omega^{-1}$.

Remark here that over the range of particle energies for which $v_f \sim E^{-1}$, the mean energy gain per unit time $\langle \Delta E / \Delta t \rangle$ is independent of energy. This will obviously tend to produce a much flatter, and considerably harder particle spectrum than that resulting from the conventional Fermi acceleration mechanism in which the energy gain per unit time is proportional to the particle energy. Obviously a fluctuation spectrum which leads to such a novel energy gain $\langle \Delta E / \Delta t \rangle$, if physically realizable, may be of great significance in the study of non-thermal sources of hard radiation such as the extar Sco X-1. It is tempting therefore to pursue these ideas further.

Because of the integrability condition on a realizable power spectrum, $S(\omega)$ cannot maintain ω^{-1} behavior for all frequencies. This immediately implies the existence of two critical frequencies ω_1 and ω_2 (take $\omega_1 < \omega_2$): below ω_1, $S(\omega)$ must decrease slower than ω^{-1}, and above ω_2, $S(\omega)$ must decrease faster than ω^{-1}. While no further restrictions on $S(\omega)$ for $\omega < \omega_1$ appear to be of any consequence, it is found from more detailed analysis [1], that the required low-energy behavior of the UR electron energy distribution, N (see above), will be obtained if $\lim S(\omega)_{\omega \to \infty} \sim \omega^{-3}$. These arguments in no way preclude the possibility of a spectral region ω, somewhere between where $S(\omega) \sim \omega^{-1}$ and where $S(\omega) \sim \omega^{-3}$, on which $S(\omega) \sim \omega^{-2}$, or some other intermediate behavior. However, for our present purposes and especially for the sake of analytic simplicity, we ignore such details (see Figure 1).

One may now inquire into the nature of the medium, and in particular, into the

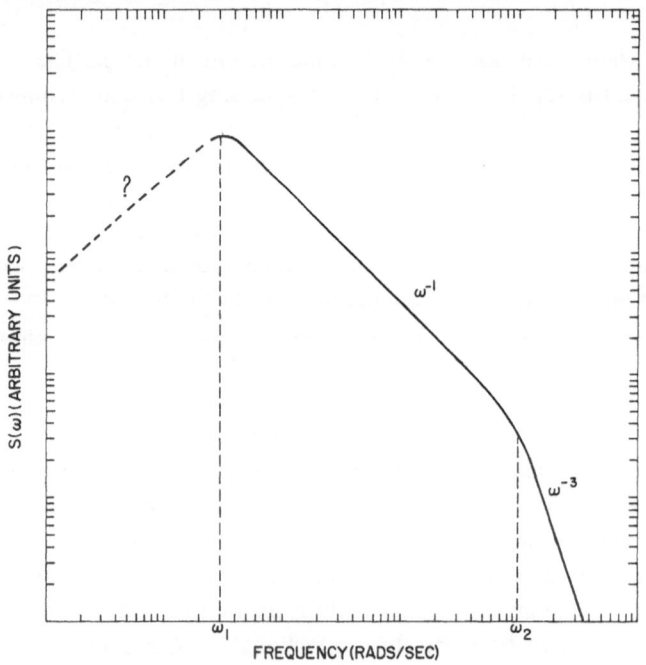

Fig. 1.   Power spectrum of magnetic fluctuations.

fluctuations therein, whose power spectrum might be just the one deduced above from our simple arguments. Of special significance appears to be the presence of the two critical frequencies ω_1 and ω_2 which endow the power spectrum $S(\omega)$ with the earmarks of band-limited noise. Now if we insist on the stochastic acceleration to arise in virtue of particle interactions with randomly moving magnetic irregularities, we should take into account that such irregularities are resolvable into a superposition of random MHD waves. Thus $S(\omega)$ may be interpreted as precisely the power spectrum of those waves (for simplicity we limit our discussion here to purely transverse Alfvén waves). There occurs then the difficulty that in an unbounded uniform domain MHD waves propagate without attenuation, thus in effect precluding the possibility of generating band-limited MHD noise.

A plausible way out of this dilemma offers itself if we abandon the idea of magneto-active plasmas being uniform on an arbitrarily large scale. In particular, theoretically useful results are obtained if plasma inhomogeneities are such as to form an aggregate of very long and thin plasmoids. For then it is evident that the finite length and thickness of these plasmoids impose long and short wavelength restrictions on transverse MHD waves. To visualize this, picture the excitation of transverse waves on a finite piece of string – on the one hand, the impossibility of sustaining identifiable vibrations with wavelengths much larger than the length $L_\parallel$ of the string itself is easily appreciated; on the other hand, it is equally easy to appreciate the difficulties in sustaining purely transverse deformations with characteristic le ngthsshorter than the diameter of the string, $L_\perp$. Since the propagation velocity of the MHD waves in the plasmoids, the Alfvén velocity u, is wavelength independent, we find that the finite dimensions of the plasmoids impose low and high frequency cut-offs, $\omega_1 \simeq u/L_\perp$ and $\omega_2 \simeq u/L_\parallel$, respectively.

Why should the plasma inhomogeneities be characterized by the two lengths $L_\parallel$ and $L_\perp$? Are these lengths uniquely relatable to any known or empirically determinable plasma properties? The answers to these questions seem to lie in the well-known fact that finite sized, optically thin plasmas which are hot enough to cool by radiation (e.g. free-free or free-bound losses) are unstable with respect to small density (and/or temperature) fluctuations [4, 5]. This radiation instability is quenched by the large, *but finite*, thermal conductivity. Dimensional analysis of the interplay between radiation losses, Q_{rad} – a volume effect, and conduction heat flow, $\kappa \nabla T$ – a surface effect, indicates the existence of a critical length, L_c, below which the growth of the radiation instability is effectively quenched. Thus

$$L_c^2 \approx \kappa T / Q_{\text{rad}},\tag{5}$$

where κ is the thermal conductivity and T is the plasma temperature. In the presence of a strong magnetic field, the flow of heat transverse to the magnetic field is strongly inhibited, while being essentially unaffected along the magnetic field ($\kappa_\parallel \approx \kappa$). Thus we expect two critical lengths – one governing the transverse size of a stable plasmoid – $L_\perp$, the other its longitudinal size – $L_\parallel \simeq L_c$. They are related by

$$\left(\frac{L_\perp}{L_\parallel}\right)^2 \simeq \frac{\kappa_\perp}{\kappa_\parallel} \simeq \frac{\nu_{\text{coll}}}{\omega_b},\tag{6}$$

where v_{coll} is the effective collision frequency in a plasma, and $\omega_b \equiv eB/m_e c$ the symbols having their conventional meaning. Here magnetic field is taken to be strong if $v_{coll}/\omega_b \ll 1$. In spite of this limitation *the plasma may not be considered as collisionless because our chosen time scale is sufficiently long for effects of radiation to be of explicit importance*. We see this in (a) having taken into account the role the radiation losses play in plasma stability and (b) noting that, for instance in the case of free-free radiation, the mean number of elastic collisions between the occurrence of radiative collisions is of the order of $m_e c^2/\alpha kT$, where α is the fine structure constant.

So far we have paid scant attention to the physical reasons for the details of the behavior of $S(\omega)$ between the two critical frequencies ω_1 and ω_2, nor have we inquired why the behavior outside of those frequencies should be as stipulated on purely phenomenological grounds. In this respect, for the present we must rest our case on purely empirical grounds – viz. the observation of magnetic noise spectra under at least two widely different conditions: interplanetary magnetic field fluctuations [6] and magnetic noise spectra generated under certain laboratory conditions [7]. In both cases the general features of the measured power spectra are found to be just those postulated by us here. Thus we feel justified in suggesting the universal nature and character of band-limited MHD noise in magneto-active plasmas and in asserting its prevalence under a wide range of astrophysical conditions. Chances are that the detailed behavior of $S(\omega)$ with frequency depends on the statistical ensemble of the plasma properties – e.g. temperature, density and magnetic field – in the region under consideration. This should be so because as we have seen, the bandwidth limitations of vibrations within individual filaments depend on their length and diameter, which in turn depend on the local temperature, density and mean magnetic field. Moreover our considerations have yielded only upper bounds on $L_\perp$ and $L_\parallel$ – as yet we have no knowledge of the rules and statistics whereby these dimensions are distributed below those bounds.

We are now in a position to write down the explicit quasi-steady state solution to the equation governing the energy dependence of the differential number density N. It can be shown that (expressing from here on all relativistic energies in terms of the Lorentz factor γ) it takes the form [1]

$$N(\gamma) = \text{const.} \, \gamma^{2-\zeta} \exp\left\{ -\left(\frac{\gamma}{\gamma_0}\right)^2 \left[1 + \frac{2}{3\zeta}\left(\frac{\gamma_m}{\gamma_0}\right)^2 \frac{\gamma}{\gamma_m} + \frac{\gamma_m}{\gamma} \right] \right\}, \tag{7}$$

where

$$\zeta \equiv \frac{10\omega_b}{3A\omega_2^2} \frac{r}{c} \tag{8a}$$

$$\gamma_m \equiv \frac{40}{3\pi} u\omega_1\omega_b/c\omega_2^2 \tag{8b}$$

$$\gamma_0^2 \equiv 3A \frac{u^2}{rc\omega_b} \tag{8c}$$

with r the classical electron radius, $m(m_p)$ the electron (proton) mass, n the mean plasmoid particle density, c velocity of light, $u \equiv (B^2/4\pi n m_p)^{1/2}$, and

$$A = \langle (\delta B/B)^2 \rangle [1 - (\omega_1/\omega_2)^2]/2\ln(\omega_2/\omega_1). \tag{9}$$

Note that $\langle (\delta B/B)^2 \rangle = \int_{-\infty}^{\infty} S(\omega) \, d\omega$ is the relative mean square amplitude of the magnetic field fluctuations and can be thought of as the ratio of the magnetic energy in fluctuations to the mean magnetic field energy. In arriving at N as given above we have modelled $S(\omega)$ in such a way as to simplify analysis, but without sacrificing in any way its principal attributes deduced from the preceding arguments. However, some of the numerical constants may be somewhat sensitive to the particular model picked. This should not affect our major conclusions.

The observed physical attributes of an extar are connected by the three parameters, ζ, γ_m, and γ_0 defined above. They may seem to arise from a rather arbitrary and not necessarily unique arrangement of various coefficients in arriving at the form of $N(\gamma)$ given above. However, a closer study reveals that they are endowed with potential physical significance hard to overlook.

Thus, on taking into account the relationships between ω_1, ω_2 and the properties of the underlying magneto-active plasma, it is found that

$$\zeta = \frac{400}{A} \frac{m_p}{m\alpha} \left(\frac{3kT}{2\pi mc^2} \right)^{3/2}. \tag{10}$$

Note that ζ governs the behavior of the low energy end of the UR electron energy distribution. It seems remarkable that this ultimate result does not depend on the plasma density and magnetic field intensity; rather it is primarily a function of the plasma temperature and the ratio of the mean energy in magnetic field fluctuations to the energy in the mean magnetic field. Of course, underlying our considerations is the assumption that the electron temperature is high enough so that the plasma is nearly fully ionized.

Next we turn to the energy, γ_m. We find that we can write it in a physically more transparent form:

$$\gamma_m = \frac{160}{3\pi^2} \frac{\omega_1 \omega_b}{\omega_2} \frac{L_\perp}{c} \tag{11}$$

from which γ_m is seen to be an energy at which the radius of gyration of an electron is guaranteed to be much smaller than $L_\perp$, i.e. smaller than the radius of the filamentary plasmoids which concern us here (this follows from the condition $\omega_1 \ll \omega_2$ implied throughout our considerations). It is perhaps even more remarkable that γ_m may be shown to be essentially a natural constant corresponding to an energy of ~ 0.5 GeV. One can conjecture that it represents a critical energy for relativistic electrons, such that when it is exceeded, the relaxation of the underlying magnetic field is governed by synchrotron emission losses of the UR electrons, rather than by plasma free-free (free-bound) emission losses.

Finally, we come to γ_0. In the limit of small γ_m/γ_0 – a limit found to be of some consequence below – γ_0 denotes a characteristic energy of the relativistic electrons described by the function $N(\gamma)$. We shall see that γ_0 plays an important role in determining the hardness of the synchrotron spectrum generated by an ensemble of electrons distributed according to $N(\gamma)$.

The function $N(\gamma)$ behaves as a power law for low electron energies, its index lying between -1 and 0, a range of importance in the study of extars provided

$$2 < \zeta < 3$$

On the other hand, $N(\gamma)$ decreases quite rapidly for $\gamma > \gamma_0$. It has been pointed out [8, 9] that if $N(\gamma)$ were to be sharply cut-off at some upper energy, the resulting synchrotron spectrum would evince an exponential behavior hardly distinguishable from that emitted by hot, optically thin plasma at an appropriate temperature. As we shall see on comparison with data, the high energy tail of our $N(\gamma)$ approaches adequately such a sharp cut-off.

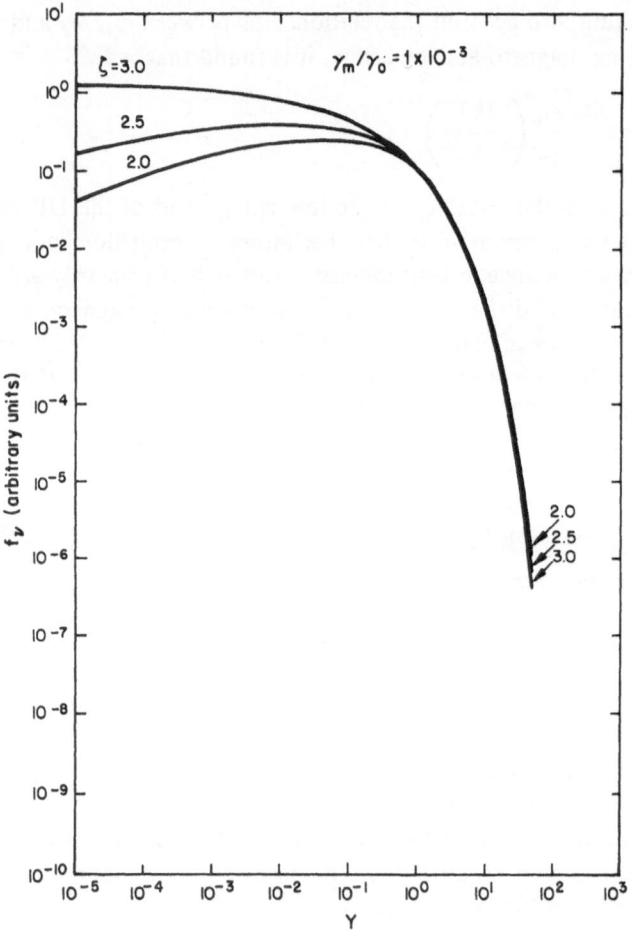

Fig. 2.   Dependence of the synchrotron spectrum, f_v, on ζ for a fixed γ_m/γ_0.

We now form the synchrotron spectrum, f_v, emitted by UR electrons – distribution with energy according to $N(\gamma)$:

$$f_v = Ber \sqrt{3} \int_0^\infty d\gamma N(\gamma) \frac{2x}{3\gamma^2} \int_{2x/3}^\infty d\eta K_{5/3}(\eta),$$ (12)

where $x \equiv 2\pi v/\omega_b$, v being the photon frequency.

In practice it turns out that the bulk of the more reliable X-ray astronomy data appears to fall in a region such that

$$0.1 \lesssim 2x/3\gamma_0^2 \lesssim 10.$$

It is found that in this frequency regime for $\gamma_m/\gamma_0 \lesssim 10^{-1}$, f_v is independent of γ_m/γ_0. f_v for several values of ζ is shown in Figure 2.

We now turn to the determination of the specific parameters describing an extar, interpreting available data in the light of the above considerations. We restrict ourselves here to data concerning Sco X-1 principally because this source is the one most extensively studied. For the sake of concreteness in the following numerical estimates we take the distance to Sco X-1 to be 300 pc [10] and hence, the X-ray luminosity (i.e. nearly the total luminosity), $L_x = 5 \times 10^{36}$ ergs/sec. Further, in the same spirit we assume Sco X-1 to be approximated by a sphere of radius R.

Recall first that the low electron energy end of the spectrum depends on the index ζ which in turn depends on the plasma temperature and mean square amplitude of magnetic fluctuations. Substitution of $T \approx 10^5$ K, and $A \approx 10^{-1}$, (i.e. $\langle (\delta B/B)^2 \rangle \approx 1$) into Equation (10) yields

$$\zeta = 2.5 \pm 0.5$$

spanning the indeterminancy in the color of the optical spectrum of Sco X-1 due to both the interstellar reddening and the observed intrinsic color fluctuations. Thus, we see that as a direct consequence of our model the observed 'flat' photon spectrum, implying a hard electron spectrum, is consistent with the observed high excitation lines requiring a plasma temperature on the order of 10^5 K. It should be noted that the parameters T, and A, selected here are probably reasonable mean values for Sco X-1 as a whole. Notwithstanding that they are subject to some temporal and spatial variations, which may account directly for the reported amplitude and color changes of the visible continuum as well as for the variability of the equivalent line widths. Of course, the possible radio emission attributed to Sco X-1 is consistent with the range of values of ζ given above.

On setting

$$B\gamma_0^2 \approx 6 \times 10^{11} \text{ G}$$

there results a reasonably good fit to the Sco X-1 X-ray spectral data, Figure 3 [11, 12, 13]. Remarkably enough, the fact that $N(\gamma)$ is actually not sharply cut-off above γ_0, serves to harden the predicted photon spectrum just enough to fit the recent high energy measurement without additional ad hoc assumptions about the structure of the extar.

Note that the recent report of a measurement of soft X-ray emission from Sco X-1 [12], is consistent with our model if the mean interstellar hydrogen density in the direction of that extar is $\sim 0.5\ \mathrm{cm}^{-3}$ (~ 2.5 mean free paths for X-ray absorption).

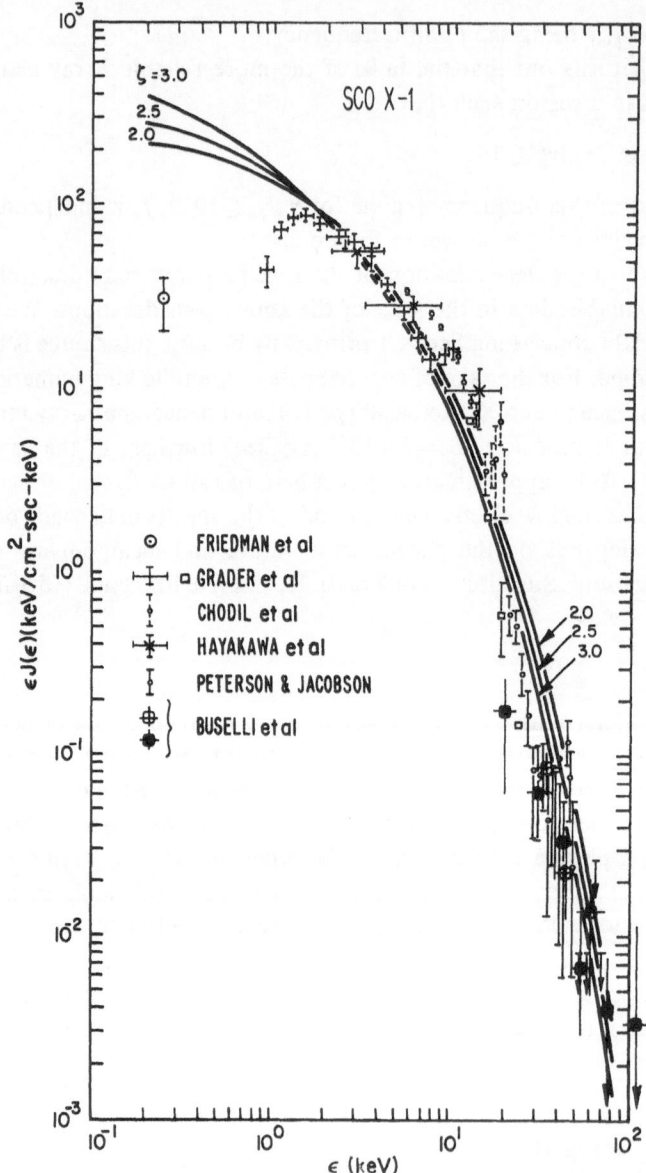

Fig. 3.   Comparison of our computations with Gould's (1967) compilation of spectral measurements of Sco X-1. Note (a) that the soft X-ray measurement reported earlier by Friedman has been amended to conform with the latest NRL measurements (Fritz *et al.*, 1968) and (b) new spectral measurements beyond 20 keV by the University of Adelaide group (Buselli *et al.*, 1968) have been added.

We find further that with the radius $R \sim 100\ AU$, $n \approx 10^6\ \mathrm{cm}^{-3}$ (consistent with absence of forbidden lines in the optical spectrum of Sco X-1) and $B \approx 2\ \mathrm{G}^*$, the relativistic electron energy density is

$$N_0 E_{\mathrm{UR}} \approx 5 \times 10^{-8}\ \mathrm{ergs/cm}^3$$

where N_0 is the mean density of UR electrons and $(E_{\mathrm{UR}} = \gamma_0\ mc^2)$. Note that for our model $N_0 E_{\mathrm{UR}} \ll nkT$

so that the dynamics of the extar is not affected directly by the relativistic electron pressure.

The spectrum of the magnetic fluctuations in our simple model spans characteristic frequencies which coincide with the observed optical and X-ray relaxation times ranging from seconds to hours. We may then assert plausibly that the observed fluctuations characterizing both the optical and X-ray ends of the Sco X-1 spectrum are a direct consequence of fluctuations in the underlying magnetic field and hence in the underlying plasma density and UR electron density. In accordance with this argument, the absence of significant polarization of the visible light emanating from Sco X-1 may well be attributed to the lack of preferred orientation of the fluctuating magnetic field. The color changes attendant on amplitude fluctuations may be evidence of sporadic hardening of the electron spectrum.

Acknowledgements

We wish to acknowledge the interest of Drs. M. Annis and R. Giacconi of American Science and Engineering, Inc. in this presentation.

This presentation was supported in part by Air Force Office of Scientific Research (OAR) Contract No. F44620-67-C-0065.

References

[1] Manley, O. P. and Olbert, S.: 1969, *Astrophys. J.* to appear.
[2] Ginzburg, V. L. and Svrovatskii, S. I.: 1964, *Origin of Cosmic Rays*, Pergamon Press, New York.
[3] Lee, Y. W.: 1960, *Statistical Theory of Communication*, J. Wiley, New York.
[4] Manley, O. P. and Olbert, S.: 1968, presented at Asilomar Conference on Plasma Instabilities in Astrophysics, California, October.
[5] Field, G. B.: 1965, *Astrophys. J.* **142**, 531.
[6] Coleman, P. J.: 1968, *Astrophys. J.* **153**, 371.
[7] Patrick, R. M. and Pugh, E. R.: 1969, *Phys. Fluids* **12**, 366.
[8] Greenstein, J.: 1964, *Astrophys. J.* **140**, 666.
[9] Manley, O. P.: 1966, *Astrophys. J.* **144**, 1253.
[10] Wallerstein, G.: 1967, *Astrophys. Letters* **1**, 31.
[11] Gould, R. J.: 1967, *Am. J. Phys.* **35**, 376.
[12] Fritz, G., Meekins, J. F., Henry, R. C., Byram, E. T., and Friedman, H.: 1968, *Astrophys. J. Letters* **153**, L199.
[13] Buselli, G., Clancy, M. C., Davison, P. J. N., Edwards, P. J., McCracken, K. G., and Thomas, R. M.: 1968, *Nature* **219**, 5159.

* Under the assumption of no external energy input to the magnetic field this corresponds to a lifetime of ~ 40 yrs.

CONCLUDING REMARKS

L. GRATTON
University of Rome, Italy

Ladies and Gentlemen,

We have come to the end of an exceptionally successfull meeting. During three very laborious days all the most important topics of X- and γ-ray non-solar astronomy have been revised and many new results have been presented.

After Rossi's beautiful introduction, Gursky has given an admirable summary of the observational techniques and Friedman presented a rapid survey of the studies of individual sources. Individual sources have been also considered in four excellent discourses by Clark, Peterson, McCracken and Giacconi, and optical results have been summarized by Johnson and myself. I do not name the great number of contributors of short communications, simply because time is short, but all of them brought excellent contributions to the success of this meeting.

Yesterday afternoon, we reached perhaps the highest point of scientific interest of the Symposium, through the discussion of the extraordinary results on pulsars by the NRL, MIT, Columbia and Rice Group, and an interesting presentation of their theoretical interpretations by the Frascati Group.

Woltjer and Felten contributed two excellent critical summaries of the theoretical problems of the sources.

Unfortunately time began to run too fast and many valuable contributions had to be compressed in a few hours. Being the last ones, the contributions which suffered most by this shortness of time were those corresponding to the background radiation. Nevertheless, I am sure you all have appreciated Oda and Kraushaar's summaries on this subject and we must be grateful to Dr. Setti, who in accepting at a two-days notice to replace the Russian colleagues, rendered an excellent service to the meeting. I think that all of you will agree that the fascinating subject of the cosmological implications of the X- and γ-ray background observations would have deserved a Symposium by itself.

Apart from the shortness of time in relation to the large number of contributions perhaps the most serious fault in this meeting was the absence of optical astronomers. I hope, being an optical astronomer myself, that this does not mean a lack of interest on the part of my colleagues for the subject of our meeting.

Before concluding, let me ask you to be excused for the many faults of our organization for which I alone am responsible. Indeed we did not expect such a large number of attendants and this is my main excuse for them. I must thank very heartily all of you for coming to Rome, some from very far; for your contributions which constitute the great success of the meeting, and for the patience and interest shown throughout.

Finally I thank in your name the International Astronomical Union for sponsoring

L. Gratton (ed.), Non-Solar X- and Gamma-Ray Astronomy, 424–425. All Rights Reserved
Copyright © 1970 by the IAU

this Symposium, The Italian National Research Council for being our host during these days and for its financial contribution to the meeting.

Good-bye, au revoir, auf Wiedersehen, hasta la vista, arrivederci!